METHODS IN MOLECULAR BIOLOGY

Series Editor
John M. Walker
School of Life and Medical Sciences
University of Hertfordshire
Hatfield, Hertfordshire, UK

For further volumes:
http://www.springer.com/series/7651

For over 35 years, biological scientists have come to rely on the research protocols and methodologies in the critically acclaimed *Methods in Molecular Biology* series. The series was the first to introduce the step-by-step protocols approach that has become the standard in all biomedical protocol publishing. Each protocol is provided in readily-reproducible step-by-step fashion, opening with an introductory overview, a list of the materials and reagents needed to complete the experiment, and followed by a detailed procedure that is supported with a helpful notes section offering tips and tricks of the trade as well as troubleshooting advice. These hallmark features were introduced by series editor Dr. John Walker and constitute the key ingredient in each and every volume of the *Methods in Molecular Biology* series. Tested and trusted, comprehensive and reliable, all protocols from the series are indexed in PubMed.

Computational Systems Biology in Medicine and Biotechnology

Methods and Protocols

Edited by

Sonia Cortassa

Laboratory of Cardiovascular Science, National Institute on Aging, NIH, Baltimore, Maryland, USA

Miguel A. Aon

Translational Gerontology Branch; Laboratory of Cardiovascular Science, National Institute on Aging, NIH, Baltimore, Maryland, USA

Humana Press

Editors
Sonia Cortassa
Laboratory of Cardiovascular Science
National Institute on Aging, NIH
Baltimore, Maryland, USA

Miguel A. Aon
Translational Gerontology Branch
Laboratory of Cardiovascular Science
National Institute on Aging, NIH
Baltimore, Maryland, USA

ISSN 1064-3745 ISSN 1940-6029 (electronic)
Methods in Molecular Biology
ISBN 978-1-0716-1830-1 ISBN 978-1-0716-1831-8 (eBook)
https://doi.org/10.1007/978-1-0716-1831-8

This Humana imprint is published by the registered company Springer Science+Business Media, LLC, part of Springer Nature.
The registered company address is: 1 New York Plaza, New York, NY 10004, U.S.A.

Dedication

This volume is dedicated to David Lloyd (Professor Emeritus, University of Cardiff, Wales), a friend, mentor, and colleague, for his inspiration, guidance, and support of the Editors, Sonia and Miguel, since early stages of their careers.

Preface

Systems biology emerged from the development of high-throughput -omics technologies and knowledge embodiment in databases, along with bioinformatic procedures enabling their interrogation, using a panoply of computational tools, in addition to mechanistic computational modeling, signal processing, and analysis in biological systems. These developments continued to be perfected in the last two decades, during which we have witnessed unprecedented breakthroughs in the transition data → information (i.e., organized data). Today, the frontier of systems biology stands at the edge between information → knowledge (i.e., organized information) in the search for meaning, as the availability of information grows. In the pursuit of knowledge, this volume highlights the state of the art of systems biology-oriented approaches under the overarching title of Computational Systems Biology (CSB), a research field that bridges experimental with computational tools to address complex challenges in diverse areas such as Medicine and Biotechnology.

As presented in this book, *Computational Systems Biology in Medicine and Biotechnology* comprises different levels of development and integration, but, conceptually and methodologically, they all share, as a common thread, a systems biology-oriented approach. CSB involves experimentation that includes comprehensive -omics technologies combined with multivariate statistical analyses performed in isolated (e.g., genomics, proteomics, metabolomics) or integrated (e.g., transcriptomics-metabolomics, metabolomics-fluxomics) datasets. In its most developed stages, CSB incorporates computational mechanistic modeling and advanced signal processing and analysis that together enhance experimental data interpretation, leading to new hypotheses and discoveries.

At the brink of the irruption of Artificial Intelligence (AI) in the science and technology scene, the Introduction of this book foresightedly analyzes its relationship with CSB, along with the potential impact of AI on the scientific *demarche*. The present volume is divided into six parts: genome/epigenome, metabolic networks, aging/longevity, disease, spatiotemporal patterns of rhythms/morphogenesis/chaos, and genome-scale metabolic modeling in biotechnology. In all these topics the readers will find varied, multifaceted, systems biology-inspired, methodological approaches to address/tackle complex questions at different levels, from molecular, cellular, organ to the organism, from the genome to the phenome, in health and disease. To produce and analyze big datasets, the authors use time-honored techniques—ordinary and partial differential equations, genome-scale metabolic modeling, parallel computing, elastic regression models, wavelets—as well as emerging ones, such as bioinformatics analyses of circular RNAs, single-cell analysis, posttranslational modifications, machine learning, network metrics, time series analysis for rhythms detection, and agent-based modeling, among others.

Comprehensive assessment of components from complex systems at different levels of organization (e.g., molecules, cells, organs, social networks) raises several challenges at logistic, analytical, and interpretative stances. Several chapters of this volume present different approaches in distinct settings to address these challenges while underscoring the immense opportunity available to address problems about *knowing (or not knowing) what we do not know.*

We gratefully thank the Editor, John Walker (Professor Emeritus, University of Hertfordshire, UK), of the Springer Nature Series on *Methods in Molecular Biology*, for his advice, guidance, and encouragement. Patrick Marton and Anna Rakovsky from Springer Nature are also thankfully acknowledged for their support.

The support by the Intramural Research Program of the National Institute on Aging, National Institutes of Health, and all authors' excellence of their contributions to this book are very gratefully acknowledged. Sonia Cortassa and Miguel A. Aon hope that this collective effort helps in shaping new research venues and approaches, while inciting the appeal of young minds to the thrilling field of Computational Systems Biology.

Baltimore, MD, USA

Sonia Cortassa
Miguel A. Aon

Contents

Contributors

EDUARDO AGOSIN • *Laboratory of Biotechnology, Department of Chemical and Bioprocess Engineering, School of Engineering, Pontificia Universidad Católica de Chile, Santiago, Chile*

CLAUDIO ANGIONE • *Computational Systems Biology and Data Analytics Research Group, Teesside University, Middlebrough, UK; Centre for Digital Innovation, Teesside University, Middlesbrough, UK; Healthcare Innovation Centre, Teesside University, Middlesbrough, UK*

MIGUEL A. AON • *Translational Gerontology Branch, National Institute on Aging, NIH, Baltimore, MD, USA; Laboratory of Cardiovascular Science, National Institute on Aging, NIH, Baltimore, MD, USA*

MICHEL BERNIER • *Translational Gerontology Branch, National Institute on Aging, National Institutes of Health, Baltimore, MD, USA*

MICHAEL O. BRECKWOLDT • *Neuroradiology Department, University Hospital Heidelberg, Heidelberg, Germany*

MARTA CASCANTE • *Department of Biochemistry and Molecular Biomedicine, Faculty of Biology, Universitat de Barcelona, Barcelona, Spain; CIBER of Hepatic and Digestive Diseases (CIBEREHD) and Metabolomics Node at Spanish National Bioinformatics Institute (INB-ISCIII-ES-ELIXIR), Institute of Health Carlos III (ISCIII), Madrid, Spain*

KYLE R. COCHRAN • *Laboratory of Genetics and Genomics and Computational Biology and Genomics Core, National Institute on Aging—Intramural Research Program, National Institutes of Health, Baltimore, MD, USA*

SONIA CORTASSA • *Laboratory of Cardiovascular Science, National Institute on Aging, NIH, Baltimore, MD, USA*

SUPRIYO DE • *Laboratory of Genetics and Genomics, National Institute on Aging (NIA), Intramural Research Program (IRP), National Institutes of Health (NIH), Baltimore, MD, USA; Laboratory of Genetics and Genomics and Computational Biology and Genomics Core, National Institute on Aging—Intramural Research Program, National Institutes of Health, Baltimore, MD, USA*

RAFAEL DE CABO • *Translational Gerontology Branch, National Institute on Aging, National Institutes of Health, Baltimore, MD, USA*

LUIGI FERRUCCI • *Biomedical Research Centre, National Institute on Aging, NIH, Baltimore, MD, USA*

ANA GEORGINA FLESIA • *Universidad Nacional de Córdoba, Facultad de Matemática, Astronomía y Física, Córdoba, Argentina; Consejo Nacional de Investigaciones Científicas y Técnicas (CONICET), Centro de Investigaciones y Estudios de Matemática (CIEM, CONICET), Ciudad Universitaria, Córdoba, Argentina*

CARLES FOGUET • *Department of Biochemistry and Molecular Biomedicine, Faculty of Biology, Universitat de Barcelona, Barcelona, Spain; CIBER of Hepatic and Digestive Diseases (CIBEREHD) and Metabolomics Node at Spanish National Bioinformatics Institute (INB-ISCIII-ES-ELIXIR), Institute of Health Carlos III (ISCIII), Madrid, Spain*

MYRIAM GOROSPE • *Laboratory of Genetics and Genomics and Computational Biology and Genomics Core, National Institute on Aging—Intramural Research Program, National Institutes of Health, Baltimore, MD, USA*
JISU HA • *Laboratory of Genetics and Genomics, National Institute on Aging (NIA), Intramural Research Program (IRP), National Institutes of Health (NIH), Baltimore, MD, USA*
JASON M. HELD • *Division of Oncology, Department of Medicine, Washington University School of Medicine, St. Louis, MO, USA*
TOSHIAKI HISADA • *UT-Heart Inc., Tokyo, Japan*
MIKE HOLCOMBE • *Department of Computer Science, University of Sheffield, Sheffield, UK*
JACKELYN MELISSA KEMBRO • *Universidad Nacional de Córdoba, Facultad de Ciencias Exactas, Físicas y Naturales, Instituto de Ciencia y Tecnología de los Alimentos (ICTA) and Catedra de Química Biológica. Consejo Nacional de Investigaciones Científicas y Técnicas (CONICET), Instituto de Investigaciones Biológicas y Tecnológicas (IIByT, CONICET-UNC), Vélez Sarsfield 1611, Ciudad Universitaria, Córdoba, Argentina*
SEULHEE KIM • *Department of Biomedical Engineering, University of Alabama at Birmingham, Birmingham, AL, USA; Department of Medicine, University of Alabama at Birmingham, Birmingham, AL, USA*
JASON M. KO • *Department of Biological Sciences, University of Maryland, Baltimore County, Baltimore, MD, USA*
ANDREAS KREMLING • *Systems Biotechnology, Technical University of Munich, Munich, Germany*
PEI-LUN KUO • *Biomedical Research Centre, National Institute on Aging, NIH, Baltimore, MD, USA*
FELIX T. KURZ • *Neuroradiology Department, University Hospital Heidelberg, Heidelberg, Germany; German Cancer Research Center, Department of Radiology, Heidelberg, Germany*
DANIEL LOBO • *Department of Biological Sciences, University of Maryland, Baltimore County, Baltimore, MD, USA*
GIUSEPPE MAGAZZÙ • *Computational Systems Biology and Data Analytics Research Group, Teesside University, Middlebrough, UK*
KRYSTYNA MAZAN-MAMCZARZ • *Laboratory of Genetics and Genomics, National Institute on Aging (NIA), Intramural Research Program (IRP), National Institutes of Health (NIH), Baltimore, MD, USA*
SEBASTIÁN N. MENDOZA • *Systems Biology Lab, AIMMS, Vrije Universiteit, Amsterdam, The Netherlands*
RUIN MOADDEL • *Biomedical Research Centre, National Institute on Aging, NIH, Baltimore, MD, USA*
PRADIP MOON • *Computational Systems Biology and Data Analytics Research Group, Teesside University, Middlebrough, UK*
ARSHAG D. MOORADIAN • *Division of Oncology, Department of Medicine, Washington University School of Medicine, St. Louis, MO, USA*
ZENOBIA MOORE • *Biomedical Research Centre, National Institute on Aging, NIH, Baltimore, MD, USA*
REZA MOUSAVI • *Department of Biological Sciences, University of Maryland, Baltimore County, Baltimore, MD, USA*
YAROSLAV R. NARTSISSOV • *Department of Mathematical Modeling and Statistical Analysis, Institute of Cytochemistry and Molecular Pharmacology, Moscow, Russia*

PAULA SOFIA NIETO • *Universidad Nacional de Córdoba, Facultad de Matemática, Astronomía y Física, Córdoba, Córdoba, Argentina; Consejo Nacional de Investigaciones Científicas y Técnicas (CONICET), Instituto de Física Enrique Gaviola (IFEG, CONICET-UNC), Ciudad Universitaria, Córdoba, Argentina*

BRIAN O'ROURKE • *Department of Medicine, The Johns Hopkins University School of Medicine, Baltimore, MD, USA*

ANNALISA OCCHIPINTI • *Computational Systems Biology and Data Analytics Research Group, Middlebrough, UK; Centre for Digital Innovation, Teesside University, Middlesbrough, UK*

JUN-ICHI OKADA • *UT-Heart Inc., Tokyo, Japan; Future Center Initiative, The University of Tokyo, Chiba, Japan*

GERNOT PLANK • *Institute of Biophysics, Medical University of Graz, Graz, Austria*

EVA QWARNSTROM • *Department of Infection, Immunity and Cardiovascular Disease, University of Sheffield, Sheffield, UK*

PEDRO A. SAA • *Laboratory of Biotechnology, Department of Chemical and Bioprocess Engineering, School of Engineering, Pontificia Universidad Católica de Chile, Santiago, Chile*

VITALY A. SELIVANOV • *Department of Biochemistry and Molecular Biomedicine, Faculty of Biology, Universitat de Barcelona, Barcelona, Spain; CIBER of Hepatic and Digestive Diseases (CIBEREHD) and Metabolomics Node at Spanish National Bioinformatics Institute (INB-ISCIII-ES-ELIXIR), Institute of Health Carlos III (ISCIII), Madrid, Spain*

PAYEL SEN • *Laboratory of Genetics and Genomics, National Institute on Aging (NIA), Intramural Research Program (IRP), National Institutes of Health (NIH), Baltimore, MD, USA*

ROBERT W. SEYMOUR • *Division of Oncology, Department of Medicine, Washington University School of Medicine, St. Louis, MO, USA*

SOROOSH SOLHJOO • *Department of Medicine, The Johns Hopkins University School of Medicine, Baltimore, MD, USA; Department of Medicine, F. Edward Hébert School of Medicine, Bethesda, MD, USA*

STEVEN J. SOLLOTT • *Laboratory of Cardiovascular Science, National Institute on Aging, NIH, Baltimore, MD, USA*

SEIRYO SUGIURA • *UT-Heart Inc., Tokyo, Japan*

TOSHIKO TANAKA • *Biomedical Research Centre, National Institute on Aging, NIH, Baltimore, MD, USA*

BAS TEUSINK • *Systems Biology Lab, AIMMS, Vrije Universiteit, Amsterdam, The Netherlands*

RAVI THARAKAN • *Biomedical Research Centre, National Institute on Aging, NIH, Baltimore, MD, USA*

CEEREENA UBAIDA-MOHIEN • *Biomedical Research Centre, National Institute on Aging, NIH, Baltimore, MD, USA*

SJOERD VAN DER POST • *Division of Oncology, Department of Medicine, Washington University School of Medicine, St. Louis, MO, USA*

SUPREETA VIJAYAKUMAR • *Computational Systems Biology and Data Analytics Research Group, Teesside University, Middlebrough, UK*

PIERRE VILLON • *Département de Génie Mécanique, Université de Technologie de Compiègne, Compiègne, France*

TAKUMI WASHIO • *UT-Heart Inc., Tokyo, Japan; Future Center Initiative, The University of Tokyo, Chiba, Japan*
OLGA A. ZAGUBNAYA • *Department of Mathematical Modeling and Statistical Analysis, Institute of Cytochemistry and Molecular Pharmacology, Moscow, Russia*
LUFANG ZHOU • *Department of Biomedical Engineering, University of Alabama at Birmingham, Birmingham, AL, USA; Department of Medicine, University of Alabama at Birmingham, Birmingham, AL, USA*

Chapter 1

Computational Systems Biology and Artificial Intelligence

Miguel A. Aon

Abstract

Aware of the rapid evolution of computational systems biology (CSB), which is the focus of this book, we address the emergence of artificial intelligence (AI). Consequently, one of the main purposes of this Introduction is to assess where the relationship between CSB and AI stands today, and to venture a vision for CSB.

Key words Algorithms, Correlation, Causation, Simulation, Prediction, Understanding

1 Introduction

Aware of the rapid evolution of computational systems biology (CSB), which is the focus of this book, we address the emergence of artificial intelligence (AI). Consequently, one of the main purposes of this introduction is to assess where the relationship between CSB and AI stands today, and to venture a vision for CSB.

2 Is There a Place for AI in CSB? Or for CSB in AI?

The answer to this question is not obvious because, at present, it is unclear whether the future of CSB will be dominated by AI. Our foresight is that rather than dominance, this topic can be visualized as a Venn diagram of two ensembles, CSB and AI, with some overlapping between them as well as unique traits. Among similarities, both are computationally driven, work with big data, and seek to learn/extract new information/understand function. A major difference is given by the way of learning, since in AI (so far) it is algorithmic whereas CSB is not because it is driven by human intelligence. The algorithmic nature of AI determines its purpose by circumscribing the search to a specific, task-oriented, purpose that demands the use of databases for training the

Sonia Cortassa and Miguel A. Aon (eds.), *Computational Systems Biology in Medicine and Biotechnology: Methods and Protocols*, Methods in Molecular Biology, vol. 2399, https://doi.org/10.1007/978-1-0716-1831-8_1,

algorithms. This does not necessarily restrain the domain of application of AI since, for example, pattern recognition can be done in, for example, pictures which can go from face recognition to images obtained with a broad range of microscopic techniques at different degrees of resolution that can be color coded, such as in heat maps. Otherwise stated, pattern recognition can be done in any dataset that can be translated into an image, and for that AI is already powerful.

Although both AI and CSB seek for the unknown, the way of achieving it is different. In principle, any task involving predetermined rules like, say, chess and Go games, the AI approach has shown that it can be superior to humans, given extensive training of the algorithms is available. However, the ruling laws of Nature, to a great extent, are unknown, which implies no training thus no AI. In this vein, and for now, CSB remains in the domain of human intelligence, and AI is another tool in the CSB toolkit for specific interrogations.

3 Understanding Through Simulation, Explanation, and Prediction

> "Again, and again, along our itinerary, three partners—understanding, theory and computation—had to dialogue to match the experimental response, or to become predictive. This weaving of theory, computation and understanding, in constant confrontation with experiment . . . ." –Hoffmann and Malrieu, 2020 [1–3]

The relationship between *simulation-understanding* and, their product, *prediction,* is germane for CSB. In this book, one definition of *knowledge* is *organized information* that, at a fundamental level, implies *understanding*, in turn essential for venturing new hypotheses and predictions.

However, facing the bewildering complexity and the tortuous networking of, for example, neurons in brain function, many scientists seem captivated, and gratified, with the idea of *emulating* rather than *understanding* the brain's amazing capabilities. Have we, scientists, capitulated in our quest of understanding Nature, and just be satisfied with its simulation/emulation (no matter how good or convincing the copy) without knowing how it works? Motivated by the recent irruption of AI, this question is relevant. The black box nature of the results obtained with either machine learning or deep learning convolutional networks is at odds with understanding.

The importance of this point is that, if a scientist would use training data, for example, essentially a collection of known input (x) and output(y) values, generated through observations or controlled experiments, to identify a function that is able to predict the output value for a new set of input data, no matter how accurate and reliable the prediction, the algorithm, implemented either through machine learning or a convolutional network, does not understand chemistry or physics, or of any other discipline as a matter of fact [1, 2]. Predicting without understanding leads to a black box paradox which conflicts with understanding.

Clearly, the way a phenomenon or behavior is simulated/predicted scientifically matters. It is not the same if a simulation comes from a model based on correlations (i.e., a function that maps a set of input data x to a predicted outcome y) compared to one in which, for example, mechanistic links are implicitly incorporated in the model. In the latter, an iterative process of experimentation $\leftarrow \rightarrow$ simulation is triggered, enabling the assessment of whether the outcome of the model's simulation, not only qualitatively but also quantitatively, is compatible with the experimental results, opening the way for understanding. Overall, during the model's validation process, if we are unable to simulate/explain/predict (i.e., understand) a given experimental result, this gives us a clue about what is missing, that is, *what we do not know*, thus suggesting new hypotheses and experimental tests.

Correlations, even based on solid statistics, are not causation, meaning that whether two, or more, entities are causally related will have to be demonstrated by experimentation. The possibility exists that two factors, X and Y, may be correlated because of a hidden factor Z, which triggers X and Y along two independent chains of causality. In this case, X or Y may be considered a cause when it is a side-effect or symptom [2].

4 A Way Ahead for Computational Systems Biology

Unlike reductionist approaches, systems biology reasons in terms of patterns and meta-patterns (patterns of patterns), looking for consistency among them while taking advantage of comprehensiveness. Accordingly, unlike classical *deductive* and *inductive* logical reasoning, appropriate for reductive approaches, systems biology works with *abductive* reasoning, that is the best possible inference that explains the ensemble of facts [4].

In complex networked systems, like living cells or organisms, the causal connection between two factors is mediated by a length-dependent chain of causality, that includes feedbacks, making the outcome of such a system to appear, at first sight, counter-intuitive. This has been observed in an integrated functional model of a cell, and described as *control by diffuse loops*, in which action (e.g., by a pharmacological agent) on a network (e.g., of chemical reactions) may bring about changes in processes without obvious direct mechanistic link between them, because they are mediated by several intermediate steps [5, 6]. In mechanistic models, the chain of causality, including feedbacks between their components, can be dissected, and understood, to interpret the experimental results in conjunction with model's simulations.

Networks of molecules, organelles, cells, and organs have the capacity to rewire themselves following challenges such as genetic or pharmacological modifications, interventions (e.g., caloric restriction, time-restricted feeding, exercise), disease, and aging, among others. Remodeling of metabolic networks in response to these interventions or natural causes constitutes a representative example. In this context, elementary flux modes (EFMs) [7] enable a precise quantification of a change in capacity of an organ or cell to metabolically rewire itself in response to a change in a key edge of the network [8]. The richness in rewiring capacity, quantified by the number of EFMs, measures the availability of ways (modes) a network can use to overcome a blockage (e.g., pharmacological inhibition, gene knock out) or raise the readiness of pathways to respond to, for example, gene editing or overexpression, noxious stimuli. From the correlation–causation perspective, a consequence of networks remodeling is that the x/y input/output of a phenomenon can be radically changed by rewiring, thus bringing forth the importance of what happens *in between* x and y. At present, AI, performed by deep learning of convolutional networks, is blind to the black box "that happens in between," that is, the causes of a phenomenon.

Comprehensive assessment of components from complex systems at different levels of organization (e.g., molecules, cells, and organs) poses several challenges at logistic, analytical, and interpretative stances. Several chapters of this volume present different approaches in distinct settings to address these challenges while underscoring the vast opportunities available to tackle problems about *knowing (or not) what we do not know*. When experimental design contemplates a priori coordination between experimental, analytical, and computational modeling strategies, discovery and unveiling of new phenomena or mechanisms can happen. Physiological outcomes reflect macroscopic emergent behavior at the organismal level such as respiration, movement, and electrical activity. Time series from electrocardiograms (ECGs), electroencephalograms (EKGs), patterns of gene expression or metabolites/ions, and animal movement, among many others, contain crucial information in the form of temporal patterns. Detecting and understanding those temporal patterns, provide insights into underlying mechanisms such as frequency-amplitude encoding of signaling, coexistence of circadian and ultradian rhythms of feeding behavior, organismic coordination, and integration of function, all essential to understand states of health and disease (*see*, e.g., Sugiura et al. chapter 10; Sohljoo et al. chapter 11; Ko et al. chapter 14; Ubaida-Moheen et al. chapter 8; Mendoza et al. chapter 16; Kembro et al. chapter 13; in this book). Moreover, the detection and identification of bipartite networks of pathways, comprising genes and metabolites or metabolites and pathways flux, can be achieved using integrated analysis of multiomics data, such as transcriptomics and metabolomics (*see* Aon et al., chapter 9 in this book) or metabolomics and fluxomics (Cortassa et al., chapter 7 in this book).

Evolutionary biology brings about the hallmark fact of biology, that is, that the networks machinery embodied in cells and organisms have evolved over hundreds of millions of years [9]. Their embodiment in, for example, animals, plants, insects, bacteria, fungi, enable their respective ability to power themselves, to divide, differentiate, see, fly, and move. In the case of humans, components of intelligence, such as *abstraction* and *analogy*, are among the hardest problems to emulate by AI [10]. We, humans, need our bodies because they form part of the unconscious activity of our brains associated with the intuitive knowledge of the physical and psychological aspects of the world around us [4, 10, 11].

Implicitly, evolutionary biology makes clear that it is not the same to be embedded in a silicon microchip than in the body of a carbon-based wired organism shaped by evolution. Consequently, to understand the evolutionary biology aspects of, for example, health, disease, aging, and to be able to discern which traits of those systems are accessible to computation from those that are not, is a long-term endeavor for CSB.

Acknowledgments

This work was supported by the Intramural Research Program of the National Institute on Aging, National Institutes of Health. The critical reading of the manuscript by Dr. Sonia Cortassa and Dr. Michel Bernier is gratefully acknowledged.

References

1. Hoffmann R, Malrieu JP (2020) Simulation vs. understanding: a tension, in quantum chemistry and beyond. Part A. Stage setting. Angew Chem Int Ed Engl 59(31): 12590–12610. https://doi.org/10.1002/anie.201902527
2. Hoffmann R, Malrieu JP (2020) Simulation vs. understanding: a tension, in quantum chemistry and beyond. Part B. The march of simulation, for better or worse. Angew Chem Int Ed Engl 59(32): 13156–13178. https://doi.org/10.1002/anie.201910283
3. Hoffmann R, Malrieu JP (2020) Simulation vs. understanding: a tension, in quantum chemistry and beyond. Part C. Toward consilience. Angew Chem Int Ed Engl 59(33):13694–13710. https://doi.org/10.1002/anie.201910285
4. Koch K (2019) The feeling of life itself. Why consciousness is widespread but can't be computed. MIT Press, Cambridge, MA
5. Aon MA, Cortassa S (2012) Mitochondrial network energetics in the heart. Wiley Interdiscip Rev Syst Biol Med 4(6):599–613. https://doi.org/10.1002/wsbm.1188
6. Cortassa S, O'Rourke B, Winslow RL, Aon MA (2009) Control and regulation of mitochondrial energetics in an integrated model of cardiomyocyte function. Biophys J 96(6):

2466–2478. https://doi.org/10.1016/j.bpj.2008.12.3893

7. Schuster S, Dandekar T, Fell DA (1999) Detection of elementary flux modes in biochemical networks: a promising tool for pathway analysis and metabolic engineering. Trends Biotechnol 17(2):53–60. https://doi.org/10.1016/s0167-7799(98)01290-6
8. Cortassa S, Sollott SJ, Aon MA (2018) Computational modeling of mitochondrial function from a systems biology perspective. Methods Mol Biol 1782:249–265. https://doi.org/10.1007/978-1-4939-7831-1_14
9. Sejnowski TJ (2018) The deep learning revolution. MIT Press, Cambridge, MA
10. Mitchell M (2019) Artificial intelligence. A guide for thinking humans. Farrar, Strauss and Giroux, New York
11. Hofstadter DR (1979) Godel, Escher, Bach: an eternal golden braid. Basic Books, New York

Part I

Systems Biology of the Genome, Epigenome, and Redox Proteome

Chapter 2

Bioinformatic Analysis of CircRNA from RNA-seq Datasets

Kyle R. Cochran, Myriam Gorospe, and Supriyo De

Abstract

Circular RNAs (circRNAs) are a vast class of covalently closed, noncoding RNAs expressed in specific tissues and developmental stages. The molecular, cellular, and pathophysiologic roles of circRNAs are not fully known, but their impact on gene expression programs is beginning to emerge, as circRNAs often associate with RNA-binding proteins and nucleic acids. With rising interest in identifying circRNAs associated with disease processes, it has become particularly important to identify circRNAs in RNA sequencing (RNA-seq) datasets, either generated by the investigator or reported in the literature. Here, we present a methodology to identify and analyze circRNAs in RNA-seq datasets, including those archived in repositories. We elaborate on the unique features of circRNAs that require specialized attention in RNA-seq datasets, the software packages designed for circRNA identification, the ongoing efforts to reconstruct the body of circRNAs starting from unique circularizing junctions, and the interacting factors that can be proposed from putative circRNA body sequences. We discuss the advantages and limitations of the current approaches for high-throughput circRNA analysis from RNA-sequencing datasets and identify areas that would benefit from the development of superior bioinformatic tools.

Key words RNA-seq, Circular RNA (circRNA), Bioinformatics, Backsplicing, Gene expression

1 Introduction

Circular (circ)RNAs originate from transcribed linear RNAs in which 5′ and 3′ ends become covalently closed, often via a backsplicing reaction catalyzed by the spliceosome machinery. CircRNAs may contain sequences from exons, introns, or combinations of both, and they may arise from messenger (m)RNAs or from noncoding RNAs [1]. Given their ability to modulate gene expression programs through interaction with a range of proteins and RNAs (particularly microRNAs), and in some cases through their partial translation, circRNAs are increasingly recognized as regulators of gene expression programs [2]. Although incompletely annotated, circRNAs comprise a vast class of transcripts, numbering tens of thousands of different molecules. Some circRNAs are highly abundant and ubiquitous, but

Sonia Cortassa and Miguel A. Aon (eds.), *Computational Systems Biology in Medicine and Biotechnology: Methods and Protocols*, Methods in Molecular Biology, vol. 2399, https://doi.org/10.1007/978-1-0716-1831-8_2,

most circRNAs appear to be expressed in low levels and only in certain tissues. Accordingly, many circRNAs are increasingly recognized as being informative about developmental stages of cells, tissues, and organs. Moreover, the covalently closed structure of circRNAs renders them relatively stable and hence they are believed to be valuable biomarkers of organ function, dysfunction, and disease [3–6].

The past few years have seen an escalation in interest in identifying circRNAs that could serve as indicators of pathophysiologic states. These efforts have been realized primarily by high-throughput identification of circRNAs expressed in a given tissue or organ, either by sequencing all RNA (both circRNA and linear RNA) or by first enriching for circRNAs [7, 8]. Each of these strategies has advantages and drawbacks. An important benefit of sequencing both circRNA and linear RNA is that the relative abundance of circRNAs and the linear counterparts can be compared, while enriching for circRNAs by first digesting the linear RNA (using RNase R among other methods [7, 8]) helps to identify low-abundance circRNAs and gain deeper knowledge of circRNAs implicated in a specific biological process.

This chapter focuses on the methodology to identify circRNAs from total RNA sequencing (RNA-seq) datasets generated by the investigator or published by other groups. As these datasets were generally designed with linear RNA analysis in mind, the key strategy is to identify circRNA junctions among the sequencing reads and to create datasets from these junction sequences. This step-by-step guideline to analyze circRNAs from RNA-seq datasets offers valuable new insight into biological processes from data that were initially generated to investigate linear RNAs, and thus representing important savings in time and resources.

2 Materials

1. Desktop with Windows, Mac OSX, and Linux operating systems with high memory.
2. Large data-storage device.
3. Microsoft Internet Explorer or Mozilla Firefox or Google Chrome browsers.

3 Methods

3.1 Identify RNA-seq Datasets to Analyze

These datasets may originate from the investigator's own RNA-seq studies in a given cell line, tissue type, healthy organ, or pathology specimen. Alternatively, the datasets may have been obtained and made publicly available by other groups who conducted RNA-seq analysis with a different goal in mind (e.g., identifying mRNAs), and deposited the RNA-seq data in public databases (e.g., Gene Expression Omnibus). (*see* **Note 1**)

3.2 Obtain the FASTQ Files from These Datasets, Containing Unprocessed RNA-seq Reads

This step is critical because the junction sequence reads are typically eliminated from subsequent analysis and are precisely the reads needed for circRNA identification.

3.3 Align the FASTQ Files to the Human Genome

This step can be performed using an alignment program such as the STAR aligner, TopHat2, or BWA-MEM [9–11]. These aligners will yield either Sequence Alignment Map (SAM) files, or their binary counterpart, Binary Alignment Map (BAM) files. This alignment is necessary to discern where on the human genome the sequences are located so that parent genes can be associated with specific circRNAs.

3.4 Use a circRNA-Identifying Software Such as CIRCexplorer2 to Generate Annotated circRNA Junction Reads

A key distinguishing feature of a circular RNA is the junction sequence, where the 3′ and 5′ ends meet to form a circular structure (Fig. 1). This feature differentiates circRNA from linear RNA. There are many widely used programs that identify and categorize circRNAs; the programs most used are listed in Table 1. We summarize the advantages and limitations, reviewed recently [12–14], stating only the competency level (precision, sensitivity) and computational ability (efficiency, memory usage, and disc space).

3.5 Construct Bioinformatically the Body of the circRNAs

While CIRCexplorer2 and similar programs can detect the single junction segments of each circRNA, it does not inform about the body of the circRNA, as sequenced reads from the body of the circRNA will be shared by sequenced reads from the parent linear RNA. Some programs (e.g., CIRCexplorer2 or CIRI) [15, 16] are capable of assembling an approximate body of each circRNA starting from the junction point and working its way outward using sequenced reads in the dataset (Fig. 2). The main reasons for performing this assembly are to identify the likely sequence of the circRNA and to characterize alternative isoforms of a circRNA; Table 2 lists software packages with de novo assembly features. Column 3 refers to the ability of the algorithms to detect circRNAs based on their genomic position (e.g., exonic circRNA versus intergenic circRNA) (*see* **Note 2**).

3.6 Analyze Bioinformatically the Levels of circRNAs

The next step is to quantify the expression levels of circRNAs and identify those circRNAs with significantly different abundance between comparison groups. The expression levels of circRNAs can also be compared to the expression levels of their parent linear RNA. The most commonly used R packages for this analysis are

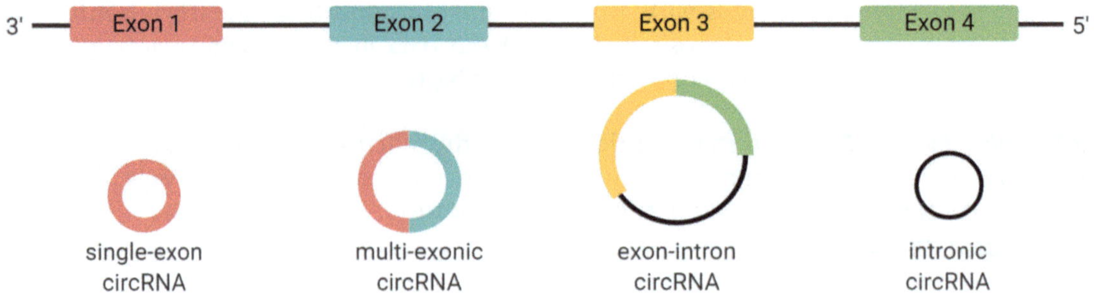

Fig. 1 Schematic representation of typical circular (circ)RNA types. CircRNAs generally arise from linear pre-mRNAs that undergo backsplicing leading to the ligation of 5′ and 3′ ends of a complete or partial exon (single-exon circRNA), two or more exons (multiexonic circRNA), exonic and intronic sequences (exon–intron circRNA), or only intronic sequences (intronic circRNA). Created at BioRender.com

Table 1
Widely used programs to identify and categorize circRNAs. Commonly used programs are evaluated for sensitivity, precision, processing speed, memory usage, and disc space requirements (*see* Note 3)

Name	Sensitivity	Precision	Run time	RAM usage	Disc space
CIRI	High	High	Fast	Average	High
CIRCexplorer	High	High	Fast	Average	Low
circRNA_finder	Low	High	Fast	Average	Low
DCC	Low	High	Fast	Average	Low
find_circ	Low	Low	Fast	Low	Low
KNIFE	High	High	Average	High	High
MapSplice	Low	High	Slow	Average	High
NCLScan	Low	High	Slow	Low	High
PTESFinder	High	High	Average	High	High
Segemehl	High	Low	Slow	High	High
UROBORUS	Low	Low	Average	Low	Low

edgeR and DESeq2 [17, 18], and the outputs are lists of circRNAs with the respective changes in abundance change and corrected *p*-values for statistical significance. This analysis identifies select circRNAs differentially expressed (more abundant or less abundant) in specific disease conditions, developmental stages, responses to immune agents or damage, etc.

3.7 Propose Functions for circRNAs Differentially Abundant

The final step is to begin to consider and possibly explore the function of the specific circRNAs of interest (Fig. 3). Unfortunately, this task is complicated for several reasons. One is that at present there is no universal nomenclature for circRNAs, so examining if other groups have reported functions for specific circRNAs

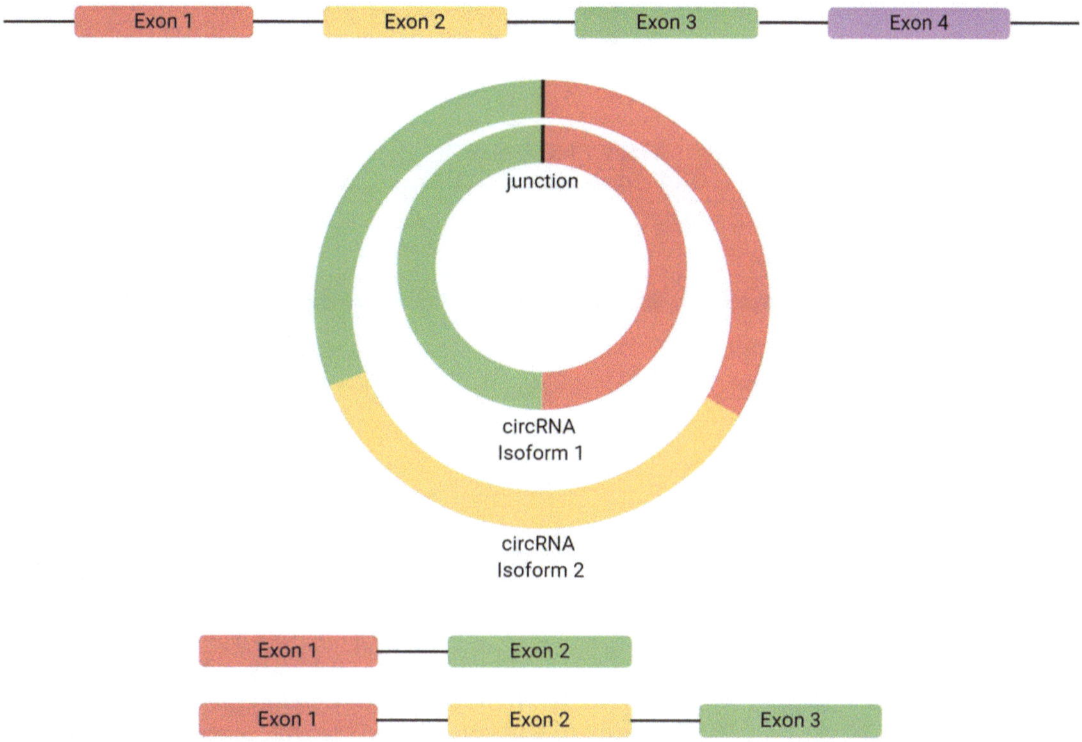

Fig. 2 CircRNA isoforms. It is possible for several circRNAs to share a junction sequence but have distinct body sequences. Since circRNA identification software programs parse out circRNAs by the backspliced junction sequence, this overlap could miss certain circRNAs. To avoid this error, de novo assembly options predict likely circRNA body sequences. Created at BioRender.com

Table 2
CircRNA analysis packages. Table specifies whether the program performs de novo assembly of the circRNAs and the type of circRNAs that it can analyze

Software	De novo assembly	Genomic position
CIRI	Yes	Exonic, Intronic, Intergenic
CIRCexplorer	Yes	Exonic, Intronic
circRNA_finder	No	Exonic, Intronic, Intergenic
DCC	Ambiguous	Exonic, Intronic, Intergenic
find_circ	Yes	Exonic, Intronic, Intergenic
KNIFE	Ambiguous	Exonic
MapSplice	Ambiguous	Exonic, Intronic
NCLScan	No	Exonic
PTESFinder	No	Exonic
Segemehl	Yes	Exonic, Intronic, Intergenic
UROBORUS	No	Exonic

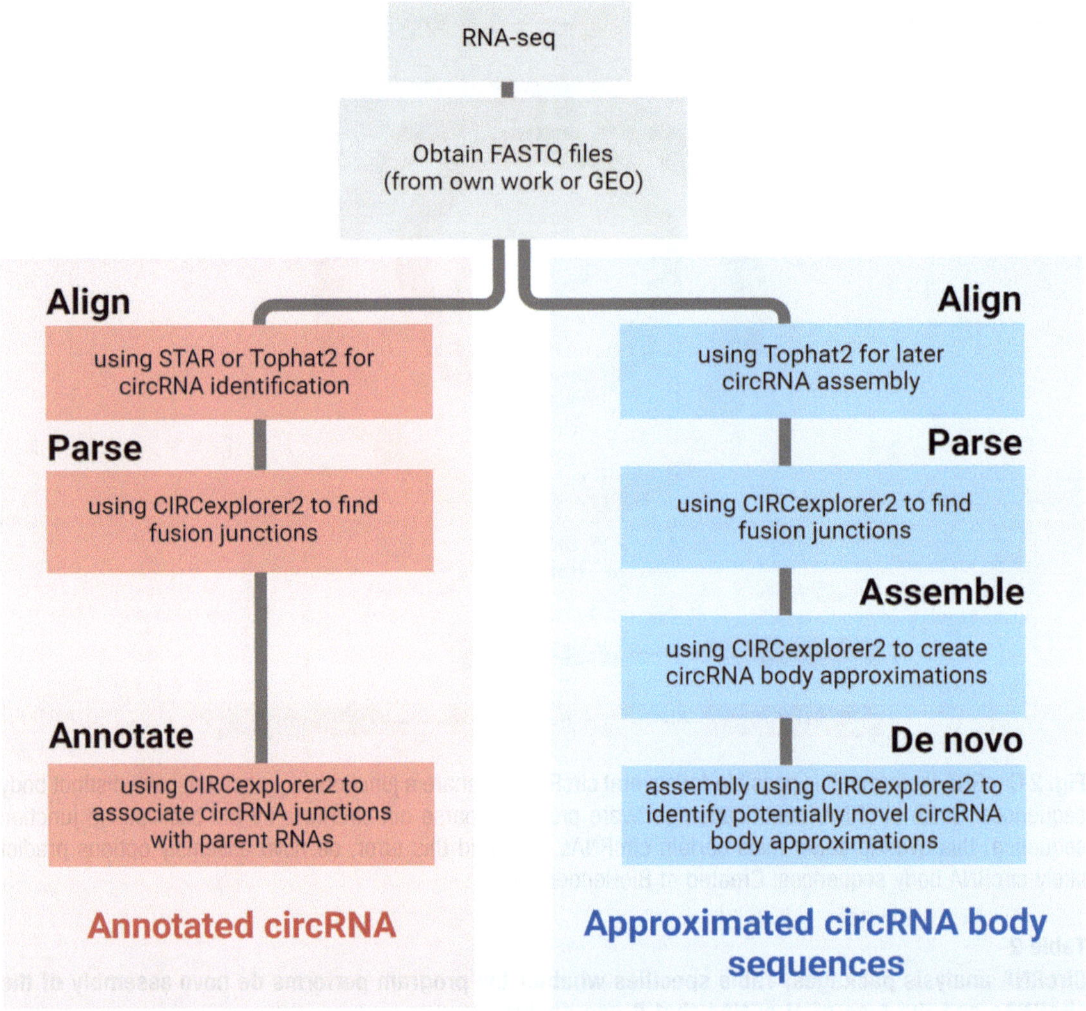

Fig. 3 CIRCexplorer2 workflow. CIRCexplorer2 takes aligned sequencing data in the form of BAM files or BED (Browser Extensible Data) files and provides annotated circRNA as output. This software also offers a de novo assembly option which constructs a circRNA body approximation based on reference annotations provided by the user. Created at BioRender.com

may be challenging if their names cannot be recognized; in this regard, the nomenclature used in circBase (http://www.circbase.org/) appears to be the most popular [19]. Another important obstacle is that little is known at present about the function of the circRNAs. One can begin by looking at the function of the linear RNA, which may offer clues as to the spatial and temporal expression of the circRNA. Alternatively, one can focus on the molecules interacting with the circRNA of interest. MicroRNAs and for RNA-binding proteins (RBPs) interacting with a circRNA of interest can be identified through a variety of methods, although this approach is tedious and can typically be done for only a handful of

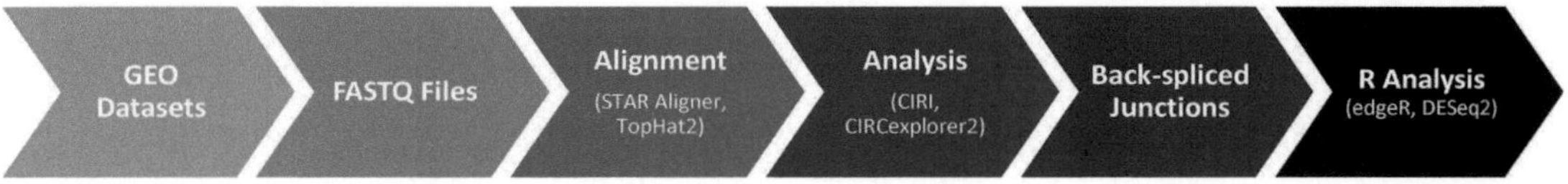

Fig. 4 Full circRNA-seq analysis workflow. The goal of the method described here is to take raw RNA-seq data (FASTQ files) from the Gene Expression Omnibus, align and analyze them, identify the junctions, and establish comparisons among samples

circRNAs at a time. Programs such as Circular RNA Interactome (circInteractome: https://circinteractome.nia.nih.gov/) [20] or ENCORI (The Encyclopedia of RNA Interactomes, formerly starBase: http://starbase.sysu.edu.cn/) [21] identify putative interactions of circRNA sequences with RBPs and microRNAs. A vast array of molecular approaches can then be utilized to study the molecular function of the circRNAs of interest (Fig. 4).

4 Example to Illustrate Workflow

1. To demonstrate how this workflow operates, we selected mouse RNA-seq datasets from the Gene Expression Omnibus (GEO) repository and passed it through the analysis pipeline. We identified four samples from two studies that used a high-purity circRNA isolation methods and deep RNA-seq analysis to ensure a robust and large circRNA sample size (GSE92632 and GSE136004); RNA was collected from mouse myoblasts cultured in proliferation medium (growth medium, "GM") (samples GSM2433793/SRR5122015 and GSM2433794/SRR5122016) and from myoblasts differentiated in culture into myotubes (differentiation medium, "DM") (samples GSM4039265/SRR10004192 and GSM4039266/SRR10004193). Figure 5 depicts the specific Linux commands and bioinformatic software used, and the corresponding output directories from one of the samples (GSM2433793/SRR5122015). The two files required for further statistical analysis are "circularRNA_known.txt" and "circularRNA_full.txt" for circRNA junction analysis and de novo circRNA body approximation analysis, respectively.
2. Once these files were generated, we used the bioinformatic analysis packages mentioned above to measure differential gene expression between DM (differentiated) and GM (undifferentiated) myoblast populations using EdgeR. To measure changes in the abundance of expressed circRNAs, we calculated the log CPM (Counts Per Million) transformation of the read counts across the chosen samples. Additionally, we combined parent RNA and isoform names to create a unique tracking name for each circRNA in order to facilitate later functional analysis.

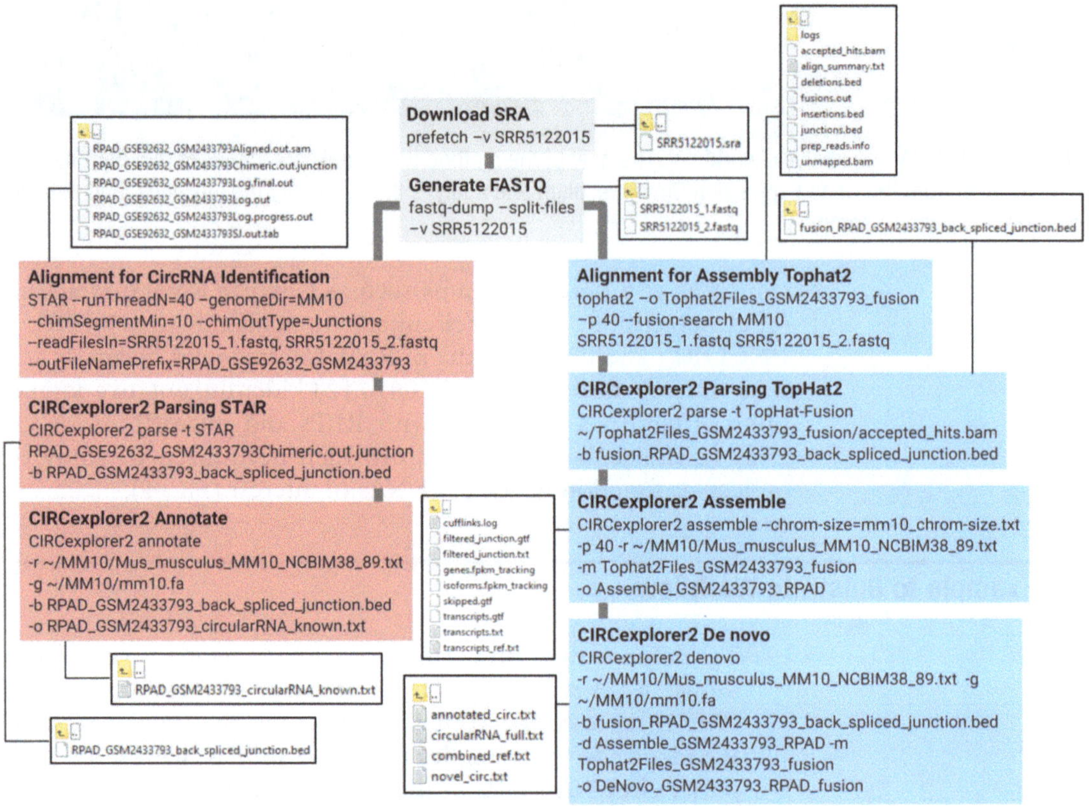

Fig. 5 Command/output workflow. Detailed depiction of the data acquisition and processing before statistical analysis. Each command has a corresponding box showing the files created when this command is run. The left branch of the diagram represents the circRNA junction analysis and the right branch represents the de novo approximation assembly of the circRNA body. Created at BioRender.com

3. The volcano plot in Fig. 6 represents each circRNA and their expression level. The final step is to identify which circRNAs are most significantly changed and how they are associated with the pathology or genes of interest using tools like circInteractome and circBase. Using the specific circRNA name, we can investigate the function of circRNAs changing significantly.

5 Notes

1. When selecting data to analyze, it is important to know whether the dataset was generated by single-read or paired-read RNA-sequencing, as this influences the choice of aligner and circRNA identification software. In fact, consulting the aligner and circRNA identification software manuals is strongly advised. It is worth mentioning that the CIRCexplorer2 de novo assembly option requires the TopHat2-Fusion aligner [22]. Additionally, to prepare the data for efficient analysis

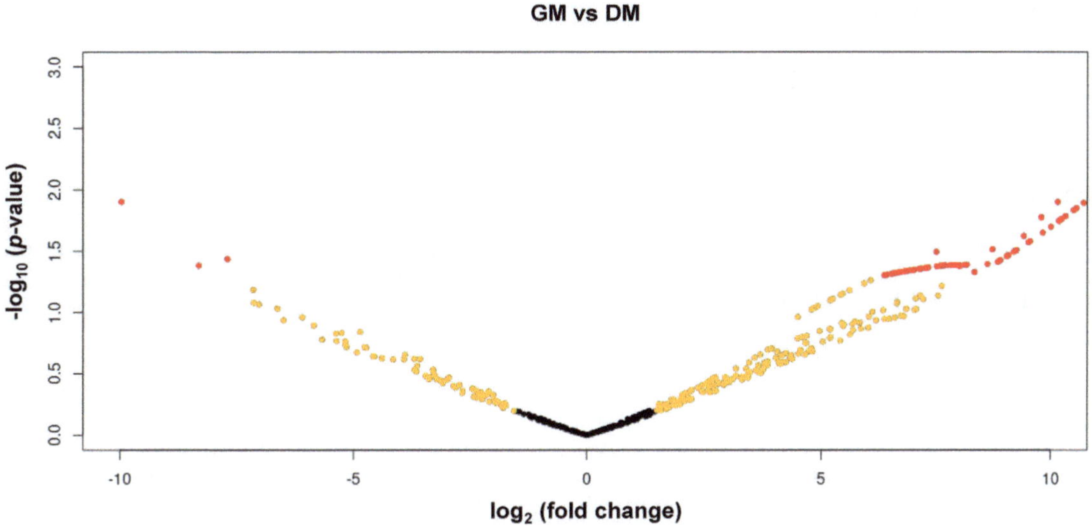

Fig. 6 CircRNA expression volcano plot. Each point on the plot represents a different circRNA with the fold-change ($\log_2$) on the *x*-axis and the *p*-value ($-\log_{10}$) on the *y*-axis. The yellow points represent circRNAs showing robust changes in abundance but not statistically significant; the red points represent circRNAs showing robust changes in abundance and statistically significant (p-value < 0.05)

in R, all reads should be amalgamated into a single data table with the user-provided gene annotations. The software JMP (https://www.jmp.com/en_us/home.html) is remarkably useful for this task, as it allows the user to match columns of tables and update datasets based on the rows they have in common, and can work with exceedingly large datasets.

2. Algorithms that constructed the body of a circRNA without using intron–exon annotations produced many more false-positive sequences than those algorithms that used exon–intron annotations [23]. Additionally, each algorithm requires a parameter that specifies the distance allowed between splice sites, ranging from 100 nucleotides to hundreds of kilobases. Algorithms that allowed distances shorter than 200 nucleotides had increased rates of false positives [14].

3. In overall performance, software packages CIRCexplorer, CIRI, KNIFE, and PTESFinder [15, 16, 24, 25] were strong in almost every category, promoting a balanced performance. CIRI and PTESFinder incur substantial computational cost, so low-budget experiments might consider other software packages. Additionally, NCLScan scored similarly high on precision, but somewhat less sensitive [26]. Other programs did not score as highly on speed and accuracy [12, 27–30]. In the burgeoning field of circRNA bioinformatics, there is still much need to develop low-cost tools that perform with high precision, speed, and sensitivity.

Acknowledgments

This work was supported in full by the National Institute on Aging Intramural Research Program, National Institutes of Health.

References

1. Eger N, Schoppe L, Schuster S, Laufs U, Boeckel JN (2018) Circular RNA splicing. Adv Exp Med Biol 1087:41–52
2. Yu CY, Kuo HC (2019) The emerging roles and functions of circular RNAs and their generation. J Biomed Sci 26:29
3. Kumar L, Shamsuzzama HR, Baghel T, Nazir A (2017) Circular RNAs: the emerging class of non-coding RNAs and their potential role in human neurodegenerative diseases. Mol Neurobiol 54:7224–7234
4. Hu W, Bi ZY, Chen ZL et al (2018) Emerging landscape of circular RNAs in lung cancer. Cancer Lett 427:18–27
5. Qu S, Yang X, Li X et al (2015) Circular RNA: a new star of noncoding RNAs. Cancer Lett 365:141–148
6. Chen B, Huang S (2018) Circular RNA: an emerging non-coding RNA as a regulator and biomarker in cancer. Cancer Lett 418:41–50
7. Xiao MS, Wilusz JE (2019) An improved method for circular RNA purification using RNase R that efficiently removes linear RNAs containing G-quadruplexes or structured 3′ ends. Nucleic Acids Res 47:8755–8769
8. Panda AC, De S, Grammatikakis I et al (2017) High-purity circular RNA isolation method (RPAD) reveals vast collection of intronic circRNAs. Nucleic Acids Res 45:e116
9. Dobin A, Davis CA, Schlesinger F, Drenkow J, Zaleski C, Jha S et al (2013) STAR: ultrafast universal RNA-seq aligner. Bioinformatics 29:15–21
10. Kim D, Pertea G, Trapnell C, Pimentel H, Kelley R, Salzberg SL (2013) TopHat2: accurate alignment of transcriptomes in the presence of insertions, deletions and gene fusions. Genome Biol 14:R36
11. Li H (2013) Aligning sequence reads, clone sequences and assembly contigs with BWA-MEM. arXiv preprint arXiv:13033997
12. Zeng X, Lin W, Guo M, Zou Q (2017) A comprehensive overview and evaluation of circular RNA detection tools. PLoS Comput Biol 13:e1005420
13. Cheng J, Metge F, Dieterich C (2016) Specific identification and quantification of circular RNAs from sequencing data. Bioinformatics 32:1094–1096
14. Hansen TB, Venø MT, Damgaard CK, Kjems J (2016) Comparison of circular RNA prediction tools. Nucleic Acids Res 44:e58
15. Zhang XO, Dong R, Zhang Y, Zhang JL, Luo Z, Zhang J, Chen LL, Yang L (2016) Diverse alternative back-splicing and alternative splicing landscape of circular RNAs. Genome Res 2016. https://doi.org/10.1101/gr.202895.115
16. Gao Y, Wang J, Zhao F (2015) CIRI: an efficient and unbiased algorithm for de novo circular RNA identification. Genome Biol 16:4
17. Robinson MD, McCarthy DJ, Smyth GK (2010) edgeR: a bioconductor package for differential expression analysis of digital gene expression data. Bioinformatics 26:139–140
18. Love MI, Huber W, Anders S (2014) Moderated estimation of fold change and dispersion for RNA-seq data with DESeq2. Genome Biol 15:550
19. Glazar P, Papavasileiou P, Rajewsky N (2014) circBase: a database for circular RNAs. RNA 20:1666–1670
20. Dudekula DB, Panda AC, Grammatikakis I, De S, Abdelmohsen K, Gorospe M (2016) CircInteractome: a web tool for exploring circular RNAs and their interacting proteins and microRNAs. RNA Biol 13:34–42
21. Li JH, Liu S, Zhou H, Qu LH, Yang JH (2014) starBase v2.0: decoding miRNA-ceRNA, miRNA-ncRNA and protein-RNA interaction networks from large-scale CLIP-Seq data. Nucleic Acids Res 42(Database issue):D92–D97
22. Kim D, Salzberg SL (2011) TopHat-Fusion: an algorithm for discovery of novel fusion transcripts. Genome Biol 12:R72
23. Hansen TB (2018) Improved circRNA identification by combining prediction algorithms. Front Cell Dev Biol 6:20
24. Szabo L, Morey R, Palpant NJ, Wang PL, Afari N, Jiang C, Parast MM, Murry CE, Laurent LC, Salzman J (2015) Statistically based splicing detection reveals neural enrichment and tissue-specific induction of circular RNA

during human fetal development. Genome Biol 16:126

25. Izuogu OG, Alhasan AA, Alafghani HM, Santibanez-Koref M, Elliott DJ, Jackson MS (2016) PTESFinder: a computational method to identify post-transcriptional exon shuffling (PTES) events. BMC Bioinformatics 17:31
26. Chuang TJ, Wu CS, Chen CY, Hung LY, Chiang TW, Yang MY (2016) NCLscan: accurate identification of non-co-linear transcripts (fusion, trans-splicing and circular RNA) with a good balance between sensitivity and precision. Nucleic Acids Res 44:e29
27. Hoffmann S, Otto C, Doose G, Tanzer A, Langenberger D, Christ S et al (2014) A multi-split mapping algorithm for circular RNA, splicing, trans-splicing and fusion detection. Genome Biol 15:R34
28. Memczak S, Jens M, Elefsinioti A, Torti F, Krueger J, Rybak A et al (2013) Circular RNAs are a large class of animal RNAs with regulatory potency. Nature 495:333–338
29. Song X, Zhang N, Han P, Moon BS, Lai RK, Wang K et al (2016) Circular RNA profile in gliomas revealed by identification tool UROBORUS. Nucleic Acids Res 44:e87
30. Wang K, Singh D, Zeng Z, Coleman SJ, Huang Y, Savich GL et al (2010) MapSplice: accurate mapping of RNA-seq reads for splice junction discovery. Nucleic Acids Res 38:e178

Chapter 3

Single-Cell Analysis of the Transcriptome and Epigenome

Krystyna Mazan-Mamczarz, Jisu Ha, Supriyo De, and Payel Sen

Abstract

Epigenome regulation has emerged as an important mechanism for the maintenance of organ function in health and disease. Dissecting epigenomic alterations and resultant gene expression changes in single cells provides unprecedented resolution and insight into cellular diversity, modes of gene regulation, transcription factor dynamics and 3D genome organization. In this chapter, we summarize the transformative single-cell epigenomic technologies that have deepened our understanding of the fundamental principles of gene regulation. We provide a historical perspective of these methods, brief procedural outline with emphasis on the computational tools used to meaningfully dissect information. Our overall goal is to aid scientists using these technologies in their favorite system of interest.

Key words Single-cell, Epigenome, Transcriptome, Multiomics

1 Introduction

With the completion of the human genome in 2003 [1], it was anticipated that in the following decade, scientists would have solutions to most major diseases afflicting humans. Although much was gleaned from the sequencing information, the returns on investment were relatively modest. This was primarily due to the faulty assumption that common genetic variants caused all human diseases. Nevertheless, it ushered in the search for single nucleotide polymorphisms (SNPs) in individuals, accelerated genome-wide association studies (GWAS) which attributed SNPs to disease risk, and whole-exome sequencing that sequenced the coding exons of all genes. Unfortunately, despite the assembly of large SNP atlases such as the SNP Consortium [2], the International HapMap Project [3] and the 1000 Genomes Project [4], only a few SNPs could be associated causally with human disease with <10% explaining heritability and almost none the underlying biology. Furthermore,

Krystyna Mazan-Mamczarz and Jisu Ha contributed equally with all other contributors.

Sonia Cortassa and Miguel A. Aon (eds.), *Computational Systems Biology in Medicine and Biotechnology: Methods and Protocols*, Methods in Molecular Biology, vol. 2399, https://doi.org/10.1007/978-1-0716-1831-8_3,

SNPs largely localized to intergenic regions, many kilobases away from a gene and almost nothing at the time was known about the function of nongenic "junk" DNA. It became clear that human diseases were complex.

To explain this apparent disconnect between genetic variants and disease causation, work during the early/mid 2000s, also focused on epigenetic regulation of genetic information, a way to link environmental exposures to phenotypic traits. Epigenetics comprised the study of chemical modifications on DNA and histone proteins, chromatin accessibility, and RNA transcript measurements. Epigenetic mechanisms were found to be transgenerational, lending some credence to Lamarckism. The NIH Roadmap Epigenomics Mapping Consortium [5], the Encyclopedia of DNA Elements (ENCODE) project [6], the related model organism ENCODE (modENCODE) project [7] and the International Human Epigenome Consortium (IHEC) [8] were thus launched to provide reference epigenomes generated using standardized protocols.

The challenge to deciphering useful information from ENCODE/modENCODE studies was to relate data from cell lines or model organisms to actual human tissues. Given the inter-individual variability in human populations, it was difficult to match timing, staging and cell types. Furthermore, the ENCODE project constituted bulk or average measurements and it was realized that ensemble averages can distort the understanding of individual cell responses. Human tissues themselves present a highly heterogenous environment that is further altered with disease. These gaps in understanding fueled the launch of the Human Cell Atlas (HCA) Consortium in 2017 [9], that in the next few years will provide comprehensive reference maps of all cells of the human body. In part, revolutionary advances in single cell preparation protocols, sensitive detection of biomolecules, sophisticated fluidics for single cell capture, clever multiplexing strategies and importantly, computational pipelines for rapid parsing of sparse data has greatly catalyzed this initiative. For single cell preparation methods, we direct the readers to a comprehensive review by the Regev lab [10]. In this chapter, we focus on summarizing the various methods for single-cell transcriptomic and epigenomic approaches and outline the computational workflow and tools for analysis. *See* Table 1 for a glossary of terms used in this chapter.

1.1 Single-Cell Transcriptomic Approaches

The single-cell revolution started with strategies to analyze the transcriptome in individual cells. The first reports of mRNA sequencing in single cells came from the profiling of individual mouse oocytes and blastomeres manually picked under a microscope. Individual cells were lysed, the RNA reverse-transcribed, the cDNA amplified by PCR, and sequenced on the SOLiD platform [11, 12].

Table 1
Glossary of terms

Term	Description
10× Genomics	A biotechnology company that designs and manufactures single-cell fluidics and kits for genomics, transcriptomics, epigenomics etc.
3C	Chromosome Conformation Capture.
Alevin	A pseudoalignment-based RNA-seq quantification program.
ArchR	An end-to-end R package for analysis of scATAC-seq data.
ArchR projects	Objects in ArchR to associate Arrow files together into a single analytical framework in R.
Arrow files	The base unit of an analytical project in ArchR that stores all of the metadata associated with an individual sample.
ASAP-seq	Select Antigen Profiling by sequencing.
ATAC-RNA-seq	Simultaneous profiling of chromatin accessibility by ATAC-seq and transcriptome by scRNA-seq in single cells.
ATAC-seq	Assay for Transposase Accessible Chromatin sequencing.
Autoencoder	A type of artificial neural network-based learning.
AutoImpute	Python package for analysis and implementation of imputation methods.
BAM	Binary Alignment Map, a compressed binary format for storing sequence data.
BRIE	Bayesian Regression for Isoform Estimation.
CCA	Canonical Correlation Analysis.
CCAN	Cis-coaccessibility networks which are hubs of coaccessibility identified utilizing Cicero.
Cell Hashing	A sample multiplexing method with oligo tagged antibodies directed against cell surface proteins based on the concept of hash functions in computer science to index datasets with specific features.
Cell Ranger	A set of analysis pipelines for processing 10× Genomics data.
CEL-seq	Cell Expression by Linear amplification and sequencing.
ChromVAR	Chromatin Variation Across Regions, a method to assess TF dynamics from scATAC-seq data.
Cicero	An algorithm that identifies coaccessible pairs of DNA elements using scATAC-seq data and makes predictions about promoter–enhancer pairs.
cisTopic	A probabilistic framework used to simultaneously discover coaccessible cis-regulatory elements and derive cell states using TOPIC modeling.
CiteFuse	A streamlined package consisting tools for preprocessing analysis and web-based visualization of CITE-seq data.
CITE-seq	Cellular Indexing of Transcriptomes and Epitopes by sequencing.
Clonealign	A method that assigns gene expression states to cancer clones using single-cell data.
coupleNMF	coupled Nonnegative Matrix Factorizations.

(continued)

Table 1
(continued)

Term	Description
CRISP-seq	A reverse genetics method that allows for analysis of thousands of CRISPR-mediated perturbations within a single experiment by combining pooled CRISPR screen to single-cell RNA sequencing. Similar techniques are Perturb-seq and CROP-seq.
CROP-seq	CRISPR droplet-sequencing, a reverse genetics method that allows for analysis of thousands of CRISPR-mediated perturbations within a single experiment by combining pooled CRISPR screen to single-cell RNA sequencing. Similar techniques are Perturb-seq and CRISP-seq.
CSV	A Comma-Separated Values file is a delimited text file format.
CUT&RUN	Cleavage Under Targets and Release Using Nuclease.
CUT&Tag	Cleavage Under Targets and Tagmentation.
CytoSeq	Gene expression cytometry.
DBSCAN	Density-Based Spatial Clustering of Applications with Noise.
DecontX	A Bayesian method to estimate and remove contamination in individual cells.
Dip-C	Method to obtain high-resolution contact maps in single diploid cells.
DoubletDecon	R package that uses deconvolution to identify and remove doublets in scRNA-seq data.
DoubletFinder	R package that can interface with Seurat and can predict and remove doublets in scRNA-seq data.
Drop-ChIP	Chromatin-immunoprecipitation in droplets.
Drop-seq	Method to profile mRNA transcripts from nanoliter-sized droplets of individual cells.
Dynamo	A tool that predicts cell states over time periods (RNA velocity), and incorporates not only splicing information but also promoter state switching, translation and RNA/protein degradation by taking advantage of scRNA-seq and combined transcriptomics and proteomics.
ENCODE	Encyclopedia of DNA Elements.
ENCODE blacklist regions	A comprehensive set of regions in the human, mouse, worm, and fly genomes that have artifactual high signal in next-generation sequencing experiments.
Expedition	A computational framework consisting of outrigger, anchor and bonvoyage, algorithms to detect alternative splicing, assign modalities and visualize results, respectively.
Feature matrix	A matrix listing the number of UMIs associated with a feature (row) and a barcode (column) created for analysis of scRNA-seq or scATAC-seq data.
Fluidigm C1	Commercially available, automated single-cell isolation and preparation system for genomic analysis from Fluidigm.
FRiP	FRaction of fragments in Peaks
fuzzy C-means	A clustering algorithm similar to K-means clustering that computes centroids and clusters cells.

(continued)

Table 1
(continued)

Term	Description
GBR	Gradient Boosting Regression.
GEM	Gel Beads in Emulsion.
GFA	Group Factor Analysis.
Graph-based clustering	An unsupervised classification algorithm that first identifies read similarities by performing pairwise comparisons and then constructs a graph in which the vertices correspond to sequence reads, the edges are overlapping reads and their similarity score is an edge weight.
Graphical LASSO	Least Absolute Shrinkage and Selection Operator
GWAS	Genome-wide Association Studies.
Harmony	A unified framework for data integration, visualization, analysis, and interpretation of single-cell genomics data across discrete timepoints.
HCA	Human Cell Atlas.
HDF5	Hierarchical Data Format version 5 is a file format that stores and organizes large data.
Hi-C	all-to-all 3C method.
Hierarchical clustering	Clustering based on similarities in objects.
HTML	HyperText Markup Language is a standard markup language for documents designed to be displayed in a web browser.
ICA	Independent Component Analysis.
iCELL8	A single-cell system from Takara Bio with open platforms that enable the processing of hundreds of single cells or nuclei.
IHEC	International Human Epigenome Consortium.
inDrop	indexing Droplets.
iNMF	Nonnegative Matrix Factorization.
IVT	In vitro transcription.
Kallisto/Bustools	A pseudoalignment-based RNA-seq quantification program.
kBET	k-nearest-neighbor Batch-Effect Test.
K-means clustering	A clustering algorithm that computes the centroids and clusters cells.
Knee plot	A standard single-cell RNA-seq ranked-ordered UMI plot that is used to determine a threshold for considering cells valid for analysis in an experiment.
LDA	Latent Dirichlet Allocation, an example of TOPIC modeling.
LIGER	Linked Inference of Genomic Experimental Relationships, a computational pipeline for integrating and analyzing multiple single-cell datasets.
Louvain algorithm	An algorithm to extract communities from large networks utilizing the greedy optimization method.
LSI	An indexing and retrieval method that uses SVD to identify patterns in the relationships between the terms and concepts.

(continued)

Table 1
(continued)

Term	Description
MACS	Model-based Analysis of ChIP-Seq, a popular peak calling algorithm.
MAGIC	Markov Affinity-based Graph Imputation of Cells.
Mandalorion	A tool to detect isoforms in scRNA-seq data.
MARS-seq	MAssively parallel RNA Single-cell sequencing.
MATCHER	Manifold Alignment to CHaracterize Experimental Relationships.
mcImpute	Matrix Completion based Imputation for scRNA-seq data.
MDS	MultiDimensional Scaling.
MEX	Market EXchange (MEX) is a format that is used to represent the gene-barcode matrix output by Cell Ranger.
MIMOSCA	Multiple Input Multiple Output Single Cell Analysis, a method to analyze single-cell Perturb-seq data.
MISC	Missing Imputation for Single-Cell RNA-seq.
MISO	Mixture-of-Isoforms, a model that estimates expression of alternatively spliced exons and isoforms.
MNN	Mutual Nearest Neighbors.
modENCODE	Model Organism Encyclopedia of DNA Elements.
Monocle	An analysis toolkit for single-cell data that performs clustering, differential analysis and trajectory construction.
mtscATAC-seq	Mitochondrial single-cell Assay for Transposase-Accessible Chromatin with sequencing.
MuSiC	Multisubject single cell deconvolution method that utilizes scRNA-seq to estimate cell type proportions in bulk RNA-seq data.
netImpute	Imputation of cell types from scRNA-seq data by integrating multiple types of biological networks.
Nucleosome banding pattern	A specific DNA fragment size banding pattern produced by transposition/ digestion on chromatin that corresponds to subnucleosome, mononucleosome, dinucleosome, and so on.
Nucleus hashing	A sample multiplexing method (like cell hashing) with oligo tagged antibodies directed against nuclear pore complex proteins based on the concept of hash functions in computer science to index datasets with specific features.
Nystrom	A sampling technique that generates the low rank embedding for large-scale dataset. It first builds a low-dimensional embedding with a subset of cells and then projects the rest of the data to the embedding structure.
Pacbio	Pacific Biosciences, a biotechnology company that pioneered long-read sequencing.
PAGA	PArtition-based Graph Abstraction.
PBAT	Post-Bisulfite Adaptor Tagging.
PBMC	Peripheral Blood Mononuclear Cells.

(continued)

Table 1
(continued)

Term	Description
PCA	Principal Component Analysis.
Perturb-seq	A reverse genetics method that allows for analysis of thousands of CRISPR-mediated perturbations within a single experiment by combining pooled CRISPR screen to single-cell RNA sequencing. Similar techniques are CRISP-seq and CROP-seq.
Pseudobulk	A sum of counts from single-cell data to imitate bulk data.
Pseudotime	A measure of how far cells have progressed along a biological process.
REAP-seq	RNA Expression And Protein sequencing.
RNA velocity	A high-dimensional vector that predicts the future state of individual cells based on time-dependent relationship between the abundance of precursor and mature mRNA.
Salmon	A pseudoalignment-based RNA-seq quantification program.
SC3	Single Cell Consensus Clustering.
Scanpy/EpiScanpy	Single-Cell ANalysis in Python, a toolkit for end-to-end analysis of single-cell datasets.
Scasat	Single Cell ATAC-Seq Analysis Tool, an end-to-end pipeline for analysis of scATAC-seq data.
scATAC-pro	An end-to-end pipeline for analysis of scATAC-seq data.
scATAC-seq	Single-cell ATAC-seq.
Scater	A Bioconductor package for analyses of single-cell RNA-seq gene expression data, with a focus on quality control and visualization.
scBS-seq	Single-cell bisulfite sequencing.
scCAT-seq	Single-cell Chromatin Accessibility and Transcriptome sequencing.
scCOOL-seq	Single-cell Chromatin Overall Omic-scale Landscape sequencing.
scGAIN	scRNA-seq data imputation using Generative Adversarial Networks.
scHi-C	Single-cell Hi-C.
scImpute	A statistical method to impute the dropouts in scRNA-seq data.
scLVM/f-scLVM	(factorial) single-cell latent variable model, a modeling framework for unraveling sources of heterogeneity and removing confounding factors for downstream analysis.
scM&T-seq	Single-cell Methylome and Transcriptome sequencing.
scMethyl-HiC	Single-cell Methyl-HiC.
scMT-seq	Single-cell methylome and transcriptome sequencing.
scNOMeRe-seq	Single-cell nucleosome, methylome, and transcriptome sequencing.
scNOMe-seq	Single-cell Nucleosome Occupancy and Methylome sequencing.
scPipe	A bioconductor package for single cell RNA-seq data analysis.

(continued)

Table 1
(continued)

Term	Description
scRRBS-seq	Single cell Reduced Representation Bisulfite Sequencing.
Scrublet	Single-Cell Remover of DoUBLETs, a python code for identifying doublets in scRNA-seq data.
Scruff	Single Cell RNA-Seq UMI Filtering Facilitator, a Bioconductor package that is used to preprocess scRNA-seq data.
scTrio-seq	Single-cell genome, methylome, and transcriptome (trio) sequencing.
scVelo	An improved estimate of RNA velocity that utilizes a likelihood-based dynamical model.
scWGBS	Single-Cell Whole Genome Bisulfite Sequencing.
Seurat	An end-to-end R package for analyses of scRNA-seq data.
Signac	An extension of Seurat for analysis of single-cell chromatin data.
SingleCellExperiment object	A lightweight Bioconductor container for storing and manipulating single-cell genomics data
SingleSplice	An algorithm for detecting alternative splicing in a population of single cells.
Slingshot	A method for inferring cell lineages and pseudotimes from scRNA-seq data.
SMART-seq	Switching Mechanism At 5′ end of RNA Template sequencing.
SnapATAC	An end-to-end pipeline for analysis of scATAC-seq data.
snm3C-seq	Single-nucleus methyl Chromatin Conformation Capture sequencing.
snmCT-seq	Single-nucleus methylome and transcriptome sequencing.
SNP	Single Nucleotide Polymorphism.
SOLiD	Small Oligonucleotide Ligation and Detection System.
Solo	A neural network framework to classify doublets in single-cell data.
souporcell	A toolkit for robust clustering, doublet detection and ambient RNA estimation in scRNA-seq data.
SoupX	An R package for the estimation and removal of ambient mRNA contamination in droplet-based scRNA-seq data.
STAR	Spliced Transcripts Alignment to a Reference, a popular alignment tool.
STREAM	Single-cell Trajectories Reconstruction, Exploration And Mapping, a comprehensive single-cell trajectory analysis pipeline.
STRT-seq	Single-cell Tagged Reverse Transcription sequencing.
SVD	Singular Value Decomposition
Tagmentation	Transposon-mediated cleaving and simultaneous tagging of DNA.
TF-IDF	Term Frequency-Inverse Document Frequency
TOPIC modeling	A type of statistical modeling for discovering the abstract “topics” that occur in a collection of documents.
t-SNE	t-distributed Stochastic Neighbor Embedding.

(continued)

Table 1
(continued)

Term	Description
TSO	Template Switch Oligo.
TSS enrichment score	Ratio of fragments centered at the TSS to fragments in TSS-flanking regions.
UMAP	Uniform Manifold Approximation and Projection.
UMI	Unique Molecular Identifier.
Variational Bayes	A family of ensemble learning techniques used in machine learning.
Velocyto	A tool to estimate RNA velocities in R or python.
VIPER	Variability-Preserving ImPutation for Expression Recovery.

Then came single-cell tagged reverse transcription sequencing (STRT-seq), which used an oligo-dT primer for first strand synthesis followed by incorporation of a cytosine string at the 3′ end of the first strand. A template switch oligo (TSO) that had complementary Gs allowed for barcoding of individual cells. The cDNA was then amplified using a single primer PCR, bound to beads, fragmented, A-tailed, and annealed to Illumina P2 adapter. The P1 adapter was introduced during library PCR step for short-read sequencing from the P1 side on an Illumina platform [13]. STRT-seq introduced one of the first multiplexing approaches for single-cell RNA-sequencing and was later adapted to the Fluidigm C1 microfluidics system for a more high-throughput version of the protocol (STRT-seq C1) [14]. Around the same time, unique molecular identifiers (UMIs) were introduced to serve as molecular barcodes during the reverse transcription step enabling absolute quantification of transcripts and distinguishing molecules from notorious PCR duplicates, a problem in low-input amplification protocols [15]. It also became routine to introduce exogenous spike-in controls to monitor technical reproducibility.

These early approaches, however, resulted in transcript read bias in the 3′ end [11, 12] due to their use of oligo-dT primers or 5′ end (STRT-seq) to incorporate barcodes for multiplexing. SMART-seq (Switching Mechanism At 5′ end of RNA Template sequencing) was introduced as a method for full length cDNA sequencing that allowed for improved read coverage across transcripts, significantly enhancing alternative transcript isoform discovery, dissection of allele-specific expression and identification of SNPs. SMART-seq used an oligo dT primer, a special TSO primer and a reverse transcriptase with terminal transferase activity. The first strand synthesis and template switch generated a full-length cDNA with an anchor for second strand synthesis which was followed by cDNA amplification. Libraries were generated by either ultrasonication followed by Illumina adapter ligation or direct tagmentation [16]. The SMART-seq kit was commercialized by Takara Clontech as the SMARTer Ultra Low RNA Kit for Illumina sequencing. SMART-seq2 was developed to further refine the reverse transcription, template switching, and preamplification steps and increase yield and length of cDNA libraries in a cost-effective way [17]. However, the SMART-seq/SMART-seq2 library preparation protocols lacked the ability to incorporate UMIs. Recently, SMART-seq3 was developed and can be used to construct UMI barcoded RNA libraries with the same 5′ end but different 3′ ends. The full-length transcript can be confidently reconstructed in silico as verified with PacBio long-read sequencing [18].

Right around the time the first SMART-seq protocol was developed, another method called CEL-seq (Cell Expression by Linear amplification and sequencing) was demonstrated to work on single cells. CEL-seq's protocol made use of linear amplification of pooled barcoded RNA samples by in vitro transcription (IVT) made possible by the use of an oligo-dT primer that also had a T7 promoter sequence [19]. CEL-seq2 was modified to increase sensitivity and throughput by primer redesign, ligation-free library preparation and use of the Fluidigm platform [20]. CEL-seq2 but not CEL-seq also incorporated UMI molecular barcoding.

Massively parallel RNA single-cell sequencing (MARS-seq) is a method similar to CEL-seq2, and uses a three-tier barcoding system at the molecular (via UMIs), cellular and plate level prior to pooling and IVT amplification of RNA from single cells in a plate format [21]. MARS-seq2.0 builds upon MARS-seq improving throughput, reducing noise and contamination, increasing yield, and providing a user-friendly computational pipeline for data analysis [22].

CytoSeq was introduced as a highly scalable, automation-free method to capture any transcribed RNA from a single cell. Individual cells are deposited in picoliter wells, lysed, and hybridized to beads containing cellular barcodes and UMIs. The diversity of available cellular barcodes and UMIs ensured that each cell and each mRNA molecule was assigned a unique tag. The cells from each well are then pooled for reverse transcription, amplification, and sequencing [23].

Drop-seq [24], inDrop (indexing Droplets) [25] and 10× Genomics [26] are newer methods that use microfluidic partitioning to generate droplets that capture single cells. They also operate on similar principles where bead embedded primers containing an oligo-dT stretch, unique cell barcodes, UMIs, and PCR handles are cocaptured in the same droplet and hybridize to the mRNAs from a single cell. A few key differences include the composition of the bead itself and the location of the reverse transcription (RT) reaction. In Drop-seq, the bead is a functionalized microparticle and the primers are synthesized on the surface. The RT reaction for Drop-seq occurs after the droplets are broken by addition of perfluorooctanol. In inDrop, the bead is a hydrogel carrying photocleavable combinatorially barcoded primers. 10× Genomics uses a Gel Beads in EMulsion (GEM) technology. For both inDrop and 10× Genomics, the RT reaction occurs inside droplets and the products are released either by photocleavage (inDrop) or dissolution of beads (10× Genomics). The 10× Genomics method with ~65% capture efficiency far outruns the other two methods which are 5–7%. Thus, 10× Genomics has become the method of choice for most laboratories although the kits and instrument are costly.

1.2 Single-Cell Epigenomic Approaches

Sometimes, steady state transcript levels alone do not reveal the nuances of gene regulation within individual cells. Thus, recent advances in single-cell technology have evolved to analyze the epigenome in detail and include DNA accessibility, transcription factor (TF) binding, DNA and histone modifications, enhancer–promoter contact maps and 3D genome organization. DNA methylation analysis for example can provide insights into promoter regulation of gene expression being anticorrelated to transcription. Due to the incredible stability of DNA methyl marks and the ability to perform absolute quantification makes methylation analysis invaluable in aging studies. For example, methylation at some CpG sites serve as epigenetic clocks that predict biological age [27]. Aged somatic cells undergo global hypomethylation at repeat regions and enhancers, and focal hypermethylation at genes that direct age-related decline and disease [28]. Chromatin accessibility patterns also offer deep insights into biological function. By measuring cell-to-cell variability of TF binding under various perturbing conditions, it is possible to identify the key TFs responsive to those perturbations [29]. Furthermore, coaccessibility patterns at enhancer promoter pairs in conjunction with single-cell chromosome conformation studies can reveal 3D genome organization in unprecedented detail.

The first single-cell epigenome technology called scRRBS-seq was a method detecting DNA methylation in single cells using a reduced representation bisulfite sequencing (RRBS) which digests DNA with restriction enzymes to enrich for CpGs. In this method, cells are lysed, the genomic DNA digested, and ligated to adapters before bisulfite conversion, all in a single tube. This is followed by DNA purification and PCR amplification for sequence-ready libraries [30]. While RRBS significantly reduces sequencing cost, it suffers from poor coverage and exclusion of many regulatory regions. Furthermore, because bisulfite conversion which fragments DNA is performed after adapter ligation, this leads to further reduction in representation.

To counter this deficiency, a method called Post-Bisulfite Adaptor Tagging (PBAT) was developed where bisulfite treatment precedes adaptor tagging [31]. Single-cell bisulfite sequencing (scBS-seq) adopted the PBAT strategy with additional PCR amplification for the absolute quantification of 5-methyl cytosine (5mC) in single cells [32]. In a related method called scWGBS (single-cell Whole Genome Bisulfite Sequencing), PBAT was used with all steps from lysis to library cleanup being performed in a single tube [33].

The first single-cell chromatin immunoprecipitation sequencing (ChIP-seq) method called Drop-ChIP was published by the Bernstein lab [34] and used an in-house microfluidic device which formed coflowing drops of cells, micrococcal nuclease (MNase) and detergent on one side, and barcoded adapters on the other side. Cell lysis, chromatin fragmentation and unique adapter ligation were done inside the drops. The drops were then broken, indexed chromatin fragments pooled, and ChIP performed with carrier chromatin before restriction digestion to remove excess adapter followed by PCR amplification. A second restriction digest created overhangs to ligate yet another sample barcode allowing for further multiplexing and Illumina sequencing-ready libraries. Given that labs had to build their own microfluidic device, this method did not gain as much popularity. Single-cell ChIP-seq remained a challenge due to the small amounts of immunoprecipitated material and low signal to noise ratio. Recently, major advancements in ChIP technology including CUT&RUN (Cleavage Under Targets and Release Using Nuclease) and CUT&Tag (Cleavage Under Targets and Tagmentation) have improved the signal-to-noise ratio by using tethered enzymes that specifically release antibody bound chromatin fragments. Indeed, CUT&Tag was used to profile H3K27me3 and H3K4me2 in single K562 and H1 cells using Takara's iCELL8 nanodispenser system and the SMARTer iCELL8 350v chip [35]. Subsequently, CUT&Tag was adapted to the 10× Genomics platform to profile histone modifications in multiple complex tissues [36, 37].

The most widely used epigenomic investigation in single cells is a technique called ATAC-seq (Assay for Transposase Accessible Chromatin sequencing). ATAC-seq works on the principle that open regions of chromatin (generally promoters, enhancers, DNA flanking TF bound sites, etc.) are more accessible for Tn5 transposome-mediated tagmentation. ATAC-seq at single cell resolution (scATAC-seq) was first performed using a highly scalable combinatorial indexing method where nuclei are first distributed in wells with a unique barcode, pooled, diluted and then distributed to another set of wells with a different set of barcodes [38]. The sequential and randomized passage of cells through two different 96-well plates at low concentration minimizes the probability of two cells receiving the same set of barcodes. The first barcode is introduced during tagmentation using uniquely barcoded transposome complexes. In the second well, the cells are lysed and the released tagmented DNA is PCR-amplified by priming off the adapters thereby introducing the second barcode. Although this was a highly scalable method, the median number of reads per cell was ~2500. In an advancement of the scATAC-seq technology, the Fluidigm platform was adopted with an automated workflow to tagment individual cells, PCR-amplify with dual index primers and pool for sequencing [29]. This method had the same throughput but increased the median number of fragments per cell to ~73,000. Currently, scATAC-seq has been adopted by the commercial 10× Genomics platform which performs the tagmentation in bulk followed by droplet capture of single cells [39].

A relatively new frontier of epigenomics is the study of the spatial organization of chromatin in 3D. Chromosome Conformation Capture (3C) techniques assess millions of contacts in the genome and together with statistical modeling can give rise to high-resolution structural views of chromosomes [40]. scHi-C (single-cell Hi-C) [41] is an "all-to-all" 3C method that performs chromatin cross-linking, restriction enzyme digestion, biotin fill-in, and proximity ligation in nuclei (also known as in situ Hi-C [42]). Individual nuclei are then placed in wells, cross-links reversed, and ligation products purified on streptavidin beads. The products are then further digested with another restriction enzyme, ligated to adapters, PCR-amplified, and sequenced. A variant method called Dip-C [43], obtained high-resolution contact maps in single diploid cells (which have many more contacts) by omitting the biotin pulldown and coupling to a whole genome amplification and sequencing.

1.3 Single-Cell Multiomics Approaches

Interrogation of single modalities such as transcriptome or accessibility or DNA/histone modifications provides only a partial picture of biological events in single cells. With technological advances, several clever modifications have enabled the simultaneous profiling of multiple modalities from a single cell.

scM&T-seq (single-cell methylome and transcriptome sequencing) is made possible by the physical separation of poly-A RNA by magnetic beads with oligo-dT and DNA from the same cell. This is followed by parallel processing of RNA using the SMART-seq2 strategy and DNA using the scBS-seq strategy [44]. Related methods include scMT-seq (methylome and transcriptome) [45], snmCT-seq (single-nucleus methylome and transcriptome) [46], and scTrio-seq (genome, methylome, and transcriptome) [47].

In addition to joint profiling of genome, methylome, and transcriptome, nucleosome occupancy mapping was adopted in single cells. scNOMe-seq (single-cell nucleosome occupancy and methylome sequencing) is performed by incubating cells with a GpC methyltransferase that methylates accessible regions while being unable to methylate nucleosome protected regions and then FACS sorting cells into individual wells. The DNA is then subjected to bisulfite conversion and sequencing enabling the measurement of accessibility, TF binding, nucleosome occupancy, and phasing [48]. This nucleosome mapping was combined with transcriptome profiling in scNMT-seq (nucleosome, methylome, and transcriptome) [49] and scNOMeRe-seq (nucleosome, methylome, and transcriptome) [50]. scCOOL-seq (single-cell Chromatin Overall Omic-scale Landscape sequencing) is a method that simultaneously measures nucleosome positioning, DNA methylation, copy number variation and ploidy from single cells by combining scNOMe-seq, PBAT-seq, and spike-ins [51].

Perhaps the most popular and commercialized multiomics approach is the simultaneous assessment of chromatin accessibility and transcriptome. Two methods that accomplish this are scCAT-seq (single-cell chromatin accessibility and transcriptome sequencing) [52] and ATAC-RNA-seq [53]. In scCAT-seq, after single cell isolation, a mild lysis method separates the cytoplasmic components (mRNA) from the nuclei (DNA). The SMART-seq2 strategy is used to prepare gene expression libraries while the nucleus is subjected to Tn5 transposition and ATAC-seq library preparation. In ATAC-RNA seq, cells are first fixed with dithiobis(succinimidyl propionate), permeabilized and transposed in bulk. The cells are then stained with DAPI and individual cells sorted into wells with lysis buffer. The mRNA component undergoes SMART-seq2 library preparation in a separate plate while the genomic DNA undergoes indexed PCR to amplify the transposed fragments. Paired scRNA-seq and scATAC-seq from the same cell has recently been adapted to the 10× Genomics platform (Chromium single-cell multiome ATAC and gene expression) enabling unprecedented scalability in this kind of multi-omics approach. In this method, nuclei are extracted, permeabilized, transposed in bulk, and captured in droplets. The nuclear mRNA and DNA components are captured by the gel bead oligos and get tagged with the same barcode identifying the cell of origin. The gene expression and ATAC components then get separated and undergo their separate library preparations based on 10× Genomics protocols.

Finally, scMethyl-HiC (single-cell Methyl-HiC) [54] and snm3C-seq (single-nucleus methyl Chromatin Conformation Capture sequencing) [55] are two methods that combine methylomics and 3D genomics in single cells. Both protocols combine in situ Hi-C followed by bisulfite conversion, library construction, and paired-end sequencing.

1.4 Single-Cell Multiplexing Approaches

RNA Expression And Protein sequencing (REAP-seq) and Cellular Indexing of Transcriptomes and Epitopes by sequencing (CITE-seq) were two methods developed independently for multimodal single cell immunophenotyping and transcriptome analysis

[56, 57]. These methods used oligo-tagged antibodies against cell surface markers that were captured in most oligo-dT-based scRNA-seq protocols thus receiving the same cell-barcode and helping to identify the cell type. REAP-seq and CITE-seq differed in how the oligo was conjugated to the antibody. In REAP-seq the aminated oligo is covalently bound to the antibody while in CITE-seq, the biotinylated oligo is bound to the streptavidin tagged antibody. Following cDNA synthesis, a size-selection separates the cDNA from cell transcripts and antibody oligos for parallel library preparations. A variant of CITE-seq applied to scATAC-seq data is ATAC with Select Antigen Profiling by sequencing (ASAP-seq) [58]. The introduction of a single bridge oligo makes ASAP-seq compatible with simultaneous phenotyping and chromatin accessibility profiling. Importantly, ASAP-seq (unlike CITE-seq) can be used to detect intracellular epitopes. On the sample preparation side, ASAP-seq utilizes the newly developed mitochondrial single-cell assay for transposase-accessible chromatin with sequencing (mtscATAC-seq) in which Tn5 is added to fixed, permeabilized whole cells rather than nuclei and both nuclear DNA and mitochondrial DNA are transposed [59].

A clever modification of the CITE-seq protocol allows for multiplexing and superloading of scRNA-seq samples in a method called Cell Hashing [60]. This multiplexing strategy dramatically reduces experimental costs. Cell Hashing uses barcoded oligo-tagged antibodies (hashtags) directed against ubiquitously expressed proteins. Every sample is uniquely tagged and then pooled before droplet capture and scRNA-seq library preparation. Downstream deconvolution of the oligo sequences can identify individual samples. In addition to multiplexing samples, Cell Hashing can be used to overload microfluidic channels and then remove intersample cell multiplets computationally. Similar multiplexing strategies have also been applied to scATAC-seq preparations. The first, called Nucleus Hashing, uses hashtags directed to the nuclear pore complex [61]. The second, is an expanded utility of ASAP-seq with the bridge oligo capturing hashtags against nuclear epitopes [58].

1.5 Single-Cell Functional Genomics Approaches

An evolving new area of single-cell technology is the development of CRISPR-based functional genomic screens combined with single-cell transcriptomics readout. Perturb-seq [62] is one such method where pools of lentiviruses containing designer barcoded guide RNA (gRNA) constructs are transduced into cells. While the gRNA component knocks out target genes, the associated barcode is captured with the rest of the cellular mRNA during droplet generation thus acting as a molecular identifier of the perturbation. The readout is simply the gene expression profile of the perturbed cell. By fine-tuning the multiplicity of infection, it is possible to study either single gene deletions or epistatic relationships between

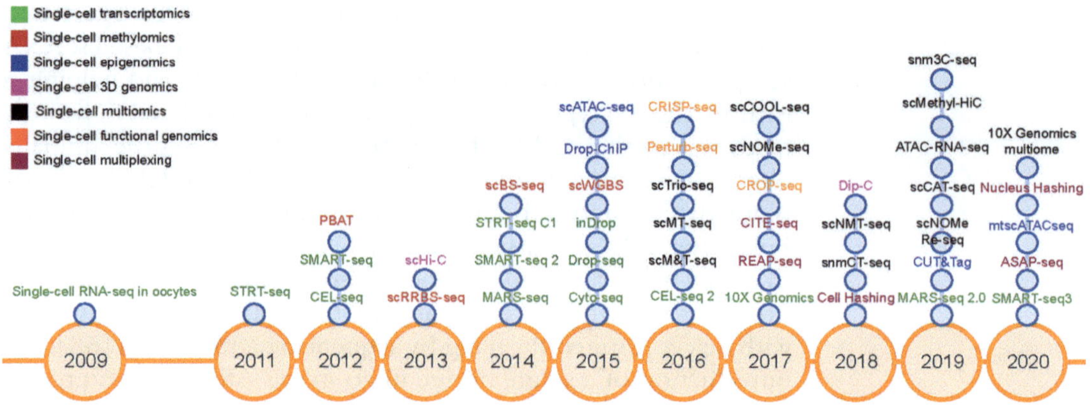

Fig. 1 A timeline of the evolution of single-cell genomics technologies. The transformative technological evolution of single-cell methods to probe the transcriptome and epigenome are outlined. The methods are colored based on the precise "omics" used

multiple genes. CRISP-seq [63] and CROP-seq (CRISPR droplet-sequencing [64]) are two alternative versions of CRISPR-screening in single cells with different vector designs. Furthermore, it was possible to use Perturb-seq and related methods with other CRISPR manipulations such as CRISPR inhibition rather than gene knockout [65, 66] as well as multiplex several gRNAs in the same vector [65]. Perturb-seq was further improved by developing a method to directly sequence the gRNA eliminating the need for barcodes [67].

A timeline of the key method developments in single-cell genomics is depicted in Fig. 1. Also, *see* **Notes 1** and **2** for some key considerations when performing the wet-lab procedures to generate single-cell genomics data.

2 Methods

2.1 Computational Methods to Analyze Single-Cell Transcriptomics Data

Rapid progress in single-cell experimental protocols (described in Subheading 1), and the great complexity of the obtained results, prompted development of new analytical tools and resources for accurate interpretation of scRNA-seq data. Although, there is no standardized computational pipeline for scRNA-seq data analysis, multiple bioinformatics tools are continuously evolving [68]. The overall workflow entails preprocessing and alignment of raw sequencing data to create count matrices of mRNA molecules. The count matrices are further evaluated for quality, normalized, and corrected to remove any technical variation and elucidated for biological function. Finally, the top highly variable genes are applied for dimensionality reduction and clustering to visualize cell heterogeneity and run desired downstream analysis. *See* **Note 3** for software version control when reporting data from single-cell analysis.

A practical example of scRNA-seq analysis covering the fundamental principles with 1K peripheral blood mononuclear cells (PBMCs) from a healthy human donor (dataset available for download from the 10× Genomics website) is outlined in Fig. 2 and discussed in detail below. Figure 2a illustrates the basic workflow.

2.1.1 Data Preprocessing

Raw scRNA-seq data generated by a sequencer is first preprocessed for cell barcode identification and deconvolution of UMIs to demultiplex cell-specific reads. Subsequently, reads from each cell are aligned to a reference genome with implementation of bioinformatics tools developed for bulk RNA-seq. The most commonly used software for initial scRNA-seq data alignment is STAR [69] that exploits feature counts, or pseudoalignment-based programs such as Kallisto/Bustools [70], Salmon [71], or Alevin [72]. 10× Genomics's Cell Ranger pipeline automates many of these preprocessing steps including STAR alignment and generating output files in BAM (Binary Alignment Map), MEX (Market EXchange), CSV (Comma Separated Values), HDF5 (Hierarchical Data Format version 5), and HTML (HyperText Markup Language) formats. A typical plot generated by Cell Ranger is an UMI vs barcode "knee plot" (Fig. 2b) that provides visual cues of a good separation between cell-containing or empty droplets. A sharp drop in UMI (knee) signals a good separation. In addition, Bioconductor offers a broad range of freely available R-packages that are continuously being enriched by the scientific community [73]. Among them, scPipe [74] and scruff [75] can be used for scRNA-seq preprocessing.

2.1.2 Quality Control

A fundamental component of scRNA-seq data analysis is a cell quality control (QC) procedure that aims to accurately identify and remove outlier barcodes corresponding to either low-quality cells or doublets, thus ensuring investigation of real biological differences in cells [76, 77]. Multiple computational tools for scRNA-seq QC use counts and gene expression features for cell quality analysis (such as Scater [78] and Seurat [79]), while other methods apply machine learning techniques [80] or use gene coverage distribution [77].

Metrics that are commonly used for assessment of cell quality involve number of counts per cell, number of genes per cell and the percent of reads in cells that comes from mitochondrial genome. Barcodes that are characterized by a low number of counts, few detected genes, and a high fraction of mitochondrial counts, are qualified as low-quality cells. These can include damaged cells with leakage of cytoplasmic RNA or cells undergoing apoptosis. On the other hand, when two or more cells (doublets or multiplets) are captured and labeled in the same droplet, the resulting barcode will contain extremely large number of detected genes and very high counts. A typical example of how a scRNA-seq dataset might look

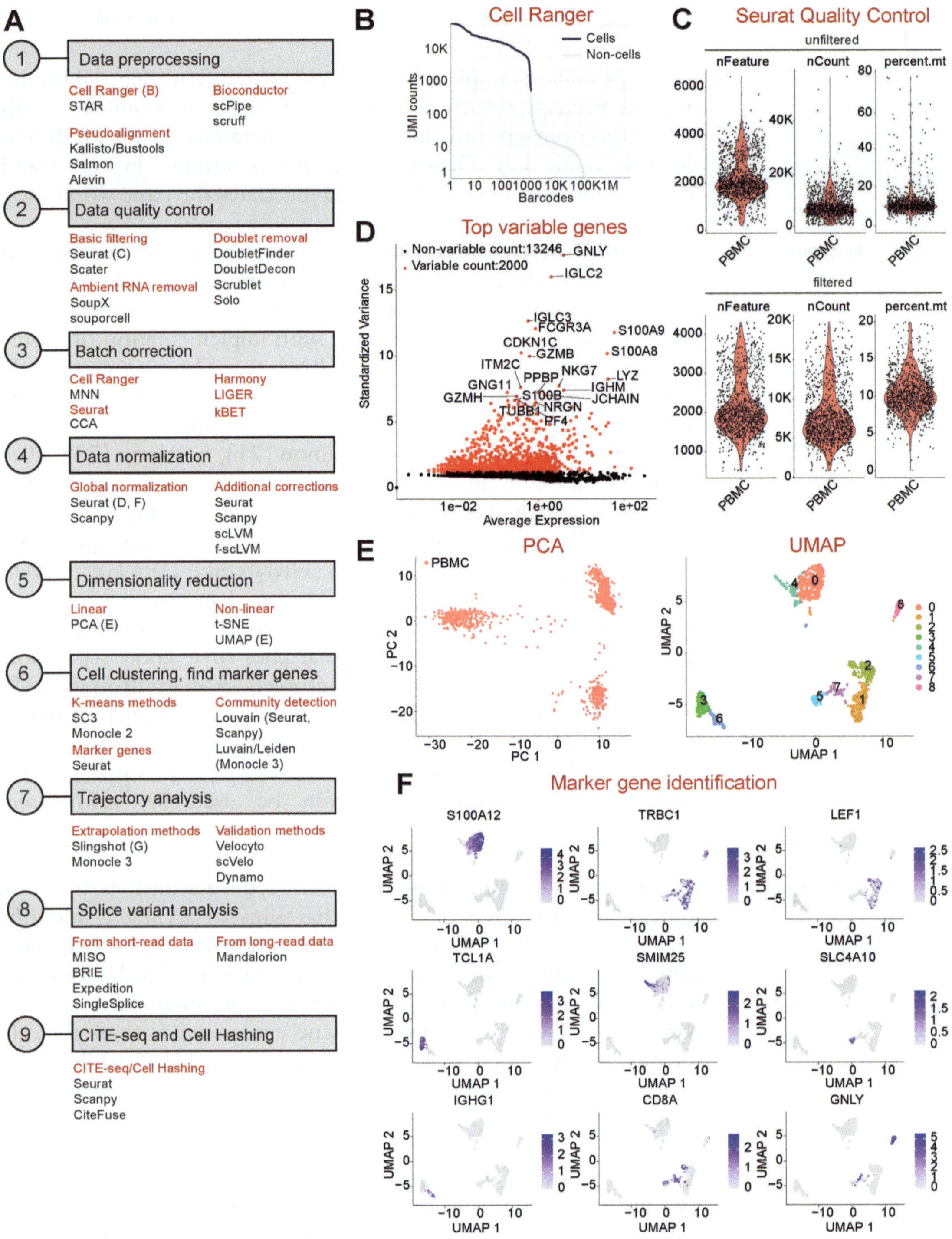

Fig. 2 Preliminary analysis of 10× Genomics scRNA-seq data on 1000 human PBMCs. (**A**) Typical workflow and tools used for analysis of scRNA-seq data. (**B**) Knee plot showing the distinction between barcodes identified as cells or noncells in Cell Ranger. (**C**) nFeature (number of genes with at least one UMI), nCount (number of UMI counts taken across all cells) and percent.mt (percent mitochondrial genes) before (top) and after (bottom) filtering in Seurat. (**D**) Top 2000 variable genes identified by mean-variance modeling in Seurat is shown in red. (**E**) PCA (left) and UMAP (right) representation of clusters. (**F**) Visualization of marker gene expression on a UMAP

before and after filtering for empty droplets, doublets or high mitochondrial reads is illustrated in Fig. 2c. Setting a universal QC threshold applicable for various types of scRNA-seq data is challenging since thresholds can depend on cell-type, cell-state and other variables. Additionally, each of the metrics used for cell filtering can illustrate other cellular characteristics, such as, lower counts can indicate smaller or quiescent cells, and a high proportion of mitochondrial genes can represent cells with high energy needs such as cardiomyocytes. Therefore, assessment of cell quality should consider all QC metrics in parallel, as using these parameters alone may be misinformative for downstream data analysis and results. Computational packages such as Seurat or Scater [78] apply a manual strategy that allows for visual evaluation of the sample before and after filtering. This provides a quick and effective way for setting dataset-specific thresholds, particularly if the results obtained from an initial setting are not clear. In addition, for more rigid cell filtering, many other computational tools are available, such as SoupX [81], souporcell [82], DecontX [83] for removal of ambient RNA contamination; and DoubletFinder [84], DoubletDecon [85], Scrublet [86], and Solo [87] for detection of doublet barcodes. *See* **Note 4** highlighting the importance of QC steps.

2.1.3 Batch Correction

Batch effect is a major driver of data variation that results from differences in experimental processing of sample sets (*see* **Note 5**). Consequently, by interfering with true biological variation, it can confound interpretation of results. Methods such as Seurat Canonical Correlation Analysis (CCA) [88] and mutual nearest neighbors (MNN) [89] are considered the most successful algorithms for removing batch effect from scRNA-seq datasets [90]. MNN algorithm has been incorporated into the Cell Ranger pipeline to enable correction of different 10× Genomics chemistries. Other widely accepted algorithms for quantification of batch effects in scRNA-seq data are Harmony [91], LIGER [92], and the recently developed k-nearest-neighbor Batch-Effect Test (kBET) [93].

2.1.4 Data Normalization

The gene counts of individual cells in a scRNA-seq dataset may show differences that are associated with technical variation created by sample processing and sequencing depth. This technical noise can confound true sample heterogeneity and must be corrected before downstream analysis. Thus, the count data are normalized to make the counts comparable across cells, and to ensure elimination of cell-specific biases. Over the years, several methods of normalization have been adapted from bulk RNA-seq analysis pipelines or developed specifically for single cells [94]. The most popular Seurat [95] and Scanpy [96] platforms employ global scaling normalization method, in which each gene expression in a cell is divided by the total gene expression in that cell, multiplied by scale factor 10,000, and finally natural log transformed. This ensures that highly expressed genes do not dominate the genes with low expression, thereby masking real biological differences. Additional data correction can be performed for regressing out genes related to unwanted biological processes such as cell cycle phase, mitochondrial, or ribosomal genes. These corrections are available in some scRNA-seq analysis platforms, such as Scanpy or Seurat but also can be performed separately with more specified tools such as scLVM [97] or f-scLVM [98]. However, given that cell processes are tightly connected, removing genes associated to specific biological functions can bias data interpretation and therefore such corrections need to be monitored with caution.

2.1.5 Dimensionality Reduction

The main goal of scRNA-seq is to uncover cell-to-cell variability within a sample, thus focusing on genes that drive heterogeneity inside the cell population. For this reason, only a small subset of genes, with the highest variation in expression across the dataset is informative and must be selected and used for downstream analysis. Most software packages implement algorithms for selection of highly variable genes [99] (usually top 500–2000 genes), that depend on dataset complexity [100], and ultimately generate a high-dimensional expression profile for each cell (Fig. 2d). Therefore, this high-dimensional data complexity requires further reduction to allow data visualization in a 2D plot and clustering of cells for biological interpretation. Several dimensional reduction algorithms have been developed [101, 102] that, in combination with different clustering models, can effectively extract biological insights of cells. This includes investigation of (1) clusters of cells with similar gene expression profile to uncover cell states or types, (2) groups of cells with coexpressing genes to study coregulatory relationships between genes, or (3) cells demonstrating continuous gene expression patterns for tracing cellular dynamic processes.

The most common, unsupervised linear dimensionality reduction technique, principal component analysis (PCA), is widely applied in many scRNA-seq software, including Seurat and Scanpy, as an internal algorithm that contributes for multiple analyses with different informational use, such as assessment of data QC and detection of batch effects (Fig. 2e, left). In addition, scores of principal components (PCs) can be exploited for other nonlinear dimensionality reduction methods, cell clustering or trajectory inferences [101, 103]. Besides PCA, other dimensionality reduction methods are often used for targeting different biological approaches in scRNA-seq analysis in combination with multiple clustering and projection models [101]. Among them, the most popular, are nonlinear dimensionality reduction methods such as multidimensional scaling (MDS) that is implemented in Scater, t-distributed stochastic neighbor embedding (t-SNE) [104], and uniform manifold approximation and projection (UMAP; Fig. 2e, right) [105]. They are used to generate 2D or 3D plots for dimension-reducing visualization of relationships between cells. For exploratory data visualization in Cell Ranger, both t-SNE and UMAP are implemented.

2.1.6 Cell Clustering, Find Marker Genes

Clustering of cells with similar gene expression into separate groups is the main strategy that elucidates different cell states or types. Clustering algorithms use intercell distance matrices to assemble subgroups of cells with similar gene expression patterns. Among the variety of clustering methods, hierarchical clustering, K-means, fuzzy C-means, DBSCAN, and Louvain algorithm are the models often used for single cell clusters establishment [106]. The most common K-means method, that selects k-clusters and assigns cells to closest cluster, is used by the SC3 package [107] and Monocle 2 [108]. Alternatively, community detection-based Louvain algorithm was shown to outperform other clustering algorithms [106, 109, 110]. Among the commonly implemented platforms, Scanpy and Seurat perform clustering by Louvain community on a single-cell K-Nearest Neighbor (KNN) graph, while Monocle 3 [111] uses Louvain/Leiden algorithm [112]. Once the cell clusters are established, they are further characterized by identification of gene markers that are uniquely expressed in each cluster, therefore defining cell state or function. This can be very challenging for single-cell data due to the low counts, high dropout rate and amplification biases in mRNA. However, several programs address these challenges and report marker genes. For example, Seurat identifies marker genes based on differential expression in a cluster relative to average expression in all other clusters (*see* **Note 6** for additional considerations). In Fig. 2f, marker genes identified were granulocytes/monocytes (cluster 0), T cells (clusters 1, 2, 5, 7 and 8), B cells and Dendritic cells (cluster 3), NK cells (cluster 4) and Plasma cells (cluster 6). Since this was a dataset of only 1000 cells, rarer populations such as hematopoietic stem cells were not detectable.

2.1.7 Trajectory Analysis

Another type of downstream analysis, that relies on results from dimensional reduction methods, is a machine-learning approach called cell trajectory inference. This analysis aims to order cells along a trajectory as a function of pseudotime (an abstract unit of progress), to trace divergence of cell lineages that can indicate multiple dynamic cellular processes such as development, differentiation, activation, or tumor progression. Among the multiple tools developed for trajectory extrapolation [113], Slingshot [114], Monocle 3 [111], and PAGA [115] are popular methods of choice. Trajectory analysis can be unsupervised, that is, it uses no prior information on genes that are important for the dynamic biological process or semisupervised, where some important marker genes are part of the input. A significant challenge of trajectory extrapolation is a necessity for supplementary validation with supportive evidence, to ensure that created trajectories truly reflect biological processes. An example of a trajectory confirmation method is RNA velocity [116] in which single-cell RNA transcription and splicing information is integrated into a trajectory, to verify directions of the pseudotime lineages. Recently developed, Velocyto [116], scVelo [117], and Dynamo [118] algorithms can be applied to recapitulate transient cell states and the dynamics of different biological processes.

2.1.8 Splice-Variant Analysis Using SMART-Seq

Given that alternative splicing is a key driver of protein diversity, scRNA-seq is a target technology for uncovering splice variant expression and transcriptomic variation that characterize individual cell functionality in physiology and disease. However, most microfluidic or droplet-based scRNA-seq platforms (such as 10× Genomics, CEL-seq, and STRT-seq) generate short UMI tagged transcripts. Since only a fragment of the transcript is sequenced, the sensitivity is too low for investigation of transcript isoforms [119]. Instead, plate-based, low-throughput methods such as SMART-seq and SMART-seq2, allow full length transcriptome sequencing that is sufficiently sensitive for gene, isoform and allele expression detection and can be applicable for single-cell alternative splicing analysis. Moreover, recently introduced SMART-seq3 method, with improved sensitivity and UMI incorporation promises to accelerate discovery of transcript splice isoforms in a high-throughput manner.

Currently, several computational strategies have been developed to perform splice variant detection and quantitation, using data obtained from existing short-read scRNA-seq technologies [120, 121]. MISO [122], BRIE [123], and Expedition [124] packages can be used to assess transcript splice variant expression at the exon level, using reads aligned to splice junctions. Single-Splice [125] applies statistical model to detect genes with different isoform usage across a set of single cells. In addition, the integrated informatics pipeline Mandalorion [126], has been introduced for single-cell isoform detection from long-read Nanopore sequencing data. However, this technology awaits further improvement for wide application to single-cell data.

2.1.9 CITE-seq and Cell-Hashing

As mentioned in Subheading 1.4, CITE-seq enables parallel quantification of RNA profiles and surface protein expression at the single cell level [56] while Cell Hashing is an application of CITE-seq that enables sample indexing for multiplexing [60]. CITE-seq analysis can be done using Seurat and Scanpy pipelines. The recently developed CiteFuse [127] package offers a more comprehensive set of tools that include doublet identification, cell hashing information, and transcriptome analysis.

2.2 Computational Methods to Analyze Single-Cell ATAC-seq Data

Compared to scRNA-seq data analysis, tools for scATAC-seq analysis are relatively limited but rapidly evolving. The workflow involves similar preprocessing and QC steps as scRNA-seq data with the parsing of cellular barcodes, generation of fragment files and a feature matrix (usually peaks). An important difference between scRNA-seq and scATAC-seq data is the inherent sparsity of the latter. The total number of accessible sites in the genome far exceed the total number of reads obtained per cell. Additionally, scATAC-seq data is binary, with each locus having mostly 0 or 1 reads and maximally two reads [29]. Thus, standard computational tools cannot accurately reconstruct cellular activities. As a remedy, most scATAC-seq-based workflows include some statistical framework that measures activities in ensemble either from bulk or single-cell (pseudobulk) data. Downstream analysis then focuses on cell-type specific peaks, gene activity deduction, underlying TF motif identification, estimation of TF dynamics, enhancer function, and so on. *See* **Note 3** for software version control when reporting data from single-cell analysis.

A practical example of scATAC-seq analysis covering the fundamental principles with 1K PBMCs from a healthy human donor (dataset available for download from the 10× Genomics website) is outlined in Fig. 3 and discussed in detail below. Figure 3a illustrates the basic workflow.

2.2.1 Data Preprocessing

The upstream data preprocessing steps such as alignment to reference genome and peak-by-cell count matrix generation from BAM files are similar to scRNA-seq data analysis and utilizes the same tools packaged into specialized scATAC-seq software. One additional step in scATAC analysis is peak calling and most software use MACS2 [128]. 10× Genomics's Cell Ranger ATAC pipeline automates these preprocessing steps and generates fragment files and peak matrices. It also outputs basic plots such as the knee plot to separate cells and empty droplets (Fig. 3b). Alternatively, Bioconductor packages such as scPipe-ATAC can preprocess data and generate SingleCellExperiment objects that can then be used as input to the many other R packages in Bioconductor. Other handy end-to-end scATAC analysis software include Scasat [129], scATAC-pro [130], EpiScanpy [131], SnapATAC [132], and ArchR [133], and can be used directly. Among these, SnapATAC and ArchR are two methods that are fast and use less memory due to optimization and parallelization of methods. Instead of using a predetermined peak set from bulk/pseudobulk data, SnapATAC and ArchR tile the genome into 5kb (SnapATAC) or 500bp (ArchR) bins. This allows for a more unbiased element representation that is otherwise skewed toward most abundant cell types. Furthermore, a higher resolution 500bp binning in ArchR ensures proper capture of singular regulatory elements that tend to be 300–500bp in size. ArchR attains an additional dimension of

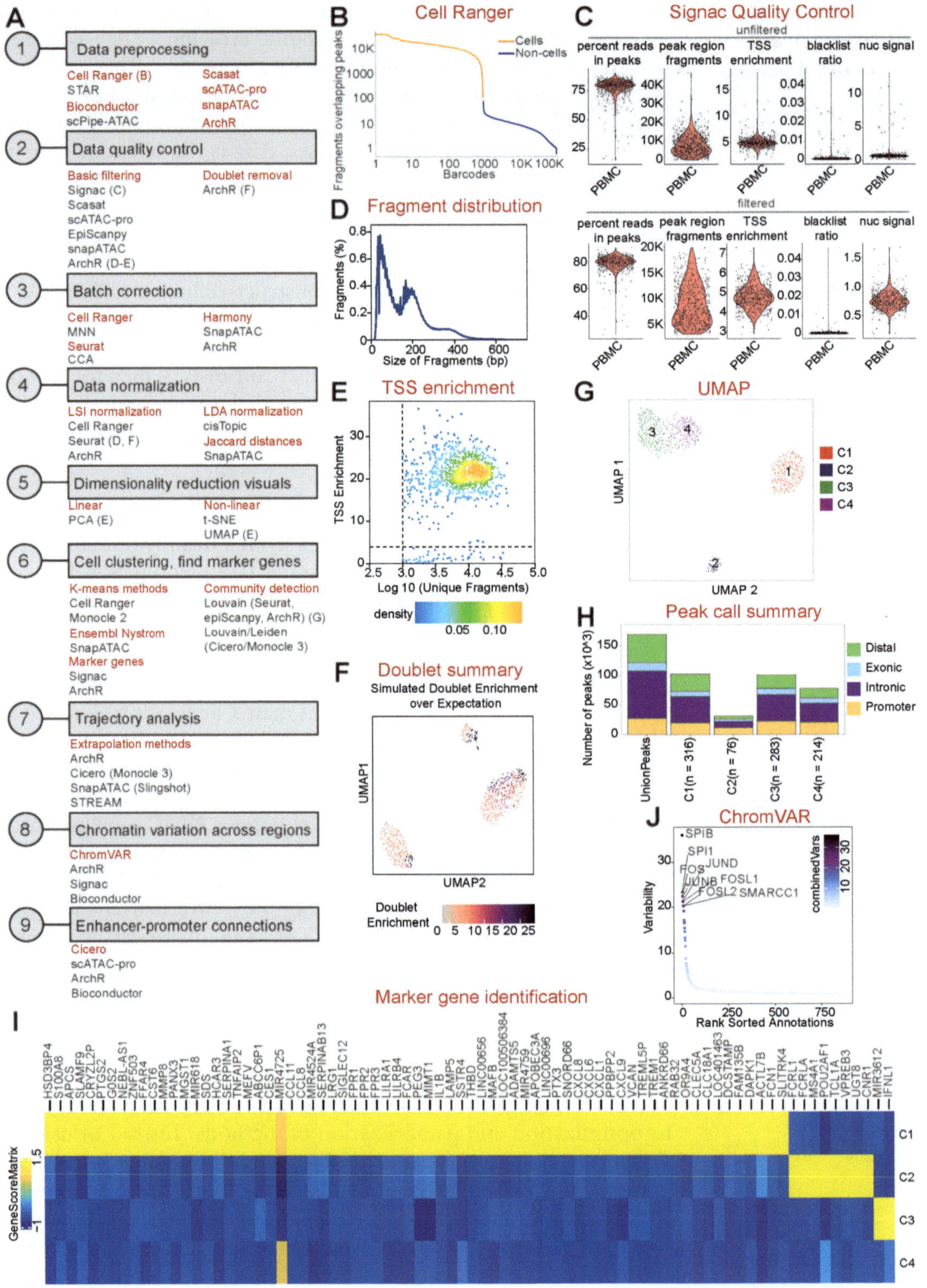

Fig. 3 Preliminary analysis of 10× Genomics scATAC-seq data on 1000 human PBMCs. (**A**) Typical workflow and tools used for analysis of scATAC-seq data. (**B**) Knee plot showing the distinction between barcodes identified as cells or noncells in Cell Ranger. (**C**) QC metrics such as percent reads in peaks (fraction of all

computational efficiency by parsing fragment files into small chunks per chromosome and then storing them in compressed random-access HDF5 files. These HDF5 files form the constituent pieces of "Arrow" files which are then grouped into "ArchR Projects" for specific downstream analysis.

2.2.2 Quality Control

The QC for scATAC data determines several quantitative parameters to determine success of transposition. Qualitative plots generated from Cell Ranger ATAC pipeline include nucleosome banding pattern, TSS enrichment score, FRiP (FRaction of fragments in Peaks), and ratio of reads in ENCODE blacklist regions, and can be used to ascertain whether majority of reads originate from expected accessible regions in the genome. Alternatively, Signac, an extension of the Seurat R toolkit, takes fragment files as input and also offers well organized QC metrics to assess quality [134]. Figure 3c illustrates how the 1K PBMC scATAC-seq data might look like before and after filtering out poor quality nuclei by Signac. Figure 3d and e shows fragment distribution and TSS enrichment of reads as determined by ArchR. ArchR also uses an innovative doublet removal technique that synthesizes synthetic doublets by mixing cells in silico in thousands of combinations and then embedding in an UMAP. A nearest neighbor analysis is then used to identify nuclei that behave like doublets and are removed (Fig. 3f, purple dots). *See* **Note 4** highlighting the importance of QC steps.

2.2.3 Batch Correction

Batch correction of scATAC-seq data is often carried out as part of preprocessing steps or dimensionality reduction, sometimes using tools already used for bulk or scRNA-seq data (*see* Subheading 2.1.3). These tools (such as Harmony used in SnapATAC and ArchR) scan for variable peaks (or noise) in the data and effectively eliminate them. *See* **Note 5** highlighting the importance of batch correction steps.

Fig. 3 (continued) fragments that fall within ATAC-seq peaks), peak region fragments (measure of cellular sequencing depth/complexity), TSS enrichment (ratio of fragments centered at the TSS to fragments in TSS-flanking regions as defined by ENCODE), blacklist ratio (proportion of reads mapping to ENCODE blacklist regions), and nucleosome signal (ratio of mononucleosomal to nucleosome-free fragments) before and after filtering in Signac. (**D**) Histogram of DNA fragment sizes from the paired-end sequencing reads showing strong nucleosome banding pattern determined in ArchR. (**E**) ArchR derived TSS enrichment scores which is the average accessibility in a 50bp region centered at the TSS divided by the average accessibility of the TSS flanking positions 2 kb. (**F**) Doublet inference with ArchR projected to a UMAP plot. Cells with high doublet enrichment (purple) are computationally removed for downstream analysis. (**G**) UMAP representation of clusters. (**H**) Compartment analysis of high-quality peaks in each cluster. (**I**) Heatmap representing marker genes identified in each cluster. (**J**) ChromVAR analysis showing the top TFs with high variability

2.2.4 Data Normalization

Cell Ranger, Signac and ArchR perform Latent Semantic Indexing (LSI), a method originally used for language processing and first applied to scATAC-seq data by Cusanovich et al. [38]. LSI performs term frequency-inverse document frequency (TF-IDF) transformation where a "term" is a peak and the "document" is a sample. TF-IDF normalizes across cells to correct for sequencing depth, and then across peaks to give more weight to rare peaks. Finally, a singular value decomposition (SVD) is applied so that the most valuable information across samples is identified and represented in a lower dimensional space. Another algorithm derived from text mining is Latent Dirichlet Allocation (LDA) which is used by cisTopic to group cells by similarities in accessibility profiles [135]. SnapATAC uses a slightly different approach: chromatin accessibility profiles of every cell are represented as binary vectors, the lengths of which correspond to the number of 5 kb bins used to segment the genome. Normalized Jaccard Indices are then calculated where the value of each element corresponds to the fraction of overlapping bins between every pair of cells. An eigenvector decomposition is then performed on this normalized similarity matrix for dimensionality reduction.

2.2.5 Dimensionality Reduction Visuals

Linear dimensionality reduction methods like PCA are generally less preferred to visualize scATAC-seq data because the sparsity results in high cell-to-cell similarity. Instead, after LSI/LDA normalization, UMAP (Fig. 3g) or t-SNE embeddings are used.

2.2.6 Cell Clustering, Find Marker Genes

Clustering of scATAC-seq data is performed using the same tools as scRNA-seq. For example, Cell Ranger uses either a k-means or a graph-based clustering method, the latter also used by Seurat and ArchR and utilizing Louvain community detection algorithms. SnapATAC utilizes a sampling method called Nystrom which first performs a low dimension embedding from a subsample of cells and then projects the remainder cells to the embedding structure. However, unlike Seurat/ArchR's deterministic clustering algorithm, Nystrom based clustering can be stochastic. Thus, to improve the robustness of the clustering method, SnapATAC uses an ensemble Nystrom, which generates a mixture of Nystrom approximations which tend to provide better clustering stability. Each cluster is then annotated based on marker gene identification techniques founded on gene score models, a variety of which are used. For example, in ArchR, a gene score model based on accessibility within gene bodies, activity of nearby regulatory elements and strict gene boundaries is used to identify marker genes for each cluster (*see* **Note 6** for additional considerations). In Fig. 3i, marker genes identified were granulocytes (cluster 1), B cells (cluster 2), T/NK/Dendritic cells (cluster 3) while cluster 4 did not show any prominent markers. Compared to scRNA-seq, marker gene identification was less informative for scATAC-seq data which can be due to the inherent challenges of clustering from sparse data, or the fact that chromatin accessibility does not fully correlate to gene expression. Finally, peak distribution may be assessed in each cluster to detect where most of the biological changes are occurring. In Fig. 3h, the PBMC dataset shows that majority of high-quality peaks are located in introns or intergenic (distal) regions.

2.2.7 Trajectory Analysis

As with scRNA-seq analysis, cellular trajectories can also be constructed with scATAC-seq data to identify nondiscrete progressive changes in chromatin accessibility along developmental or differentiation pseudotimes. As before, if there are multiple outcomes, the trajectory will show a branched structure, indicative of critical cellular decision-making points. Once the cells have been ordered, differential peak analysis can be used to identify the exact genomic regions that drive the progressive changes. Commonly used tools for trajectory generation with scATAC-seq data include Cicero (which uses Monocle 3), SnapATAC (which uses Slingshot), ArchR, and STREAM [136].

2.2.8 Chromatin Variation Across Regions

Chromatin accessibility data is highly variable across individual cells due to heterogenous cell distribution in a sample, and cell type-, time-, and sex-specific TF activities within samples. ChromVAR (Chromatin Variation Across Regions) is an R package that assesses the functionality of trans-acting factors and cis-regulatory elements from this highly variable scATAC-seq data by measuring gain and loss of accessibility within peaks [29, 137]. ChromVAR takes three input files: aligned sequencing reads, pseudobulk/bulk peak information, and a set of chromatin features representing either position weight matrices (PWMs) of TF motifs or other user-determined genomic features such as enhancer locations, ChIP-seq peaks, and GWAS annotations. ChromVAR outputs include bias-corrected "deviation" values which reflect the difference between the observed number of fragments that map to peaks containing a particular motif and the expected number of mapped fragments based on bulk/pseudobulk data and a "z-score." Variability, which is the standard deviation of the z-scores is then calculated, the expected value of which is 1 if the motif peak sets are no more variable than the background peak sets for that motif. High variability is correlated to TF activity in a biological state or in response to a perturbation. In the example PBMC dataset, a number of TFs known to correlate with immune cell activity and division show high variability by ChromVAR analysis (Fig. 3j). ChromVAR thus provides biologically relevant, comprehensive views of TF activity in cell subsets from scATAC-seq data.

2.2.9 Enhancer–Promoter Looping Predictions by Cicero

The goal of Cicero is to detect connections between regulatory DNA elements and their putative target genes, using chromatin coaccessibility patterns from scATAC-seq data [138]. For accurate estimation, cells are first aggregated (in groups of 50 based on clustering or trajectory space) using a KNN algorithm to obtain high density, high depth counts. Nearby peaks are then assessed for their coaccessibility structure using covariance analysis. Regularized covariance matrices are generated by Graphical LASSO estimation with distance penalties. Finally, the coaccessibility scores (between −1 and 1) between each pair of accessible sites within a user-defined distance is reported. Additionally, positive coaccessibility scores can detect larger cis-coaccessibility networks (CCANs or chromatin accessibility hubs) by establishing graph structures. The nodes are the peaks of accessibility, and the edges of the graph are coaccessibility scores above a user defined threshold. Communities within this genome-wide graph identify CCANs and can be found using the Louvain algorithm. Cicero also provides an extension toolkit for differential peak analysis, clustering, visualization, trajectory analysis, and so on of scATAC-seq experiments using Monocle 3.

2.3 Computational Methods to Analyze Single-Cell DNA Multiomics Data

Rapid development of single-cell multiomics experimental protocols has prompted coevolution of innovative computational methods to integrate heterogeneous data from multiple datasets. This heterogeneity is compounded by missing data from the various technologies. In theory, all measurements should come from the same cell (paired or matched data) but to take advantage of already published datasets, more often than not, unpaired data from different cells from the same sample or different samples are used. One of the approaches for integration uses a common latent space to generate a combined reference [139]. In many cases, scRNA-seq is used as the common reference as it is the most developed single-cell application and given the reputation of RNA in the central dogma. To fill in the missing values, imputing algorithms have been developed based on heuristic model (MAGIC [140]), regression model (MISC [141], scImpute [142], VIPER [143]), matrix completion (mcImpute [144]), network-based (netImpute [145], scGAIN [146]) or using autoencoder (AutoImpute [147]).

The initial scRNA-seq preprocessing steps remain as mentioned before (Subheading 2.1.1) and the data is further integrated by CCA using latent space projection. This can be effectively used for integrating expression with epigenome data (scM&T-seq, scMT-seq, snmCT-seq, etc.). The Variational Bayes (VB) method is based on Bayesian modeling and is best suited for integrating single-cell genome sequencing with expression. LASSO (Least Absolute Shrinkage and Selection Operator) and GBR (Gradient Boosting Regression) are regression-based methods used mainly for expression and chromatin accessibility data (scCAT-seq, ATAC-RNA-seq, etc.). scRNA and scATAC-seq data have also been analyzed using matrix factorization methods such as integrative Nonnegative Matrix Factorization (iNMF), coupled Nonnegative Matrix Factorizations (coupleNMF), Group Factor Analysis (GFA), and Independent Component Analysis (ICA). TOPIC modeling has been used for integrating scRNA-seq with single-cell CRISPR. For integrating >2 single-cell modalities, Deep Learning (Autoencoder) has been used. Several published software tools (e.g., Seurat, MOFA+ [148], MATCHER [149], MuSiC [150], MIMOSCA [151], LIGER, clonealign [152]) are used to handle single-cell multiomics data [153].

2.4 Conclusions

The advent of single-cell genomics has revolutionized the world of big data, adding volume, complexity, resolution, and dimension. While this revolution promises conceptual insights into biological processes and disease development, it presents significant analytical and statistical challenges. In this chapter, we summarize the variety of single-cell methodologies available till date focusing on the transcriptome and epigenome. We also systematically outline and exemplify how scientists embarking on the single-cell journey might approach the analysis using a small publicly available PBMC dataset for each category of single-cell experiment. We discuss the pros and cons of popularly used packages and suggest alternatives when necessary. Overall, we hope this chapter will encourage scientists to consider single-cell applications to answer their favorite biological questions.

3 Notes

1. Sample quality and dead cell removal: sample quality is one of the key factors that determines experimental success. For scRNA-seq, it is necessary to ensure that the single-cell suspension has high viability and if not, dead cells must be removed prior to droplet production. Poor cell viability may result in increased ambient RNA that makes it difficult to distinguish cells from empty droplets. For scATAC-seq, it is important to use intact nuclei which can be especially challenging to obtain from frozen tissues.
2. Number of cells and read depth: the number of cells queried, and individual cell read depth greatly influences biological interpretations. For example, enough cells per sample must be targeted for droplet generation to ensure proper sampling of both abundant and rare cell populations (tissue-resident stem cells, cancer subclones, etc.). Similarly, sufficient read depth is required to detect low-abundance transcripts and for proper detection of differentially expressed genes and cell annotation.

3. Software versions: due to the constant evolution of software dedicated to single-cell data analysis, and some stochasticity in the performance of dimensionality reduction, clustering or imputation algorithms, it is imperative that researchers record the software versions for every experiment and adequately report in publications.
4. Filtering: filtering out low quality cells from the analysis is perhaps the most important upstream step in single-cell analysis as it greatly influences biological interpretations. QC steps outlined in Subheadings 2.1.2 and 2.2.2 are used to remove empty droplets, multiplets, apoptotic or lysing cells with high mitochondrial reads, cells with poor transposition or other technical artifacts.
5. Batch effects: systemic variations in single-cell datasets are produced due to differences in the source/lab generating the data, technologies used to make the single-cell libraries, sequencing platforms, scale of the data, and so on. Additionally, researchers may choose to integrate multiple modalities produced by different labs. For these purposes, efficacious batch correction techniques (*see* Subheadings 2.1.3 and 2.2.3) need to be applied to dissect true biological variation.
6. Cell annotation: marker gene identification of cell clusters give clues to cell identity. While several automatic cell classification methods have been identified, we propose that researchers always check several marker genes manually for correct identification of cell types. This is because many annotation tools perform poorly for rare cell populations. Additionally, clustering itself may not be optimal and may have mixed populations of cells.

References

1. International Human Genome Sequencing C (2004) Finishing the euchromatic sequence of the human genome. Nature 431(7011): 931–945. https://doi.org/10.1038/nature03001
2. Sachidanandam R, Weissman D, Schmidt SC, Kakol JM, Stein LD, Marth G, Sherry S, Mullikin JC, Mortimore BJ, Willey DL, Hunt SE, Cole CG, Coggill PC, Rice CM, Ning Z, Rogers J, Bentley DR, Kwok PY, Mardis ER, Yeh RT, Schultz B, Cook L, Davenport R, Dante M, Fulton L, Hillier L, Waterston RH, JD MP, Gilman B, Schaffner S, Van Etten WJ, Reich D, Higgins J, Daly MJ, Blumenstiel B, Baldwin J, Stange-Thomann-N, Zody MC, Linton L, Lander ES, Altshuler D, International SNPMWG (2001) A map of human genome sequence variation containing 1.42 million single nucleotide polymorphisms. Nature 409(6822): 928–933. https://doi.org/10.1038/35057149
3. International HapMap C (2005) A haplotype map of the human genome. Nature 437(7063):1299–1320. https://doi.org/10.1038/nature04226
4. Genomes Project C, Abecasis GR, Altshuler D, Auton A, Brooks LD, Durbin RM, Gibbs RA, Hurles ME, GA MV (2010) A map of human genome variation from population-scale sequencing. Nature 467(7319):1061–1073. https://doi.org/10.1038/nature09534
5. Bernstein BE, Stamatoyannopoulos JA, Costello JF, Ren B, Milosavljevic A, Meissner A, Kellis M, Marra MA, Beaudet AL, Ecker JR,

Farnham PJ, Hirst M, Lander ES, Mikkelsen TS, Thomson JA (2010) The NIH roadmap epigenomics mapping consortium. Nat Biotechnol 28(10):1045–1048. https://doi.org/10.1038/nbt1010-1045

6. Consortium EP, Birney E, Stamatoyannopoulos JA, Dutta A, Guigo R, Gingeras TR, Margulies EH, Weng Z, Snyder M, Dermitzakis ET, Thurman RE, Kuehn MS, Taylor CM, Neph S, Koch CM, Asthana S, Malhotra A, Adzhubei I, Greenbaum JA, Andrews RM, Flicek P, Boyle PJ, Cao H, Carter NP, Clelland GK, Davis S, Day N, Dhami P, Dillon SC, Dorschner MO, Fiegler H, Giresi PG, Goldy J, Hawrylycz M, Haydock A, Humbert R, James KD, Johnson BE, Johnson EM, Frum TT, Rosenzweig ER, Karnani N, Lee K, Lefebvre GC, Navas PA, Neri F, Parker SC, Sabo PJ, Sandstrom R, Shafer A, Vetrie D, Weaver M, Wilcox S, Yu M, Collins FS, Dekker J, Lieb JD, Tullius TD, Crawford GE, Sunyaev S, Noble WS, Dunham I, Denoeud F, Reymond A, Kapranov P, Rozowsky J, Zheng D, Castelo R, Frankish A, Harrow J, Ghosh S, Sandelin A, Hofacker IL, Baertsch R, Keefe D, Dike S, Cheng J, Hirsch HA, Sekinger EA, Lagarde J, Abril JF, Shahab A, Flamm C, Fried C, Hackermuller J, Hertel J, Lindemeyer M, Missal K, Tanzer A, Washietl S, Korbel J, Emanuelsson O, Pedersen JS, Holroyd N, Taylor R, Swarbreck D, Matthews N, Dickson MC, Thomas DJ, Weirauch MT, Gilbert J, Drenkow J, Bell I, Zhao X, Srinivasan KG, Sung WK, Ooi HS, Chiu KP, Foissac S, Alioto T, Brent M, Pachter L, Tress ML, Valencia A, Choo SW, Choo CY, Ucla C, Manzano C, Wyss C, Cheung E, Clark TG, Brown JB, Ganesh M, Patel S, Tammana H, Chrast J, Henrichsen CN, Kai C, Kawai J, Nagalakshmi U, Wu J, Lian Z, Lian J, Newburger P, Zhang X, Bickel P, Mattick JS, Carninci P, Hayashizaki Y, Weissman S, Hubbard T, Myers RM, Rogers J, Stadler PF, Lowe TM, Wei CL, Ruan Y, Struhl K, Gerstein M, Antonarakis SE, Fu Y, Green ED, Karaoz U, Siepel A, Taylor J, Liefer LA, Wetterstrand KA, Good PJ, Feingold EA, Guyer MS, Cooper GM, Asimenos G, Dewey CN, Hou M, Nikolaev S, Montoya-Burgos JI, Loytynoja A, Whelan S, Pardi F, Massingham T, Huang H, Zhang NR, Holmes I, Mullikin JC, Ureta-Vidal A, Paten B, Seringhaus M, Church D, Rosenbloom K, Kent WJ, Stone EA, Program NCS, Baylor College of Medicine Human Genome Sequencing C, Washington University Genome Sequencing C, Broad I, Children's Hospital Oakland Research I, Batzoglou S, Goldman N, Hardison RC, Haussler D, Miller W, Sidow A, Trinklein ND, Zhang ZD, Barrera L, Stuart R, King DC, Ameur A, Enroth S, Bieda MC, Kim J, Bhinge AA, Jiang N, Liu J, Yao F, Vega VB, Lee CW, Ng P, Shahab A, Yang A, Moqtaderi Z, Zhu Z, Xu X, Squazzo S, Oberley MJ, Inman D, Singer MA, Richmond TA, Munn KJ, Rada-Iglesias A, Wallerman O, Komorowski J, Fowler JC, Couttet P, Bruce AW, Dovey OM, Ellis PD, Langford CF, Nix DA, Euskirchen G, Hartman S, Urban AE, Kraus P, Van Calcar S, Heintzman N, Kim TH, Wang K, Qu C, Hon G, Luna R, Glass CK, Rosenfeld MG, Aldred SF, Cooper SJ, Halees A, Lin JM, Shulha HP, Zhang X, Xu M, Haidar JN, Yu Y, Ruan Y, Iyer VR, Green RD, Wadelius C, Farnham PJ, Ren B, Harte RA, Hinrichs AS, Trumbower H, Clawson H, Hillman-Jackson J, Zweig AS, Smith K, Thakkapallayil A, Barber G, Kuhn RM, Karolchik D, Armengol L, Bird CP, de Bakker PI, Kern AD, Lopez-Bigas N, Martin JD, Stranger BE, Woodroffe A, Davydov E, Dimas A, Eyras E, Hallgrimsdottir IB, Huppert J, Zody MC, Abecasis GR, Estivill X, Bouffard GG, Guan X, Hansen NF, Idol JR, Maduro VV, Maskeri B, McDowell JC, Park M, Thomas PJ, Young AC, Blakesley RW, Muzny DM, Sodergren E, Wheeler DA, Worley KC, Jiang H, Weinstock GM, Gibbs RA, Graves T, Fulton R, Mardis ER, Wilson RK, Clamp M, Cuff J, Gnerre S, Jaffe DB, Chang JL, Lindblad-Toh K, Lander ES, Koriabine M, Nefedov M, Osoegawa K, Yoshinaga Y, Zhu B, de Jong PJ (2007) Identification and analysis of functional elements in 1% of the human genome by the ENCODE pilot project. Nature 447 (7146):799–816. https://doi.org/10.1038/nature05874

7. Celniker SE, Dillon LA, Gerstein MB, Gunsalus KC, Henikoff S, Karpen GH, Kellis M, Lai EC, Lieb JD, MacAlpine DM, Micklem G, Piano F, Snyder M, Stein L, White KP, Waterston RH, modENCODE Consortium (2009) Unlocking the secrets of the genome. Nature 459(7249):927–930. https://doi.org/10.1038/459927a

8. Stunnenberg HG, International Human Epigenome C, Hirst M (2016) The International Human Epigenome Consortium: a Blueprint for scientific collaboration and discovery. Cell 167 (5):1145–1149. doi:https://doi.org/10.1016/j.cell.2016.11.007

9. Rozenblatt-Rosen O, Stubbington MJT, Regev A, Teichmann SA (2017) The Human Cell Atlas: from vision to reality. Nature 550(7677):451–453. https://doi.org/10.1038/550451a

10. Slyper M, Porter CBM, Ashenberg O, Waldman J, Drokhlyansky E, Wakiro I, Smillie C, Smith-Rosario G, Wu J, Dionne D, Vigneau S, Jane-Valbuena J, Tickle TL, Napolitano S, Su MJ, Patel AG, Karlstrom A, Gritsch S, Nomura M, Waghray A, Gohil SH, Tsankov AM, Jerby-Arnon L, Cohen O, Klughammer J, Rosen Y, Gould J, Nguyen L, Hofree M, Tramontozzi PJ, Li B, Wu CJ, Izar B, Haq R, Hodi FS, Yoon CH, Hata AN, Baker SJ, Suva ML, Bueno R, Stover EH, Clay MR, Dyer MA, Collins NB, Matulonis UA, Wagle N, Johnson BE, Rotem A, Rozenblatt-Rosen O, Regev A (2020) A single-cell and single-nucleus RNA-Seq toolbox for fresh and frozen human tumors. Nat Med 26(5):792–802. https://doi.org/10.1038/s41591-020-0844-1

11. Tang F, Barbacioru C, Nordman E, Li B, Xu N, Bashkirov VI, Lao K, Surani MA (2010) RNA-Seq analysis to capture the transcriptome landscape of a single cell. Nat Protoc 5(3):516–535. https://doi.org/10.1038/nprot.2009.236

12. Tang F, Barbacioru C, Wang Y, Nordman E, Lee C, Xu N, Wang X, Bodeau J, Tuch BB, Siddiqui A, Lao K, Surani MA (2009) mRNA-Seq whole-transcriptome analysis of a single cell. Nat Methods 6(5):377–382. https://doi.org/10.1038/nmeth.1315

13. Islam S, Kjallquist U, Moliner A, Zajac P, Fan JB, Lonnerberg P, Linnarsson S (2011) Characterization of the single-cell transcriptional landscape by highly multiplex RNA-seq. Genome Res 21(7):1160–1167. https://doi.org/10.1101/gr.110882.110

14. Pollen AA, Nowakowski TJ, Shuga J, Wang X, Leyrat AA, Lui JH, Li N, Szpankowski L, Fowler B, Chen P, Ramalingam N, Sun G, Thu M, Norris M, Lebofsky R, Toppani D, Kemp DW 2nd, Wong M, Clerkson B, Jones BN, Wu S, Knutsson L, Alvarado B, Wang J, Weaver LS, May AP, Jones RC, Unger MA, Kriegstein AR, West JA (2014) Low-coverage single-cell mRNA sequencing reveals cellular heterogeneity and activated signaling pathways in developing cerebral cortex. Nat Biotechnol 32(10):1053–1058. https://doi.org/10.1038/nbt.2967

15. Islam S, Zeisel A, Joost S, La Manno G, Zajac P, Kasper M, Lonnerberg P, Linnarsson S (2014) Quantitative single-cell RNA-seq with unique molecular identifiers. Nat Methods 11(2):163–166. https://doi.org/10.1038/nmeth.2772

16. Ramskold D, Luo S, Wang YC, Li R, Deng Q, Faridani OR, Daniels GA, Khrebtukova I, Loring JF, Laurent LC, Schroth GP, Sandberg R (2012) Full-length mRNA-Seq from single-cell levels of RNA and individual circulating tumor cells. Nat Biotechnol 30(8):777–782. https://doi.org/10.1038/nbt.2282

17. Picelli S, Faridani OR, Bjorklund AK, Winberg G, Sagasser S, Sandberg R (2014) Full-length RNA-seq from single cells using Smart-seq2. Nat Protoc 9(1):171–181. https://doi.org/10.1038/nprot.2014.006

18. Hagemann-Jensen M, Ziegenhain C, Chen P, Ramskold D, Hendriks GJ, Larsson AJM, Faridani OR, Sandberg R (2020) Single-cell RNA counting at allele and isoform resolution using Smart-seq3. Nat Biotechnol 38(6):708–714. https://doi.org/10.1038/s41587-020-0497-0

19. Hashimshony T, Wagner F, Sher N, Yanai I (2012) CEL-Seq: single-cell RNA-Seq by multiplexed linear amplification. Cell Rep 2(3):666–673. https://doi.org/10.1016/j.celrep.2012.08.003

20. Hashimshony T, Senderovich N, Avital G, Klochendler A, de Leeuw Y, Anavy L, Gennert D, Li S, Livak KJ, Rozenblatt-Rosen O, Dor Y, Regev A, Yanai I (2016) CEL-Seq2: sensitive highly-multiplexed single-cell RNA-Seq. Genome Biol 17:77. https://doi.org/10.1186/s13059-016-0938-8

21. Jaitin DA, Kenigsberg E, Keren-Shaul H, Elefant N, Paul F, Zaretsky I, Mildner A, Cohen N, Jung S, Tanay A, Amit I (2014) Massively parallel single-cell RNA-seq for marker-free decomposition of tissues into cell types. Science 343(6172):776–779. https://doi.org/10.1126/science.1247651

22. Keren-Shaul H, Kenigsberg E, Jaitin DA, David E, Paul F, Tanay A, Amit I (2019) MARS-seq2.0: an experimental and analytical pipeline for indexed sorting combined with single-cell RNA sequencing. Nat Protoc 14(6):1841–1862. https://doi.org/10.1038/s41596-019-0164-4

23. Fan HC, Fu GK, Fodor SP (2015) Expression profiling. Combinatorial labeling of single cells for gene expression cytometry. Science 347(6222):1258367. https://doi.org/10.1126/science.1258367

24. Macosko EZ, Basu A, Satija R, Nemesh J, Shekhar K, Goldman M, Tirosh I, Bialas AR, Kamitaki N, Martersteck EM, Trombetta JJ, Weitz DA, Sanes JR, Shalek AK, Regev A, McCarroll SA (2015) Highly parallel genome-wide expression profiling of individual cells using nanoliter droplets. Cell 161(5):

1202–1214. https://doi.org/10.1016/j.cell.2015.05.002

25. Klein AM, Mazutis L, Akartuna I, Tallapragada N, Veres A, Li V, Peshkin L, Weitz DA, Kirschner MW (2015) Droplet barcoding for single-cell transcriptomics applied to embryonic stem cells. Cell 161(5):1187–1201. https://doi.org/10.1016/j.cell.2015.04.044
26. Zheng GX, Terry JM, Belgrader P, Ryvkin P, Bent ZW, Wilson R, Ziraldo SB, Wheeler TD, McDermott GP, Zhu J, Gregory MT, Shuga J, Montesclaros L, Underwood JG, Masquelier DA, Nishimura SY, Schnall-Levin-M, Wyatt PW, Hindson CM, Bharadwaj R, Wong A, Ness KD, Beppu LW, Deeg HJ, McFarland C, Loeb KR, Valente WJ, Ericson NG, Stevens EA, Radich JP, Mikkelsen TS, Hindson BJ, Bielas JH (2017) Massively parallel digital transcriptional profiling of single cells. Nat Commun 8:14049. https://doi.org/10.1038/ncomms14049
27. Horvath S, Raj K (2018) DNA methylation-based biomarkers and the epigenetic clock theory of ageing. Nat Rev Genet 19(6):371–384. https://doi.org/10.1038/s41576-018-0004-3
28. Sen P, Shah PP, Nativio R, Berger SL (2016) Epigenetic mechanisms of longevity and aging. Cell 166(4):822–839. https://doi.org/10.1016/j.cell.2016.07.050
29. Buenrostro JD, Wu B, Litzenburger UM, Ruff D, Gonzales ML, Snyder MP, Chang HY, Greenleaf WJ (2015) Single-cell chromatin accessibility reveals principles of regulatory variation. Nature 523(7561):486–490. https://doi.org/10.1038/nature14590
30. Guo H, Zhu P, Wu X, Li X, Wen L, Tang F (2013) Single-cell methylome landscapes of mouse embryonic stem cells and early embryos analyzed using reduced representation bisulfite sequencing. Genome Res 23(12):2126–2135. https://doi.org/10.1101/gr.161679.113
31. Miura F, Enomoto Y, Dairiki R, Ito T (2012) Amplification-free whole-genome bisulfite sequencing by post-bisulfite adaptor tagging. Nucleic Acids Res 40(17):e136. https://doi.org/10.1093/nar/gks454
32. Smallwood SA, Lee HJ, Angermueller C, Krueger F, Saadeh H, Peat J, Andrews SR, Stegle O, Reik W, Kelsey G (2014) Single-cell genome-wide bisulfite sequencing for assessing epigenetic heterogeneity. Nat Methods 11(8):817–820. https://doi.org/10.1038/nmeth.3035
33. Farlik M, Sheffield NC, Nuzzo A, Datlinger P, Schonegger A, Klughammer J, Bock C (2015) Single-cell DNA methylome sequencing and bioinformatic inference of epigenomic cell-state dynamics. Cell Rep 10(8):1386–1397. https://doi.org/10.1016/j.celrep.2015.02.001
34. Rotem A, Ram O, Shoresh N, Sperling RA, Goren A, Weitz DA, Bernstein BE (2015) Single-cell ChIP-seq reveals cell subpopulations defined by chromatin state. Nat Biotechnol 33(11):1165–1172. https://doi.org/10.1038/nbt.3383
35. Kaya-Okur HS, Wu SJ, Codomo CA, Pledger ES, Bryson TD, Henikoff JG, Ahmad K, Henikoff S (2019) CUT&Tag for efficient epigenomic profiling of small samples and single cells. Nat Commun 10(1):1930. https://doi.org/10.1038/s41467-019-09982-5
36. Wu SJ, Furlan SN, Mihalas AB, Kaya-Okur H, Feroze AH, Emerson SN, Zheng Y, Carson K, Cimino PJ, Keene CD, Holland EC, Sarthy JF, Gottardo R, Ahmad K, Henikoff S, Patel AP (2020) Single-cell analysis of chromatin silencing programs in developmental and tumor progression. bioRxiv:2020.2009.2004.282418. https://doi.org/10.1101/2020.09.04.282418
37. Bartosovic M, Kabbe M, Castelo-Branco G (2020) Single-cell profiling of histone modifications in the mouse brain. bioRxiv:2020.2009.2002.279703. https://doi.org/10.1101/2020.09.02.279703
38. Cusanovich DA, Daza R, Adey A, Pliner HA, Christiansen L, Gunderson KL, Steemers FJ, Trapnell C, Shendure J (2015) Multiplex single cell profiling of chromatin accessibility by combinatorial cellular indexing. Science 348(6237):910–914. https://doi.org/10.1126/science.aab1601
39. Satpathy AT, Granja JM, Yost KE, Qi Y, Meschi F, McDermott GP, Olsen BN, Mumbach MR, Pierce SE, Corces MR, Shah P, Bell JC, Jhutty D, Nemec CM, Wang J, Wang L, Yin Y, Giresi PG, Chang ALS, Zheng GXY, Greenleaf WJ, Chang HY (2019) Massively parallel single-cell chromatin landscapes of human immune cell development and intratumoral T cell exhaustion. Nat Biotechnol 37(8):925–936. https://doi.org/10.1038/s41587-019-0206-z
40. Bonev B, Cavalli G (2016) Organization and function of the 3D genome. Nat Rev Genet 17(12):772. https://doi.org/10.1038/nrg.2016.147
41. Nagano T, Lubling Y, Stevens TJ, Schoenfelder S, Yaffe E, Dean W, Laue ED, Tanay A, Fraser P (2013) Single-cell Hi-C reveals cell-to-cell variability in chromosome

structure. Nature 502(7469):59–64. https://doi.org/10.1038/nature12593

42. Rao SS, Huntley MH, Durand NC, Stamenova EK, Bochkov ID, Robinson JT, Sanborn AL, Machol I, Omer AD, Lander ES, Aiden EL (2014) A 3D map of the human genome at kilobase resolution reveals principles of chromatin looping. Cell 159(7):1665–1680. https://doi.org/10.1016/j.cell.2014.11.021
43. Tan L, Xing D, Chang CH, Li H, Xie XS (2018) Three-dimensional genome structures of single diploid human cells. Science 361(6405):924–928. https://doi.org/10.1126/science.aat5641
44. Angermueller C, Clark SJ, Lee HJ, Macaulay IC, Teng MJ, Hu TX, Krueger F, Smallwood S, Ponting CP, Voet T, Kelsey G, Stegle O, Reik W (2016) Parallel single-cell sequencing links transcriptional and epigenetic heterogeneity. Nat Methods 13(3):229–232. https://doi.org/10.1038/nmeth.3728
45. Hu Y, Huang K, An Q, Du G, Hu G, Xue J, Zhu X, Wang CY, Xue Z, Fan G (2016) Simultaneous profiling of transcriptome and DNA methylome from a single cell. Genome Biol 17:88. https://doi.org/10.1186/s13059-016-0950-z
46. Luo C, Liu H, Wang B-A, Bartlett A, Rivkin A, Nery JR, Ecker JR (2018) Multiomic profiling of transcriptome and DNA methylome in single nuclei with molecular partitioning. bioRxiv:434845. https://doi.org/10.1101/434845
47. Hou Y, Guo H, Cao C, Li X, Hu B, Zhu P, Wu X, Wen L, Tang F, Huang Y, Peng J (2016) Single-cell triple omics sequencing reveals genetic, epigenetic, and transcriptomic heterogeneity in hepatocellular carcinomas. Cell Res 26(3):304–319. https://doi.org/10.1038/cr.2016.23
48. Pott S (2017) Simultaneous measurement of chromatin accessibility, DNA methylation, and nucleosome phasing in single cells. Elife 6. https://doi.org/10.7554/eLife.23203
49. Clark SJ, Argelaguet R, Kapourani CA, Stubbs TM, Lee HJ, Alda-Catalinas C, Krueger F, Sanguinetti G, Kelsey G, Marioni JC, Stegle O, Reik W (2018) scNMT-seq enables joint profiling of chromatin accessibility DNA methylation and transcription in single cells. Nat Commun 9(1):781. https://doi.org/10.1038/s41467-018-03149-4
50. Wang Y, Yuan P, Yan Z, Yang M, Huo Y, Nie Y, Zhu X, Yan L, Qiao J (2019) Single-cell multiomics sequencing reveals the functional regulatory landscape of early embryos. bioRxiv:803890. https://doi.org/10.1101/803890
51. Guo F, Li L, Li J, Wu X, Hu B, Zhu P, Wen L, Tang F (2017) Single-cell multi-omics sequencing of mouse early embryos and embryonic stem cells. Cell Res 27(8):967–988. https://doi.org/10.1038/cr.2017.82
52. Liu L, Liu C, Quintero A, Wu L, Yuan Y, Wang M, Cheng M, Leng L, Xu L, Dong G, Li R, Liu Y, Wei X, Xu J, Chen X, Lu H, Chen D, Wang Q, Zhou Q, Lin X, Li G, Liu S, Wang Q, Wang H, Fink JL, Gao Z, Liu X, Hou Y, Zhu S, Yang H, Ye Y, Lin G, Chen F, Herrmann C, Eils R, Shang Z, Xu X (2019) Deconvolution of single-cell multi-omics layers reveals regulatory heterogeneity. Nat Commun 10(1):470. https://doi.org/10.1038/s41467-018-08205-7
53. Reyes M, Billman K, Hacohen N, Blainey PC (2019) Simultaneous profiling of gene expression and chromatin accessibility in single cells. Adv Biosyst 3(11). https://doi.org/10.1002/adbi.201900065
54. Li G, Liu Y, Zhang Y, Kubo N, Yu M, Fang R, Kellis M, Ren B (2019) Joint profiling of DNA methylation and chromatin architecture in single cells. Nat Methods 16(10):991–993. https://doi.org/10.1038/s41592-019-0502-z
55. Lee DS, Luo C, Zhou J, Chandran S, Rivkin A, Bartlett A, Nery JR, Fitzpatrick C, O'Connor C, Dixon JR, Ecker JR (2019) Simultaneous profiling of 3D genome structure and DNA methylation in single human cells. Nat Methods 16(10):999–1006. https://doi.org/10.1038/s41592-019-0547-z
56. Stoeckius M, Hafemeister C, Stephenson W, Houck-Loomis B, Chattopadhyay PK, Swerdlow H, Satija R, Smibert P (2017) Simultaneous epitope and transcriptome measurement in single cells. Nat Methods 14(9):865–868. https://doi.org/10.1038/nmeth.4380
57. Peterson VM, Zhang KX, Kumar N, Wong J, Li L, Wilson DC, Moore R, McClanahan TK, Sadekova S, Klappenbach JA (2017) Multiplexed quantification of proteins and transcripts in single cells. Nat Biotechnol 35(10):936–939. https://doi.org/10.1038/nbt.3973
58. Mimitou EP, Lareau CA, Chen KY, Zorzetto-Fernandes AL, Takeshima Y, Luo W, Huang T-S, Yeung B, Thakore PI, Wing JB, Nazor KL, Sakaguchi S, Ludwig LS, Sankaran VG, Regev A, Smibert P (2020) Scalable, multimodal profiling of chromatin accessibility and

protein levels in single cells. bioRxiv:2020.2009.2008.286914. https://doi.org/10.1101/2020.09.08.286914

59. Lareau CA, Ludwig LS, Muus C, Gohil SH, Zhao T, Chiang Z, Pelka K, Verboon JM, Luo W, Christian E, Rosebrock D, Getz G, Boland GM, Chen F, Buenrostro JD, Hacohen N, Wu CJ, Aryee MJ, Regev A, Sankaran VG (2021) Massively parallel single-cell mitochondrial DNA genotyping and chromatin profiling. Nat Biotechnol 39(4):451–461. https://doi.org/10.1038/s41587-020-0645-6
60. Stoeckius M, Zheng S, Houck-Loomis B, Hao S, Yeung BZ, Mauck WM 3rd, Smibert P, Satija R (2018) Cell Hashing with barcoded antibodies enables multiplexing and doublet detection for single cell genomics. Genome Biol 19(1):224. https://doi.org/10.1186/s13059-018-1603-1
61. Gaublomme JT, Li B, McCabe C, Knecht A, Drokhlyansky E, Wittenberghe NV, Waldman J, Dionne D, Nguyen L, Jager PD, Yeung B, Zhao X, Habib N, Rozenblatt-Rosen O, Regev A (2018) Nuclei multiplexing with barcoded antibodies for single-nucleus genomics. bioRxiv:476036. https://doi.org/10.1101/476036
62. Dixit A, Parnas O, Li B, Chen J, Fulco CP, Jerby-Arnon L, Marjanovic ND, Dionne D, Burks T, Raychowdhury R, Adamson B, Norman TM, Lander ES, Weissman JS, Friedman N, Regev A (2016) Perturb-Seq: dissecting molecular circuits with scalable single-cell RNA profiling of pooled genetic screens. Cell 167(7):1853–1866.e1817. https://doi.org/10.1016/j.cell.2016.11.038
63. Jaitin DA, Weiner A, Yofe I, Lara-Astiaso D, Keren-Shaul H, David E, Salame TM, Tanay A, van Oudenaarden A, Amit I (2016) Dissecting immune circuits by linking CRISPR-pooled screens with single-cell RNA-Seq. Cell 167(7):1883–1896.e1815. https://doi.org/10.1016/j.cell.2016.11.039
64. Datlinger P, Rendeiro AF, Schmidl C, Krausgruber T, Traxler P, Klughammer J, Schuster LC, Kuchler A, Alpar D, Bock C (2017) Pooled CRISPR screening with single-cell transcriptome readout. Nat Methods 14(3):297–301. https://doi.org/10.1038/nmeth.4177
65. Adamson B, Norman TM, Jost M, Cho MY, Nunez JK, Chen Y, Villalta JE, Gilbert LA, Horlbeck MA, Hein MY, Pak RA, Gray AN, Gross CA, Dixit A, Parnas O, Regev A, Weissman JS (2016) A multiplexed single-cell CRISPR screening platform enables systematic dissection of the unfolded protein response. Cell 167(7):1867–1882.e1821. https://doi.org/10.1016/j.cell.2016.11.048
66. Xie S, Duan J, Li B, Zhou P, Hon GC (2017) Multiplexed engineering and analysis of combinatorial enhancer activity in single cells. Mol Cell 66(2):285–299.e285. https://doi.org/10.1016/j.molcel.2017.03.007
67. Replogle JM, Norman TM, Xu A, Hussmann JA, Chen J, Cogan JZ, Meer EJ, Terry JM, Riordan DP, Srinivas N, Fiddes IT, Arthur JG, Alvarado LJ, Pfeiffer KA, Mikkelsen TS, Weissman JS, Adamson B (2020) Combinatorial single-cell CRISPR screens by direct guide RNA capture and targeted sequencing. Nat Biotechnol 38(8):954–961. https://doi.org/10.1038/s41587-020-0470-y
68. Rostom R, Svensson V, Teichmann SA, Kar G (2017) Computational approaches for interpreting scRNA-seq data. FEBS Lett 591(15):2213–2225. https://doi.org/10.1002/1873-3468.12684
69. Dobin A, Davis CA, Schlesinger F, Drenkow J, Zaleski C, Jha S, Batut P, Chaisson M, Gingeras TR (2013) STAR: ultrafast universal RNA-seq aligner. Bioinformatics 29(1):15–21. https://doi.org/10.1093/bioinformatics/bts635
70. Melsted P, Booeshaghi AS, Gao F, Beltrame E, Lu L, Hjorleifsson KE, Gehring J, Pachter L (2019) Modular and efficient pre-processing of single-cell RNA-seq. bioRxiv:673285. https://doi.org/10.1101/673285
71. Patro R, Duggal G, Love MI, Irizarry RA, Kingsford C (2017) Salmon provides fast and bias-aware quantification of transcript expression. Nat Methods 14(4):417–419. https://doi.org/10.1038/nmeth.4197
72. Srivastava A, Malik L, Smith T, Sudbery I, Patro R (2019) Alevin efficiently estimates accurate gene abundances from dscRNA-seq data. Genome Biol 20(1):65. https://doi.org/10.1186/s13059-019-1670-y
73. Amezquita RA, Lun ATL, Becht E, Carey VJ, Carpp LN, Geistlinger L, Marini F, Rue-Albrecht K, Risso D, Soneson C, Waldron L, Pages H, Smith ML, Huber W, Morgan M, Gottardo R, Hicks SC (2020) Orchestrating single-cell analysis with bioconductor. Nat Methods 17(2):137–145. https://doi.org/10.1038/s41592-019-0654-x
74. Tian L, Su S, Dong X, Amann-Zalcenstein D, Biben C, Seidi A, Hilton DJ, Naik SH, Ritchie ME (2018) scPipe: a flexible R/Bioconductor

preprocessing pipeline for single-cell RNA-sequencing data. PLoS Comput Biol 14(8): e1006361. https://doi.org/10.1371/journal.pcbi.1006361

75. Wang Z, Hu J, Johnson WE, Campbell JD (2019) scruff: an R/bioconductor package for preprocessing single-cell RNA-sequencing data. BMC Bioinformatics 20(1):222. https://doi.org/10.1186/s12859-019-2797-2
76. Jiang P (2019) Quality control of single-cell RNA-seq. Methods Mol Biol 1935:1–9. https://doi.org/10.1007/978-1-4939-9057-3_1
77. Abugessaisa I, Noguchi S, Cardon M, Hasegawa A, Watanabe K, Takahashi M, Suzuki H, Katayama S, Kere J, Kasukawa T (2020) Quality assessment of single-cell RNA sequencing data by coverage skewness analysis. bioRxiv:2019.2012.2031.890269. https://doi.org/10.1101/2019.12.31.890269
78. McCarthy DJ, Campbell KR, Lun AT, Wills QF (2017) Scater: pre-processing, quality control, normalization and visualization of single-cell RNA-seq data in R. Bioinformatics 33(8):1179–1186. https://doi.org/10.1093/bioinformatics/btw777
79. Butler A, Hoffman P, Smibert P, Papalexi E, Satija R (2018) Integrating single-cell transcriptomic data across different conditions, technologies, and species. Nat Biotechnol 36(5):411–420. https://doi.org/10.1038/nbt.4096
80. Ilicic T, Kim JK, Kolodziejczyk AA, Bagger FO, McCarthy DJ, Marioni JC, Teichmann SA (2016) Classification of low quality cells from single-cell RNA-seq data. Genome Biol 17:29. https://doi.org/10.1186/s13059-016-0888-1
81. Young MD, Behjati S (2020) SoupX removes ambient RNA contamination from droplet based single-cell RNA sequencing data. bioRxiv:303727. https://doi.org/10.1101/303727
82. Heaton H, Talman AM, Knights A, Imaz M, Gaffney D, Durbin R, Hemberg M, Lawniczak M (2019) Souporcell: robust clustering of single cell RNAseq by genotype and ambient RNA inference without reference genotypes. bioRxiv:699637. https://doi.org/10.1101/699637
83. Yang S, Corbett SE, Koga Y, Wang Z, Johnson WE, Yajima M, Campbell JD (2020) Decontamination of ambient RNA in single-cell RNA-seq with DecontX. Genome Biol 21(1):57. https://doi.org/10.1186/s13059-020-1950-6
84. McGinnis CS, Murrow LM, Gartner ZJ (2019) DoubletFinder: doublet detection in single-cell RNA sequencing data using artificial nearest neighbors. Cell Syst 8(4): 329–337.e324. https://doi.org/10.1016/j.cels.2019.03.003
85. DePasquale EAK, Schnell DJ, Van Camp PJ, Valiente-Alandi I, Blaxall BC, Grimes HL, Singh H, Salomonis N (2019) DoubletDecon: deconvoluting doublets from single-cell RNA-sequencing data. Cell Rep 29(6): 1718–1727.e1718. https://doi.org/10.1016/j.celrep.2019.09.082
86. Wolock SL, Lopez R, Klein AM (2019) Scrublet: computational identification of cell doublets in single-cell transcriptomic data. Cell Syst 8(4):281–291.e289. https://doi.org/10.1016/j.cels.2018.11.005
87. Bernstein NJ, Fong NL, Lam I, Roy MA, Hendrickson DG, Kelley DR (2020) Solo: doublet identification in single-cell RNA-Seq via semi-supervised deep learning. Cell Syst 11(1):95–101.e105. https://doi.org/10.1016/j.cels.2020.05.010
88. Hardoon DR, Szedmak S, Shawe-Taylor J (2004) Canonical correlation analysis: an overview with application to learning methods. Neural Comput 16(12):2639–2664. https://doi.org/10.1162/0899766042321814
89. Haghverdi L, Lun ATL, Morgan MD, Marioni JC (2018) Batch effects in single-cell RNA-sequencing data are corrected by matching mutual nearest neighbors. Nat Biotechnol 36(5):421–427. https://doi.org/10.1038/nbt.4091
90. Tran HTN, Ang KS, Chevrier M, Zhang X, Lee NYS, Goh M, Chen J (2020) A benchmark of batch-effect correction methods for single-cell RNA sequencing data. Genome Biol 21(1):12. https://doi.org/10.1186/s13059-019-1850-9
91. Korsunsky I, Millard N, Fan J, Slowikowski K, Zhang F, Wei K, Baglaenko Y, Brenner M, Loh PR, Raychaudhuri S (2019) Fast, sensitive and accurate integration of single-cell data with Harmony. Nat Methods 16(12): 1289–1296. https://doi.org/10.1038/s41592-019-0619-0
92. Welch JD, Kozareva V, Ferreira A, Vanderburg C, Martin C, Macosko EZ (2019) Single-cell multi-omic integration compares and contrasts features of brain cell identity. Cell 177(7):1873–1887.e1817. https://doi.org/10.1016/j.cell.2019.05.006
93. Buttner M, Miao Z, Wolf FA, Teichmann SA, Theis FJ (2019) A test metric for assessing single-cell RNA-seq batch correction. Nat

Methods 16(1):43–49. https://doi.org/10.1038/s41592-018-0254-1

94. Lytal N, Ran D, An L (2020) Normalization methods on single-cell RNA-seq data: an empirical survey. Front Genet 11:41. https://doi.org/10.3389/fgene.2020.00041
95. Hafemeister C, Satija R (2019) Normalization and variance stabilization of single-cell RNA-seq data using regularized negative binomial regression. Genome Biol 20(1): 296. https://doi.org/10.1186/s13059-019-1874-1
96. Wolf FA, Angerer P, Theis FJ (2018) SCANPY: large-scale single-cell gene expression data analysis. Genome Biol 19(1):15. https://doi.org/10.1186/s13059-017-1382-0
97. Buettner F, Natarajan KN, Casale FP, Proserpio V, Scialdone A, Theis FJ, Teichmann SA, Marioni JC, Stegle O (2015) Computational analysis of cell-to-cell heterogeneity in single-cell RNA-sequencing data reveals hidden subpopulations of cells. Nat Biotechnol 33(2):155–160. https://doi.org/10.1038/nbt.3102
98. Buettner F, Pratanwanich N, McCarthy DJ, Marioni JC, Stegle O (2017) f-scLVM: scalable and versatile factor analysis for single-cell RNA-seq. Genome Biol 18(1):212. https://doi.org/10.1186/s13059-017-1334-8
99. Yip SH, Sham PC, Wang J (2019) Evaluation of tools for highly variable gene discovery from single-cell RNA-seq data. Brief Bioinformat 20(4):1583–1589. https://doi.org/10.1093/bib/bby011
100. Townes FW, Hicks SC, Aryee MJ, Irizarry RA (2019) Feature selection and dimension reduction for single-cell RNA-Seq based on a multinomial model. Genome Biol 20(1): 295. https://doi.org/10.1186/s13059-019-1861-6
101. Sun S, Zhu J, Ma Y, Zhou X (2019) Accuracy, robustness and scalability of dimensionality reduction methods for single-cell RNA-seq analysis. Genome Biol 20(1):269. https://doi.org/10.1186/s13059-019-1898-6
102. Heiser CN, Lau KS (2020) A quantitative framework for evaluating single-cell data structure preservation by dimensionality reduction techniques. Cell Rep 31(5): 107576. https://doi.org/10.1016/j.celrep.2020.107576
103. Tsuyuzaki K, Sato H, Sato K, Nikaido I (2020) Benchmarking principal component analysis for large-scale single-cell RNA-sequencing. Genome Biol 21(1):9. https://doi.org/10.1186/s13059-019-1900-3
104. vanDerMaaten L, Hinton G (2008) Visualizing data using t-SNE. J Machine Learning Res 9:2579–2605
105. McInnes L, Healy J, Melville J (2018) UMAP: uniform manifold approximation and projection for dimension reduction. arXiv:1802.03426
106. Feng C, Liu S, Zhang H, Guan R, Li D, Zhou F, Liang Y, Feng X (2020) Dimension reduction and clustering models for single-cell RNA sequencing data: a comparative study. Int J Mol Sci 21(6):2181. https://doi.org/10.3390/ijms21062181
107. Kiselev VY, Kirschner K, Schaub MT, Andrews T, Yiu A, Chandra T, Natarajan KN, Reik W, Barahona M, Green AR, Hemberg M (2017) SC3: consensus clustering of single-cell RNA-seq data. Nat Methods 14(5):483–486. https://doi.org/10.1038/nmeth.4236
108. Qiu X, Mao Q, Tang Y, Wang L, Chawla R, Pliner HA, Trapnell C (2017) Reversed graph embedding resolves complex single-cell trajectories. Nat Methods 14(10):979–982. https://doi.org/10.1038/nmeth.4402
109. Duo A, Robinson MD, Soneson C (2018) A systematic performance evaluation of clustering methods for single-cell RNA-seq data. F1000Res 7:1141. https://doi.org/10.12688/f1000research.15666.2
110. Freytag S, Tian L, Lonnstedt I, Ng M, Bahlo M (2018) Comparison of clustering tools in R for medium-sized 10x Genomics single-cell RNA-sequencing data. F1000Res 7:1297. https://doi.org/10.12688/f1000research.15809.2
111. Cao J, Spielmann M, Qiu X, Huang X, Ibrahim DM, Hill AJ, Zhang F, Mundlos S, Christiansen L, Steemers FJ, Trapnell C, Shendure J (2019) The single-cell transcriptional landscape of mammalian organogenesis. Nature 566(7745):496–502. https://doi.org/10.1038/s41586-019-0969-x
112. Traag VA, Waltman L, van Eck NJ (2019) From Louvain to Leiden: guaranteeing well-connected communities. Sci Rep 9(1):5233. https://doi.org/10.1038/s41598-019-41695-z
113. Saelens W, Cannoodt R, Todorov H, Saeys Y (2019) A comparison of single-cell trajectory inference methods. Nat Biotechnol 37(5): 547–554. https://doi.org/10.1038/s41587-019-0071-9
114. Street K, Risso D, Fletcher RB, Das D, Ngai J, Yosef N, Purdom E, Dudoit S (2018)

Slingshot: cell lineage and pseudotime inference for single-cell transcriptomics. BMC Genomics 19(1):477. https://doi.org/10.1186/s12864-018-4772-0

115. Wolf FA, Hamey FK, Plass M, Solana J, Dahlin JS, Gottgens B, Rajewsky N, Simon L, Theis FJ (2019) PAGA: graph abstraction reconciles clustering with trajectory inference through a topology preserving map of single cells. Genome Biol 20(1):59. https://doi.org/10.1186/s13059-019-1663-x

116. La Manno G, Soldatov R, Zeisel A, Braun E, Hochgerner H, Petukhov V, Lidschreiber K, Kastriti ME, Lonnerberg P, Furlan A, Fan J, Borm LE, Liu Z, van Bruggen D, Guo J, He X, Barker R, Sundstrom E, Castelo-Branco G, Cramer P, Adameyko I, Linnarsson S, Kharchenko PV (2018) RNA velocity of single cells. Nature 560(7719):494–498. https://doi.org/10.1038/s41586-018-0414-6

117. Bergen V, Lange M, Peidli S, Wolf FA, Theis FJ (2020) Generalizing RNA velocity to transient cell states through dynamical modeling. Nat Biotechnol. https://doi.org/10.1038/s41587-020-0591-3

118. Qiu X, Zhang Y, Yang D, Hosseinzadeh S, Wang L, Yuan R, Xu S, Ma Y, Replogle J, Darmanis S, Xing J, Weissman JS (2019) Mapping vector field of single cells. bioRxiv:696724. https://doi.org/10.1101/696724

119. Mereu E, Lafzi A, Moutinho C, Ziegenhain C, McCarthy DJ, Alvarez-Varela-A, Batlle E, Sagar GD, Lau JK, Boutet SC, Sanada C, Ooi A, Jones RC, Kaihara K, Brampton C, Talaga Y, Sasagawa Y, Tanaka K, Hayashi T, Braeuning C, Fischer C, Sauer S, Trefzer T, Conrad C, Adiconis X, Nguyen LT, Regev A, Levin JZ, Parekh S, Janjic A, Wange LE, Bagnoli JW, Enard W, Gut M, Sandberg R, Nikaido I, Gut I, Stegle O, Heyn H (2020) Benchmarking single-cell RNA-sequencing protocols for cell atlas projects. Nat Biotechnol 38(6):747–755. https://doi.org/10.1038/s41587-020-0469-4

120. Wen WX, Mead AJ, Thongjuea S (2020) Technological advances and computational approaches for alternative splicing analysis in single cells. Comput Struct Biotechnol J 18:332–343. https://doi.org/10.1016/j.csbj.2020.01.009

121. Arzalluz-Luque A, Conesa A (2018) Single-cell RNAseq for the study of isoforms-how is that possible? Genome Biol 19(1):110. https://doi.org/10.1186/s13059-018-1496-z

122. Katz Y, Wang ET, Airoldi EM, Burge CB (2010) Analysis and design of RNA sequencing experiments for identifying isoform regulation. Nat Methods 7(12):1009–1015. https://doi.org/10.1038/nmeth.1528

123. Huang Y, Sanguinetti G (2017) BRIE: transcriptome-wide splicing quantification in single cells. Genome Biol 18(1):123. https://doi.org/10.1186/s13059-017-1248-5

124. Song Y, Botvinnik OB, Lovci MT, Kakaradov B, Liu P, Xu JL, Yeo GW (2017) Single-cell alternative splicing analysis with expedition reveals splicing dynamics during neuron differentiation. Mol Cell 67(1):148–161.e145. https://doi.org/10.1016/j.molcel.2017.06.003

125. Welch JD, Hu Y, Prins JF (2016) Robust detection of alternative splicing in a population of single cells. Nucleic Acids Res 44(8):e73. https://doi.org/10.1093/nar/gkv1525

126. Byrne A, Beaudin AE, Olsen HE, Jain M, Cole C, Palmer T, DuBois RM, Forsberg EC, Akeson M, Vollmers C (2017) Nanopore long-read RNAseq reveals widespread transcriptional variation among the surface receptors of individual B cells. Nat Commun 8:16027. https://doi.org/10.1038/ncomms16027

127. Kim HJ, Lin Y, Geddes TA, Yang JYH, Yang P (2020) CiteFuse enables multi-modal analysis of CITE-seq data. Bioinformatics 36(14):4137–4143. https://doi.org/10.1093/bioinformatics/btaa282

128. Zhang Y, Liu T, Meyer CA, Eeckhoute J, Johnson DS, Bernstein BE, Nusbaum C, Myers RM, Brown M, Li W, Liu XS (2008) Model-based analysis of ChIP-Seq (MACS). Genome Biol 9(9):R137. https://doi.org/10.1186/gb-2008-9-9-r137

129. Baker SM, Rogerson C, Hayes A, Sharrocks AD, Rattray M (2019) Classifying cells with Scasat, a single-cell ATAC-seq analysis tool. Nucleic Acids Res 47(2):e10. https://doi.org/10.1093/nar/gky950

130. Yu W, Uzun Y, Zhu Q, Chen C, Tan K (2020) scATAC-pro: a comprehensive workbench for single-cell chromatin accessibility sequencing data. Genome Biol 21(1):94. https://doi.org/10.1186/s13059-020-02008-0

131. Danese A, Richter ML, Fischer DS, Theis FJ, Colomé-Tatché M (2019) EpiScanpy: integrated single-cell epigenomic analysis. bioRxiv:648097. https://doi.org/10.1101/648097

132. Fang R, Preissl S, Li Y, Hou X, Lucero J, Wang X, Motamedi A, Shiau AK, Zhou X,

Xie F, Mukamel EA, Zhang K, Zhang Y, Behrens MM, Ecker JR, Ren B (2020) SnapATAC: a comprehensive analysis package for single cell ATAC-seq. bioRxiv:615179. https://doi.org/10.1101/615179

133. Granja JM, Corces MR, Pierce SE, Bagdatli ST, Choudhry H, Chang HY, Greenleaf WJ (2020) ArchR: an integrative and scalable software package for single-cell chromatin accessibility analysis. bioRxiv:2020.2004.2028.066498. https://doi.org/10.1101/2020.04.28.066498

134. Stuart T, Srivastava A, Lareau C, Satija R (2020) Multimodal single-cell chromatin analysis with Signac. bioRxiv:2020.2011.2009.373613. https://doi.org/10.1101/2020.11.09.373613

135. Bravo Gonzalez-Blas C, Minnoye L, Papasokrati D, Aibar S, Hulselmans G, Christiaens V, Davie K, Wouters J, Aerts S (2019) cisTopic: cis-regulatory topic modeling on single-cell ATAC-seq data. Nat Methods 16(5):397–400. https://doi.org/10.1038/s41592-019-0367-1

136. Chen H, Albergante L, Hsu JY, Lareau CA, Lo Bosco G, Guan J, Zhou S, Gorban AN, Bauer DE, Aryee MJ, Langenau DM, Zinovyev A, Buenrostro JD, Yuan GC, Pinello L (2019) Single-cell trajectories reconstruction, exploration and mapping of omics data with STREAM. Nat Commun 10(1):1903. https://doi.org/10.1038/s41467-019-09670-4

137. Schep AN, Wu B, Buenrostro JD, Greenleaf WJ (2017) chromVAR: inferring transcription-factor-associated accessibility from single-cell epigenomic data. Nat Methods 14(10):975–978. https://doi.org/10.1038/nmeth.4401

138. Pliner HA, Packer JS, McFaline-Figueroa JL, Cusanovich DA, Daza RM, Aghamirzaie D, Srivatsan S, Qiu X, Jackson D, Minkina A, Adey AC, Steemers FJ, Shendure J, Trapnell C (2018) Cicero Predicts cis-Regulatory DNA interactions from single-cell chromatin accessibility data. Mol Cell 71(5):858–871. e858. https://doi.org/10.1016/j.molcel.2018.06.044

139. Efremova M, Teichmann SA (2020) Computational methods for single-cell omics across modalities. Nat Methods 17(1):14–17. https://doi.org/10.1038/s41592-019-0692-4

140. van Dijk D, Sharma R, Nainys J, Yim K, Kathail P, Carr AJ, Burdziak C, Moon KR, Chaffer CL, Pattabiraman D, Bierie B, Mazutis L, Wolf G, Krishnaswamy S, Pe'er D (2018) Recovering gene interactions from single-cell data using data diffusion. Cell 174(3):716–729.e727. https://doi.org/10.1016/j.cell.2018.05.061

141. Yang MQ, Weissman SM, Yang W, Zhang J, Canaann A, Guan R (2018) MISC: missing imputation for single-cell RNA sequencing data. BMC Syst Biol 12(Suppl 7):114. https://doi.org/10.1186/s12918-018-0638-y

142. Li WV, Li JJ (2018) An accurate and robust imputation method scImpute for single-cell RNA-seq data. Nat Commun 9(1):997. https://doi.org/10.1038/s41467-018-03405-7

143. Chen M, Zhou X (2018) VIPER: variability-preserving imputation for accurate gene expression recovery in single-cell RNA sequencing studies. Genome Biol 19(1):196. https://doi.org/10.1186/s13059-018-1575-1

144. Mongia A, Sengupta D, Majumdar A (2019) McImpute: matrix completion based imputation for single cell RNA-seq data. Front Genet 10:9. https://doi.org/10.3389/fgene.2019.00009

145. Qi Y, Guo Y, Jiao H, Shang X (2020) A flexible network-based imputing-and-fusing approach towards the identification of cell types from single-cell RNA-seq data. BMC Bioinformatics 21(1):240. https://doi.org/10.1186/s12859-020-03547-w

146. Gunady MK, Kancherla J, Bravo HC, Feizi S (2019) scGAIN: single cell RNA-seq data imputation using generative adversarial networks. bioRxiv:837302. https://doi.org/10.1101/837302

147. Talwar D, Mongia A, Sengupta D, Majumdar A (2018) AutoImpute: autoencoder based imputation of single-cell RNA-seq data. Sci Rep 8(1):16329. https://doi.org/10.1038/s41598-018-34688-x

148. Argelaguet R, Arnol D, Bredikhin D, Deloro Y, Velten B, Marioni JC, Stegle O (2020) MOFA+: a statistical framework for comprehensive integration of multi-modal single-cell data. Genome Biol 21(1):111. https://doi.org/10.1186/s13059-020-02015-1

149. Welch JD, Hartemink AJ, Prins JF (2017) MATCHER: manifold alignment reveals correspondence between single cell transcriptome and epigenome dynamics. Genome Biol 18(1):138. https://doi.org/10.1186/s13059-017-1269-0

150. Wang X, Park J, Susztak K, Zhang NR, Li M (2019) Bulk tissue cell type deconvolution with multi-subject single-cell expression

reference. Nat Commun 10(1):380. https://doi.org/10.1038/s41467-018-08023-x

151. Duan B, Zhou C, Zhu C, Yu Y, Li G, Zhang S, Zhang C, Ye X, Ma H, Qu S, Zhang Z, Wang P, Sun S, Liu Q (2019) Model-based understanding of single-cell CRISPR screening. Nat Commun 10(1): 2233. https://doi.org/10.1038/s41467-019-10216-x

152. Campbell KR, Steif A, Laks E, Zahn H, Lai D, McPherson A, Farahani H, Kabeer F, O'Flanagan C, Biele J, Brimhall J, Wang B, Walters P, Consortium I, Bouchard-Cote A, Aparicio S, Shah SP (2019) clonealign: statistical integration of independent single-cell RNA and DNA sequencing data from human cancers. Genome Biol 20(1):54. https://doi.org/10.1186/s13059-019-1645-z

153. Ma A, McDermaid A, Xu J, Chang Y, Ma Q (2020) Integrative methods and practical challenges for single-cell multi-omics. Trends Biotechnol 38(9):1007–1022. https://doi.org/10.1016/j.tibtech.2020.02.013

Chapter 4

Automating Assignment, Quantitation, and Biological Annotation of Redox Proteomics Datasets with ProteoSushi

Sjoerd van der Post, Robert W. Seymour, Arshag D. Mooradian, and Jason M. Held

Abstract

Redox proteomics plays an increasingly important role characterizing the cellular redox state and redox signaling networks. As these datasets grow larger and identify more redox regulated sites in proteins, they provide a systems-wide characterization of redox regulation across cellular organelles and regulatory networks. However, these large proteomic datasets require substantial data processing and analysis in order to fully interpret and comprehend the biological impact of oxidative posttranslational modifications. We therefore developed ProteoSushi, a software tool to biologically annotate and quantify redox proteomics and other modification-specific proteomics datasets. ProteoSushi can be applied to differentially alkylated samples to assay overall cysteine oxidation, chemically labeled samples such as those used to profile the cysteine sulfenome, or any oxidative posttranslational modification on any residue.

Here we demonstrate how to use ProteoSushi to analyze a large, public cysteine redox proteomics dataset. ProteoSushi assigns each modified peptide to shared proteins and genes, sums or averages signal intensities for each modified site of interest, and annotates each modified site with the most up-to-date biological information available from UniProt. These biological annotations include known functional roles or modifications of the site, the protein domain(s) that the site resides in, the protein's subcellular location and function, and more.

Key words Redox, Proteomics, Cysteines, Bioinformatics, Systems biology, Posttranslational modifications, Protein inference, ProteoSushi, Reactive oxygen species

1 Introduction

It is now possible to quantify the redox state of thousands, or even tens of thousands, of cysteines in the proteome of cells or tissues using quantitative proteomics [1, 2]. Conceptually, each cysteine in the proteome can be considered a unique sentinel monitoring distinct aspects of redox biology since each cysteine has distinct properties including subcellular localization, solvent accessibility, reactivity, function, and localization in specific protein domains. Large scale profiling of the cysteine redoxome, the set of all

Sonia Cortassa and Miguel A. Aon (eds.), *Computational Systems Biology in Medicine and Biotechnology: Methods and Protocols*, Methods in Molecular Biology, vol. 2399, https://doi.org/10.1007/978-1-0716-1831-8_4,

oxidized cysteines in the proteome, therefore casts a systems-wide net to monitor numerous aspects of cellular redox biology and regulation [3]. For example, subcellularly localized redox changes are indicated by coordinated redox regulation of many cysteines located in organelle-specific proteins and can be discerned without the use of a microscope or organelle-targeted redox probes [1, 3].

Redox proteomics workflows consist of three main components: (1) sample preparation, (2) liquid-chromatography mass spectrometry (LC-MS) data acquisition, and (3) data analysis. The technical details of the first two components are well documented [4–6] and are primarily dictated by the type(s) of redox modification to measure, the LC-MS instrumentation, the number of samples to be analyzed, and the desired coverage depth. These will be discussed briefly. We primarily focus on the third component in this chapter, redox proteomic data analysis, which has been described in less detail.

One feature of most redox proteomics datasets is the use of 'bottom-up' proteomics, in which proteins are proteolyzed into peptides prior to LC-MS analysis [7]. Inferring which gene(s) and protein(s) that each peptide is derived from requires careful consideration. Subsequent assignment of cysteine residue number and automated annotation of publicly available molecular and cellular knowledge for each cysteine and redox regulated protein is also critical to fully evaluate and interpret the cysteine redoxome and cellular redox state. This includes features such as known function (s) and redox regulation, or the protein domain and subcellular localization of the modified sites and proteins.

We have developed a new tool, ProteoSushi [8], that takes peptide-centric, posttranslational modification (PTM)-focused proteomic data, such as redox proteomics results, and performs assignment of peptides to genes and proteins, simplifies quantitation, and annotates each redox regulated site with up-to-date information via real time query of UniProt. In this chapter we detail its application to a published cysteine redox proteomics dataset that identified and quantified ~4000 cysteines using DIA-MS [1]. We then describe how to perform several common statistical analyses to identify significantly regulated cysteines sites in this dataset and enriched biological annotations across the redoxome using ANOVA, multiple hypothesis correction of p-values using the Benjamini–Hochberg method, as well as the Fisher Exact test and Monte Carlo simulation.

Cellular redox regulation by growth factor stimulation is a common experimental model system that has delineated numerous regulatory paradigms of redox signaling [3]. The example redox proteomic dataset quantifying nearly 4000 cysteines at 5 timepoints after EGF treatment of A431 cells [1] serves as an example for how to extract systems-level knowledge from this type of data. We briefly describe the cellular, biochemical, and mass spectrometry methods, all of which are published, focusing primarily on data processing and analysis using ProteoSushi as well as downstream statistics and enrichment analysis of biological annotations.

2 Materials

2.1 Sample Preparation and Mass Spectrometry Analysis

This is described briefly below for completeness. *See* [1] for complete details.

2.2 Mass Spectrometry Data Availability

The mass spectrometry data and processed files analyzed in this protocol are published [1] and available from the ProteomeXchange with the identifier PXD010880. http://www.proteomexchange.org/. The files used for this tutorial are included in the ProteoSushi installation.

2.3 Computational Resources and Required Software

1. Hardware requirements: ProteoSushi is operating system independent and requires at least 8 GB of RAM (12 GB on a Windows-based system). A stable internet connection required for real time annotation retrieval from UniProt.
2. Python v3.8 or higher: https://www.python.org/downloads/. Check the current python version by executing the command *python --version*. The version number should appear.
3. ProteoSushi and example files: https://github.com/HeldLab/ProteoSushi.
4. R for statistical analysis: https://www.r-project.org/.

3 Methods

3.1 Redox Proteomics Sample Preparation

Epidermal growth factor (EGF) stimulation is well known to activate NADPH oxidases which endogenously produce reactive oxygen species (ROS) [9, 10]. A431 cells are the most common model system for these studies since they express high levels of the EGF receptor EGFR. The example EGFR dataset focused on investigating the temporal dynamics of cysteine oxidation upon growth factor stimulation; thus, A431 cells were stimulated with EGF and lysed at various timepoints afterward (*see* Fig. 1a) as described in detail in reference [1].

Sample preparation for redox proteomics can be broadly categorized into two categories: (1) indirect methods based on differential alkylation or (2) direct labeling. Differential alkylation (*see* ref. 5 for review) is based on covalently labeling free, nonoxidized, cysteine thiols prior to reduction, followed by labeling after reduction with a different alkylating reagent that can be distinguished from the first during the mass spectrometry analysis. The reductant

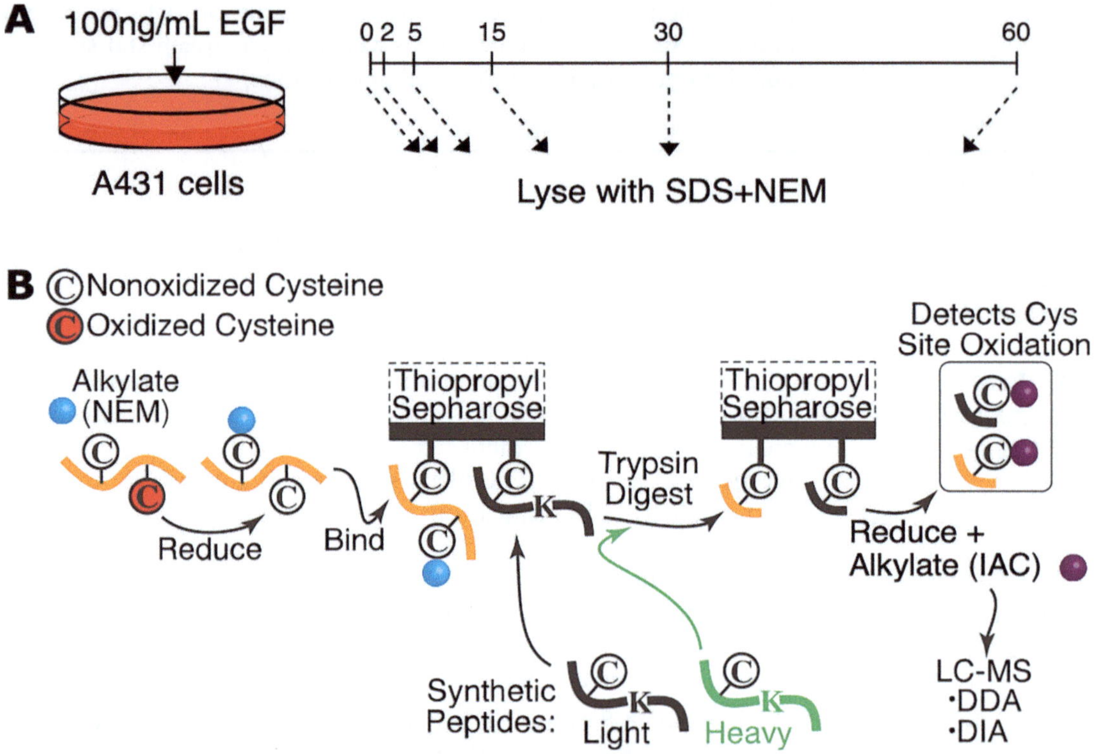

Fig. 1 The OxRAC workflow to globally profile cysteine oxidation. (**a**) Serum-starved A431 cells were left untreated (0 min) or stimulated with EGF (100 ng/ml) for the times indicated before lysis. (**b**) OxRAC workflow schematic in which free cysteine residues are trapped with NEM, and oxidized thiols are enriched by thiopropyl Sepharose resin and trypsin digested on-resin. The oxidized cysteine residues remain bound during washing, then are eluted by reduction, and labeled with iodoacetamide (IAC) to differentiate oxidized (IAC-labeled) from nonoxidized (NEM-labeled) cysteine residues. Peptides are analyzed by data-dependent acquisition (DDA) to identify peptides and data-independent acquisition (DIA) mass spectrometry for quantification purposes based on high-resolution MS2 scans. From *Science Signaling 2020* 13(615) eaay7315, doi: 10.1126/scisignal.aay7315. Reprinted with permission from AAAS

used is chosen based on whether total reversible oxidation is to be measured, or after selective reduction of glutathionylation via glutaredoxin treatment [11], selective reduction of S-nitrosothiols (SNOs) using copper and ascorbate [12], or detection of persulfides with ProPerDP [13]. In contrast, direct labeling covalently conjugates a tag to trap a specific cysteine oxoform. The probe typically has a fluorescent, biotinylated, or clickable handle for downstream analysis. Common examples include labeling of the unstable cysteine sulfenic acid using Dyn-2 [10] or a beta-ketoester [14], or SNOs using PBZyn [15], all of which contain an alkyne tag that can be leveraged for click chemistry to conjugate a wide array of azide-based moieties for enrichment purposes [16].

For the example EGFR dataset, differential alkylation was performed using two common reagents, N-ethylmaleimide (NEM) and iodoacetamide (IAC), to label free thiols before and after

TCEP reduction of all reducible modifications (*see* Fig. 1b). While quantitative cysteine redox proteomics often employs stable isotope variants of chemically related probes, such as unlabeled and deuterated NEM [17, 18], these relatively small mass shifts can confound downstream analysis when using DIA-MS acquisition by being acquired in the same 'SWATH window'. The large mass shift between NEM and IAC labeled cysteines is therefore ideal when performing DIA-MS for redox proteomics.

Thiopropyl Sepharose 6B is a commonly utilized resin to enrich for cysteines with detailed technical notes for redox proteomics found in reference [6]. For the EGFR dataset, we adapted this resin to a new workflow we termed oxidation resin-assisted capture "OxRAC" (*see* Fig. 1b). Previously oxidized cysteines in proteins were bound and enriched using thiopropyl Sepharose prior to alkylation with iodoacetamide, leading to a proteomic dataset primarily enriched with carbamidomethylated cysteines [+57.021 Da] that represent the level of oxidation for each cysteine and were the focus of the analysis. In contrast, NEM labeled cysteines represent the nonoxidized form of the cysteine.

3.1.1 Overview of Liquid Chromatography-Mass Spectrometry (LC-MS): Data-Dependent Acquisitions (DDA) and Data-Independent Acquisitions (DIA)

The example EGFR dataset employs two different types of LC-MS methods, DDA and DIA, sometimes called SWATH (Reviewed in [19]). DDA LC-MS is focused on peptide identification, first assaying intact peptides in an MS1 scan that reports their mass-to-charge (m/z) ratio. The peptides present are then isolated one by one in the gas phase using quadrupoles, fragmented by collision with gas, and the resulting fragment ions are analyzed in an MS2 scan. In a typical LC-MS analysis, hundreds of thousands or more MS2 spectra are acquired and need to be assigned to peptides as discussed in *see* Subheading 2.3. DIA, in contrast, fragments ions in an MS1 scan with a wide m/z range (typically 10 m/z units), and typically does so sequentially, starting at ~400 m/z up to 1200 m/z, the m/z range of typical peptides. For example, the first MS1 scan is from 400–410 m/z, then 410–420 m/z, etc. DIA data is not typically used to identify peptides, but rather to quantify peptides. Importantly, by constantly scanning the full m/z range, DIA can consistently quantify all peptides present in a sample at sufficient signal intensity to be detected without any missing datapoints. In a typical workflow, a sample is first analyzed by a DIA to detect peptides to determine their elution time and MS2 fragmentation characteristics, followed by two DDA analyses to quantify peptides [20–22] as was performed for the EGFR dataset [1] (*see* **Note 1**).

3.1.2 Liquid Chromatography–Mass Spectrometry (LC-MS)

Samples were analyzed by reverse-phase HPLC on a nano-LC 2D HPLC system (Eksigent) directly connected to a quadrupole time-of-flight (QqTOF) TripleTOF 5600 mass spectrometer (AB SCIEX) in direct injection mode. Peptide mixtures are separated on self-packed (ReproSil-Pur C18-AQ, 3 μm, Dr. Maisch,

Germany) nanocapillary HPLC column (75 μm I.D. × 22 cm column) and eluted at a flow rate of 250 nl/min using the following gradient: 2% solvent B in A (from 0 to 7 min), 2–5% solvent B in A (from 7.1 min), 5–30% solvent B in A (from 7.1 to 130 min), 30–80% solvent B in A (from 130 to 145 min), isocratic 80% solvent B in A (from 145 to 149 min) and a gradient 80-2% solvent B in A (from 149 to 150 min), with a total runtime of 180 min including mobile phase equilibration. Solvents were prepared as follows: mobile phase A, 0.1% formic acid (v/v) in water, and mobile phase B, 0.1% formic acid (v/v) in acetonitrile (*see* **Note 2**).

Mass spectra were recorded in positive-ion mode. For DDA, the mass window for precursor ion selection of the quadrupole mass analyzer was set to ±0.7 *m/z*. MS1 scans ranged from 380 to 1250 *m/z* at a resolution of 30,000 with an accumulation time of 250 ms. The 50 most abundant parent ions were selected for MS2 following each survey MS1 scan. Dynamic exclusion features were based on value MH^+ not *m/z* and were set to an exclusion mass width 50 mDa for a duration of 30 s. MS2 scans ranged from 100 to 1500 *m/z* with a maximum accumulation time of 50 ms. For DIA-MS, a wider first quadrupole (Q1) window of 10 *m/z* is passed in incremental steps over the full mass range *m/z* 400 to 1250 with 85 SWATH segments, 63 ms accumulation time each.

3.2 MS2 Database Searches to Generate Peptide Spectral Matches (PSMs)

Many database search tools are available (*see* **Note 3**). For this study, mass spectral data sets were analyzed and searched with both MaxQuant [23] and Mascot [24] against the UniProt Human reference proteome. MaxQuant search parameters included: First peptide search tolerance of 0.07 Da and main peptide search tolerance of 0.0006 Da, and variable methionine oxidation, protein N-terminal acetylation, carbamidomethyl, and NEM modifications with a maximum of 5 modifications per peptide, 2 missed cleavages and trypsin/P protease specificity. Razor protein false discovery rate (FDR) was utilized and the maximum expectation value for accepting individual peptides was 0.01 (1% FDR). For all Mascot searches, parameters were the same except for mass tolerance of 25 ppm and 0.1 Da for MS1 and MS2 spectra, respectively, and decoy searches were performed choosing the Decoy checkbox within the search engine. For all further data processing, peptide expectation values were filtered to keep the FDR rate at 1%.

3.3 Label Free Data-Independent Acquisition (DIA) Quantitation Using Skyline

Skyline [25], a freely available and open-source software tool that runs on the Windows platform, was used for peak integration of the resulting DIA-MS data. Detailed technical notes, webinars, and an active support forum for using Skyline are included at the Skyline website (*see* **Note 4**). Spectral libraries from peptides identified by MaxQuant and Mascot were generated in Skyline. Raw files were directly imported into Skyline in their native file format, and only cysteine-containing peptides were quantified.

3.4 Processing Peptide-Centric, PTM-Focused Proteomic Results Using ProteoSushi

ProteoSushi is software written in Python that is designed to be easy to use and aided by a GUI (*see* Fig. 2). Download the most recent version and example files from GitHub and confirm that Python is installed (*see* Subheading 2.3). ProteoSushi has a few common, minor hardware requirements. (*See* **Note 5** for additional details.)

3.4.1 ProteoSushi Data Requirements

Raw data from the example EGFR dataset are deposited on the ProteomeXchange repository with the identifier PXD010880, but relevant output files are included in the ProteoSushi installation under the examples folder. In order to run ProteoSushi, there are several required files needed in specific formats:

1. *Mascot output (if using as input)*: the CSV file output from Mascot using default CSV export settings.

   The file must have the header lines with the information from the search included, such as the protease used to generate peptides and the maximum number of missed cleavages (see example file GitHub).
2. *MaxQuant output (if using as input)*: the output txt folder.

   This folder must have the summary.txt and evidence.txt files. Other files from the output are not used.
3. *Other search engines*: The CSV file must have a column containing peptide sequences to be analyzed with the header "peptide sequence," and a second column including the modified peptide sequence with the header "peptide modified sequence" (specify PTMs between brackets or parenthesis after the modified residue). Optional: if you elect to use the quantitation values in the analysis, there must be at least 1 column with the header "Intensity" or "Intensities." All column names with "intensity" or "intensities" will be used by default to allow multiplexed analysis (e.g., "Intensity light" and "Intensity heavy").
4. *Protein sequence file in FASTA format*: Typically, this is the same reference proteome FASTA file used in the MS2 database search.
5. *Optional files*: A list of gene names in a TXT file, one gene name per line, to prioritize the user-provided genes whenever there are multiple matches once ProteoSushi performs a search.

3.4.2 ProteoSushi Installation and Data Analysis

1. Open a terminal window (or command prompt in Windows) and run the command:

```
pip install proteosushi
```

A

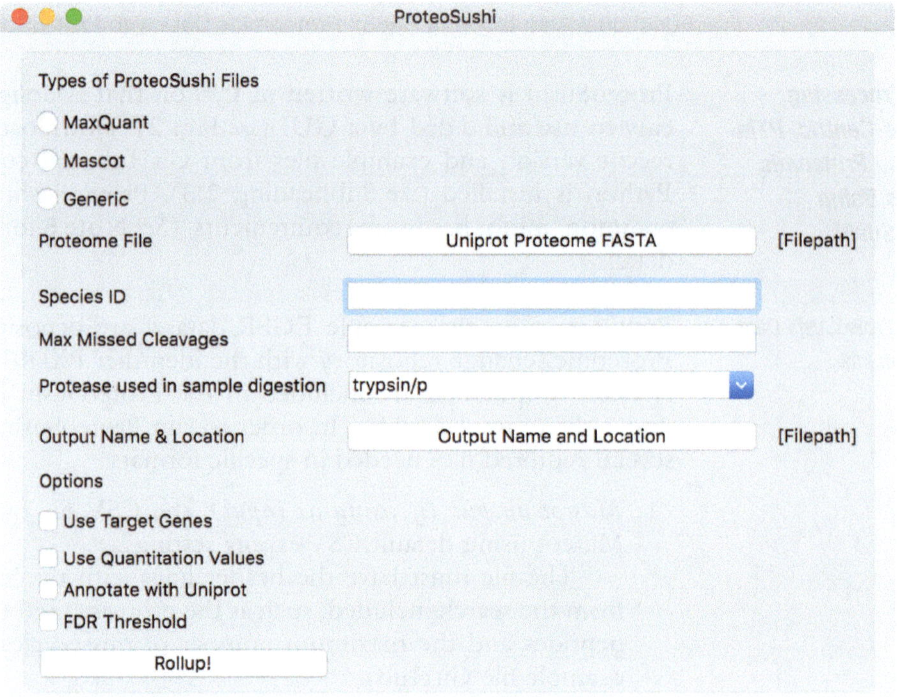

B

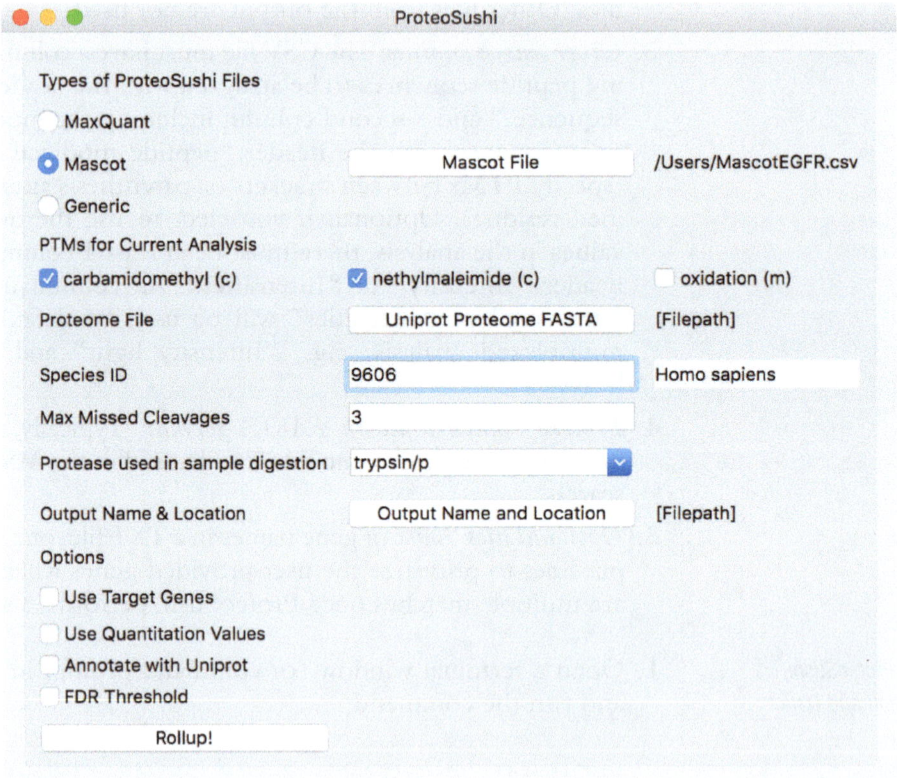

Fig. 2 ProteoSushi graphical user interface (GUI). (**a**) The ProteoSushi GUI includes options to analyze generic peptide lists, or directly process the output files from MaxQuant or Mascot. (**b**) The user is prompted to load the proteome FASTA file and specify variable options to tailor processing based on species, quantitation, peptide false discovery rate filtering and the protease used for protein digestion

While in the terminal, run ProteoSushi with the command.

```
python -m proteosushi
```

Alternatively, download the files directly from Heldlab GitHub repository. Once unpacked, navigate the terminal window to the ProteoSushi folder and run the following command to start the GUI (*see* Fig. 2a).

```
python run_proteosushi.py
```

2. Using the example data on GitHub from the EGFR dataset, choose the search engine used and navigate to the output file to analyze. For a given search engine, there one of the following files is required:
   (a) *Mascot*. Choose the annotated Mascot output file. This should be a CSV file. The example file is called MascotEGFR.csv.
   (b) *MaxQuant*. Choose the MaxQuant txt output folder containing the evidence.txt and summary.txt files.
   (c) *Generic*. Choose the peptide output from any search engine, supplementary table or Skyline analysis in CSV format. The peptide list must include the peptide sequence and peptide modified sequence as separate columns, and optionally intensity columns used for quantification. The example file is EGFR_Skyline_data.csv.
3. ProteoSushi will then parse the file to populate the maximum allowed missed cleavages, protease field and PTMs available for analysis (*see* Fig. 2b). The PTM options are dynamic and will change based on the file provided. If there are many different PTMs in the file, you may need to expand the ProteoSushi window horizontally as they will all be on the same line. Next, choose the protein sequence FASTA UniProt file to use with ProteoSushi.

   The following options are additional settings that can be modified or omitted. First, choose whether to use a prioritized gene list (*see* **Note 6**). If so, choose the file to be used. These genes will be used for peptide assignment if there is a tie in annotation score between multiple matches for a PTM site of a peptide. If one of the PTM sites is included in this list, it will be chosen as the assigned protein/gene.

   Second, choose whether to use the quantitation values. You will need to specify whether to sum or average values that will be combined. The "Intensity" column must be provided in the input file to do this (*see* **Note 7**).

If not already populated, specify the number of maximum allowed missed cleavages for a given peptide (typically 2). It is recommended to stay consistent with what was selected for the original database search. If the protease field is empty, specify the one used in the sample digestion step. Use the drop-down menu to select the protease used for protein digestion. These include: trypsin/p, trypsin!p, lys-c, asp-n, asp-nc, and lys-n (*see* **Note 8**).

Specify the threshold for FDR or posterior error probability (PEP), if using Mascot or MaxQuant (*see* **Note 3**) and reference [26]. This value can be left blank if you do not want to specify a threshold. Once all of the necessary options are included, click on the *"Rollup!"* button to start the analysis.

Peptides can be shared between multiple proteins and genes. To minimize redundancy, ProteoSushi uses the UniProt annotation score to prioritize assignment, using the annotation from those shared proteins with the highest UniProt annotation score. This helps achieves the best quality and in-depth annotations of the modified sites identified. For a detailed flow chart of peptide assignment, annotation and optional quantification steps *see* Fig. 3.

4. Results will be returned as a CSV spreadsheet with the filename and location that the user chose in the GUI

   The resulting file will include information for each modified residue of interest such as gene, UniProt accession number, peptide, modified cysteine site, active site annotation, known PTM and subcellular localization (*see* Table 1).

5. ProteoSushi will annotate each cysteine site spanning peptide with the following data columns from Uniprot:

   Gene: Gene associated with the modified site.

   Site: Position number of the modified site within the protein.

   Protein_Name: Name of the protein with the modified site.

   Shared_Genes: Additional genes that contain the peptide sequence and have the same annotation score.

   Target_Genes: Indicates if the assigned genes are in the supplied target gene list (if provided).

   Peptide_Sequence: Amino acid sequence of the unmodified peptide.

   Peptide_Modified_Sequence: Amino acid sequence of the peptides indicating the modification(s).

   Annotation_Score: UniProt annotation score for the protein.

   Uniprot_Accession_ID: UniProt accession number for the protein.

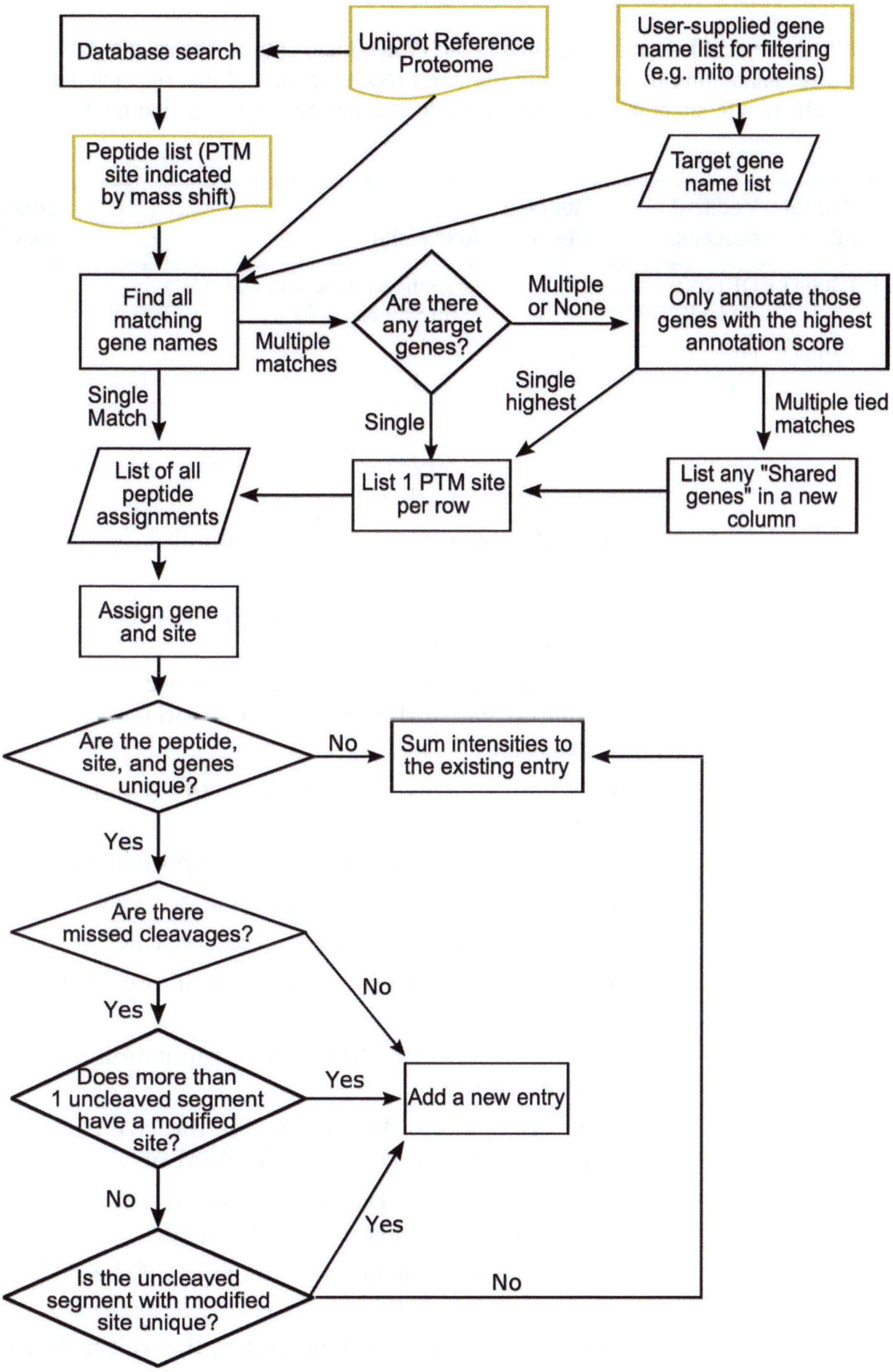

Fig. 3 Flowchart of ProteoSushi's peptide assignment and merging multiple peptide forms. Flowchart detailing how ProteoSushi assigns peptides to shared proteins and genes as well as combines multiple forms of a peptide sharing the same modification(s)

Table 1

Example of the biological features annotated in the ProteoSushi output. The results table will include 34 columns with annotation retrieved from the most recent version of UniProt including common identifiers, cysteine site specific annotation, domain and protein region assignment, and protein annotation

Gene	UniProt AC	Modified sequence	Modified site	Active site	Subcellular location
PRDX6	P30041	DFTPVC (ca)TTEIGR	47	Cysteine sulfenic acid (-SOH) intermediate; for peroxidase activity	Cytoplasm, lysosome
PRDX6	P30041	DINAYNC(ca) EEPTEK	91		Cytoplasm, lysosome

*[*Intensit*]or[*intensit*]* (optional): Column(s) with the quantitation values for the site.

Length_Of_Sequence: Number of amino acids in the protein sequence.

Range_of_Interest: Amino acid position of the regions of interest in the following column.

Region_of_Interest: Amino acid sequence of the regions of interest where the site is located (domains, binding sites, etc.).

Subcellular_Location: Location(s) of the protein within the cell.

Enzyme_Class: Type of enzyme (if the protein is an enzyme).

Rhea: Hyperlinks to the RHEA database (www.rhea-db.org).

Secondary_Structure: Protein secondary structure at the site.

Active_Site_Annotation: Annotation if the PTM site is an active site.

Alternative_Sequence_Annotation: Annotation related to protein isoforms.

Chain_Annotation: Annotation for the chain in the mature protein after processing at the PTM site.

Compositional_Bias_Annotation: Annotation indicating overrepresentation of certain amino acids.

Disulfide_Bond_Annotation: Annotation if the PTM site part of a disulfide bond.

Domain_Extent_Annotation: Description of the domain.

Lipidation_Annotation: Annotation if the residue is known to be lipidated.

Metal_Binding_Annotation: Type of metal binding at the PTM site.

Modified_Residue_Annotation: Known type of PTM at site.

Motif_Annotation: Description of a short, conserved sequence motif of biological significance.

Mutagenesis_Annotation: Mutations and known effect on protein function at the site.

Natural_Variant_Annotation: Known natural variant at the PTM site (if there is one).

NP_Binding_Annotation: If the site is a known nucleotide binding site.

Other: If there is an annotation different than the other _Annotation columns.

Region_Annotation: Annotation related to the "Region_of_Interest" column.

Repeat_Annotation: The types of repeated sequence motifs or repeated domains.

Topological_Domain_Annotation: Orientation in the plasma membrane (cytosolic or extracellular).

Zinc_Finger_Annotation: If the modified site is within a zinc finger domain.

More detailed information on any of the above annotations is available on the help section of the UniProt website (https://www.uniprot.org/help).

3.5 Statistical Analysis of Redox Regulated Cysteine Sites: Multiple Hypothesis Correction

Two questions naturally arise after processing redox proteomics datasets with ProteoSushi. First, which cysteines are redox regulated? This is the focus of Subheading 3.6. Second, are there certain types of proteins, protein types, subcellular locations, or other trends in the data or annotations that are preferentially redox regulated? This is the focus of Subheading 3.7. These analyses both require statistical hypothesis testing to evaluate the likelihood of observing changes in the redox state between samples.

Conventionally, an alpha of 0.05 is used to set the threshold for determining statistical significance, for example a *t*-test. However, using an alpha cutoff of 0.05 for an unadjusted *p*-value is only valid for a single independent hypothesis test. If multiple hypotheses are tested, correction is necessary to preserve the original error rate cutoff of 5%. Without multiple-hypothesis correction there is an increased chance for type I errors (false positives) beyond what is suggested by the alpha cutoff. Since typical redox proteomics dataset contains thousands of cysteines, it is especially important that statistical evaluation includes multiple hypothesis correction.

A stringent multiple hypothesis correction method is the Bonferroni method. The Bonferroni method simply divides the alpha cutoff by the number of hypothesis tests performed in the analysis to set a new threshold for significance. The Bonferroni correction controls the family-wise error rate, which is the likelihood of finding at least one false positive among the hypothesis tests performed, and can therefore be overly conservative when performing thousands of hypothesis tests to discover and identify sites of redox regulation. This method is far too stringent for proteomics datasets due to their relatively high experimental variance and fewer samples per condition than is ideal from a statistics standpoint.

The most common multiple hypothesis correction used for proteomics data is the Benjamini–Hochberg (BH) method. In contrast to the Bonferroni method, the Benjamini–Hochberg correction controls the false discovery rate, which is the proportion of false positives expected in a population of hypothesis tests. This method is less stringent than the Bonferroni method and is weighted, correcting less stringently for the most significant *p*-values. It is typically calculated using software tools as demonstrated here. To BH correct *p*-values, now termed *q*-values, first sort the *p*-values from lowest to highest. Next, use the equation $(r/N)\alpha$ where r is the rank of the *p*-value (1 being the lowest *p*-value), N is the number of *p*-values to be corrected, and α is the alpha value (also known as the *p*-value cutoff, usually 0.05).

1. To calculate the false discovery rate (BH corrected *p*-values) for a set of hypothesis tests, to test for example "which peptides are oxidized by a given condition?" the following command can be used in R:

```
df <- readxl::read_xlsx("experiment.xlsx")
df$qval <- p.adjust(df$value, method = "BH")
qval <- p.adjust(value, method = "BH")
```

where 'pval' is a numeric vector containing unadjusted *p*-values from a hypothesis test.

3.5.1 Analyses of Variance (ANOVA)

To determine whether cysteine redox sites are differentially oxidized between multiple treatments or time points, one-way analyses of variance (ANOVA) can be performed. ANOVA compares two or more groups of data against each other to determine the likelihood of observing a difference in any of the groups' means by chance. In an example of a time course experiment, like the EGFR dataset, ANOVAs can be run for each cysteine site, where the experimental groups are time points representing the duration that cells were treated with a growth factor:

1. ANOVA can be performed in R with the following command:

```
anova_results <- summary(aov(value ~ as.factor(time_
point), data = df))
```

Where df is the data frame containing the data of interest. *See* (Table 3) for an example of the test results.

2. For an experiment containing thousands of modified sites, *p*-values for each site can be calculated in R with the following commands:

```
library(dplyr)
p_value_df <- df %>%
group_by(Modified_Sequence) %>%
 do(pvalue = summary(aov(value ~ as.factor(time_point),
data =.))[[1]]$'Pr(>F)'[1]) %>%
data.frame()
```

Because thousands of ANOVAs may be performed this way for a single experiment, multiple hypothesis correction should be performed to convert the ANOVA *p*-values to false discovery rate *q*-values (detailed above).While the ANOVA can indicate that one or more groups' means differ from one another, no indication is given as to which groups are different. Identifying which specific group means differ can be accomplished using post hoc tests for cysteine sites where the null hypothesis of the initial ANOVA is rejected ($q < 0.05$). In the example above, the experimental design includes a set of triplicate observations at time 0 which serves as the control. Dunnett's test can be performed to determine whether any treatment groups' means differs from the control group's mean (Table 4):

3. For a single site, as in the example in Table 2:

```
library(DescTools)
 DunnettTest(x = df$value, g = as.factor(df$time_point),
control = "0")
```

4. For all pairwise comparisons between experimental groups, Tukey's Honest Significant Difference can be used (Table 5):

```
TukeyHSD(aov(value ~ as.factor(time_point), data = df))
```

3.6 Statistical Analysis of Biological Annotations

A common investigative interest about large datasets is whether certain features, such as protein domains, or functional annotations, or subcellular locations are enriched by a certain situation such as a treatment. For example, "Are small GTPases preferentially oxidized by EGF treatment?" Two approaches to assess statistical significance of a group of values include the Fisher's exact test (Subheading 3.6.1) and Monte Carlo simulation (Subheading

Table 2
Example of a peptide containing an oxidized cysteine from an EGF treatment time course experiment. Samples were analyzed in triplicate

Modified_Sequence	Time_point	Value
LTVVDTPGYGDAINC[+57]R	0	−0.04486
LTVVDTPGYGDAINC[+57]R	0	0.075969
LTVVDTPGYGDAINC[+57]R	0	−0.08259
LTVVDTPGYGDAINC[+57]R	2	0.066154
LTVVDTPGYGDAINC[+57]R	2	0.511476
LTVVDTPGYGDAINC[+57]R	2	−0.04662
LTVVDTPGYGDAINC[+57]R	5	0.179514
LTVVDTPGYGDAINC[+57]R	5	0.093883
LTVVDTPGYGDAINC[+57]R	5	0.15436
LTVVDTPGYGDAINC[+57]R	15	0.507034
LTVVDTPGYGDAINC[+57]R	15	0.711176
LTVVDTPGYGDAINC[+57]R	15	0.591411
LTVVDTPGYGDAINC[+57]R	30	0.609015
LTVVDTPGYGDAINC[+57]R	30	0.854675
LTVVDTPGYGDAINC[+57]R	30	0.73025
LTVVDTPGYGDAINC[+57]R	60	−0.25985
LTVVDTPGYGDAINC[+57]R	60	0.137838
LTVVDTPGYGDAINC[+57]R	60	0.089007

Table 3
Results of the ANOVA calculations for the peptide LTVVDTPGYGDAINC[+57]R at the different time points

	Df	Sum Sq	Mean Sq	*F* value	Pr(>*F*)	Significance
time_point	5	1.5304	0.30608	10.9	0.000394	***
Residuals	12	0.3371	0.02809			

3.6.2). Fisher's exact test is based on contingency tables, uses categorical data such as protein domains, and can be used when sample sizes are small. Monte Carlo simulation, on the other hand, simulates a distribution using the actual results that is then used for hypothesis testing.

Table 4
Results for of Dunnett's test to compare the means of each sample to control. The table displays the differences in means (diff), lower and upper end points of the 95% confidence intervals (lwr.ci, upr.ci), and *p*-values after correction for multiple comparisons (pval)

Comparison	diff	lwr.ci	upr.ci	pval
15-0	0.620367	0.223257	1.017478	0.00268
2-0	0.194165	−0.20295	0.591275	0.50713
30-0	0.748474	0.351363	1.145584	0.00069
5-0	0.159746	−0.23736	0.556857	0.66788
60-0	0.006159	−0.39095	0.40327	1

Table 5
Results of the Tukey Honest Significant Difference test which makes pairwise comparisons between the means of all samples tested. This table displays the differences in means (diff), lower and upper end points of the 95% confidence intervals (lwr, upr), and *p*-values after adjustment for multiple comparisons (*p* adj)

Comparison	diff	lwr	upr	*p* adj
15-0	0.620367	0.160725	1.08001	0.006941
2-0	0.194165	−0.26548	0.653807	0.716322
30-0	0.748474	0.288832	1.208116	0.001535
5-0	0.159746	−0.2999	0.619388	0.843682
60-0	0.006159	−0.45348	0.465801	1
15-2	−0.4262	−0.88584	0.03344	0.075056
30-15	0.128106	−0.33154	0.587749	0.929211
15-5	−0.46062	−0.92026	−0.00098	0.049404
60-15	−0.61421	−1.07385	−0.15457	0.007479
30-2	0.554309	0.094667	1.013951	0.015566
2–5	−0.03442	−0.49406	0.425223	0.999822
60-2	−0.18801	−0.64765	0.271637	0.740944
30-5	−0.58873	−1.04837	−0.12909	0.010201
60-30	−0.74231	−1.20196	−0.28267	0.001647
60-5	−0.15359	−0.61323	0.306055	0.863111

3.6.1 Peptide Annotation Enrichment Analysis: Fisher Exact Test

Enrichment analysis can be used to determine if certain annotations are overrepresented in a sample group or a subset of differentially regulated peptides in comparison to the control group. Here we apply the commonly used Fisher exact test to calculate significance of this categorical analysis. Enrichment analysis can be performed at the site level, protein region, domain, or gene- or protein-level annotations. In this example, we perform enrichment analysis to test if "cysteine sites identified using the OxRAC method are more often the active site in a protein?" and in addition "In which domains are the cysteines that are differentially regulated in response to EGF found?"

1. Input for analysis is a list of significantly differentially regulated peptides annotated using ProteoSushi (i.e., the output from **step** 3.5).
2. The dataset will be compared to a reference list containing the frequency of the annotation of interest in the proteome of the analyzed species. The method of creating this list depends on the type of annotation. In principle, most ProteoSushi outputs can be used for this type of analysis (*see* **Note 9**).
3. For the comparison and statistical analysis of the experimental results toward the reference proteome list, start by creating a 2 × 2 matrix. The first column will hold the frequency of the annotation active site in the sample (S1, 90) and the total number of unique cysteine sites found in the datasets (S2, 5476). The second column holds the same information for the reference list R1 (625, cysteine active sites) and R2 (261,464, number of cysteines in the human proteome). Create the data frame (df) in R and conduct the Fisher exact test using the following commands.

```
df <- matrix(c(90,5476,625,261464), nrow=2, byrow=T)
fisher.test(df)
Fisher's Exact Test for Count Data
data: df
p-value < 2.2e-16
```

4. The probability that the annotation of interest is enriched in the experimental dataset is reported as a *p*-value. This analysis can be performed on multiple annotations terms at once using a row-based matrix, for example testing all domain annotations in parallel. Use the following command to perform a Fisher exact test for multiple annotations in parallel. The required input file should contain one term per row, followed by S1, R1, S2 and R2 in the subsequent columns. Import the dataset into R and perform a Fisher exact test for each row with the following commands.

```
ft.res <- apply(df, 1, function(x){
t1 <- fisher.test(matrix(x, nrow = 2))
 data.frame(p_value = t1$p.value, odds_ratio = t1$esti-
mate)})
cbind(df, do.call(rbind, ft.res))
```

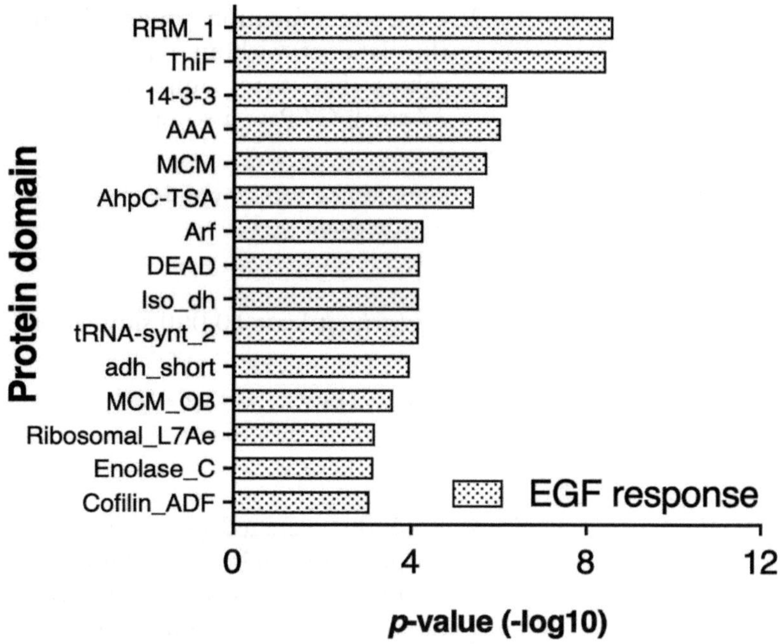

Fig. 4 Fisher exact test results domain annotation. Top 15 most enriched protein domains containing cysteine sites that are regulated in response to EGF stimulation, as determined by Fisher exact test

5. The *p*-value are generally log10 converted for visualization and interpretation purposes. This can be presented simply as ranked bar graphs or as enrichment maps with annotated nodes highlighting the most enriched pathway in a sample set [27]. An example figure showing the domains enriched in which the cysteine-sites are found that respond to EGF stimulation (*see* Fig. 4).

3.6.2 Monte Carlo Simulation

In the EGFR dataset, 37 peptides are assigned to small GTPases which have an average *q* value of 0.132. To determine if ATPases have a statistically significant enrichment of oxidized cysteines, use R to perform the Monte Carlo simulation.

1. Import the FDR adjusted *q*-values for every peptide in the dataset, this is included in the ProteoSushi Github repository in a list called ATPasesMonteCarlo (https://tinyurl.com/y2yct4tb).

```
experiment_qvals <- read.table("ATPasesMonteCarlo.txt",
header = FALSE, col.names = c("qvals"))
```

2. Generate a function "one.trial" that samples 37 q-values from the data set without replacement, and then averages the q-values that were sampled. 37 is chosen because there are 37 small GTPase associated peptides in the EGFR dataset.

```
one.trial  <-  function(qvals){mean(sample(qvals,37,
replace=FALSE))}
```

3. Perform at least 10,000 simulations. Here we do 100,000.

```
output.list <- replicate(100000,one.trial(experiment_qvals
$qvals))
```

4. Rank the averaged q-value results from the simulated trials in output.list from lowest to highest. Compare this to the average of the 37 q-values of the small GTPase assigned peptides. The rank, divided by the number of samplings, 100,000 in this case, is the p-value for enrichment. For example, if the real data has an average q-value lower than all simulations the p-value would be $1/100{,}000 = 0.00001$. In this case, the average q-value for small GTPases is 0.132, which is ranked #892 of 100,000 samplings for a p-value of 0.00892.

3.7 Conclusions

Regulation of the cysteine redoxome provides a systems-level view of cellular redox biology. The increasing numbers of redox regulated sites identified by redox proteomics due to new reagents and advances in LC-MS technology mean that it is poised to provide unparalleled insights into the dynamics of redox signaling networks. ProteoSushi facilitates the processing, analysis, and evaluation of these modification-specific proteomics datasets. When coupled with the downstream statistical analysis demonstrated here, redox proteomics data provides a broad characterization of redox systems biology as well as focused analysis of individual proteins, modification sites, and redox trends across protein domains and families, and subcellular locations.

4 Notes

1. There are two common quantitative approaches in proteomics, each having advantages and disadvantages. First, data-independent mass spectrometry (DIA-MS, also known as "SWATH") is label free, and thus does not require purchasing tagging reagents which can be expensive. In addition, the resulting data has no missing values. However, accurately

selecting chromatographic peaks for thousands of peptides across many samples remains challenging. In addition, since each sample must be run individually it is also not practical to prefractionate peptides prior to LC-MS analysis which decreases peptide coverage depth. Second, isobaric tagging and isotopic labeling, such as tandem mass tag (TMT) labeling followed by DDA LC-MS, can now routinely quantify 10 or more samples in parallel. However, this approach typically uses proprietary reagents and often has missing data, 20–30%, since the peptides chosen for MS2 in each sample pool are not the same. However, isobaric tagging enables offline prefractionation of peptides which increases the number of peptides identified. In addition, quantitation is highly automated.

2. LC parameters for peptide-centric LC-MS are relatively similar across instrumentation and samples due to the relatively consistent chemistry of peptides. However, if a hydrophobic modification is being monitored, increasing the percent of organic solvent can be beneficial. In addition, low flow rates (100–300 nl/min) improve sensitivity.

3. Many MS2 database search software tools exist and all can be used with ProteoSushi with minimal formatting. The protein database search engine takes the thousands, or even millions, of MS2 fragmentation spectra acquired for a sample set and assign them to a peptide sequence match. This is known as a peptide spectral match, or PSM. MS3 data acquisition, in which MS2 fragment ions are fragmented again, can also be performed [28]. The most common open source MS2 database search software is MaxQuant [23], but development of new tools continues and is reviewed in [29]. For redox proteomics, one major difference between search algorithms is the requirement that the user specify a priori which modifications are present, versus de novo algorithms that use the underlying spectral features to assign both a peptide sequence as well as modifications without prior specification. Site assignment, specifically determining the accuracy of assigning a modification to a specific residue in the peptide, is also an important consideration [30] and is also an area of continued tool development [29].

   Once peptides are assigned to MS2 spectra, a false discovery rate (FDR) is determined based on the quality of the assignment. This is typically performed with target-decoy approaches [31] and set to 1%. For a detailed review of false discovery rate versus posterior error probabilities *see* reference [26].

4. For large scale DIA experiments, automated peak picking is essential. Skyline is an open-source option, and options for improving automated peak picking are detailed here [32].

Spectronaut is an alternative software tool for automated analysis of large DIA datasets [33].

5. There are only two requirements for a computer to run ProteoSushi: (1) 16 GB of memory is recommended, but at least 8 GB of RAM (macOS) or 12 GB (Windows) are needed. (2) A stable internet connection is needed to retrieve data from UniProt. Optional hardware that will allow ProteoSushi to process results faster include a fast processor and a solid-state drive (as opposed to a hard disk drive).
6. Inputting a custom genes list allow the user to prioritize a group of specific proteins to assign a shared peptide and site (s) to. This feature is useful for users who wish to focus their analysis on a single organelle (such as the mitochondrion), for example, where only a subset of homologous isoforms or proteins are expected to be present and are prioritized for peptide assignment.
7. ProteoSushi accepts any type of quantitative value, including intensity or ratios. Summing quantitative values is appropriate for intensity measurements, such as DIA results or reporter ion heights from isobarically labeled samples. Using intensity will weight based on abundance. Alternatively, if reporter ion ratios are being analyzed, from isotopically labeled samples for example, averaging quantitative values is a more appropriate option and will equally weight all forms of the modified peptide.
8. There are six possible proteases to choose from in ProteoSushi and should match the protease used to prepare the sample for mass spectrometry. These are trypsin/p, trypsin!p, lys-c, asp-n, asp-nc, and lys-n. Trypsin cleaves after arginine (R) and lysine (K). The "/p" indicates that trypsin cleaves when proline (P) follows R or K. By contrast, "trypsin!p" does not cleave when a P follows an R or K. "lys-c" cleaves after lysine, whereas "lys-n" cleaves before lysine. "asp-n" cleaves before aspartic acid (D) and "asp-nc" cleaves before both aspartic acid (D) and glutamic acid (E).
9. The two options available to determine overrepresentation is to compare the significantly regulated cysteine sites against either (1) all identified sites in the dataset, or (2) versus all cysteines in the complete proteome of the species of interest. In the first case, the reference values can be calculated directly from the ProteoSushi results. For the second comparison, the values can typically be obtained using the UniProt website by browsing the complete proteome section for the species analyzed and selecting the map to UniProtKB function to retrieve a table with all proteins available. Additional columns with the protein information to be evaluated can be added from this page, for

example domain annotation as used in the example. When calculating the reference values, one should ideally exclude non–cysteine-containing protein regions and domains from analysis as these are negatively enriched by default in most redox proteomics sample preparation methods and will skew the analysis.

References

1. Behring JB, van der Post S, Mooradian AD et al (2020) Spatial and temporal alterations in protein structure by EGF regulate cryptic cysteine oxidation. Sci Signal 13:eaay7315. https://doi.org/10.1126/scisignal.aay7315
2. Xiao H, Jedrychowski MP, Schweppe DK et al (2020) A quantitative tissue-specific landscape of protein redox regulation during aging. Cell 180:968–983.e24. https://doi.org/10.1016/j.cell.2020.02.012
3. Held JM (2019) Redox systems biology: harnessing the sentinels of the cysteine redoxome. Antioxid Redox Signal 32:659–676. https://doi.org/10.1089/ars.2019.7725
4. Held JM, Gibson BW (2012) Regulatory control or oxidative damage? Proteomic approaches to interrogate the role of cysteine oxidation status in biological processes. Mol Cell Proteomics 11:R111.013037. https://doi.org/10.1074/mcp.R111.013037
5. Wojdyla K, Rogowska-Wrzesinska A (2015) Differential alkylation-based redox proteomics—lessons learnt. Redox Biol 6:240–252. https://doi.org/10.1016/j.redox.2015.08.005
6. Guo J, Gaffrey MJ, Su D et al (2014) Resin-assisted enrichment of thiols as a general strategy for proteomic profiling of cysteine-based reversible modifications. Nat Protoc 9:64–75. https://doi.org/10.1038/nprot.2013.161
7. Aebersold R, Mann M (2003) Mass spectrometry-based proteomics. Nature 422:198–207. https://doi.org/10.1038/nature01511
8. Seymour RW, van der Post S, Mooradian AD, Held JM (2021) ProteoSushi: A Software Tool to Biologically Annotate and Quantify Modification-Specific, Peptide-Centric Proteomics Data Sets. J Proteome Res acs.jproteome.1c00203. https://doi.org/10.1021/acs.jproteome.1c00203
9. Bae YS, Kang SW, Seo MS et al (1997) Epidermal growth factor (EGF)-induced generation of hydrogen peroxide. J Biol Chem 272:217–221. https://doi.org/10.1074/jbc.272.1.217
10. Paulsen CE, Truong TH, Garcia FJ et al (2012) Peroxide-dependent sulfenylation of the EGFR catalytic site enhances kinase activity. Nat Chem Biol 8:57–64. https://doi.org/10.1038/nchembio.736
11. Lind C, Gerdes R, Hamnell Y et al (2002) Identification of S-glutathionylated cellular proteins during oxidative stress and constitutive metabolism by affinity purification and proteomic analysis. Arch Biochem Biophys 406(2):229–240. https://doi.org/10.1016/S0003-9861(02)00468-X
12. Wang X, Kettenhofen NJ, Shiva S et al (2008) Copper dependence of the biotin switch assay: modified assay for measuring cellular and blood nitrosated proteins. Free Radic Biol Med 44(7):1362–1372. https://doi.org/10.1016/j.freeradbiomed.2007.12.032
13. Doka E, Pader I, Biro A et al (2016) A novel persulfide detection method reveals protein persulfide- and polysulfide-reducing functions of thioredoxin and glutathione systems. Sci Adv 2:e1500968. https://doi.org/10.1126/sciadv.1500968
14. Qian J, Wani R, Klomsiri C et al (2012) A simple and effective strategy for labeling cysteine sulfenic acid in proteins by utilization of beta-ketoesters as cleavable probes. Chem Commun 48:4091–4093. https://doi.org/10.1039/c2cc17868k
15. Clements JL, Pohl F, Muthupandi P et al (2020) A clickable probe for versatile characterization of S-nitrosothiols. Redox Biol 37:101707. https://doi.org/10.1016/j.redox.2020.101707
16. Pickens CJ, Johnson SN, Pressnall MM et al (2018) Practical considerations, challenges, and limitations of bioconjugation via azide-alkyne cycloaddition. Bioconjug Chem 29:686–701. https://doi.org/10.1021/acs.bioconjchem.7b00633
17. Held JM, Danielson SR, Behring JB et al (2010) Targeted quantitation of site-specific cysteine oxidation in endogenous proteins using a differential alkylation and multiple reaction monitoring mass spectrometry approach.

Mol Cell Proteomics 9:1400–1410. https://doi.org/10.1074/mcp.m900643-mcp200

18. Danielson SR, Held JM, Oo M et al (2011) Quantitative mapping of reversible mitochondrial complex I cysteine oxidation in a parkinson disease mouse model. J Biol Chem 286: 7601–7608. https://doi.org/10.1074/jbc.M110.190108
19. Gillet LC, Leitner A, Aebersold R (2016) Mass spectrometry applied to bottom-up proteomics: entering the high-throughput era for hypothesis testing. Annu Rev Anal Chem 9: 449–472. https://doi.org/10.1146/annurev-anchem-071015-041535
20. Held JM, Schilling B, D'Souza AK et al (2013) Label-free quantitation and mapping of the ErbB2 tumor receptor by multiple protease digestion with data-dependent (MS1) and data-independent (MS2) acquisitions. Int J Proteomics 2013:1–11. https://doi.org/10.1155/2013/791985
21. Zawadzka AM, Schilling B, Held JM et al (2014) Variation and quantification among a target set of phosphopeptides in human plasma by multiple reaction monitoring and SWATH-MS2 data-independent acquisition. Electrophoresis 35:3487–3497. https://doi.org/10.1002/elps.201400167
22. Collins BC, Hunter CL, Liu Y et al (2017) Multi-laboratory assessment of reproducibility, qualitative and quantitative performance of SWATH-mass spectrometry. Nat Commun 8: 291 https://doi.org/10.1038/s41467-017-00249-5
23. Cox J, Mann M (2008) MaxQuant enables high peptide identification rates, individualized p.p.b.-range mass accuracies and proteome-wide protein quantification. Nat Biotechnol 26:1367–1372. https://doi.org/10.1038/nbt.1511
24. Perkins DN, Pappin DJC, Creasy DM, Cottrell JS (1999) Probability-based protein identification by searching sequence databases using mass spectrometry data. Electrophoresis 20: 3551–3567. https://doi.org/10.1002/(SICI)1522-2683(19991201)20:18<3551::AID-ELPS3551>3.0.CO;2-2
25. MacLean B, Tomazela DM, Shulman N et al (2010) Skyline: an open source document editor for creating and analyzing targeted proteomics experiments. Bioinformatics 26(7): 966–968. https://doi.org/10.1093/bioinformatics/btq054
26. Käll L, Storey JD, MacCoss MJ, Noble WS (2008) Posterior error probabilities and false discovery rates: two sides of the same coin. J Proteome Res 7:40–44. https://doi.org/10.1021/pr700739d
27. Reimand J, Isserlin R, Voisin V et al (2019) Pathway enrichment analysis and visualization of omics data using g:Profiler, GSEA, Cytoscape and EnrichmentMap. Nat Protoc 14(2): 482–517. https://doi.org/10.1038/s41596-018-0103-9
28. McAlister GC, Nusinow DP, Jedrychowski MP et al (2014) MultiNotch MS3 enables accurate, sensitive, and multiplexeddetection of differential expression across cancer cell line proteomes. Anal Chem 86:7150–7158. https://doi.org/10.1021/ac502040v
29. Chen C, Hou J, Tanner JJ, Cheng J (2020) Bioinformatics methods for mass spectrometry-based proteomics data analysis. Int J Mol Sci 21:. https://doi.org/10.3390/ijms21082873
30. Beausoleil SA, Villén J, Gerber SA et al (2006) A probability-based approach for high-throughput protein phosphorylation analysis and site localization. Nat Biotechnol 24:1285–1292. https://doi.org/10.1038/nbt1240
31. Elias JE, Gygi SP (2010) Target-decoy search strategy for mass spectrometry-based proteomics. Methods Mol Biol. https://doi.org/10.1007/978-1-60761-444-9_5
32. Egertson JD, MacLean B, Johnson R et al (2015) Multiplexed peptide analysis using data-independent acquisition and Skyline. Nat Protoc 10:887–903. https://doi.org/10.1038/nprot.2015.055
33. Bruderer R, Bernhardt OM, Gandhi T et al (2015) Extending the limits of quantitative proteome profiling with dataindependent acquisition and application to acetaminophen-treated three-dimensional liver microtissues. Mol Cell Proteomics 14:1400–1410. https://doi.org/10.1074/mcp.M114.044305

Part II

Systems Biology of Metabolic Networks

Chapter 5

A Practical Guide to Integrating Multimodal Machine Learning and Metabolic Modeling

Supreeta Vijayakumar, Giuseppe Magazzù, Pradip Moon, Annalisa Occhipinti, and Claudio Angione

Abstract

Complex, distributed, and dynamic sets of clinical biomedical data are collectively referred to as multimodal clinical data. In order to accommodate the volume and heterogeneity of such diverse data types and aid in their interpretation when they are combined with a multi-scale predictive model, machine learning is a useful tool that can be wielded to deconstruct biological complexity and extract relevant outputs. Additionally, genome-scale metabolic models (GSMMs) are one of the main frameworks striving to bridge the gap between genotype and phenotype by incorporating prior biological knowledge into mechanistic models. Consequently, the utilization of GSMMs as a foundation for the integration of multi-omic data originating from different domains is a valuable pursuit towards refining predictions. In this chapter, we show how cancer multi-omic data can be analyzed via multimodal machine learning and metabolic modeling. Firstly, we focus on the merits of adopting an integrative systems biology led approach to biomedical data mining. Following this, we propose how constraint-based metabolic models can provide a stable yet adaptable foundation for the integration of multimodal data with machine learning. Finally, we provide a step-by-step tutorial for the combination of machine learning and GSMMs, which includes: (i) tissue-specific constraint-based modeling; (ii) survival analysis using time-to-event prediction for cancer; and (iii) classification and regression approaches for multimodal machine learning. The code associated with the tutorial can be found at https://github.com/Angione-Lab/Tutorials_Combining_ML_and_GSMM.

Key words Multi-omics, Multimodal, Metabolic modeling, Flux balance analysis, Machine learning, Data integration, Cancer survival prediction

1 Introduction

With the rapid development of next-generation sequencing technologies (also known as high-throughput sequencing), we are currently observing an unprecedented growth in data from various sources, accelerated by the decreased cost of sequencing [1–3]. In parallel, the expansion of patient data has enabled the growth of personalized healthcare in the clinical field. The increasing volume and availability of samples that are accessible for analysis and testing

Sonia Cortassa and Miguel A. Aon (eds.), *Computational Systems Biology in Medicine and Biotechnology: Methods and Protocols*, Methods in Molecular Biology, vol. 2399, https://doi.org/10.1007/978-1-0716-1831-8_5,

have widespread applications in improving disease management, cohort discovery for clinical trials and early disease diagnosis [4].

One of the most eagerly anticipated applications of biomedical data analysis has been its influence on personalized medicine in cancer treatment [5, 6]. Ever since cancer was identified as having its roots in genetics [7], the first studies of cell replication assumed that different gene variants were solely responsible for the development of cancer cells [8]. Following the advancement of sequencing technologies, it was recognized that errors in cell replication could also occur during the transcription or translation phase, and that healthy cells were gradually malformed into cancer cells owing to damage that arises in the genome [9]. During the multiple stages of cellular mitosis, it was realized that many factors could influence the gene expression process and produce errors in replication [7]. Any number of gene mutations could affect the translation of proteins further downstream [8, 10], which was why particular attention was paid to changes in codons that encode proteins directly involved in metabolism and the regulation of cellular growth.

Understanding the underlying causes of the processes leading to cellular growth malfunctions is important for devising more effective strategies to halt disease progression. Recent computational studies have focused on mapping molecular features for individual tumors and mutations that influenced the growth of various cancers, including a renewed interest in cancer metabolism [11, 12]. Computational approaches have aided companion diagnostics, supported drug development, and guided selection of personalized treatments [5, 13]. The data accumulating from these studies has also helped to answer many questions concerning clinical and biological research, for instance, how mutations came to be and how they progressed to causing cancer. In particular, it has been well established that cancer cells possess high metabolic flexibility in terms of substrate utilization, which enables their sustained proliferation and survival under fluctuating nutrient availability [14].

Within each cell, metabolism is the network of biochemical reactions that determines function [15, 16]. With recent advances in mathematical and computational methods, we are now able to reconstruct all known pathways of these reactions as complete networks spanning the entirety of metabolic functions in living organisms [15–17]. These reconstructions are known as genome-scale metabolic models (GSMMs), and they are now becoming further enriched following the application of condition- or patient-specific experimental data, as well as various levels of omic and splice isoform information [18, 19]. In order to accommodate the volume and heterogeneity of such diverse data types and aid in their interpretation when they are combined with a multi-scale predictive model, machine learning is a useful tool that can be wielded to deconstruct biological complexity and extract relevant outputs for clinical biomedicine. The complementary characteristics of GSMMs and machine learning make them suitable to be used in combination. As the features introduced by GSMMs are fully informative, a combined approach also helps to address the lack of interpretability associated with machine learning models in biology.

In this chapter, we process multi-omic cancer data via metabolic modeling and multimodal machine learning. Firstly, we focus on the merits of adopting an integrative systems biology approach to biomedical data mining. Following this, we propose how constraint-based metabolic models can provide a stable yet adaptable foundation for the integration of multimodal data with machine learning. Finally, we provide a step-by-step tutorial for the combination of machine learning and GSMMs, which includes: (i) tissue-specific constraint-based modeling; (ii) survival analysis using time-to-event prediction for cancer; and (iii) classification and regression approaches for multimodal machine learning. We envisage that this pipeline (summarized in Fig. 1) can be further adapted by users to suit their own analyses of clinical data, with the goal of building more informative predictive models that combine the quantitative power of machine learning with biomedical interpretability provided by data-enriched GSMMs.

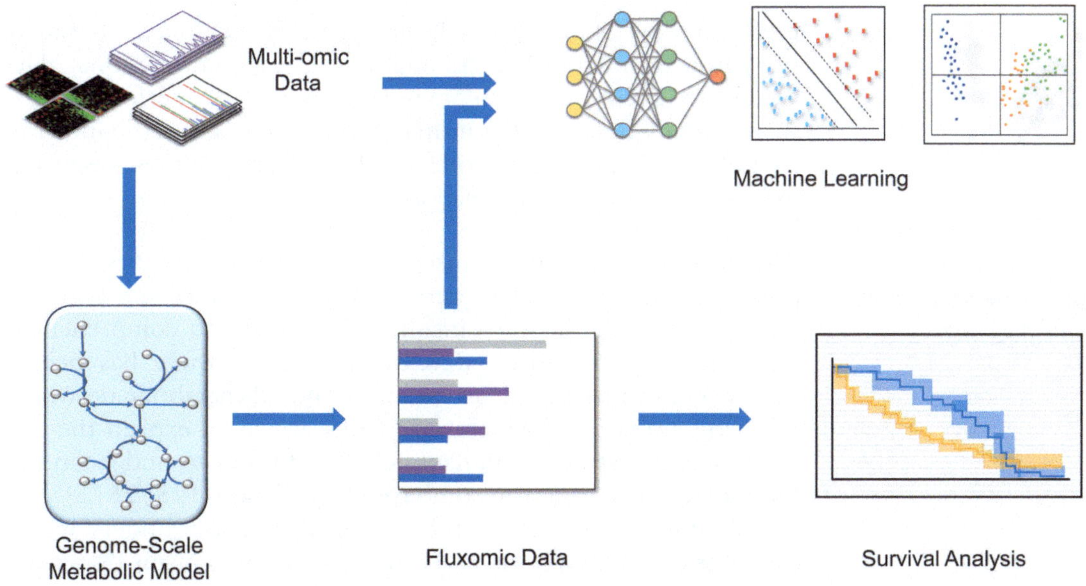

Fig. 1 Pipeline for the integration of metabolic modeling with multimodal machine learning and survival analysis. A summary of the tutorials presented in Subheading 3. Multi-omic data (transcriptomic, proteomic, metabolomic, etc.) can directly be used as input for multimodal machine learning or fed into genome-scale models to generate context-specific metabolic fluxes, e.g. for different cancer cell lines. Furthermore, the resulting fluxomic profiles can be used as input for survival (time-to-event) prediction or machine learning analysis, using data that has been informed by alterations in metabolic pathways

2 Materials

2.1 Data Mining in Biomedicine

Initially, primary genomic analyses were characterized by the examination of DNA or RNA sequence reads [3, 20]. Following this were secondary analyses targeting raw sequenced data, i.e. outputs from next-generation sequencing that filtered and aligned these reads to a reference genome in order to locate gene mutations [3, 20]. Ultimately, tertiary analyses were regarded as the most meaningful since they targeted pre-processed data with an emphasis on learning how genomic features were represented by different genomic regions and how they interacted [20]. Nonetheless, all three of these stages in genomic data analysis were necessary for scientists to develop the current approach that gives broader insights into the development of diseases such as cancer.

Data analysis in biomedicine is currently driven by the need to understand the underlying mechanisms of diseases. However, the key problem is how to extract desirable knowledge from large and complex datasets. The challenge of managing and integrating large, multi-dimensional datasets is still an open problem, and new analytical tools are required to utilize these data to their full potential. Manual analysis of data is considered to be difficult, largely ineffective and inefficient even with the support of statistical methods [21, 22]. These problems are being alleviated by the development of machine learning techniques that seek to interactively develop an understanding of datasets by building predictive models that identify patterns in data [23]. An analyst can use computational algorithms to decide how an investigation will be performed and subsequently develop an automated process to assess the information obtained. Thus, the model can learn from prior data as well as from the outcomes of its analysis [24].

2.2 Constraint-Based Reconstruction and Modeling

Although genomics and transcriptomics provide insights into the presence and expression of genes, pattern discovery and comparison only between genes is insufficient to gather a comprehensive understanding of disease development; thus, there has been a recent focus on the study of metabolism and metabolic networks. Studying alterations in metabolic pathways could explain the dysfunctional growth of malignant cells in addition to understanding the molecular mechanisms underpinning diseases [25, 26].

Systems biology has its focus on analyzing and understanding different components and reactions in living organisms by regarding biological processes from a global perspective, in order to observe connections at the level of the individual cell, organism, or community [26, 27]. In this view, all molecules that comprise a living cell interact to cohesively perform physiological functions [28]. Systems biology can thus be viewed as a collaborative, interdisciplinary venture to examine changes in multiple systems simultaneously and under different conditions [29]. Ultimately, the aim is to improve understanding of how an organism's genetically inherited characteristics (genotype) affects its externally observable characteristics (phenotype) under a given set of conditions. The need to develop new mathematical approaches to model relations between the components of a system was recognized due to the complexity of metabolism, where metabolites are often involved in numerous chemical reactions [30, 31]. Besides, integrating mathematical models or biological networks with additional data has been recognized as an important tool for gaining novel insights from large amounts of new biological information [32].

Constraint-based methods are currently the most commonly used to estimate the flow of metabolites through a metabolic network. They have also proved to be suitable for the interpretation of omics data [33, 34] and the generation of hypotheses that can be experimentally validated to support the improvement of precision medicine [35]. Flux balance analysis (FBA) is currently the most popular tool used to estimate the flow of metabolites in metabolic networks and identify the range of feasible flux values [36, 37]. FBA calculates the rate at which compounds are consumed or produced during metabolic reactions. This approach is based on constraints and is extensively used for the study of GSMMs [38, 39]. The underlying idea is that constraints are imposed on the flow of metabolites through the network by assigning a lower and upper bound to each reaction in the network, which regulates the minimum and maximum amount of flux that would be allowed. This feature also allows for flux predictions based on various experimental conditions [40]. FBA also adopts two other important principles. It is led by the assumption that the total amount of compounds being produced must be equal to the amount of the compounds being consumed, known as the steady-state assumption [36]. For the purpose of mass conservation, this theory also assumes that the fluxes are temporally invariant (do not change over time) and spatially homogeneous.

Most applications of FBA are suitable for the analysis of fluxes in steady-state since they cannot accurately predict the concentration of metabolites in the chemical reactions without introducing additional data. In order to develop more accurate models to record changes in metabolism over time, several dynamic flux balance analysis (dFBA) techniques have been designed to increase the accuracy of predictions [41]. Within their computational frameworks, these approaches incorporate kinetic parameters and changes in the concentrations of specific metabolites over time. Other methods modeling unique solutions that are more consistent with observed metabolic states may consider conditional dependencies or thermodynamic uncertainties within the metabolic network [42–45].

2.3 Machine Learning for Multi-Omic Data Integration

Machine learning can be described as a set of algorithms that improve predictive accuracy through experience, given a certain processable input from which they are able to learn and generalize. Beyond their predictive power, their widespread usage in bioinformatics and computational biology is also due to the limited assumptions they require compared to other statistical or computational approaches. This makes them essential in a number of tasks, ranging from the understanding of RNA folding to estimating the impact of mutations on splicing, and from the exploration of gene expression profiles to reconstructing phylogenetic trees [46–49].

In recent years, machine and deep learning are increasingly becoming regarded as fundamental tools for the inspection, interpretation, and exploitation of multi-omic data, driving it several leaps forward and expected to increasingly drive it in the near future [50, 51]. Additionally, there has been an increase in the application of machine learning to glean more information from biological systems. This abundance of data can only prove to be useful if utilized in a suitable manner, with the aid of appropriate statistical techniques for analysis. A number of recent developments in the application of machine learning to biological problems have been critically analyzed elsewhere [52].

Large volumes of high-throughput biological data are not easily interpretable and analyzable with classical statistical techniques alone. Relying on modern powerful computing architectures, machine and deep learning algorithms have become the go-to suite of methods in almost every field of data analysis. In biology, the most used data are omics, which represent experimental profiles with large coverage over multiple biological domains. The principal "-omic" technologies [53] include genomics (total gene content), transcriptomics (total mRNA transcribed from the genome), proteomics (all expressed proteins), lipidomics (complete lipid profile), metabolomics (all compounds participating in metabolic reactions), epigenomics (complete set of heritable phenotypic changes that do not modify DNA), and many more. The most common techniques used to acquire these data include DNA sequencing (genomics), microarrays and RNA sequencing (transcriptomics), DNA methylation and histone modification assays (epigenomics), and protein or metabolite mass spectrometry (proteomics and metabolomics). Fluxomics, stemming from metabolic modeling or metabolic flux measurements, aims to capture the *in-vivo* activity of the compounds involved in metabolic pathways [54]. Since each of these data types is inextricably linked with the others in a coherent, mechanical way, it is of paramount importance that they can be combined in order to achieve a better understanding of the entire biological system. The more data types included in models, the more information is available to trace molecular components across multiple functional states [55].

2.4 Multimodal Machine Learning

In Subheading 2.3, we discussed how omic data provide direct and convenient access to genetic variability and cellular activity. However, these datasets can only be useful if processed and deciphered through appropriate analytical tools. It has also been stressed that the quality of data generated must also be high—with a sufficient number of replicates to take account of biological variability, reduce the batch effect, and provide sufficient statistical power [56]. Consequently, the quality of available experimental data can become a limiting factor for the predictive power of the model. Multimodal (or multi-view) learning is a branch of machine learning that combines multiple aspects of a common problem in a single setting, in an attempt to offset their limitations when used in isolation [57, 58]. This could prove to be an effective strategy when dealing with multi-omic datasets, as all types of omic data are interconnected.

Data values may be directly concatenated as single sample matrices into one large matrix, transformed into a common intermediate format, or analyzed separately as multiple models with different training sets for each data type. The stage at which data integration is carried out for meta-dimensional analysis must be carefully considered, as this has an impact on the transformation of data. The initial pooling of all samples (early integration) results in increased noise unless regularization is performed. Therefore, it is often preferable to build a similarity matrix between data types (intermediate integration) or analyze each data type separately (late integration) prior to the application of machine learning [59].

Due to their versatility, multimodal algorithms could be also be utilized in a wide range of biomedical applications involving multi-omic data. Correspondingly, the preprocessing of features coupled with late or intermediate fusion should be preferred to early fusion [60]. This approach, based on a combination of computational systems biology and machine learning tools, has been shown to provide key mechanistic insights into neurological disorders [61] and yeast cellular growth [62].

2.5 Multi-Omic Data Integration with Survival Analysis

Multimodal learning has been successfully applied in a multi-omic data integration setting also for predicting cancer survival [63]. Survival analysis, also known as failure time analysis or time-to-event analysis, is a subfield of statistics used to analyze and model data where the outcome is the expected duration of time until events occur. This is commonly used in metabolic modeling to predict the effects of metabolic changes on survival rates [64]. Survival analysis is one of the most significant advancements in the mathematical statistics in the last quarter of the 20th century and is widely used in the fields of biomedical data analysis, economics, finance, engineering, and medicine [65–68].

In recent years, machine learning methods have been applied to survival analysis owing to their capability of modeling non-linear relationships and achieving high-quality predictions. Jhajharia et al.

recognized the difficulties associated with predicting statistics for cancer survival [64]; as a consequence, they calculated survival rates based on a sample of 683 cases by applying five different data mining algorithms (Naive Bayes classification, decision trees, IBK, support vector machine with sequential minimal optimization, and OneR). They compared the performance of these methods and reported on their precision, concluding that the multifactorial modality of breast cancer necessitated a separate predictive factor study for assessing survivability.

The main challenge associated with combining survival analysis with machine learning is when some instances do not experience any event after a certain time period. This is known as censoring and can be resolved by using data transformation techniques such as uncensoring or calibration [69]. Time estimation of the model can also prove to be unreliable. In fact, during the study of a survival analysis problem, it is possible that events of interest are not observed in some instances. This might occur because of the limited observation time-window or because the subject left the study prior to experiencing the event.

2.5.1 Evaluation Metrics for Survival Analysis

Machine learning is effective when the dataset contains a large number of instances in a reasonable dimensional feature space, which is not always the case when working with survival data [70]. For this reason, machine learning methods and performance evaluation metrics must be carefully tailored to develop an accurate model for survival prediction. Due to the presence of censoring in survival data, the standard evaluation metrics such as root of mean squared error and R^2 are not suitable for measuring the model performance in survival analysis. Instead, more specialized metrics need to be applied such as the Concordance index (C-index), Brier Score, and Mean Absolute Error.

C-Index

A common way to evaluate a survival model is to consider the relative risk of an event for different instances rather than the absolute survival times for each instance. This can be achieved by using the C-index [71], which calculates the percentage of "concordant" (i.e. correctly ordered) pairs among all feasible evaluation pairs. A comparable pair is said to be concordant if the sample with a higher estimated risk score within each pair has a shorter actual survival time. Considering both the observation and prediction values of two instances, $(y_A, \hat{y}_A)$ and $(y_B, \hat{y}_B)$, where y_i and $\hat{y}_i$ represent the actual observation time and the predicted value, respectively, the concordance probability between them is computed as:

$$c = Pr(\hat{y}_A > \hat{y}_B | y_A > y_B) \tag{1}$$

In this instance, if patient A's actual event happens before that of patient B and the predicted event time for A is before that of B, this pair is considered a concordant pair (regardless of how long before B the event A actually happened).

Brier Score

The Brier score [72] was initially developed to predict the inaccuracy of probabilistic forecasts. It can only be used to evaluate prediction models that have probabilistic outcomes (i.e. where the outcome is in the range [0, 1] and the sum of all the possible outcomes for a certain individual is equal to 1). The empirical definition of the Brier score at the specific time t can be given by:

$$BS(t) = \frac{1}{N}\sum_{i=1}^{N}[\hat{y}(t)_i - y(t)_i]^2 \quad (2)$$

where $\hat{y}(t)_i$ is the predicted outcome and $y(t)_i$ is the actual outcome at time t.

Brier score can be adapted as a performance metric for survival analysis by including a weight variable $w_i(t)$ defined as:

$$w_i(t) = \begin{cases} \delta_i/G(y_i) & \text{if } y_i \le t \\ 1/G(y_i) & \text{if } y_i > t \end{cases} \quad (3)$$

where δ_i is the binary event indicator (i.e. $\delta_i = 1$ if instance i is uncensored and $\delta_i = 0$ otherwise). G is the Kaplan–Meier estimator of the censoring distribution using the actual outcome y_i, $i = 1, \ldots, N$. With this distribution, the weights of censored instances before t will be 0. However, they will contribute indirectly to the calculation of the Brier score since they are used to calculate G.

Using the weight variable w in Eq. 3, the individual contributions to the empirical Brier score can be re-weighted according to the censored information as follows:

$$BS_w(t) = \frac{1}{N}\sum_{i=1}^{N} w_i(t)[\hat{y}(t)_i - y(t)_i]^2 \quad (4)$$

Mean Absolute Error

In survival analysis, the mean absolute error (MAE) is defined as an average of the differences between the predicted time values and the observed time values. MAE can be calculated using the formula below:

$$MAE = \frac{1}{N}\sum_{i=1}^{N}(\delta_i|y_i - \hat{y}_i|) \quad (5)$$

where y_i and $\hat{y}_i$ represent the actual observation time and the predicted time, respectively. Since $\delta_i = 0$ for censored instances, only the samples for which the event occurs are being considered in this metric. For this reason, MAE can only be used to evaluate models that provide the same event time as the predicted target value (such as accelerated failure time models, which describe the survival time as a function of covariates [73]).

2.6 Multi-Omic Analysis Using Deep Neural Networks

Inspired by neuroscience, neural networks are weighted graphs in which each input node is connected to one or more output nodes via a hidden layer that enables the prediction of non-linear relationships between the input and output variable(s) [74]. Deep neural networks contain multiple hidden layers comprising of many units, making them more applicable for representing functions of increasing complexity [75]. Among the most common and successful integrative deep learning architectures are the autoencoders: they are artificial neural networks composed of two parts—an encoder and a decoder. The encoder tries to compress the information onto a lower-dimensional space in order to retain only the relevant features of the data, while the decoder controls this process indirectly. This is because the decoder has to reconstruct the input data starting from their compressed representation, which is only possible when the important information regarding them has been retained. Autoencoders have been used successfully in classification settings [76, 77] (for a more complete overview on integration with variational autoencoders, the reader is referred to recent reviews [78, 79]).

Following technological advances in computing power, deep neural networks are becoming more commonly used and currently represent the state-of-the-art performance for many different tasks. For instance, tasks such as integrative clustering [80], drug response prediction [81], cancer survival prediction [82], cancer subtyping [83], and psychiatric disease risk prediction [84] have been successfully tackled with deep neural networks. The flexibility of neural networks means that, in theory, they can approximate any arbitrary function—provided that they contain a sufficient number of computational units (nodes) and network depths (layers). However, this often results in networks exploiting high computational predictive power at the expense of interpretability. Although some steps have been taken towards the development of an interpretative framework [62, 85], a definitive approach has not yet taken hold.

2.7 Multimodal GSMMs—Merging Metabolic Analyses with Machine Learning

Although results from integrating machine learning into biological models seem promising, there is still much leeway for improvements with regards to refining phenotypic predictions. With this in mind, we argue that GSMMs can be used both as a foundation for the integration of multi-omic data originating from different domains and as a source of interpretable features for machine learning algorithms. Cuperlovic [86] summarized the main prerequisites essential for the successful implementation of machine learning as follows: the proper selection of learning attributes, construction of training and test sets, selection of the appropriate learning algorithm(s), careful design of the learning approach, and an accurate evaluation of predictive performance. Consequently, it is important to consider how these key attributes can be adhered to when considering the application of machine learning to GSMMs.

GSMMs are one of the main frameworks striving to bridge the gap between genotype and phenotype by incorporating prior biological knowledge into mechanistic models, but they require additional quantitative power to refine parameters within feasible limits and increase predictive performance [87]. On the other hand, machine learning algorithms suffer from a lack of biological interpretability in spite of their optimal predictive power. It is important to note that biological interpretability does not only depend on the data-driven models but extends to the preprocessing of their inputs. It must be emphasized that machine learning algorithms are highly context-specific and they are efficient when they can be tailored to test a particular hypothesis in its applicable setting.

When machine learning is being applied to clinical data, all biological knowledge derived from the training samples must be carefully selected, such that the algorithms are able to ascertain relationships between inputs and responses and generalize them to make further predictions for new inputs [88]. In order to exploit the advantages and complementarity of both frameworks, we argue that constraint-based GSMMs and machine learning can be formulated as a joint problem by embedding stoichiometric constraints in a learning task [60]. Thus, we can envisage the ideal model as one that can leverage the quantitative power of machine learning with the biological interpretability provided by data-enriched GSMMs.

GSMMs are sufficiently flexible to accommodate data corresponding to specific cell types, processes, or environments within the human body. For example, context-specific metabolic profiles have been generated using the integration of gene expression measurements with drug-specific small intestine epithelial cell metabolic models [89]. In this instance, gene expression and flux measurements were subsequently used as features for a multi-label support vector machine to predict the occurrence of gastrointestinal side effects and to cluster drugs according to their latent similarities. A metabolic network could be also represented as a neural network that incorporates a number of hidden layers comprising multiple omic data types [90].

A range of machine learning algorithms have been successfully applied to settings where the features originate from different omic domains, including recent attempts where mechanistic features were generated by GSMMs [62, 91–98]. These approaches have achieved the goal of improving the performance achieved by single-omic datasets for various predictive tasks, while also ensuring biological interpretability of the machine learning predictions. In the following section, we provide a step-by-step tutorial centered on the integration of machine learning and metabolic modeling.

3 Methods

We here describe a series of tutorials for the analysis of cancer-related clinical data within a joint framework that combines metabolic modeling and machine learning. Firstly, we demonstrate how tissue-specific gene expression values measured for cancer cells can be integrated within a human GSMM to yield context-specific flux predictions (Subheading 3.1). Subsequently, we show how both the initial transcriptomic dataset and the newly generated fluxomic dataset can be used as input data for either survival analysis (Subheading 3.2) or the application of multimodal machine learning via classification or regression tasks (Subheading 3.3). The updated code associated with these tutorials can be found at https://github.com/Angione-Lab/Tutorials_Combining_ML_and_GSMM. We note that the example analyses shown here are simply one of many potential implementations of machine learning, and the reader is encouraged to select the type of method most applicable to their data.

3.1 Integrating Gene Expression Data into Flux Balance Analysis

This tutorial describes how gene expression data (e.g. for breast cancer patients) can be integrated within a human GSMM to perform tissue-specific flux balance analysis and yield context-specific metabolic fluxes (Fig. 2).

3.1.1 System Requirements

If required, ensure that the latest versions of MATLAB (R2020b Update 2 at the time of writing), git, and the COBRA Toolbox are installed before beginning. The MATLAB programming language can be downloaded from https://uk.mathworks.com/downloads/.

Instructions for configuring git and downloading/testing the COBRA Toolbox and its compatible solvers are provided here: https://opencobra.github.io/cobratoolbox/stable/installation.html. All installations can be run using Linux, Mac, or Windows operating systems, but in this tutorial we describe analyses run using the Windows platform and the Gurobi Optimizer https://www.gurobi.com/products/gurobi-optimizer/.

3.1.2 Flux Balance Analysis

Within the MATLAB computing environment, add the COBRA Toolbox and code directories to the file path to ensure that they are accessible:

```
% Add the path to COBRA Toolbox
addpath(genpath('C:\Program Files\MATLAB\R2020b\toolbox\cobra'));
```

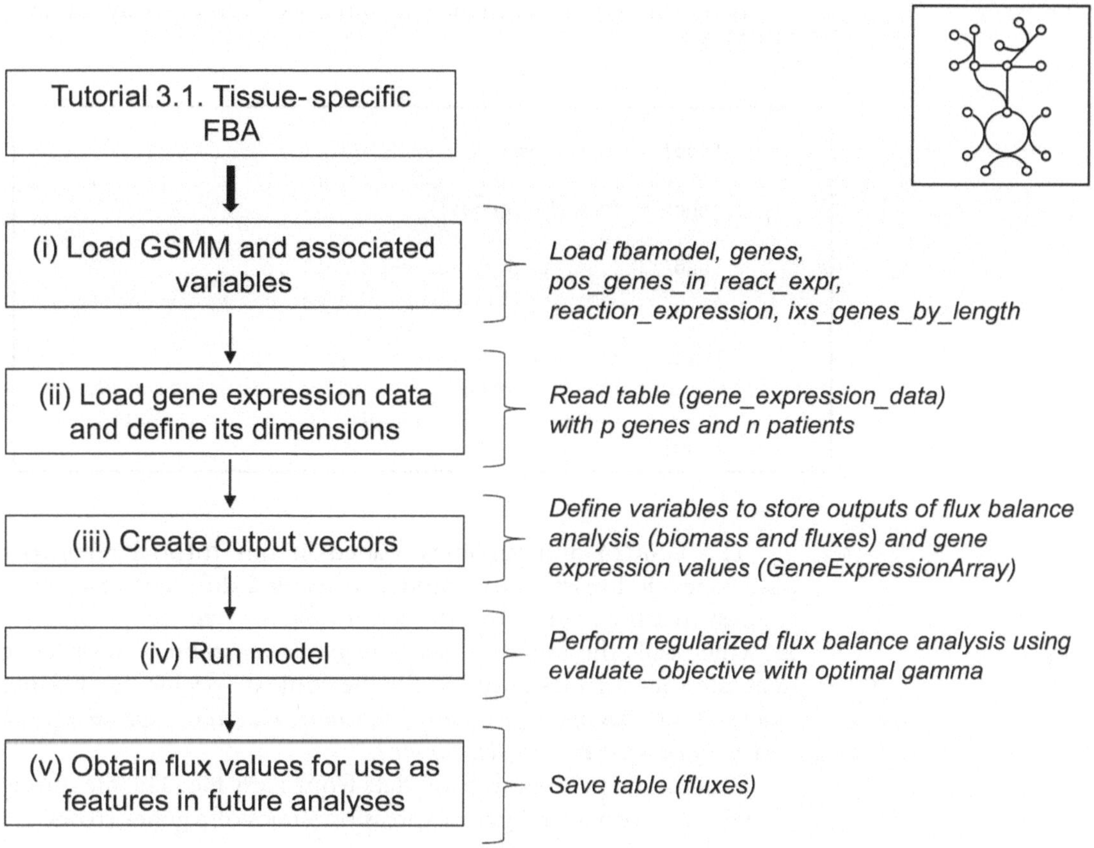

Fig. 2 Tutorial 3.1. Tissue-specific FBA

Initialize the COBRA Toolbox (this step only needs to be completed once):

```
% Initialize the COBRA Toolbox
initCobraToolbox
```

Select the quadratic optimization solver:

```
% Select Gurobi as the quadratic solver
changeCobraSolver('gurobi','QP');
% Avoid solver feasibility error by adjusting the optimization
    parameter
changeCobraSolverParams('QP', 'method', 1);
```

Load the required variables within a local directory available to MATLAB:

```
% Human genome-scale model of breast cancer
load('fbamodel.mat');
% Specify genes as a separate variable
genes = fbamodel.genes;
% Array indexing the position of genes within reactions
load('pos_genes_in_react_expr.mat');
% Array defining the connection between genes and reactions
load('reaction_expression.mat');
% Array indexing genes by their length (required when replacing genes with their expression values)
load('ixs_genes_sorted_by_length');
```

The GSMM and variables loaded in this tutorial refer to a tissue-specific breast cancer model. If using a different model, we suggest that the reader runs the script *compute_reaction_expression, m*, which calls the function *associate_genes_reactions.m* in order to generate the variables required for the context-specific flux balance analysis—i.e. *pos_genes_in_react_expr.mat*, *reaction_expression.mat*, and *ixs_genes_sorted_by_length.mat.*

Read in the gene expression data from a .csv file. The file should contain the normalized gene expression values of *p* genes (rows) for *n* samples (columns). Here, every sample represents a gene expression value associated with a single patient:

```
% Load gene expression data
Data = readtable('gene_expression_data.csv');
% Load the list of gene IDs from the transcriptomic data file
genes_in_dataset = string(table2cell(Data(:, 1)));
% Load gene expression profiles from the data
Data = table2array(Data(:,2:end));
% Convert log2FC gene expression into fold change values centered around 1
Data = 2.^(Data);
% Define the dimensions of the data
% p genes (rows) and n samples (columns)
p = size(Data,1);
n = size(Data,2);
```

Define variables to store the biomass outcomes and flux outputs:

```
% Create vector to store the optimal biomass flux for
    each sample, if of interest
biomass = zeros(1,n);
% Create cell array to store the total fluxes obtained
    for each sample
fluxes{n,size(fbamodel.rxns,1)} = [];
% Create vector to change gene expression values
GeneExpressionArray = ones(numel(genes),1);
```

The flux balance analysis is carried out using the function *evaluate_objective*, which regularizes the flux distribution by minimizing norm-2 for the user-defined objective function (f). The regularized optimization is formulated by subtracting a concave function from the objective, where $\sigma = 10^{-6}$ [99]:

$$\begin{aligned} \max \quad & f^{\mathrm{T}}\nu - \frac{\sigma}{2}\nu^{\mathrm{T}}\nu \\ \text{such that} \quad & S\nu = 0, \\ & \nu_{\min}\varphi(\Theta) \leq \nu \leq \nu_{\max}\varphi(\Theta) \end{aligned} \tag{6}$$

Here, S is the stoichiometric matrix recording all reactions and metabolites within the GSMM, ν is the vector of reaction flux rates, and f is a Boolean vector of weights selecting the objective reaction in ν, while $\nu_{\min}$ and $\nu_{\max}$ are vectors representing the lower- and upper-limits for the flux rates in ν for the unconstrained model. The gene set expression of reactions associated with the fluxes in ν is represented by the vector Θ.

φ is a vector-valued function mapping the expression level of each gene set to a coefficient for the lower- and upper-limits of the corresponding reaction, where γ represents the strength of gene expression mapped to each reaction in the model (all the operations involving Θ are to be intended as element-wise operations):

$$\varphi(\Theta) = [\mathbf{1} + \gamma|log(\Theta)|]^{\mathrm{sgn}(\Theta - 1)} \tag{7}$$

Here, the strength of gene expression can be defined as the detectability of specific gene effects informed by the ability of gene expression values to fine tune the flux rates. In other words, varying orders of magnitude for γ can magnify the level of gene up-regulation or down-regulation in cases where the values are very low. This results in higher metabolic sensitivity, ensuring that experimentally feasible flux values are yielded across different patients.

In this instance, the function employed is logarithmic [100] but several alternatives could be used, including linear functions that can also be set in a reaction-specific fashion [101]. For each variable in *pos_genes_in_react_expr*, each substring of *reaction_expression* is replaced with a value representing their gene expression. For reactions whose genes are not differentially expressed (e.g. exchange reactions), this value is set to 1—i.e. equivalent to a gene expressed normally without up- or down-regulation.

The value of γ can be determined using a grid-search within a range of different values to select the optimum value providing the highest model accuracy ($\gamma = 10$ has been used in this example).

The code below shows how to run the model for each patient and save the flux outputs in a .csv file:

```
% Set the gamma value
gamma = 10;
% Specify an expression profile corresponding to each
    sample in the dataset
for i=1:n
expr_profile = Data(:,i);
% Filter the gene expression data for the genes that are
    present in the dataset
% For those genes that are absent, leave the expression
    value as 1, i.e. normally expressed
pos_genes_in_dataset = zeros(numel(genes),1);
for j=1:length(pos_genes_in_dataset)
position = find(strcmp(genes{j},genes_in_dataset),1);
if ~isempty(position)
pos_genes_in_dataset(j) = position(1);
GeneExpressionArray(j) = expr_profile(
    pos_genes_in_dataset(j));
end
end
% Calculate the flux distribution by optimizing the flux
    objective f
[v_out, f_out] = evaluate_objective(GeneExpressionArray,
    fbamodel,genes,reaction_expression,
    pos_genes_in_react_expr,ixs_genes_sorted_by_length,
    gamma);
% Store biomass values if of interest
% biomass(i) = f_out;
% Store total flux distribution
fluxes(i,:) = num2cell(v_out);
end
% Save flux values for all samples into a .csv file to
    be used for survival analysis and machine learning
    applications
writetable(cell2table(fluxes),'fluxes.csv');
```

The function *evaluate_objective* uses the instruction below to solve the flux balance analysis problem (from the COBRA Toolbox):

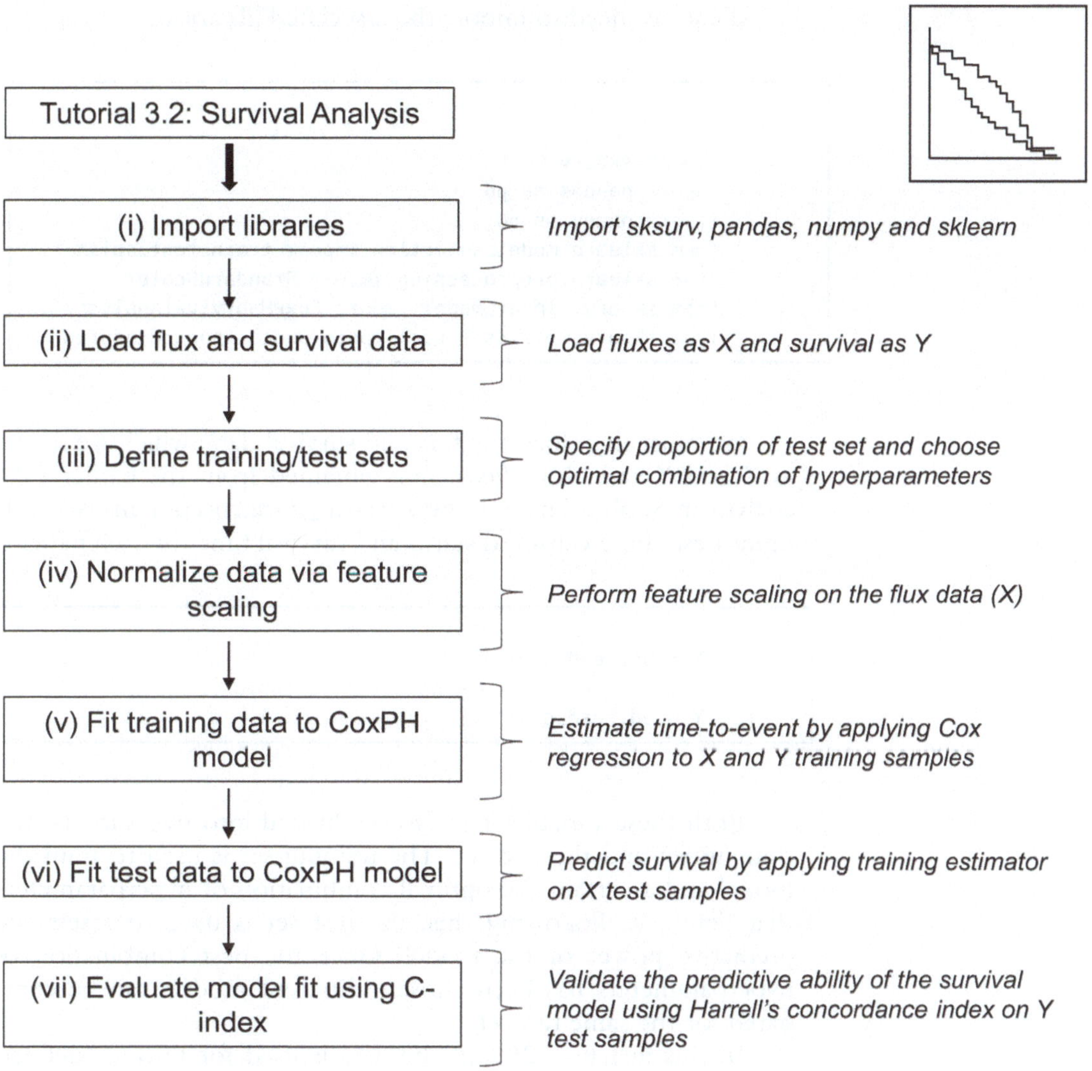

Fig. 3 Tutorial 3.2. Survival analysis

```
% This command is integrated within evaluate_objective
    and does not need to be run as part of the tutorial
[solution] = optimizeCbModel(fbamodel,'max', 1e-6);
f_out = solution.f;
v_out = solution.v;
```

3.2 Survival Analysis

The following tutorial describes a multimodal time-to-event and survival prediction model for cancer that takes metabolic fluxes and survival data as inputs (Fig. 3).

First, we need to import the associated libraries:

```
# Import libraries for feature scaling and selection, fitting and evaluation of the survival model
import sksurv
import pandas as pd
import numpy as np
from sklearn.model_selection import train_test_split
from sklearn.preprocessing import StandardScaler
from sksurv.linear_model import CoxPHSurvivalAnalysis
from sksurv.metrics import concordance_index_censored
```

We must then load the flux and survival data, which we define as X and Y variables. *fluxes.csv* is obtained from the earlier FBA analysis in Subheading 3.1 and *survival_data.csv* contains two columns describing censored status and survival time for each patient:

```
# Load flux data as the X variable
X = pd.read_csv('fluxes.csv')
# Load survival data as the Y variable
Y = pd.read_csv('survival_data.csv')
```

Both these variables need to be divided into two subsets: the training set and the test set. The training set is used to train the model and ascertain the optimal combination of hyperparameters that define it. Following this, the test set is used to assess the predictive power of the model. Once the best combination of hyperparameters has been chosen, different models can be compared for the same test set.

In this instance, 20% of the data is used for testing (defined using the parameter test_size = 0.2):

```
# Define the training and test sets and specify the proportion of data to be used as the test set
X_train, X_test, Y_train, Y_test = train_test_split(X, Y, test_size = 0.2, random_state = 1)
```

Prior to analysis, feature scaling must be performed on training and test sets of the X variable:

```
# Perform feature scaling on the flux data
sc_X = StandardScaler()
X_train = sc_X.fit_transform(X_train)
X_test = sc_X.transform(X_test)
```

We use Cox regression for analyzing our training samples as follows:

```
# Fit the training data to the Cox proportional hazards
    model
estimator = CoxPHSurvivalAnalysis()
estimator.fit(X_train, Y_train)
```

This is followed by fitting the same estimator on the test set to make our final time-to-event prediction:

```
# Fit the test data to the Cox proportional hazards
    model
prediction = estimator.predict(X_test)
```

Finally, we check the concordance index on the test set to validate the predictive ability of the survival model:

```
# Evaluate the fit of the survival model using Harrell`s
     concordance index
result = concordance_index_censored(Y_test['Event'],
    Y_test['Time'], prediction)
```

3.3 Multi-Omic Data Integration and Machine Learning Analyses

Classification approaches comprise many instance-based machine learning algorithms that solve problems based on instances seen in training. Such algorithms include support vector machines (SVMs), k-nearest neighbor (kNN), self-organizing maps, and locally weighted learning. On the other hand, regression-based approaches can prove useful in analyzing relationships between variables, e.g. between metabolic fluxes (x) and their corresponding growth rates (y).

In the following tutorial, instances of early- and late-stage data integration of gene expression and metabolic flux values are demonstrated, according to the type of machine learning analyses that are performed. Herein, in the case of classification using SVM and kNN, data integration is performed at an early stage, whereas late-stage integration is adopted in the case of neural networks (Fig. 4).

3.3.1 System Requirements

We suggest that the reader installs Python and all the required packages via the data science platform Anaconda, which can be downloaded from https://www.anaconda.com/. Anaconda installs

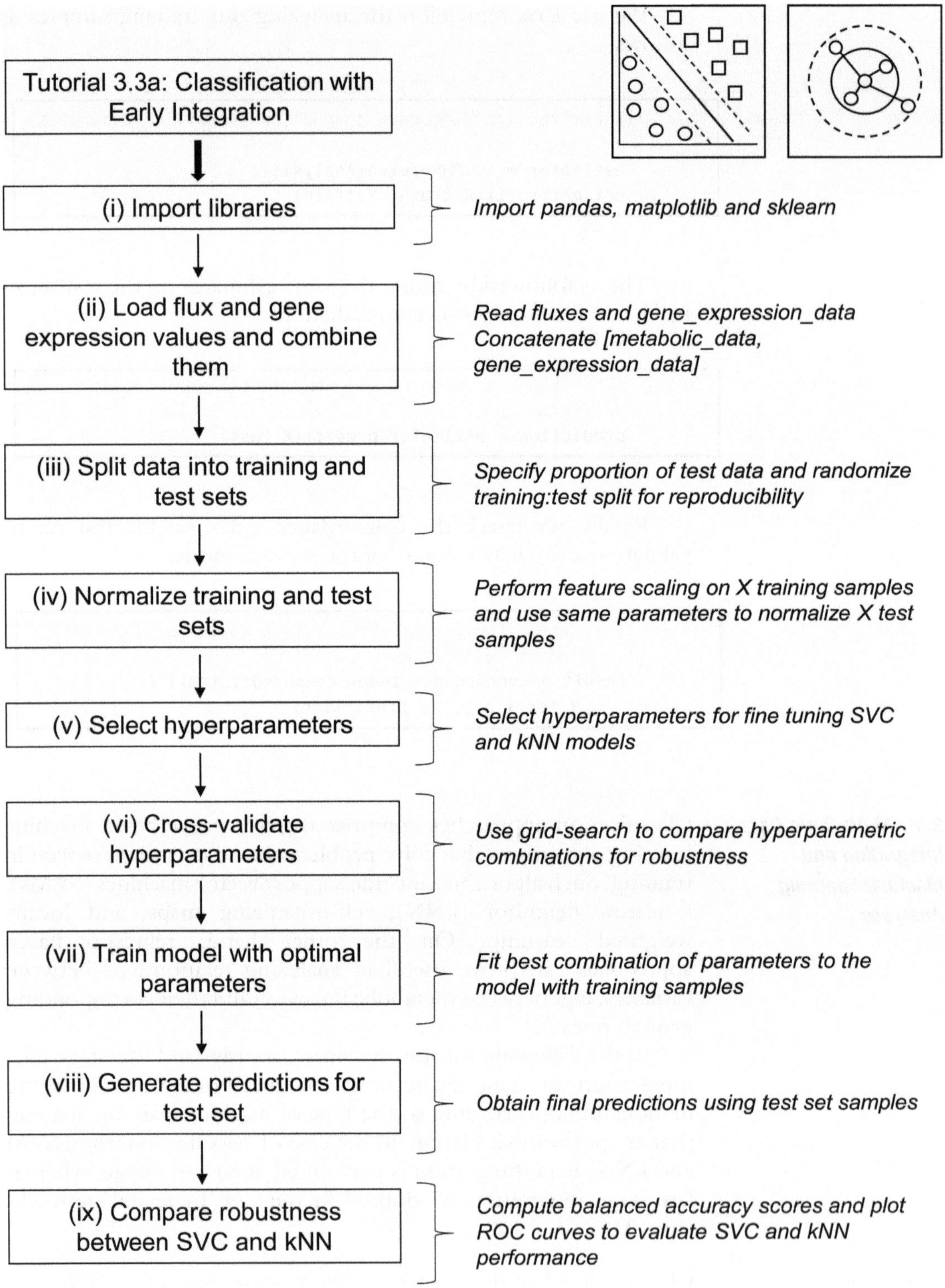

Fig. 4 Tutorial 3.3a. Classification with early integration

all the necessary additional libraries for the tutorial by default, with the exception of Keras (with TensorFlow backend) which can be downloaded like any other package using the Conda installer in Anaconda.

3.3.2 Classification Task with Early Data Integration

In this tutorial, two methods for classification are adopted: Support Vector Machine and k-Nearest Neighbor classifiers. The first step is to import all the necessary libraries:

```
# Import libraries for feature scaling, model selection,
    SVC, kNN and comparison metrics
import pandas as pd
import matplotlib.pyplot as plt
from sklearn.model_selection import train_test_split,
    GridSearchCV
from sklearn.preprocessing import LabelEncoder,
    StandardScaler
from sklearn.svm import SVC
from sklearn.neighbors import KNeighborsClassifier
from sklearn.metrics import balanced_accuracy_score,
    plot_roc_curve
```

Following this, we import *fluxes.csv*, which contains the metabolic fluxes for each sample in our dataset and *gene_expression_data.csv* which contains the respective gene expression levels and the class information.

```
# Import flux data
metabolic_data = pd.read_csv('fluxes.csv', encoding='utf
    -8')
# Disregard the null fluxes
metabolic_data = metabolic_data.loc[:, (metabolic_data.
    abs() >= 1e-7).any(axis=0)]
# Import gene expression data
gene_expression_data = pd.read_csv('gene_expression_data
    .csv',
encoding='utf-8')
```

As part of the early data integration, we combine the gene expression and flux data values and ensure that all data labels are numerical and normalized:

```
# Concatenate the gene expression and flux data matrices
dataset = pd.concat([metabolic_data, gene_expression_data],
axis=1)
# Define the X and Y variables
X = dataset[dataset.columns[:-1]]
Y = dataset[dataset.columns[-1]]
# Transform all labels to numerical normalized labels
Y = LabelEncoder().fit_transform(Y)
```

Similar to the survival analysis, we partition the concatenated data into training and test sets. In this instance, we specify that 30% of the initial data will be used as the test set:

```
# Specify the proportion of data to be used as the test
    set
percent_test = 0.3
# Ensure reproducibility of the results by setting the
    random_state parameter
rndm_state = 0
X_train, X_test, Y_train, Y_test = train_test_split(X, Y,
test_size=percent_test, shuffle=True,
random_state=rndm_state)
```

Following this, the data have to be normalized so that all the feature are comparable and the algorithms can learn from them without effects due to magnitude:

```
# Perform feature scaling to normalize the training data
stdscaler = StandardScaler()
X_train = stdscaler.fit_transform(X_train)
# Normalize the test set with the same parametric values
    as the training set
X_test = stdscaler.transform(X_test)
```

In order to optimize the model for our data, we need to select the best hyperparameter combination for the SVC. This is achieved by trying all possibilities according to the parametric values defined below. Here, the main function of the kernel is to define the high-dimensional space in which the data would be linearly separable by a hyperplane whereas C is a penalty term that controls the tradeoff between building a smooth decision boundary and classifying the training points correctly without overfitting:

```
# Perform parametric tuning for SVC by creating a
    dictionary containing parameters for kernels and C
svc_params = {
        'kernel': ['poly', 'rbf', 'sigmoid'],
        'C':[1, 5, 10]
}
```

A cross-validation approach is used to robustly compare the hyperparametric combinations, where in each split the proportion of classes is maintained:

```
# Create a GridSearchCV object and fit it to the
    training data
grd_search_scv = GridSearchCV(estimator=SVC(),
    param_grid=svc_params, cv=10)
grd_search_scv.fit(X_train, Y_train)
```

The best hyperparametric combination is then selected to train the model from scratch and test it on the test set:

```
# Find the optimal parameters and use them to fit the
    model
svc = SVC(C=grd_search_scv.best_params_['C'],
kernel=grd_search_scv.best_params_['kernel'])
svc.fit(X_train, Y_train)
```

The same process is repeated for the kNN, but with different hyperparameters. Here, n_neighbors represents the number of neighbors used for the KNeighbors queries whereas p determines the metric used to compute distances between data points:

```
# Perform parametric tuning for kNN by creating a dictionary containing parameters for n_neighbors and p
knn_params = {
      'n_neighbors': [5, 7, 9],
      'p':[1, 2]
}
# Create a GridSearchCV object and fit it to the training data
grd_search_knn = GridSearchCV(estimator=KNeighborsClassifier(),
param_grid=knn_params, cv=10)
grd_search_knn.fit(X_train, Y_train)
# Find the optimal parameters and use them to fit the model
kNN = KNeighborsClassifier(
n_neighbors=grd_search_knn.best_params_['n_neighbors'],
p=grd_search_knn.best_params_['p'])
kNN.fit(X_train, Y_train)
```

Finally, we compute balanced accuracy scores and plot ROC curves to compare the robustness between the two methods:

```
# Compute balanced accuracy score for SVC
Y_pred = svc.predict(X_test)
svc_acc = balanced_accuracy_score(Y_test, Y_pred)
# Compute balanced accuracy score for kNN
Y_pred = kNN.predict(X_test)
knn_acc = balanced_accuracy_score(Y_test, Y_pred)
# Print balanced accuracy scores for both methods
print('SVC accuracy: {}, kNN accuracy:{}'.format(svc_acc, knn_acc))
# Get the current axes, creating one if necessary
ax = plt.gca()
# Plot receiver operating characteristic (ROC) curves for SVC and kNN and display them in the same figure
svc_roc = plot_roc_curve(svc, X_test, Y_test, ax=ax)
knn_roc = plot_roc_curve(kNN, X_test, Y_test, ax=ax)
knn_roc.figure_.suptitle(ROC curves)
plt.show()
```

3.3.3 Regression Task with Late Data Integration

In this tutorial, we employ neural networks to incorporate different types of omic data within a regression task. We achieve this by building two single-view neural networks to learn omic-specific features and combining their hidden layers as inputs of a multimodal neural network to learn multi-omic features (Fig. 5).

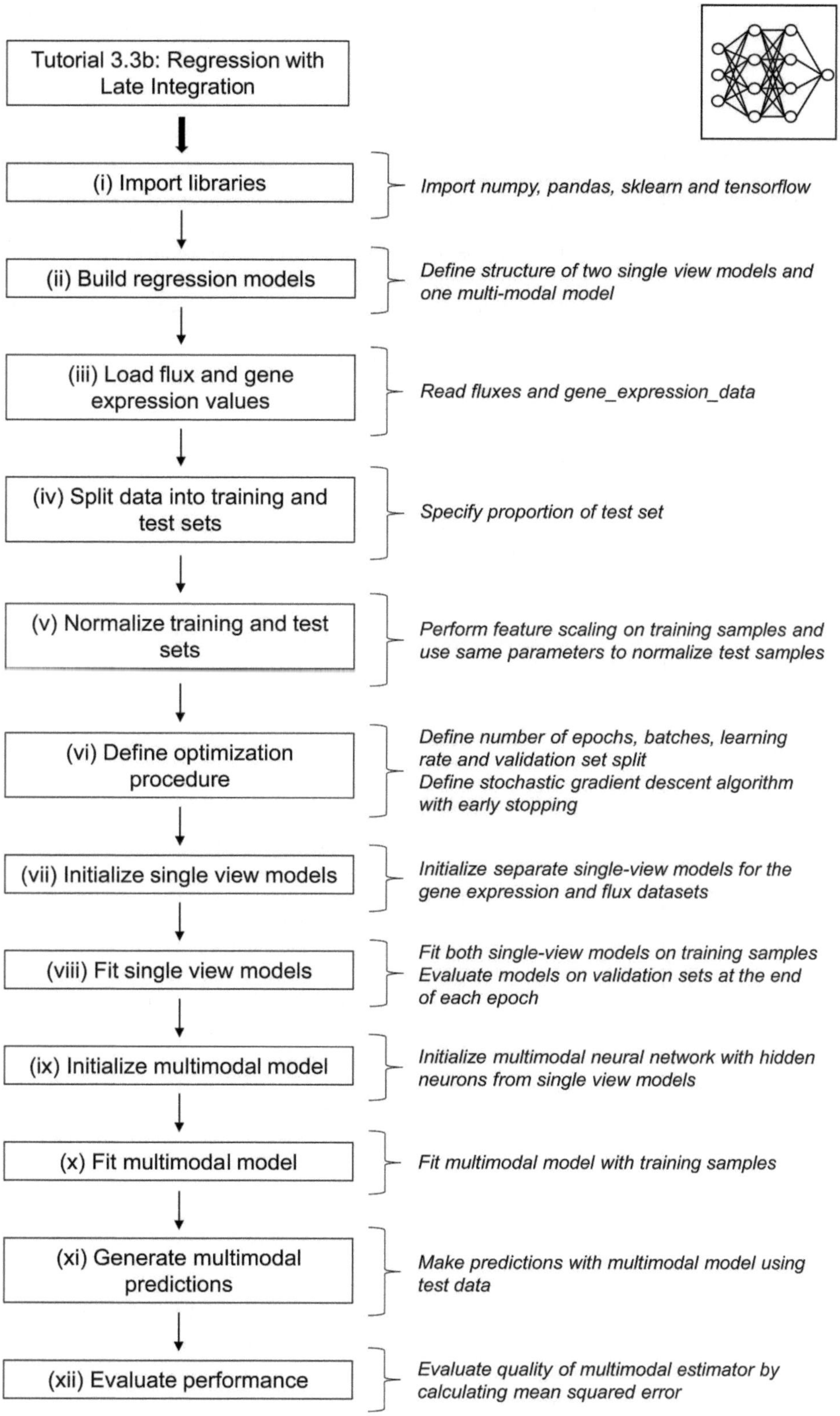

Fig. 5 Tutorial 3.3b. Regression with late integration

As in the previous tutorials, we begin by importing libraries as required:

```
# Import pandas and numpy libraries for data analysis
import numpy as np
import pandas as pd
from sklearn.preprocessing import StandardScaler
# Import the metric to calculate mean squared error
from sklearn.metrics import mean_squared_error
# Import Tensorflow libraries for deep learning
from tensorflow.keras.callbacks import EarlyStopping
from tensorflow.keras.models import Model
from tensorflow.keras.layers import Dense, Dropout,
    Input, Concatenate
from tensorflow.keras.optimizers import SGD
from tensorflow.keras.constraints import max_norm
```

We now need to define the neural network models. Firstly, we build a 4-layer feed-forward network using Keras for the single-view model:

```
# Define the neural network model to use with only one
    view
def init_model(input_dim, learning_rate, epochs,
    momentum, neurons,
        trainable=True):
        input = Input(shape=(input_dim,))
        layer = Dense(neurons, activation='sigmoid',
        kernel_constraint=max_norm(3)) (input)
        layer = Dropout(rate=0.6) (layer)
        layer = Dense(neurons, activation='sigmoid',
        kernel_constraint=max_norm(3)) (layer)
        layer = Dropout(rate=0.6) (layer)
        predictions = Dense(1, activation='linear') (layer)
        model = Model(inputs=input, outputs=predictions)
        rms = SGD(lr=learning_rate, decay=learning_rate / epochs
        momentum=momentum)
        model.trainable = trainable
        if (trainable):
                model.compile(loss='mean_squared_error',
                    optimizer=rms,
                metrics=['mean_squared_error'])
        return model
```

We then build a multimodal model, which uses a concatenation layer to join the high-level features representations learned by the two single-view models, one for gene expression and one for flux data:

```
# Define the neural network to use in a multi-view fashion
def init_multi_model(input_dim,input_dim2, learning_rate, epochs,
momentum, neurons, metabolic_fluxes_layer,
gene_expression_layer):
        metabolic_fluxes_input = Input(shape=(input_dim,))
        gene_expression_input = Input(shape=(input_dim2,))
        comb_layer = Concatenate()([
        metabolic_fluxes_layer(metabolic_fluxes_input),
        gene_expression_layer(gene_expression_input)])
        comb_layer = Dense(neurons, activation='sigmoid',
        kernel_constraint=max_norm(3)) (comb_layer)
        predictions = Dense(1, activation='linear') (comb_layer)
        model = Model(inputs=[
        metabolic_fluxes_input, gene_expression_input],
        outputs=predictions)
        rms = SGD(lr=learning_rate, decay=learning_rate / epochs,
        momentum=momentum)
        model.compile(loss='mean_squared_error', optimizer=rms,
        metrics=['mean_squared_error'])
        return model
```

The data are the same as in the classification tutorial, except for the last column of *gene_expression_data.csv*, which contains continue values instead of class labels:

```
# Specify the proportion of data to be used as the test set
percent_test = 0.3
# Import flux data
metabolic_data = pd.read_csv('fluxes.csv', encoding='utf-8')
# Disregard null fluxes
metabolic_data = metabolic_data.loc[:, (metabolic_data.abs() >= 1e-7).any(axis=0)]
# Import gene expression data
gene_expression_data = pd.read_csv('gene_expression_data.csv',
encoding='utf-8')
X = gene_expression_data[gene_expression_data.columns[:-1]]
Y = gene_expression_data[gene_expression_data.columns[-1]]
# Split gene expression data into training and test sets
gene_expression_train, gene_expression_test, Y_train, Y_test =
train_test_split(X, Y, test_size=percent_test,
shuffle=False)
# Split flux data into training and test sets
metabolic_fluxes_train, metabolic_fluxes_test = train_test_split(
X, test_size=percent_test, shuffle=False)
```

As in the previous tutorial, the data have to be first normalized so that the training will be easier for the models:

```
# Perform feature scaling to normalize the training data
stdscaler_f = StandardScaler()
stdscaler_g = StandardScaler()
metabolic_fluxes_train = stdscaler_f.fit_transform
   (metabolic_fluxes_train)
gene_expression_train = stdscaler_g.fit_transform
   (gene_expression_train)
# Normalize the test sets with the same parametric
    values as the training sets
metabolic_fluxes_test = stdscaler_f.transform
   (metabolic_fluxes_test)
gene_expression_test = stdscaler_g.transform
   (gene_expression_test)
```

The neural network is optimized using stochastic gradient descent and an early stopping approach is adopted to prevent the network from learning irrelevant features in the data, which would hinder generalization:

```
# Define the number of epochs, batches, learning rate
    and validation set split size
epochs = 6000
batches = 256
lrate = 0.005
validation = 0.2
lrate2 = 0.05
epochs2 = 500
# Define the stochastic gradient descent algorithm and
    the early stopping strategy to prevent overfitting
rms = SGD(lr=lrate , decay=lrate / epochs, momentum
    =0.75)
earlyStopping=EarlyStopping(monitor='val_loss', patience
    =15000,
verbose=0, mode='auto')
```

The two types of data (gene expression and flux rates) are processed by two separate neural networks, and then the trained hidden layers of each network are used as input in a third network to predict the regression values. The model is evaluated on a validation dataset at the end of each training epoch to prevent overfitting:

```
# Initialize separate single-view models for the gene expression and flux datasets
model_gene_expression = init_model(gene_expression_train.shape[1],
lrate, 3000, 0.75, 1000)
model_metabolic_fluxes = init_model(metabolic_fluxes_train.shape[1],
lrate, 3000, 0.75,1000)
# Fit both the learning models on the training data
model_gene_expression.fit(x=gene_expression_train, y=Y_train,
epochs=epochs, batch_size=batches,
validation_split=validation,
callbacks=[earlyStopping], verbose=0)
model_metabolic_fluxes.fit(x=metabolic_fluxes_train, y=Y_train,
epochs=epochs, batch_size=batches,
validation_split=validation,
callbacks=[earlyStopping], verbose=0)
```

We remove the last layers in the two single-view models in order to use the trained hidden neurons as input for the multimodal neural network:

```
# Remove the last layer from each single view model
model_gene_expression.layers.pop()
model_gene_expression.layers.pop()
model_gene_expression.outputs = [model_gene_expression.layers[-1].output]
model_metabolic_fluxes.layers.pop()
model_metabolic_fluxes.layers.pop()
model_metabolic_fluxes.outputs = [model_metabolic_fluxes.layers[-1].output]
```

Finally, we initialize and train the multimodal model to make predictions on the test set, using the mean squared error to evaluate performance:

```
        # Initialize the multi-modal model
        multi_modal_model = init_multi_model(
            metabolic_fluxes_train.shape[1],
        gene_expression_train.shape[1], lrate2, epochs2,
        0.75, 15, model_metabolic_fluxes,
        model_gene_expression)
        # Fit the multi-modal model using training samples
        multi_modal_model.fit(x=[metabolic_fluxes_train,
            gene_expression_train],
        y=Y_train, epochs=epochs2, batch_size=batches,
        validation_split=validation, verbose=0)
        # Generate predictions for the test set samples
        predictions = multi_modal_model.predict([
            metabolic_fluxes_test,
        gene_expression_test])
        # Print mean squared error
        print('MSE: ', mean_squared_error(predictions, Y_test))
```

3.4 Conclusions

Metabolic models can be adapted to suit several applications by integrating additional data, for instance, to simulate metabolism under specific genetic or environmental conditions, thus generating condition-specific GSMMs that are mechanistically and biologically interpretable. At the same time, given the difficulty in extracting information from large and complex datasets, machine learning algorithms can be harnessed to reduce dimensionality and elucidate cross-omic relationships. Constraint-based modeling is ideally poised to bridge the gap between knowledge-driven biological datasets and biologically agnostic machine learning models, thus consolidating these two computational frameworks in order to reveal novel insights relating to metabolism. Machine learning algorithms and constraint-based models share complementary characteristics and common mathematical bases which makes them compatible to be combined. On the one hand, GSMMs provide critical data in terms of stoichiometry and the genetic control of biochemical reactions. On the other hand, machine learning can deconstruct biological complexity by extracting relevant outputs from data. Together, they improve omic-based statistical and machine learning analyses by supplementing the learning process with biological knowledge and refining phenotypic predictions.

4 Notes

1. **Significance of Generating Fluxomic Data**

   In this chapter, we have provided instances of targeted hybrid approaches that can better leverage the advantages of multiple methods, whilst offsetting their respective limitations. In Subheading 2.3, we illustrated how multi-omic data integration presents the unique advantage of tracking biological processes across multiple functional states. Through simulating flux using GSMMs, an additional (fluxomic) perspective is generated, which facilitates the identification of metabolically significant fluxes and adaptive mechanisms that are not apparent when analyzing each type of data in isolation. It is expected that integrating further omic datasets, i.e. further data views in a multimodal machine learning setting could derive even more detailed insights into metabolic adaptations, or better support existing findings in the biomedical context.

2. **Multi-Omic Functional Redundancy**

   In addition to considering the quality of data when generating replicates across different omics, a balance needs to be struck between achieving sufficient statistical power and coverage of biological variability whilst reducing the batch effect. However, combining omic datasets increases the risk of introducing redundancies across different types of data, which must be resolved in order to prevent generalization of models on other data samples. This is particularly applicable in the case of fluxomic data, which are derived as a direct result of other omics incorporated within metabolic models. As a consequence, this necessitates the development of new generative techniques or regularization approaches specifically for flux outputs. Although all omics are directly interpretable from a biological perspective, their integration within machine learning tasks remains limited by the interpretability of results generated from the most complex techniques such as neural networks. This calls for the development of better, more interpretable models tailored for the analysis of multi-omic data.

3. **Future Applications**

   The prospective research directions that could be followed include the building and testing of new mathematical models that would, for instance, allow the prediction of behavior for different types of cancer under diverse environmental conditions. Further studies could also devise new ways to incorporate microbial networks and their influence on human cells. Particularly, studies incorporating signalling and regulatory networks could introduce additional information in relation to factors that are not included in the human genome but still affect cellular functions. Through this chapter, we hope that prospective readers in the biomedical field can see the merits of fusing data from multiple sources in a data-driven manner. This can be achieved by finding suitable machine learning approaches that can be applied to the data and task at hand, with the goal of building more accurate and biologically interpretable (white-box) predictive models.

Acknowledgements

We would like to acknowledge the support from UKRI Research England's THYME project, from the Children's Liver Disease Foundation, and from Earlier.org.

Author Contributions
Conceptualization: S.V. and C.A.; Data curation: S.V., G.M. and A. O.; Formal analysis: S.V., G.M., P.M. and A.O.; Funding Acquisition: C.A.; Investigation: S.V., G.M., P.M. and A.O.; Methodology: S.V., G.M., P.M., A.O. and C.A.; Project administration: S.V. and C.A.; Resources: S.V., G.M., P.M. and A.O.; Software: S. V., G.M., P.M., A.O. and C.A.; Supervision: S.V. and C.A.; Validation: S.V., G.M. and A.O.; Visualization: S.V.; Writing—original draft: S.V., G.M., P.M., A.O. and C.A.; Writing—reviewing and editing: S.V., G.M., A.O. and C.A.

Declaration of Interests The authors declare no competing interests.

References

1. Shi Y, Kim S (2014) Towards information analysis for big data. In: 2014 7th conference on Control and automation (CA). IEEE, Piscataway, pp 3–5
2. Gupta A (2015) Big data analysis using computational intelligence and Hadoop: a study. In: 2015 2nd international conference on computing for sustainable global development (INDIACom). IEEE, Piscataway, pp 1397–1401
3. Ceri S, Kaitoua A, Masseroli M, Pinoli P, Venco F (2016) Data management for heterogeneous genomic datasets. IEEE/ACM Trans Comput Biol Bioinform 14(6): 1251–1264
4. Kench A, Janeja VP, Yesha Y, Rishe N, Grasso MA, Niskar A (2015) Clinico-genomic data analytics for precision diagnosis and disease management. In: 2015 international conference on healthcare informatics (ICHI). IEEE, Piscataway, pp 263–271
5. Zieba A, Grannas K, Söderberg O, Gullberg M, Nilsson M, Landegren U (2012) Molecular tools for companion diagnostics. New Biotechnol 29(6):634–640
6. Ascolani G, Occhipinti A, Liò P (2015) Modelling circulating tumour cells for personalised survival prediction in metastatic breast cancer. PLoS Comput Biol 11(5):e1004
7. Rieger PT (2004) The biology of cancer genetics. In: Seminars in oncology nursing, vol 20. Elsevier, Amsterdam, pp 145–154
8. Moorcraft SY, Gonzalez D, Walker BA (2015) Understanding next generation sequencing in oncology: a guide for oncologists. Crit Rev Oncol/Hematol 96(3): 463–474

9. Bertram JS (2000) The molecular biology of cancer. Mol Aspects Med 21(6):167–223
10. Schatz MC, Langmead B (2013) The DNA data deluge. IEEE Spectr 50(7):28–33
11. Eyassu F, Angione C (2017) Modelling pyruvate dehydrogenase under hypoxia and its role in cancer metabolism. R Soc Open Sci 4(10): 170
12. Pavlova NN, Thompson CB (2016) The emerging hallmarks of cancer metabolism. Cell Metab 23(1):27–47
13. Pacheco MP, Bintener T, Sauter T (2019) Towards the network-based prediction of repurposed drugs using patient-specific metabolic models. EBioMedicine 43:26–27
14. Martin SD, McGee SL (2019) A systematic flux analysis approach to identify metabolic vulnerabilities in human breast cancer cell lines. Cancer Metab 7(1):12
15. Edwards LM (2017) Metabolic systems biology: a brief primer. J Physiol 595(9): 2849–2855
16. Palsson B (2015) Systems biology. Cambridge University Press, Cambridge
17. Angione C (2019) Human systems biology and metabolic modelling: a review—from disease metabolism to precision medicine. BioMed Res Int 2019:8304260
18. Ryu JY, Kim HU, Lee SY (2017) Framework and resource for more than 11,000 gene-transcript-protein-reaction associations in human metabolism. Proc Nat Acad Sci 114(45): E9740–E9749
19. Angione C (2018) Integrating splice-isoform expression into genome-scale models characterizes breast cancer metabolism. Bioinformatics 34(3):494–501
20. Montanari P, Bartolini I, Ciaccia P, Patella M, Ceri S, Masseroli M (2016) Pattern similarity search in genomic sequences. IEEE Trans Knowl Data Eng 28(11):3053–3067
21. Wang Xl, Li Jy, Liu Y, Wang Yf, Zhao Ds (2013) Building localized bioinformatics platform based on galaxy and high performance computing cluster. In: 2013 6th International Conference on Biomedical engineering and informatics (BMEI). IEEE, Piscataway, pp 712–716
22. Belgrave D, Henderson J, Simpson A, Buchan I, Bishop C, Custovic A (2017) Disaggregating asthma: big investigation versus big data. J Allergy Clin Immunol 139(2): 400–407
23. Han J, Pei J, Kamber M (2011) Data mining: concepts and techniques. Elsevier, Amsterdam
24. Kitchin R (2014) The data revolution: big data, open data, data infrastructures and their consequences. SAGE Publishing, Thousand Oaks
25. Cairns RA, Harris IS, Mak TW (2011) Regulation of cancer cell metabolism. Nat Rev Cancer 11(2):85
26. Mardinoglu A, Nielsen J (2016) The impact of systems medicine on human health and disease. Fron Physiol 7:552
27. Barrett CL, Kim TY, Kim HU, Palsson BØ, Lee SY (2006) Systems biology as a foundation for genome-scale synthetic biology. Curr Opin Biotechnol 17(5):488–492
28. Yurkovich JT, Palsson BO (2015) Solving puzzles with missing pieces: the power of systems biology. Proc IEEE 104(1):2–7
29. Palsson BØ (2011) Systems biology: simulation of dynamic network states. Cambridge University Press, Cambridge
30. Gomez-Cabrero D, Abugessaisa I, Maier D, Teschendorff A, Merkenschlager M, Gisel A, Ballestar E, Bongcam-Rudloff E, Conesa A, Tegnér J (2014) Data integration in the era of omics: current and future challenges. BMC Syst Biol 8(Suppl 2):I1
31. Ivanov O, van der Schaft A, Weissing FJ (2016) Steady states and stability in metabolic networks without regulation. J Theor Biol 401:78–93
32. Nielsen J (2017) Systems biology of metabolism: a driver for developing personalized and precision medicine. Cell Metab 25(3): 572–579
33. Joyce AR, Palsson BØ (2006) The model organism as a system: integrating 'omics' data sets. Nat Rev Mol Cell Biol 7(3):198
34. Aurich MK, Fleming RM, Thiele I (2016) Metabotools: a comprehensive toolbox for analysis of genome-scale metabolic models. Front Physiol 7:327
35. Bordbar A, Palsson BO (2012) Using the reconstructed genome-scale human metabolic network to study physiology and pathology. J Internal Med 271(2):131–141
36. Orth JD, Thiele I, Palsson BØ (2010) What is flux balance analysis? Nat Biotechnol 28(3): 245
37. O'Brien EJ, Monk JM, Palsson BO (2015) Using genome-scale models to predict biological capabilities. Cell 161(5):971–987
38. Di Filippo M, Colombo R, Damiani C, Pescini D, Gaglio D, Vanoni M, Alberghina L, Mauri G (2016) Zooming-in on cancer metabolic rewiring with tissue

specific constraint-based models. Comput Biol Chem 62:60–69

39. Vivek-Ananth R, Samal A (2016) Advances in the integration of transcriptional regulatory information into genome-scale metabolic models. Biosystems 147:1–10

40. Yilmaz LS, Walhout AJ (2017) Metabolic network modeling with model organisms. Curr Opin Chem Biol 36:32–39

41. Fernandes S, Robitaille J, Bastin G, Jolicoeur M, Wouwer AV (2016) Dynamic metabolic flux analysis of underdetermined and overdetermined metabolic networks. IFAC-PapersOnLine 49(26):318–323

42. Rügen M, Bockmayr A, Steuer R (2015) Elucidating temporal resource allocation and diurnal dynamics in phototrophic metabolism using conditional FBA. Sci Rep 5:15,247

43. Lularevic M, Racher AJ, Jaques C, Kiparissides A (2019) Improving the accuracy of flux balance analysis through the implementation of carbon availability constraints for intracellular reactions. Biotechnol Bioeng 116(9):2339–2352

44. Ataman M, Hatzimanikatis V (2015) Heading in the right direction: thermodynamics-based network analysis and pathway engineering. Curr Opin Biotechnol 36:176–182

45. Willemsen AM, Hendrickx DM, Hoefsloot HC, Hendriks MM, Wahl SA, Teusink B, Smilde AK, van Kampen AH (2015) MetDFBA: incorporating time-resolved metabolomics measurements into dynamic flux balance analysis. Mol BioSyst 11(1):137–145

46. Zhang Y, Rajapakse JC (2009) Machine learning in bioinformatics, vol 4. Wiley, London

47. Leung MK, Delong A, Alipanahi B, Frey BJ (2016) Machine learning in genomic medicine: a review of computational problems and data sets. Proc IEEE 104(1):176–197

48. Angermueller C, Pärnamaa T, Parts L, Stegle O (2016) Deep learning for computational biology. Mol Syst Biol 12(7):878

49. Min S, Lee B, Yoon S (2017) Deep learning in bioinformatics. Briefings Bioinform 18(5):851–869

50. Libbrecht MW, Noble WS (2015) Machine learning applications in genetics and genomics. Nat Rev Genet 16(6):321

51. Ching T, Himmelstein DS, Beaulieu-Jones BK, Kalinin AA, Do BT, Way GP, Ferrero E, Agapow PM, Zietz M, Hoffman MM, et al (2018) Opportunities and obstacles for deep learning in biology and medicine. J R Soc Interface 15(141):20170

52. Zeng ISL, Lumley T (2018) Review of statistical learning methods in integrated omics studies (an integrated information science). Bioinform Biol Insights 12:1177932218759

53. Horgan RP, Kenny LC (2011) 'omic' technologies: genomics, transcriptomics, proteomics and metabolomics. Obstet Gynaecol 13(3):189–195

54. Biedendieck R, Borgmeier C, Bunk B, Stammen S, Scherling C, Meinhardt F, Wittmann C, Jahn D (2011) Systems biology of recombinant protein production using bacillus megaterium. In: Methods in enzymology, vol 500. Elsevier, Amsterdam, pp 165–195

55. Fondi M, Liò P (2015) Multi-omics and metabolic modelling pipelines: challenges and tools for systems microbiology. Microbiol Res 171:52–64

56. Yurkovich JT, Palsson BO (2018) Quantitative-omic data empowers bottom-up systems biology. Curr Opin Biotechnol 51:130–136

57. Sun S (2013) A survey of multi-view machine learning. Neural Comput Appl 23(7–8):2031–2038

58. Vijayakumar S, Conway M, Lió P, Angione C (2018) Seeing the wood for the trees: a forest of methods for optimization and omic-network integration in metabolic modelling. Briefings Bioinform 19(6):1218–1235

59. Serra A, Fratello M, Fortino V, Raiconi G, Tagliaferri R, Greco D (2015) MVDA: a multi-view genomic data integration methodology. BMC Bioinform 16(1):261

60. Zampieri G, Vijayakumar S, Yaneske E, Angione C (2019) Machine and deep learning meet genome-scale metabolic modeling. PLoS Comput Biol 15(7):e1007

61. Sertbas M, Ulgen KO (2018) Unlocking human brain metabolism by genome-scale and multiomics metabolic models: relevance for neurology research, health, and disease. OMICS: J Integr Biol 22(7):455–467

62. Culley C, Vijayakumar S, Zampieri G, Angione C (2020) A mechanism-aware and multiomic machine-learning pipeline characterizes yeast cell growth. Proc Nat Acad Sci 117(31):18,869–18,879

63. Tong L, Mitchel J, Chatlin K, Wang MD (2020) Deep learning based feature-level integration of multi-omics data for breast cancer patients survival analysis. BMC Med Inform Decis Making 20(1):1–12

64. Jhajharia S, Verma S, Kumar R (2016) Predictive analytics for breast cancer survivability: a comparison of five predictive models. In:

Proceedings of the second international conference on information and communication technology for competitive strategies. ACM, New York, p 26

65. Ma Z, Krings AW (2008) Survival analysis approach to reliability, survivability and prognostics and health management (PHM). In: 2008 IEEE aerospace conference. IEEE, Piscataway, pp 1–20
66. Iuliano A, Occhipinti A, Angelini C, De Feis I, Lió P (2016) Cancer markers selection using network-based Cox regression: a methodological and computational practice. Front Physiol 7:208
67. Iuliano A, Occhipinti A, Angelini C, De Feis I, Liò P (2018) Combining pathway identification and breast cancer survival prediction via screening-network methods. Front Genet 9: 206
68. Lee C, Zame WR, Yoon J, van der Schaar M (2018) Deephit: a deep learning approach to survival analysis with competing risks. In: AAAI, pp 2314–2321
69. Wang P, Li Y, Reddy CK (2019) Machine learning for survival analysis: a survey. ACM Comput Surv 51(6):1–36
70. Zupan B, DemšAr J, Kattan MW, Beck JR, Bratko I (2000) Machine learning for survival analysis: a case study on recurrence of prostate cancer. Artif Intell Med 20(1):59–75
71. Harrell Jr FE, Lee KL, Califf RM, Pryor DB, Rosati RA (1984) Regression modelling strategies for improved prognostic prediction. Stat Med 3(2):143–152
72. Brier GW (1950) Verification of forecasts expressed in terms of probability. Mon Weather Rev 78(1):1–3
73. Kleinbaum DG, Klein M (2010) Survival analysis. Springer, Berlin
74. Nisbet R, Elder J, Miner G (2009) Basic algorithms for data mining: a brief overview. In: Handbook of statistical analysis and data mining applications, pp 121–150
75. Goodfellow I, Bengio Y, Courville A (2016) Deep learning. MIT Press, Cambridge. https://www.deeplearningbook.org
76. Xu J, Wu P, Chen Y, Meng Q, Dawood H, Dawood H (2019) A hierarchical integration deep flexible neural forest framework for cancer subtype classification by integrating multi-omics data. BMC Bioinform 20(1):1–11
77. Lemsara A, Ouadfel S, Fröhlich H (2020) Pathme: pathway based multi-modal sparse autoencoders for clustering of patient-level multi-omics data. BMC Bioinform 21:1–20
78. Kingma DP, Welling M (2013) Auto-encoding variational bayes. arXiv preprint arXiv:13126114
79. Simidjievski N, Bodnar C, Tariq I, Scherer P, Andres Terre H, Shams Z, Jamnik M, Liò P (2019) Variational autoencoders for cancer data integration: design principles and computational practice. Front Genet 10:1205
80. Liang M, Li Z, Chen T, Zeng J (2014) Integrative data analysis of multi-platform cancer data with a multimodal deep learning approach. IEEE/ACM Trans Comput Biol Bioinform 12(4):928–937
81. Sharifi-Noghabi H, Zolotareva O, Collins CC, Ester M (2019) Moli: multi-omics late integration with deep neural networks for drug response prediction. Bioinformatics 35(14):i501–i509
82. Cheerla A, Gevaert O (2019) Deep learning with multimodal representation for pancancer prognosis prediction. Bioinformatics 35(14): i446–i454
83. Chen R, Yang L, Goodison S, Sun Y (2020) Deep-learning approach to identifying cancer subtypes using high-dimensional genomic data. Bioinformatics 36(5):1476–1483
84. Wang D, Liu S, Warrell J, Won H, Shi X, Navarro FC, Clarke D, Gu M, Emani P, Yang YT, et al. (2018) Comprehensive functional genomic resource and integrative model for the human brain. Science 362(6420):eaat8464
85. Qiu S, Joshi PS, Miller MI, Xue C, Zhou X, Karjadi C, Chang GH, Joshi AS, Dwyer B, Zhu S, et al (2020) Development and validation of an interpretable deep learning framework for Alzheimer's disease classification. Brain 143(6):1920–1933
86. Cuperlovic-Culf M (2018) Machine learning methods for analysis of metabolic data and metabolic pathway modeling. Metabolites 8(1):4
87. Vijayakumar S, Conway M, Lió P, Angione C (2018) Optimization of multi-omic genome-scale models: methodologies, hands-on tutorial, and perspectives. In: Metabolic network reconstruction and modeling. Springer, Berlin, pp 389–408
88. Lawson C, Martí JM, Radivojevic T, Jonnalagadda SVR, Gentz R, Hillson NJ, Peisert S, Kim J, Simmons BA, Petzold CJ, et al (2021) Machine learning for metabolic engineering: a review. Metab Eng 63(1):34–60
89. Ben Guebila M, Thiele I (2019) Predicting gastrointestinal drug effects using contextualized metabolic models. PLoS Comput Biol 15(6):e1007,100

90. Guo W, Xu Y, Feng X (2017) Deepmetabolism: a deep learning system to predict phenotype from genome sequencing. arXiv preprint arXiv:170503094
91. Ajjolli Nagaraja A, Fontaine N, Delsaut M, Charton P, Damour C, Offmann B, Grondin-Perez B, Cadet F (2019) Flux prediction using artificial neural network (ANN) for the upper part of glycolysis. PloS One 14(5):e0216,178
92. Occhipinti A, Eyassu F, Rahman TJ, Rahman PK, Angione C (2018) In silico engineering of pseudomonas metabolism reveals new biomarkers for increased biosurfactant production. PeerJ 6:e6046
93. Yaneske E, Angione C (2018) The poly-omics of ageing through individual-based metabolic modelling. BMC Bioinform 19(14):83–96
94. Yang JH, Wright SN, Hamblin M, McCloskey D, Alcantar MA, Schrübbers L, Lopatkin AJ, Satish S, Nili A, Palsson BO, et al. (2019) A white-box machine learning approach for revealing antibiotic mechanisms of action. Cell 177(6):1649–1661
95. Vijayakumar S, Rahman PKMSM, Angione C (2020) A hybrid flux balance analysis and machine learning pipeline elucidates the metabolic response of cyanobacteria to different growth conditions. iScience 23(12):101818
96. Kavvas ES, Yang L, Monk JM, Heckmann D, Palsson BO (2020) A biochemically-interpretable machine learning classifier for microbial GWAS. Nat Commun 11(1):1–11
97. Occhipinti A, Hamadi Y, Kugler H, Wintersteiger C, Yordanov B, Angione C (2020) Discovering essential multiple gene effects through large scale optimization: an application to human cancer metabolism. IEEE/ACM Trans Comput Biol Bioinform. https://doi.org/10.1109/TCBB.2020.2973386
98. Zhang J, Petersen SD, Radivojevic T, Ramirez A, Pérez-Manríquez A, Abeliuk E, Sánchez BJ, Costello Z, Chen Y, Fero MJ, et al. (2020) Combining mechanistic and machine learning models for predictive engineering and optimization of tryptophan metabolism. Nat Commun 11(1):1–13
99. Heirendt L, Arreckx S, Pfau T, Mendoza SN, Richelle A, Heinken A, Haraldsdóttir HS, Wachowiak J, Keating SM, Vlasov V, et al. (2019) Creation and analysis of biochemical constraint-based models using the cobra toolbox v. 3.0. Nat Protoc 14(3):639–702
100. Angione C, Conway M, Lió P (2016) Multiplex methods provide effective integration of multi-omic data in genome-scale models. BMC Bioinform 17(4):257–269
101. Tian M, Reed JL (2018) Integrating proteomic or transcriptomic data into metabolic models using linear bound flux balance analysis. Bioinformatics 34(22):3882–3888

Chapter 6

MITODYN: An Open Source Software for Quantitative Modeling of Mitochondrial and Cellular Energy Metabolic Flux Dynamics in Health and Disease

Vitaly A. Selivanov, Olga A. Zagubnaya, Carles Foguet, Yaroslav R. Nartsissov, and Marta Cascante

Abstract

Mitochondrial respiratory chain (RC) transforms the reductive power of NADH or FADH2 oxidation into a proton gradient between the matrix and cytosolic sides of the inner mitochondrial membrane, that ATP synthase uses to generate ATP. This process constitutes a bridge between carbohydrates' central metabolism and ATP-consuming cellular functions. Moreover, the RC is responsible for a large part of reactive oxygen species (ROS) generation that play signaling and oxidizing roles in cells. Mathematical methods and computational analysis are required to understand and predict the possible behavior of this metabolic system. Here we propose a software tool that helps to analyze individual steps of respiratory electron transport in their dynamics, thus deepening understanding of the mechanism of energy transformation and ROS generation in the RC. This software's core is a kinetic model of the RC represented by a system of ordinary differential equations (ODEs). This model enables the analysis of complex dynamic behavior of the RC, including multistationarity and oscillations. The proposed RC modeling method can be applied to study respiration and ROS generation in various organisms and naturally extended to explore carbohydrates' metabolism and linked metabolic processes.

Key words Electron transport chain, Respiratory complexes, Respiratory chain, Reactive oxygen species, ROS generation, Central energetic metabolism, Kinetic model, Ordinary differential equations

Abbreviations

CI	Respiratory complex I
CII	Respiratory complex II
CIII	Respiratory complex III
CIV	Respiratory complex IV
ODE	Ordinary differential equation
RC	Respiratory chain
ROS	Reactive oxygen species

Sonia Cortassa and Miguel A. Aon (eds.), *Computational Systems Biology in Medicine and Biotechnology: Methods and Protocols*, Methods in Molecular Biology, vol. 2399, https://doi.org/10.1007/978-1-0716-1831-8_6,

1 Introduction

Ordinary differential equations (ODEs) are traditionally used for modeling metabolic dynamics. Since certain parameters in the equations are unknown, their estimation has been performed with optimization procedures to fit model simulations to experimentally measured time courses of metabolites. This approach is widely used in different organs such as cardiac and skeletal muscle, liver, and brain [1, 2]. The metabolic pathways of cellular energetics are highly complex. They are regulated by specific enzymes, the amounts/activities of which depend on the particular tissue or cell type [3]. Indeed, because flux balance analysis is usually expressed in the form of a linear system, the nonuniqueness of the solution can be associated naturally with the null space of the system matrix [4]. Several constraints must be fulfilled to guarantee that the considered system is physiologically meaningful and energetically feasible. The methodological aspects of the algorithms applied to respiration in mitochondria are described in this work.

Due to the accessibility of redox centers to oxygen and appropriate oxidative-reduction potential, the mitochondrial electron transport chain is one of the major generators of ROS in most cell types [5]. Mammalian mitochondria can generate superoxide or hydrogen peroxide from at least 11 different sites associated with substrate catabolism and the electron transport chain, and these different sites have very distinct properties [6]. Indeed, ROS generation plays a role in the regulation of intracellular signaling cascades in, for example, endothelium, smooth muscle cells, cardiac myocytes [7, 8]. However, overproduction of these chemicals is in the background and accompanying process for different kinds of pathologies [9], especially in neurodegenerative diseases [10, 11]. Thus, the description of fluxes of redox reactions in mitochondria, including key metabolites, will be essential for biomedical research and the development of new therapeutic approaches.

It has been established that the functional structure of the respiratory chain (RC) is prone to multistationarity and sustained oscillations [12–14]. This finding essentially extends the currently accepted concept of the oxidative stress mechanism. Validation and further investigation of this novel mechanism would be of high practical interest for biomedical research and therapy. The kinetic model, which is at the core of the herein described software tool Mitodyn, is applicable for in-depth studies of nonlinear dynamic behavior in metabolic systems linked with respiration. Moreover, the algorithms described here are also applicable for mathematical description of different metabolic systems.

2 Materials

The open-source software tool "Mitodyn" supports the analysis of mitochondrial and cellular energy metabolism dynamics. Its binary, coded in C++, was tested in Linux Ubuntu in desktop and notebook computers. The code was compiled using the g++ compiler. However, it should work in any operating system that has a C++ compiler. Mitodyn code can be freely downloaded from https://github.com/seliv55/cell_mito. This website contains instructions for compiling and running the program.

The main function of Mitodyn is numerical solving of ODEs representing a kinetic model, comprising mitochondrial electron transport, biochemical reactions providing substrates for electron transport, oxidative phosphorylation, ATP consumption, and transport of some metabolites through the cellular and inner mitochondrial membranes. It uses the Fortran implementation of DASSL method for ODEs solving [15].

The software simulates the time course of model variables, including the concentrations of metabolites and redox states of respiratory complexes, ROS generation from various sites of the electron transport chain, mitochondrial membrane potential ($\Delta\psi$), and oxygen consumption (VO_2). Comparing these output variables with respective experimental measurements provides information about the underlying processes.

Mitodyn can calculate steady states as a function of a model parameter. This feature enables finding qualitative changes in the dynamic behavior of a complex system in the parametric space, by performing a series of simulations as a function of a parameter that is changed incrementally by a small value and using the latest steady state as initial value. In this way, it can locate dynamic behaviors where a continuous small change of a parameter leads to an abrupt, discontinuous, change of the steady state. This method allows to determine hysteresis intervals between two branches of steady states obtained as a function of a continuously decreasing or increasing parameter. In the case that two branches of steady states are located in the same interval of the continuation parameter, it is assumed that the discontinuity between the two branches of steady states corresponds to a branch of unstable steady-states, but Mitodyn does not find its exact location. Since such a technique can only locate branches of stable steady states within the same interval of the continuation parameter, the strict existence of bistability (i.e., two possible steady states for a single parametric set) can only be presumed since Mitodyn does not perform stability analysis, which can determine the location and type of bifurcation. A similar reasoning applies to the detection of oscillatory behavior [16] regarding the distinction between damped and sustained oscillations with Mitodyn.

3 Methods

3.1 Mitodyn Performance: A Detailed Respiratory Chain Model

Mitodyn is a tool designed to in-deep analysis of respiration and related processes: substrates supply, ROS generation. Therefore, the electron transport in the RC is represented in the model in detail. The model considers two places where electrons can enter the RC: NADH oxidation in complex I and succinate oxidation in complex II. Both oxidations result in ubiquinone (Q) reduction to ubiquinol (QH_2), which is then oxidized in complex III. The latter passes electrons further in the RC that ultimately reduces oxygen to H_2O in complex IV. Oxidation of substrates in complexes I, III, and IV is coupled with protons translocation from the matrix into the intermembrane space. The model accounts for 4 H^+ translocation in complex I per NADH molecule oxidized [17], and 2 H^+ translocation in complex III per QH_2 oxidized [18]. Complex IV translocates 2 H^+ per 2 cytochromes c oxidized [19]. Protons and other ions translocation across the inner mitochondrial membrane change the $\Delta\psi$, which is used to drive ATP synthesis. The model accounts for the translocation of 4 H^+ back to the mitochondrial matrix per one ATP synthesized from ADP and inorganic phosphate [20].

The model accounts that each respiratory complex is a composition of electron carriers. Each one-electron carrier can be in one of two redox states: reduced (0) or oxidized (1). A combination of carrier's redox states is referred to as a redox state of the complex. The fractions of redox states of respiratory complexes are model variables. Therefore, a complex containing N1 one-electron carriers can be in 2^{N1} states so that 2^{N1} variables represent it. A complex may include two-electron transporters, such as ubiquinone. In general, a two-electron carrier can be in one of four redox states: 00, 01, 10, 11. However, in our case, states 01 and 10 are indistinguishable. Therefore, the total number of its redox states is 3 and the total number of states of a complex containing N2 two-electron carriers is $2^{N1} \cdot 3^{N2}$. Such a consideration is implemented in the model representation of complex III. It results in many ODEs, and we simplified it for the model representation of complexes I and II. The basis of this respiratory chain model is described elsewhere [12–14]. Here the modifications and additions to that model are explained.

3.1.1 Respiratory Complex I

Complex I delivers reducing equivalents from NADH to Q, thus providing fuel for complex III, and uses the free energy to translocate four protons from the matrix to the intermembrane space:

$$NADH + H^+{}_{in} + Q + 4H^+{}_{in} \rightarrow NAD^+ + QH_2 + 4H^+{}_{out}$$

This overall process consists of a series of reactions between individual carriers. NADH initially transfers two electrons to the flavin mononucleotide (FMN), creating $FMNH_2$. These electrons are then transported via a series of iron-sulfur (FeS) clusters, and finally to coenzyme Q10 (Q, ubiquinone): NADH – FMN – N3 – N1b – N4 – N5 – N6a – N6b – N2 – Q, where Nx is a conventional name for iron-sulfur clusters. This electron flow induces conformational changes of the protein, causing four protons to be pumped out of the mitochondrial matrix [21]. Q accepts two electrons thus reducing to ubiquinol (QH_2). The model represents these electron transport reactions, as described next.

Quasi-equilibrium distribution of electrons in the relay of FeS centers. FMN delivers the electrons accepted from NADH further to the chain of FeS centers of complex I: N1a, N3, N1b, N4, N5, N6a, N6b, N2. FMN is located between N1a and N3 to deliver its two electrons to these two centers, in practical terms, simultaneously. The centers from N3 to N6b in the above list constitute a chain of carriers transporting electrons from reduced FMN to N2. The electrons taken from NADH are distributed between complex I carriers according to their midpoint potentials (*see* **Note 1**).

The rate constants of electron tunneling calculated using Marcus's expression with coefficients [22] indicates that all these reactions proceed in the submillisecond scale. Thus, we assumed them to be in fast equilibrium (Eqs. 1).

$$\begin{aligned} &\mathrm{FMNS} \times \mathrm{Nr}/(\mathrm{FMNH}^- \times \mathrm{No}) \\ &\quad = \mathrm{KFHN}(\mathrm{FMNH}^- + \mathrm{No} \leftrightarrow \mathrm{FMNS} + \mathrm{Nr}) \end{aligned} \tag{1a}$$

$$\begin{aligned} \mathrm{FMN} \quad &\times \mathrm{Nr}/(\mathrm{FMNS} \times \mathrm{No}) \\ &= \mathrm{KFN}(\mathrm{FMNS} + \mathrm{No} \leftrightarrow \mathrm{FMN} + \mathrm{Nr}) \end{aligned} \tag{1b}$$

$$\begin{aligned} &\mathrm{FMN} \times \mathrm{N1ar}/(\mathrm{FMNS} \times \mathrm{N1ao}) \\ &\quad = \mathrm{KFN1a}(\mathrm{FMNS} + \mathrm{N1ao} \leftrightarrow \mathrm{FMN} + \mathrm{N1ar}) \end{aligned} \tag{1c}$$

$$\mathrm{No} \times \mathrm{N2r}/(\mathrm{Nr} \times \mathrm{N2o}) = \mathrm{KNN2}(\mathrm{Nr} + \mathrm{N2o} \leftrightarrow \mathrm{No} + \mathrm{N2r}) \tag{1d}$$

Mass balances:

$$\mathrm{FMNH}^- + \mathrm{FMNS} + \mathrm{FMN} = 1 \tag{1e}$$

$$\mathrm{N1ao} + \mathrm{N1ar} = 1 \tag{1f}$$

$$\mathrm{No} + \mathrm{Nr} = 4 \tag{1g}$$

$$\mathrm{N2o} + \mathrm{N2r} = 1 \tag{1h}$$

$$\mathrm{FMNH}^- \times 2 + \mathrm{FMNS} + \mathrm{N1ar} + \mathrm{Nr} + \mathrm{N2} = \#\mathrm{e}^- \tag{1i}$$

Here $FMNH^-$, FMNS, and FMN are the concentrations of reduced, semiquinone, and oxidized forms of FMN respectively, N1a and N2 are the concentrations of the respective centers where

Table 1
The main determinants of the distribution of redox states of complex I electron carriers: midpoint potential (Em), redox reaction where the carrier participates, ΔG of the reaction, equilibrium constant, distance between the participating carriers, forward and reverse rate constants

Center	Em (V)	Reaction	ΔG(eV)	$K = k_f/k_r$	D (Å)	k_f (s^{-1})	k_r (s^{-1})
FMN second e	−0.293	FMNH-N3	−0.043	5.35	7.6	9.12E7	1.70E7
FMN first e	−0.389	FMNS-N3	−0.139	226.58	7.6	4.64E8	2.05E6
N1a	−0.38	FMN-N1a	−0.009	1.42	12.3	6.70E4	4.71E4
N3	−0.25	N3-N1b	0	1.0	11	3.47E5	3.47E5
N1b	−0.25	N1b-N4	0	1.0	10.7	5.28E5	5.28E5
N4	−0.25	N4-N5	0	1.0	8.5	1.15E7	1.15E7
N5	−0.25	N5-N6a	0	1.0	14	5.21E3	5.21E3
N6a	−0.25	N6a-N6b	0	1.0	9.4	3.26E6	3.26E6
N6b	−0.25	N6b-N2	−0.15	348.02	10.5	9.50E6	2.73E4
N2	−0.1	N2-Q1	−0.1	49.48	12	5.23E5	1.06E4
Q first e	0	N2-Q2	0.02	0.458	12	5.77E4	1.26E5
Q second e	−0.12						

the indexes "o" and "r" mean "oxidized" and "reduced," N is the total amount of FeS centers from N3 to N6b.

Equation (1g) assumes that the chain from N3 to N6b can contain four electrons as maximum so that the reduction state varies between 0 and 4.

Equation (1i) quantifies the total amount of electrons that the chain from FMN to N2 carries. By convention, this fast equilibrium group can contain eight or less electrons. The number of electrons is referred to as redox state of this group. Since it can contain any integer number of electrons between 0 and 8, nine redox states are possible for this fast equilibrium group.

Numerically solving system Eqs. (1a)–(1i) with the equilibrium constants indicated in Table 1 gives the relative concentrations of oxidized and reduced states of the carriers within the fast equilibrium group for its various redox states (number of electrons ($\#e^-$) remaining in the group), as Table 2 shows.

The chain of complex I carriers, containing only the fast equilibrium group of centers, without bound Q is referred to as the complex I core or CI. The nine redox states of CI are the state variables of the model. CI with bound Q is referred to as CIQ. Since Q can be in any of the three states carrying 0, 1, or 2 valency electrons, the total number of redox states of complex I with bound Q is $9 \times 3 = 27$. The redox states of CIQ are referred to as Fig. 1 shows. The first nine positions in the array of redox states

Table 2
The distribution of redox states in the system of complex I electron carriers assumed to be in fast equilibrium

#e⁻	FMNH	FMNS	N1o	Nr	N2r
0	0	0	1	0	0
1	5.4E−07	0.00011	0.99983	0.10025	0.89946
2	9.3E−05	0.00148	0.9979	1.0047	0.9915
3	0.00081	0.00435	0.99382	1.9907	0.9971
4	0.00652	0.0123	0.9825	2.9581	0.99898
5	0.09896	0.04474	0.93089	3.6884	0.99975
6	0.43792	0.07126	0.829	3.8819	0.99991
7	0.82025	0.05012	0.64545	3.9548	0.99996
8	1	0	0	4	1

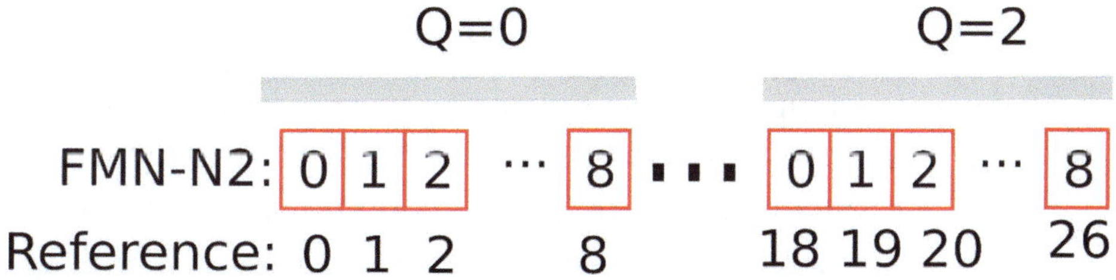

Fig. 1 A scheme of organization of the array of complex I redox states concentrations. 'fe' states for the fast equilibrium group. An explanation is in the text

correspond to the redox states of the fast equilibrium group with reduced Q, next nine positions are for the group with Q carrying 1 electron (SQ), and the last nine for the group with Q carrying 2 electrons (QH_2). In the case of CIQ, for each of the three states of bound Q, the core part of CIQ has the same probability of being reduced or oxidized. Thus, Table 2 remains valid for the states of CIQ.

The electron transport in complex I starts from transferring hydride ion from NADH to FMN [23, 24] (*see* **Note 2**). The electrons accepted by FMN are transferred through the relay of iron-sulfur centers to the center N2, where bound ubiquinone (Q) can be subsequently reduced to semiquinone radical (SQ) and then to ubiquinol (QH_2). This process is linked with four protons translocation from the matrix to the intermembrane space (*see* **Note 3**).

Reduced QH_2 dissociates from complex I and transfers the reduced equivalents further to complex III. The liberated place can be occupied by Q, oxidized by complex III (*see* **Note 4**).

3.1.2 Respiratory Complex II

The reaction catalyzed by complex II is:

$$\text{succinate} + Q \rightarrow \text{fumarate} + QH_2$$

Respiratory complex II, also referred to as succinate dehydrogenase (SDH), oxidizes succinate to fumarate, reducing Q to QH_2. Ubiquinol is an energy source for complex III that, by using this energy, translocates protons from the matrix to the intermembrane space. Complex II contains several electron carriers. Flavin adenine nucleotide (FAD) accepts two electrons from succinate and passes them one by one to the nearest iron-sulfur center, 2Fe2S. Then the electrons are tunneled subsequently to 4Fe4S and 3Fe4S centers. Cytochrome b can also accept an electron. The model accounts for these four centers to be in fast equilibrium with bound FAD representing the core of complex II (CII) that, in turn, can bind ubiquinone. The ubiquinone binding site is close to the 3Fe4S center. If ubiquinone is attached (referred below as CIIQ), it accepts two electrons one by one from the latter iron–sulfur center.

The distance between the redox centers forming the electron transport chain in complex II is less than 10 angstroms, which enables fast tunneling. The model also accounts for the distribution of redox states of the iron–sulfur (FeS) centers and cytochrome b, which are assumed in fast equilibrium. The following equations describe the fast equilibrium in the chain of centers:

$$\text{fs1o} \times \text{fs2r}/(\text{fs1r} \times \text{fs2o}) = \text{K1}(\text{fs1r} + \text{fs2o} \leftrightarrow \text{fs1o} + \text{fs2r}) \quad (2a)$$

$$\text{fs2o} \times \text{fs3r}/(\text{fs2r} \times \text{fs3o}) = \text{K2}(\text{fs2r} + \text{fs3o} \leftrightarrow \text{fs2o} + \text{fs3r}) \quad (2b)$$

$$\text{fs3o} \times \text{cytbr}/(\text{fs3r} \times \text{cytbo}) = \text{K3}(\text{fs3r} + \text{cytbo} \leftrightarrow \text{fs3o} + \text{cytbr}) \quad (2c)$$

$$\text{fs1o} + \text{fs1r} = 1; \text{fs2o} + \text{fs2r} = 1; \text{fs3o} + \text{fs3r} = 1; \text{cytbo} + \text{cytbr} = 1 \quad (2d)$$

Sum of electrons:

$$\text{fs1r} + \text{fs2r} + \text{fs3r} + \text{cytbr} = \#e- \quad (2e)$$

Here the symbols fs1, fs2, fs3, and cytb state for reduced (r) and oxidized (o) forms of centers 2Fe2S, 4Fe4S, 3Fe4S, and cytochrome b, respectively. The equilibrium constants K1 – K3 are calculated based on the midpoint potentials [25] given in Table 5.

Numerically solving the system of Eqs. (2a)–(2e) gives the relative concentrations of oxidized and reduced states of the carriers of whole complex II with bound ubiquinone for a various number of electrons (#e-) remaining in the system, as Table 6 shows.

The number of valency electrons present in the fast equilibrium group defines individual carriers' redox states. Four carriers, in principle, can keep four electrons. However, as Table 6 shows, when the number of valency electrons in this group reaches 3, the centers 2Fe2S and 3Fe4S, which interact with $FADH_2$ and Q, are almost entirely reduced. For simplicity, we limited in the model the maximal number of valency electrons to 3. Thus, the fast equilibrium group can be in one of four redox states (containing 0 to 3 electrons). Since FAD can be in one of 3 states, having 0, 1, or 2 electrons, their combination with the fast equilibrium group gives 12 possible states for CII. Binding ubiquinone multiplies the number of redox states in CIIQ by three due to combination with states 0, 1, or 2 of Q.

References to various redox states of complex II in the model. A redox state of the CIIQ consists of three parts. First, three possible redox states of FAD, which correspond to 0, 1, or 2 valency electrons. Second, four possible states of the fast equilibrium group, which correspond to 0-3, valency electrons contained in this group. Third, three possible redox states of Q correspond to 0, 1, or 2 valency electrons. Various combinations of these states constitute redox states of CIIQ. The concentrations of these states, which are state variables of the model, represented as an array of nonnegative *numbers.* All the algorithms calculating derivatives for these variables use this array. Figure 2 shows the positions of the redox states in the array of the state variables.

The first three numbers are the FAD redox states 0, 1, 2 in combination with the state 0 of the fast equilibrium group, and the state 0 of Q. Next three numbers are the same states of FAD in combination with the state 1 of the fast equilibrium group, and the state 0 of Q, and so on. In this way, the first 12 numbers represent all possible combinations of FAD's and the fast equilibrium group's redox states and the state 0 of Q. The next 12 numbers represent the same combinations of the two parts with the state 1 of Q. Ultimately, the last 12 numbers complete all the possible combinations of redox states for all three parts of the CIIQ. The redox states of CII occupy 12 positions, like CIIQ with Q in the state 0.

Oxidation of succinate and electron transport from $FADH_2$ to FeS centers. Succinate oxidation is the first of the reactions catalyzed

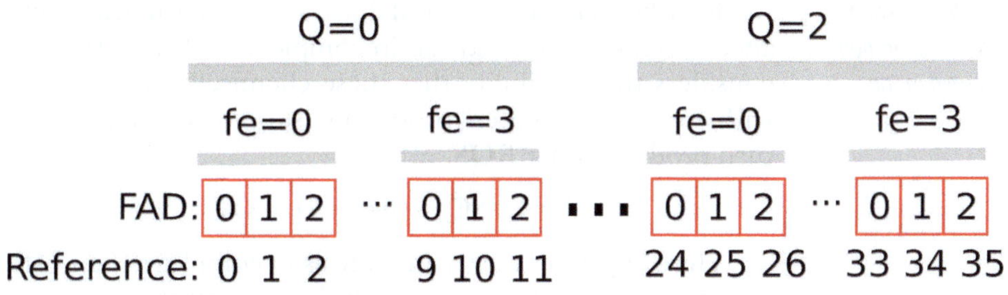

Fig. 2 A scheme of organization of the array of complex II redox states concentrations. 'fe' states for the fast equilibrium group. An explanation is in the text

by complex II (*see* **Note 5**). This reaction proceeds between free molecules, but further electron transport occurs within complex II. By performing this reaction, it changes its redox state, delivering electrons from FAD to the fast equilibrium group (*see* **Note 6**).

Electron transport from FeS centers to Q. The primary function of complex II is to reduce ubiquinone, thus providing fuel for complex III. When Q is bound, it receives electrons from the 3Fe4S center (*see* **Note 7**).

Dissociation of bound ubiquinol and binding next ubiquinone. After complete reduction of quinone, the produced quinol can dissociate from complex II (*see* **Note 8**). Then core of complex II can bind free quinone (*see* **Note 9**).

3.1.3 Respiratory Complex III

The reaction catalyzed by complex III is:

$$2QH_2 + 2Fe3^+ - \text{cyt C} + Q \rightarrow 2Fe^{2+} - \text{cyt C} + 2\ Q + QH_2$$

Implementation of this reaction in the model is described elsewhere [12–14]. The model employs in detail Mitchell's Q-cycle of ubiquinone oxidation coupled with the translocation of protons [31]. QH_2, produced by complex I or complex II, binds to the Qo site of complex III. The FeS center of the Rieske protein accepts one electron from bound QH_2 producing semiquinone anion radical (SQ) and releasing 2 H^+ to the intermembrane space. Then the Rieske protein delivers the electron accepted by the FeS center to cytochrome c1, which passes it further in the RC. The generated SQ provides its unpaired electron to cytochrome bL and dissociates, releasing the Qo site for the next QH_2. bL passes the accepted electron to bH. If Q is bound at the Qi site of complex III, it accepts the electron from bH. The SQ produced at the Qi site waits for another electron coming from another QH_2 bound at the Qo site. This process results in oxidizing two QH_2 molecules, reducing one Q molecule and two protons translocation.

However, there is a probability that the semiquinone radical at Qo site delivers its unpaired electron instead of bL directly to oxygen, producing superoxide radicals [32–34].

3.1.4 Reactive Oxygen Species (ROS) Generation as Implemented in the Model

As described above, the model accounts for the transient appearance of semiquinone radicals in complexes I, II and III. The model considers the possibility that these chemically active radicals react with molecular oxygen producing superoxide radicals, which in turn produce other ROS.

$$SQ + O_2 \rightarrow Q + O_2^-$$

Here SQ states for flavin radicals in complexes I and II, and for ubiquinone radicals formed in all three complexes.

3.2 Model Implementation

The directory containing the source code of the program with all necessary files for its compilation and running, and a brief user guide, can be downloaded (cloned) from the GitHub repository https://github.com/seliv55/cell_mito. It contains makefiles to make an executable program with g++ or equivalent.

The compiler creates an executable binary file "mitodyn.out". The repository has a script "mito.sh" that can be used to run Mitodyn. At runtime Mitodyn reads a file with the initial values of the state variables, and a file with the values of model parameters, and runs simulations of various modes (single simulation, continuation) specified in the script "mito.sh" (*see* **Note 10**). Examples of input files with the initial values of the state variables ("i1") and the model parameters ("1") are provided in the GitHub repository. An analysis can be started from the presented example files and then the parameters can be changed manually. The modified parameters can be saved in a different file and then used for subsequent analysis. The values of state variables obtained at the end of single simulation are saved as "i0" and then can be used as initial values for subsequent simulations. After performing a single simulation, Mitodyn saves the time course of variables of interest ($\Delta\psi$), and combinations of variables (sum of potentially ROS producing redox states) and functions of variables (VO_2) in a text file "dynamics." If GNUplot is installed, it plots the saved data executing the GNUplot script "gplt.p". The plot is saved in the file "./kin/dynamics.png".

Figure 3 shows an example of a plot of state variables $\Delta\psi$ (Fig. 3a), QH2 (Fig. 3b), ATP levels (Fig. 3c). The program calculates the reaction rates as functions of state variables as described above. Specifically, it calculates the electron flux (reaction rate) that flows from complex III to oxygen. This flow (taking into account that four electrons are needed to reduce O_2 to $2H_2O$) is shown in Fig. 3d.

Orange curves labeled as "high" characterizes some basal state with high rate of electron transport (Fig. 3d) that effectively maintains $\Delta\psi$ (Fig. 3a). ATP synthase compensates ATP consumption in this state (Fig. 3c). Ubiquinone in this state is mainly oxidized (QH_2 levels are very low, Fig. 3b).

The model accounts for the redox states of respiratory complexes as described above. Some states can be grouped to facilitate comparing the model output with experimental data. In this way, the fractions of the redox states that contain potentially ROS generating radicals can be shown in the output. Specifically, such radicals could be FMN semiquinone in complex I, FAD semiquinone in complex II, semiquinone at Qo site in complex III. The program calculates from the state variables the fractions of the states containing these radicals. When the radical is a part of a fast equilibrium group, the program sums the amounts of the states containing 1, 2, . . ., n electrons in the fast equilibrium group multiplied

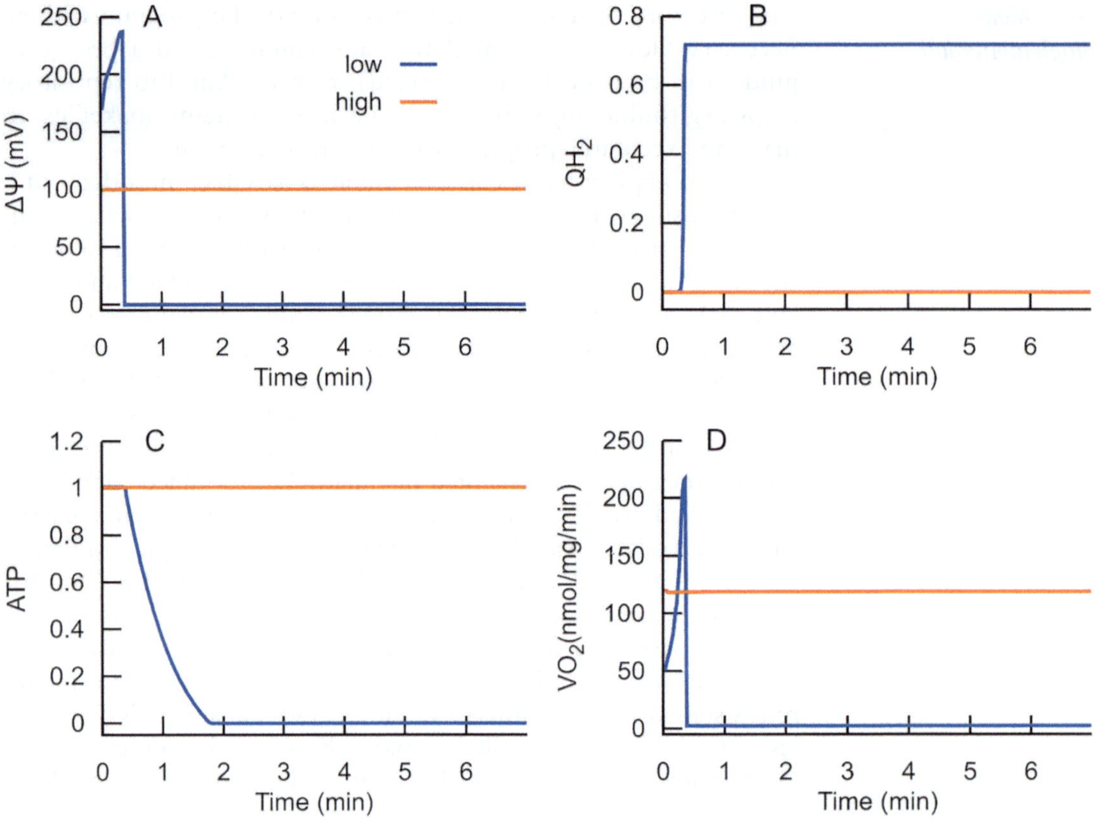

Fig. 3 Dynamics of state variables $\Delta\psi$ (**a**), ubiquinol (QH_2) (**b**), ATP (**c**), and a function of state variables, oxygen consumption (**d**). Calculations performed for some basal state when ATP synthesis compensates its consumption (orange curves labeled as "high") and ten-times decreased ATPase activity (blue curves labeled as "low"). The corresponding values of parameters and initial values of state variables are listed in the files "1" and "i1" respectively in https:/github.com/seliv55/cell_mito, where all files necessary for Mitodyn functioning are presented

by the percentage of the radical corresponding to the given number of electrons. It performs these calculations for all the states of other carriers that are not a part of the fast equilibrium group. For complex III, where the amount of all the redox states are simulated explicitly, the program sums the amounts of the states containing SQ radical at the Qo site. Figure 4a–c shows the dynamics of these potentially ROS producing radicals calculated as described above. According to this figure, the concentration of ROS generating radicals in complexes II and III is low in this basal state (orange curves), and only complex I makes a significant contribution to ROS generation. In this way, the program can show in output any redox states or their combinations.

The ten-times increase of the maximal capacity for ATPase activity compared to the basal state does not change qualitatively the characteristics shown by orange curves in Figs. 3 and 4 (data not shown).

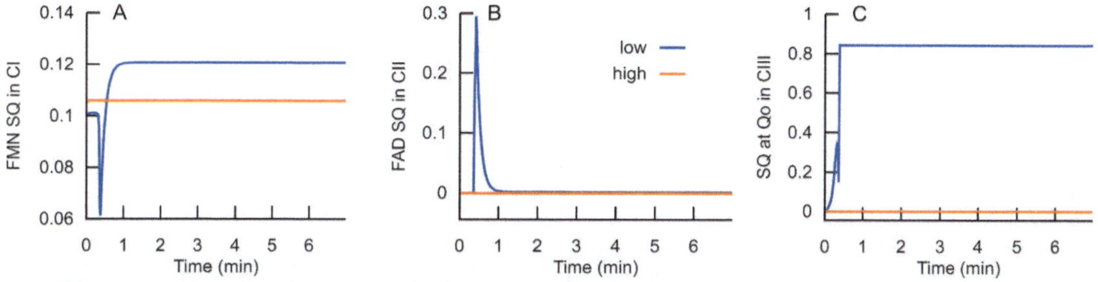

Fig. 4 Dynamics of combinations of redox states containing potentially ROS producing radicals calculated by Mitodyn. Time course of semiquinone radicals of FMN in complex I (**a**), semiquinone radicals of FAD in complex II (**b**), and semiquinone in Qo site of complex III (**c**) taken from the same simulations as shown in Fig. 3 (orange curves indicated as "high" for a basal state and blue curves indicated as "low" for decreased ATPase activity)

The blue curves labeled as "low" in Figs. 3 and 4 show the response of the system to ten-times decrease of ATP consumption compared to the basal state shown by the orange curves. A decrease of ATPase activity at an unchanged supply of substrates for the RC leads to an almost complete reduction of ubiquinol to QH_2 (Fig. 3b). The levels of the latter remain below 1 because a part of Q is bound to the respiratory complexes, but the concentration of free Q is close to 0. Such a reduction of ubiquinone results in immense augmentation of ROS generation in complex III (Fig. 4c), decrease of electron flow to oxygen (Fig. 3d) and, as a consequence, drop in $\Delta\psi$ (Fig. 3a) and ATP levels (Fig. 3c). Thus, the RC reduction, which can occur at a decreased ATPase activity and unchanged substrate supply, is more harmful to the cells than a high workload increase. Probably, there are mechanisms in living cells protecting them from such an over reduction of the RC.

A reason for the intracellular ATP levels drop is given by the presumed bistable behavior of the RC. The decrease of ATP demand results in an almost complete reduction of the ubiquinone pool and, as a consequence, Q deficiency. The latter is a reason of the RC switch from the state of high rate of electron transport (e.g., as indicated by orange curve in Fig. 3d) coupled with protons translocation and ATP synthesis to a state of low rate of electron transport (e.g., as indicated by blue curve in Fig. 3d) and significantly increased content of ROS generating radicals, especially in complex III (Fig. 4c).

Running Mitodyn in the mode of finding a series of steady states incrementing a parameter allows visualizing the branches of stable steady states existing at the same interval of ATP demand (*see* Subheading 2).

Performing iterations incrementing a parameter, it can find a point where any small change of a parameter leads to an abrupt, discontinuous, change of a steady state, that is, switch to a different branch of stable steady states. After reaching a border of a desired

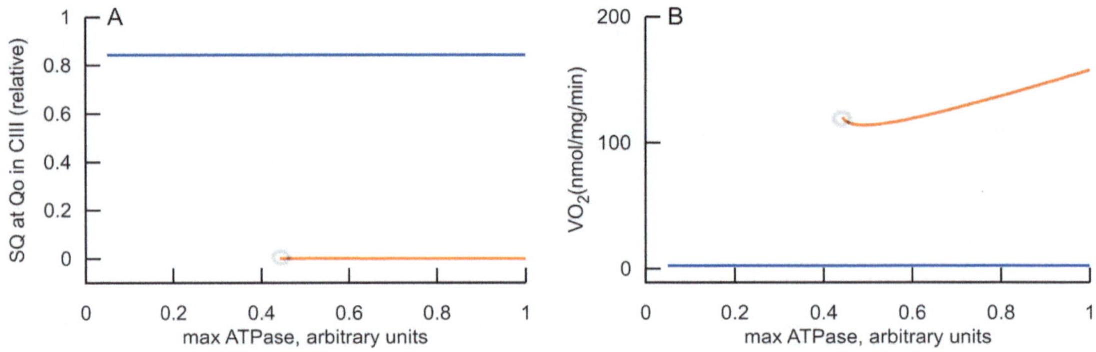

Fig. 5 Two branches of stable steady states revealed by running Mitodyn in "cont" mode. The steady states for semiquinone at Qo site of complex III (**a**) and oxygen consumption (**b**) are shown as functions of ATPase activity (parameter 48 in file "1"). The branch of stable steady states obtained by simulations gradually decreasing the parameter is shown in orange. The parameter value at which the system's dynamics shifts between branches of steady states is denoted by a circle. Further decrease of the parameter switches the system to the branch of stable steady states depicted in blue color. After subsequent increase of the parameter, the system remains in the blue branch

region of a parameter Mitodyn reverses the direction of the parameter change and finds steady states passing the same region of the parameter using the latest achieved steady state as initial value. It is possible that, in this way, Mitodyn finds a region of hysteresis, where reaching one or another stable steady-state with the same parametric set depends only on the initial point in the space of state variables. The explored parameter should be specified in the text file "cont" (*see* **Note 11**). Mitodyn saves all the stored values in the file "00000" and plots some of them in the file "cont.png" using GNUplot.

Figure 5 shows an example of such an analysis with Vmax of the ATPase reaction as a parameter subsequently decremented (incremented with a negative value) starting from some basal state with high rate of electron transport and low ROS generation rate (orange curves in Fig. 5). Decreasing the parameter to a value, corresponding to the point in the orange curve surrounded by a circle, leads to a discontinuous shift to a different branch of steady-states characterized by the sudden change in dynamic behavior (indicated by the blue curves in Fig. 5).

With further decrease and subsequent increase of the parameter the system remains in the branch of stable steady states indicated by blue curves in Fig. 5. The fact that Mitodyn did not find a parameter value where the system's dynamics return to the orange branch entails that the perturbation of other parameters will be needed. The two branches of stable steady states the Mitodyn captured are, presumably, separated by a branch of unstable steady states that Mitodyn is unable to calculate (*see* Subheading 2).

3.3 Conclusions

The RC is a subject of great interest due to its crucial role in energy metabolism, ROS signaling, and oxidative stress. Mathematical modeling is a widely used tool applied to study parts of metabolism that include the RC (e.g., [35, 36]). However, to our knowledge, Mitodyn is the first model that describes the dynamics of redox states of the first three respiratory complexes using a realistic and detailed approach. Our approach is realistic because it accounts for the real situation that the electron carriers of each unit of a complex are bound together and cannot interact with other units of the same type. Such an approach predicts possible complex nonlinear dynamic behavior of the RC, presumably bistability (Fig. 5) or oscillations [14].

Bistable dynamic behavior, involving sudden jumps in ROS generation, could underlie switching mechanisms between low and high ROS generation such as in hypoxia and reoxygenation after ischemic injury. This dynamic mechanism can give insights into therapeutic improvement of diseases related to oxidative stress, for example, the use of antioxidants. In this context, some antioxidants are biologically active reducing agents, thus, potentially, provoking switches between distinct modes of ROS generation. Moreover, because of similarities in the presence of electron carriers between respiration and photosynthetic systems, slight modifications of the model presented herein, can be used for studying energetic metabolism in plant systems and cyanobacteria.

4 Notes

1. *Midpoint potentials of complex I redox centers.* The values of midpoint potentials of complex I redox centers are summarized (reviewed in [25]) in the second column of Table 1. This table also shows ΔG for the reactions between interacting centers and values of equilibrium constants follow the corresponding midpoint potentials:

$$K = \exp\left((n \times F \times \Delta\mathrm{Em})/(R \times T)\right) \qquad (3)$$

Here ΔEm is the difference of midpoint potentials of product and substrate of reaction, n is the number of elementary charges transferred in the reaction, $R = 8.3$ J/(mol × K) is the ideal gas constant, T is the temperature in Kelvin scale, $F = 96{,}485$ C × mol^{-1} is Faraday's constant. F/RT is approximately 0.039 mV^{-1}.

The rate constants of forward (k_f) reactions are calculated using Marcus's expression with coefficients given in [22]: $\log10(k) = 13.0–0.6(D\text{-}3.6)\text{-}4.23(\Delta G + 0.7)^2/0.7$, where D is the edge to edge distance (in angstrom) between the interacting centers.

The backward constants (k_r) are calculated using the equilibrium (K) and forward constants.

2. *NADH oxidation.*

$$NADH + FMN \leftrightarrow NAD^+ + FMNH^- \quad (4)$$

The respective midpoint potentials (Em) are as follows.
for $NAD^+/NADH$ Em = − 320 mV [37];
for $FMN/FMNH^-$ Em = − 340 mV [37];

$$\Delta Em = Em(FMN/FMNH^-) - Em(NAD^+/NADH) \\ = -340 + 320 = -20 \text{ mV}.$$

Forward and reverse rate constants are linked according to the expression,

$$k_f = k_r \times \exp((n \times F \times \Delta Em)/(R \times T)) \quad (5)$$

where n is the number of electrons transported, F=96,485 c/mol is the Faraday constant, R = 8.3 J/(mol × K) is the gas constant, and T = 298°K is temperature. With the defined above ΔEm, Eq. (5) gives

$$kNADH = kNAD \times \exp(2 \times 0.039 \times (-20)) \\ \approx kNAD \times 0.21 \quad (6)$$

The distribution of redox states shown in Table 2, together with the rate constants, determine the electron transfer rate from NADH to complex I, where the electrons accepted from NADH are distributed among the above listed group of redox centers that are in fast equilibrium. It is proportional to the probability of finding the first electron acceptor, FMN, in an oxidized state (fourth column in Table 2), which depends on the number of electrons (i) already confined within the carriers' chain:

$$v[i] = kNADH \times CI[i] \times PFMN[i] \times [NADH] - kNAD \\ \times CI[i+2] \times PFMNH[i+2] \times [NAD] \quad (7)$$

Here k is the rate constant defined by Eq. (2a), i = 0, . . ., nCI-2 is the reference number of the redox state of core species (CI), which corresponds to the number of electrons confined within the chain of CI carriers. nCI is the maximal number of electrons. CI[i] is the concentration of the redox states referred to by i, PFMN[i] = 1- PFMNS[i]- PFMNH[i] is the probability to find oxidized (FMN) form when the fast equilibrium group contains i electrons. PFMNS[i] and PFMNH[i] are the probabilities to find semiquinone or completely reduced forms respectively; these probabilities are presented in the second and third columns of Table 2.

Equation (7) describes NADH oxidation rates by the core of complex I (CI). The same principle is used to describe

NADH oxidation by CIQ. Still, in this case, Eq. (7) should be applied to the species containing ubiquinone at different degrees of oxidation, with the number of electrons varying within the range 0 to 2.

$$v[i+9\times j] = \text{kNADH}\times \text{CIQ}[i+9\times j]\times \text{PFMN}[i]\times[\text{NADH}]-\text{kNAD}\times \text{CIQ}[i+2+9\times j]\times \text{PFMNH}[i+2]\times[\text{NAD}] \quad (8)$$

Here $j = 0, 1, 2$ is the number of electrons harbored by the quinone bound to CI. The other terms are the same as described above (*see* Eq. 7).

The rates defined in Eq. (7) contribute to the derivatives for the concentrations of CI[i] and CI[i + 2]:

$$\begin{aligned} &d(\text{CI}[i])_{\text{NADH}}/dt = -v[i];\\ &d(\text{CI}[i+2]_{\text{NADH}}/dt = v[i]; \end{aligned} \quad (9)$$

Here $d(\text{CI}[i])_{\text{NADH}}/dt$ represents the contribution of NADH oxidation to the total pool of CI[i].

Similarly, the rates defined in Eq. (8) contribute to the concentrations of CIQ[$i + 9 \times j$] and CIQ[$i + 2 + 9 \times j$]:

$$\begin{aligned} &d(\text{CIQ}[i+9\times j])_{\text{NADH}}/dt = -v[i+9\times j];\\ &d(\text{CIQ}[i+2+9\times j]_{\text{NADH}}/dt = v[i+9\times j]; \end{aligned} \quad (10)$$

The contribution of NADH oxidation by complex I into total NADH pool is given by the sum of all the rates defined in Eqs. (9) and (10).

$$d\,[\text{NADH}]_{\text{NADH}}/dt = -\sum_i v[i] - \sum_i\sum_j v[i+9\times j] \quad (11)$$

3. *Reduction of ubiquinone (Q) bound to complex I in the proximity of FeS cluster N2.*

   The overall process is described by:

$$\text{N2–Q} + 2e + 2H^+ \rightarrow \text{N2–QH}_2 \quad (12)$$

   The process is divided into two one-electron reduction steps of ubiquinone, first one.

$\text{Em}(Q/Q^-) = 0$ mV [26];

$\Delta\text{Em1} = \text{Em}(Q/Q^-) - \text{Em}(\text{N2o/N2r}) = 0 + 100 = 100$ mV.

Rate constants $k_{\text{N2Q1}} = k_{\text{QN21}} \times \exp(0.039 \times 100) = 50 \times k_{\text{QN21}}$

and second one

$\text{Em} = -120$ mV [26];

$\Delta\text{Em2} = \text{Em}(Q^-/\text{QH2}) - \text{Em}(\text{N2o/N2r}) = -120 + 100 = -20$ mV.

Rate constants $k_{\text{N2Q2}} = k_{\text{QN22}} \times \exp(0.039 \times (-20)) = 0.5 \times k_{\text{QN22}}$

(13)

K is the equilibrium constant calculated using Eq. (5). k_f is the rate constant in forward direction calculated using Marcus's expression with coefficients given in [22]:

$\log 10(k) = 13.0–0.6(D-3.6)-4.23(\Delta G + 0.7)^2/0.7$, where *D* is the edge to edge distance between the interacting centers.

The rates of quinone reduction depend on the probability of N2 center reduction, which, in turn, depends on the number of electrons in the fast equilibrium group.

$$v_1[i] = k_{N2Q1} \times CIQ[i] \times PN2r[i] - k_{QN21} \times CIQ[i-1+9] \times PN2o[i-1] \quad (14)$$

$$v_2[i] = k_{N2Q2} \times CIQ[i+9] \times PN2r[i] - k_{QN22} \times CIQ[i-1+9\times 2] \times PN2o[i-1] \quad (15)$$

Here $CIQ[i + 9 \times j]$ is the concentration of CIQ species containing $i = 0, \ldots, 8$ valency electrons in the fast equilibrium group, which are bound with quinone containing $j = 0, 1, 2$ valency electrons. $PN2r[i]$ is the probability to find N2 center in reduced form when the fast equilibrium group contains *i* electrons (column N2r in Table 2). $PN2o[i] = 1\text{-}PN2r[i]$ is the probability to find N2 center in oxidized form. *k* denotes the rate constants of the electron transition shown in Table 3.

Each of the two reductions of Q were assumed to couple electron transfer with translocation of 2 H^+ from matrix to the intermembrane space thus changing the mitochondrial membrane potential ($\Delta\Psi$) by four elementary charges.

$$d(\Delta\Psi)_{N2Q}/dt = \sum_i (F/C\tilde{n}4\tilde{n}(v_1[i] + v_2[i]) \quad (16)$$

Here $F = 96{,}485\ C \times mol^{-1}$ is Faraday's constant; *C* is electric capacitance of the membrane.

$F/C = 500\ mV \times nmol^{-1} \times$ mg protein. $d(\Delta\Psi)_{N2Q}$ is the contribution of quinone reduction in complex I in total derivative $d(\Delta\Psi)$

4. *Dissociation of bound ubiquinol and ubiquinone binding.*

   After complete reduction of quinone, the produced quinol can dissociate from complex I:

$$CIQH_2 \leftrightarrow core + QH_2 \quad (17)$$

Table 3
Characteristics of first and second ubiquinone reduction by N2 center implemented in the model

	Em(V)	ΔG(eV)	$K = k_f/k_r$	distance (Å)	$k_f(s^{-1})$	$k_r(s^{-1})$
First reduction (Q → Q-)	0	−0.1	49.5	12	523,000	10,500
2nd reduction(Q- → QH2)	−0.12	0.02	0.46	12	58,000	126,000

Table 4
Rate constants of quinone binding/dissociation in complex I

QH_2 dis	197,409	s^{-1}
QH_2 bnd	38,814	$(nmol/mg)^{-1} \cdot s^{-1}$
Q bnd	114,778	$(nmol/mg)^{-1} \cdot s^{-1}$
Q dis	5800	s^{-1}

The rate of this reaction is expressed as follows:

$$v_d[i] = k_{df} \times CIQ[i + 9 \times 2] - k_{dr} \times CI[i] \times [QH_2] \qquad (18)$$

Here k denotes the rate constants of QH_2 dissociation from CIQ and its binding to CI. Their values accepted in the model are shown in Table 4. CIQ[i + 9 × 2] is the concentration of complex I with bound QH_2 containing $i = 0, \ldots, 8$ valency electrons in the fast equilibrium group; CI[ì] concentration of the core containing i electrons in the fast equilibrium group, $[QH_2]$ is the concentration of free QH_2.

This rate changes the derivatives for the concentrations of $CIQ[i + 9 \times 2]$ and $CI[i]$

$$\begin{aligned} d(CIQ[i + 9 \times 2])_{CIdis}/dt = -v_d[i] \\ d(CI[i]_{CIdis}/dt = v_d[i] \end{aligned} \qquad (19)$$

The contribution of quinol dissociation in total derivative of QH_2 is:

$$d(QH_2)_{CIdis}/dt = \sum_i v_d[i] \qquad (20)$$

Ubiquinone binding. The form of complex I core can bind

$$\text{free quinonecore} + Q \leftrightarrow CIQ \qquad (21)$$

The rate of this reaction is expressed as follows:

$$v_b[i] = k_{bf} \times [Q] \times CI[i] - k_{br} \times CIQ[i] \qquad (22)$$

here k are the rate constants of Q association and dissociation from complex I, and their values are shown in Table 4. CI $[i]$ is the concentrations of core containing $i = 0, \ldots, 8$ electrons in the fast equilibrium group, $CIQ[i]$ is the concentrations of CIQ containing i electrons in the fast equilibrium group with bound Q.

The concentrations of $CIQ[i]$ and $CI[i]$ are given by:

$$\begin{aligned} d(CIQ[i])CIbin/dt = v_b[i] \\ d(CI[i]CIbin/dt = -v_b[i] \end{aligned} \qquad (23)$$

The contribution of quinone binding in total derivative of Q is:

$$d(Q)CIbin/dt = -\sum_i v_b[i] \quad (24)$$

5. *Oxidation of succinate.*

$$\text{Succinate} + \text{FAD} \leftrightarrow \text{fumarate} + \text{FADH2} \quad (25)$$

Em value for succinate/fumarate redox couple is +0.025 V (as reviewed in [25]). Em for FAD/$FADH_2$ is −0.08 V [38]. Equilibrium constant calculated based on this data is:

$$K = k_f/k_r = \text{FAD} \times \text{succinate}/(\text{FADH2} \times \text{fumarate}) = 0.017$$

The values of forward and reverse rate constants in the model correspond to this equilibrium constant. Two electron reduction of FAD leads to the change of its redox state from 0 to 2. These states appear several times combined with various redox states of the fast equilibrium group and Q if it is bound (*see* Fig. 2). Specifically, the reference numbers i0 for the state 0 are $i0 = i \times 3$, where $i = 0,1,2,3$ is the redox state (number of valency electrons) of fast equilibrium group. The reference numbers of reduced $FADH_2$ are $i \times 3 + 2$. With this definition of the reference numbers, the reaction rates for the individual redox states of complex II core are as follows.

$$v[i] = k_f \times CII[i0] \times [suc] - k_r \times CII[i0 + 2] \times [fum] \quad (26)$$

Here k_f and k_r are the rate constants of forward and reverse reactions.

Calculating the reference numbers of the redox states complex II with bound quinone, which participate in this reaction, it is necessary to account for the states 0 of FAD that appears in combination with various Q states. $i0(i, j) = 4 \times 3 \times j + i \times 3$, where $i = 0, 1, 2, 3$ are the redox states of the fast equilibrium group, $j = 0, 1, 2$ are the redox states of Q. The expression for the individual rates is similar to Eq. (26):

$$v[i,j] = k_f \times CIIQ[i0(i,j)] \times [suc] - k_r \times CIIQ[i0(i,j) + 2] \times [fum] \quad (27)$$

The rates defined in Eqs. (26) and (27) contribute to the derivatives for the concentrations of the redox states of complex II participating in the respective reactions, succinate, and fumarate. Such a contribution of succinate oxidation is as follows.

$$\begin{aligned}
&d\,CII[i0]suc/dt = -v[i];\\
&d\,CII[i0 + 2]suc/dt = v[i];\\
&d\,CIIQ[i0]suc/dt = -v[i,j];\\
&d\,CIIQ[i0 + 2]suc/dt = v[i,j];\\
&d\,[suc]_{CII}/dt = -\sum i\, v[i] - \sum i \sum j\, v[i,j];\\
&d\,[fum]_{CII}/dt = \sum i\, v[i] + \sum i \sum j\, v[i,j];
\end{aligned} \quad (28)$$

Table 5
Midpoint potentials of the redox centers of complex II

Redox center	Em7 first reduction	Em7 second reduction	Reference
FAD	– 0.127	– 0.031	[26]
2Fe2S (fs1)	0.00		[27]
4Fe4S (fs2)	– 0.260		[27]
3Fe4S (fs3)	0.06		[28]
Cyt b	−0.185		[29]
Q	0.096	0.172	[30]

Here the references i0, i, j and the rates $v[i]$, $v[i,j]$ are defined above in Eqs. (26) and (27), the index 'suc' indicates that it is a contribution to the derivative from succinate oxidation. The index 'CII' indicates a contribution to the derivative from oxidation by complex II.

6. *Electron transport from* $FADH_2$ *to FeS centers.*

The primary acceptor of $FADH_2$ electrons is the 2Fe2S center. According to the data shown in Table 5, ΔEm for the first electron transition from $FADH_2$ to 2Fe2S center is 0-(−0.031) = 0.031. For the second electron transition ΔEm = 0-(−0.127) = 0.127. These values determine the respective equilibrium constants $K_1 = k_{f1}/k_{r1} = 3.35$, $K_2 = k_{f2}/k_{r2} = 141.9$.

This electron transport changes the redox state of complex II from the state referred to as isub to the state referred to as iprod according to the convention outlined in Fig. 2. For the first electron transition FAD is in the redox state 2 and the reference isub(i) = 2 + i × nfad, where nfad = 3 is the total number of redox states of FAD and i = 0,1,. . ., nfs-2 where nfs = 4 is the number of redox states in the fast equilibrium group of FeS centers. For the second electron transition FAD is in the redox state 1 and isub(i) = 1 + i × nfad. The reference iprod(i) is isub(i) + nfad-1 because obtaining one electron by the fast equilibrium group is reflected in the model by shifting the current position in the array of concentrations by nfad positions to the right (*see* Fig. 2). Since FAD loses one electron, the position shifts one position to the left. With these designations, the rate of this electron transport for the CII forms is:

$$v[i] = k_f \times \mathrm{CII}[\mathrm{isub}(i)] \times \mathrm{Pfs2o}[i] - k_r \times \mathrm{CII}[\mathrm{iprod}(i)] \times \mathrm{Pfs2r}[i+1] \quad (29)$$

Here the reference $k_f = k_{f1}$ and $k_r = k_{r1}$ for the first electron transition, and $k_f = k_{f2}$ and $k_r = k_{r2}$ for the second electron transition. CII[isub] is the concentration of the core of

Table 6
The probabilities of complex II electron carriers to be reduced. First column shows the number of electrons that contains the fast equilibrium group of complex II electron carriers, and the other columns show the corresponding probabilities for the individual carriers to be reduced

#e⁻	2Fe2S	4Fe4S	3Fe4S	Cyt b
0	0	0	0	0
1	0.1351	1.14E-4	0.5718	0.2929
2	0.44567	5.89E-4	0.873	0.6807
3	0.97849	0.0323	0.9974	0.9918
4	1	1	1	1

complex II in the state isub. Pfs2o[i] = 1-Pfs2r[i] is a probability of the 2Fe2S center to be oxidized. Column 2 in Table 6 shows the probabilities Pfs2r[i] for the 2Fe2S center to be reduced when the whole fast equilibrium group contains various numbers of electrons shown in column 1.

The rate of this electron transport for the CIIQ is:

$$v[i,j] = k_f \times \text{CIIQ}[\text{isub}(i,j)] \times \text{Pfs2o}[i] - k_r \times \text{CIIQ}[\text{iprod}(i,j)] \times \text{Pfs2r}[i+1] \quad (30)$$

Here $j = 0, 1, 2$ is the redox state ubiuinone, the reference isub(i,j) = 2 + i × nfad + j × nfs × nfad for the first electron transition (FAD is in the redox state 2), and isub(i, j) = 1 + i × nfad + j × nfs × nfad for the second electron transition (FAD is in the redox state 1); iprod(i,j) = isub(i, j) + nfad-1; CIIQ[isub] is the concentration of the core of complex II in the state isub. Other designations as in Eq. (29).

7. *Electron transport from FeS centers to Q.*

According to the data shown in Table 6, ΔEm for the first electron transition from 3Fe4S center to Q is 0.096–0.06 = 0.03. For the second electron transition ΔEm = 0.172–0.06 = 0.112. These values determine the respective equilibrium constants $K_1 = k_{f1}/k_{r1} = 3.22$, $K_2 = k_{f2}/k_{r2} = 79.0$.

This electron transport changes the state of CIIQ from the state referred to as isub(i,j,k) = i + j × nfad+k × nfad×nfs to the state iprod(i,j,k) = isub(i,j,k) + nfs × nfad-nfad, where $i = 0, 1, 2$ is the reference to the redox state of FAD, $j = 0, 1, 2, 3$ is the reference to the redox state of the group of FeS centers, nfad = 3 is the number of FAD redox centers, k is the reference to the redox state of ubiquinone ($k = 0$ for the first electron transition and $k = 1$ for the second one), nfs = 4 is the number of redox states in the group of fast equilibrium FeS centers. The references isub and iprod reflect the fact that the transition transfers one electron to Q (shifts the state of CIIQ by nfs × nfad positions to the right), taking one electron from the FeS group (shifts the state of CIIQ by nfad positions to the left, *see* Fig. 2).

The rate of this electron transport for the complex II with bound Q is

$$v[i] = k_f \times \mathrm{CIIQ}[\mathrm{isub}(i,j,k)] \times \mathrm{Pfs3r}[i] - k_r \times \mathrm{CIIQ}[\mathrm{iprod}(i,j,k)] \times \mathrm{Pfs3o}[i-1] \quad (31)$$

Here $k_f = k_{f1}$ and $k_r = k_{r1}$ for the first electron transition, $k_f = k_{f2}$ and $k_r = k_{r2}$ for the second electron transition (Q is in the redox state 1). CIIQ[isub(i,j,k)] is the concentration of complex II in the state isub. CIIQ[iprod(i,j,k)] is the concentration of complex II in the state iprod. Pfs3o[i] = 1-Pfsrr[i] is a probability of the 3Fe4S center to be oxidized. Column 2 in Table 6 shows the probabilities Pfs3r[i] for the 3Fe4S center to be reduced when the whole fast equilibrium group contains various numbers of electrons shown in column 1.

8. *Dissociation of bound ubiquinol* ($CIIQH_2 \leftrightarrow CII + QH_2$).

   The rate of this reaction is expressed as follows.

$$v_d[i] = k_{df} \times \mathrm{CIIQ}[i + 2 \times \mathrm{nfad} \times \mathrm{nfs}] - k_{dr} \times \mathrm{CII}[i] \times [\mathrm{QH_2}] \quad (32)$$

Here k_{dr} and k_{df} are the rate constants of QH2 dissociation from and its binding to complex II, nfad = 3, and nfs = 4 are the quantities of redox states of FAD and FeS centers, i = 0,1,2, ...,nfad×nfs-1 is the reference to the redox states of complex II. i + 2 × nfad×nfs refers to the species with Q in state 2 (*see* Fig. 2). CII[i] is the concentration of respective states of complex II core, [QH_2] is the concentration of free QH_2.

The contribution of rates described in Eq. (32) into the derivatives of complex II variables is:

$$\mathrm{d\,CIIQ}[i + 2 \times \mathrm{nfad} \times \mathrm{nfs}]_{\mathrm{dis}}/\mathrm{d}t = -v_d[i]$$

$$\mathrm{d\,CII}[i]_{\mathrm{dis}}/\mathrm{d}t = v_d[i] \quad (33)$$

$$\mathrm{d\,[QH2]_{cIIdis}}/\mathrm{d}t = \sum\nolimits_i v_d[i]$$

Here the reference i = 0,1, ..., nfad×nfs-1, indexes "dis," and "cIIdis" refer to the contribution of QH_2 dissociation into the derivatives of respective variables.

9. *Binding of next ubiquinone.*

   The core of complex I can bind free quinone:

$$\mathrm{CII} + \mathrm{Q} \leftrightarrow \mathrm{CIIQ} \quad (34)$$

The rate of this reaction is expressed as follows:

$$v_b = k_{bf} \times [Q] \times CII[i] - k_{br} \times CIIQ[i] \quad (35)$$

Here k_{bf} and k_{br} are the rate constants of Q binding to complex II and dissociation from it. CII[i] is the concentrations of complex II core in the redox state referred to as $i = 0$, 1,. . ., nfad×nfs-1. CIIQ[i] is the respective concentrations of CIIQ containing Q in the state 0.

This rate contributes to the derivatives for the model variables:

$$\begin{aligned} &\mathrm{d}\ CII[i]_{Qbind}/\mathrm{d}t = -v_b \\ &\mathrm{d}\ CIIQ[i]_{Qbind}/\mathrm{d}t = v_b \\ &\mathrm{d}\ [Q]_{CIIbind}/\mathrm{d}t = -\sum\nolimits_i v_b \end{aligned} \quad (36)$$

Here $i = 0, 1, \ldots,$ nfad×nfs-1. The indexes "Qbind" and "CIIbind" refer to the contribution of Q binding to complex II.

10. *Script "mito.sh" that can be used to run Mitodyn.*

```
#!/bin/sh
edata="a011110rbm.txt" #file with experimental data
init="i1" #file with initial values
par="1" #file with set of parameters
mode="0" #
tst=yes
while getopts ":e:i:p:m:" opt; do
 case $opt in
 i) init=$OPTARG;;
 p) par=$OPTARG;;
 m) mode=$OPTARG;;
 *)
 echo "Valid options: -e edata, -i init value, -p parameters,
-m mode"
 cat help
 tst=no
 ;;
 esac
done
if [ $tst = yes ]
then
./mitodyn.out $edata $init $par $mode
fi
```

In this script the following information should be provided:

- Path to a file with initial values for all variables, an example is provided in a file "i1."
- Path to a file with a set of parameters, an example is provided in a file "1."
- Mode of desirable functioning.

If mode = "0," Mitodyn makes a single simulation and stops.

"r" also produces a single simulation but assuming inhibition of complex I by rotenone.

"cont" produces a series of simulations incrementing a parameter in some interval starting from the point achieved in the previous simulation (shortened word "continuation"). The file "cont" specifies the number of points in the series, the number of a chosen parameter in the array of parameters, the starting value of the parameter, and the interval's size. This file existing in the root directory contains an example of such specification.

"rc" is the same as "cont" but assuming inhibition of complex I by rotenone.

The command

./mito.sh.

executes Mitodyn with default values of the parameters. The script could be edited to change the default behavior, or it could be executed with different desirable options.

Example: ./mito.sh -i i0 -m cont.

Here the character "-" indicates option, which could be "i," "p," or "m." A set of characters following the option specifies a path to initial values file, parameters file or mode of execution correspondingly. The possible modes of execution are listed above.

11. *Parameters for running Mitodyn in the mode "cont".*

    A parameter chosen for continuation, the interval of its change, number of intermediate points inside the interval should be specified in the text file "cont" in the Mitodyn directory. Here is an example of such a file:

    parameter 48

    points 150

    start 0.0002

    interval -0.00019.

It indicates that the continuation parameter is #48 in the list of parameters, presented currently in the file "1," which Mitodyn reads. According to the name and comment for this parameter, also given in file "1," it is the rate constant for ATPase activity. Mitodyn should make simulations in 30 points evenly distributed inside the interval chosen for this parameter. It should start at 1. The length of the interval is 0.9. A negative sign means that, first, a parameter will be decrementing from 1 to 0.1, and then incrementing from 0,1 to 1.

Acknowledgments

This work was supported by the Spanish Ministerio de Economia y Competitividad (MINECO)-European Commission FEDER funds—"Una manera de hacer Europa" (SAF2017-89673-R and PID2020-115051RB-I00), the Agència de Gestió d'Ajuts Universitaris i de Recerca (AGAUR) Generalitat de Catalunya (2017SGR1033), CIBERehd (CB17/04/00023) (ISCIII, Spain), M. C also received support through the prize "ICREA Academia" for excellence in research, funded by ICREA foundation–Generalitat de Catalunya. Part of the work was performed according to Memorandum of Understanding between Universitat de Barcelona and Institute of cytochemistry and molecular pharmacology (04/02/2020).

Author contributions: VAS developed algorithms, analyzed the data, and wrote and edited the text; OAZ analyzed data and wrote and edited the text; CF edited the text and analyzed data; YRN and MC designed the work, wrote the text, and analyzed data.

References

1. Somersalo E, Cheng Y, Calvetti D (2012) The metabolism of neurons and astrocytes through mathematical models. Ann Biomed Eng 40 (11):2328–2344
2. Strutz J, Martin J, Greene J, Broadbelt L, Tyo K (2019) Metabolic kinetic modeling provides insight into complex biological questions, but hurdles remain. Curr Opin Biotechnol 59:24–30
3. Dash RK, Li Y, Kim J, Saidel GM, Cabrera ME (2008) Modeling cellular metabolism and energetics in skeletal muscle: large-scale parameter estimation and sensitivity analysis. IEEE Trans Biomed Eng 55(4):1298–1318
4. Calvetti D, Cheng Y, Somersalo E (2015) A spatially distributed computational model of brain cellular metabolism. J Theor Biol 376:48–65
5. Mazat JP, Devin A, Ransac S (2020) Modelling mitochondrial ROS production by the respiratory chain. Cell Mol Life Sci 77(3):455–465
6. Brand MD (2016) Mitochondrial generation of superoxide and hydrogen peroxide as the source of mitochondrial redox signaling. Free Radic Biol Med 100:14–31
7. Dröge W (2002) Free radicals in the physiological control of cell function. Physiol Rev 82 (1):47–95
8. Sies H, Berndt C, Jones DP (2017) Oxidative stress. Annu Rev Biochem 86(1):715–748
9. Sies H (2018) On the history of oxidative stress: concept and some aspects of current development. Curr Opin Toxicol 7:122–126
10. Zhao M, Zhu P, Fujino M, Zhuang J, Guo H, Sheikh I, Zhao L, Li X-K (2016) Oxidative stress in hypoxic-ischemic encephalopathy: molecular mechanisms and therapeutic strategies. Int J Mol Sci 17(12):2078
11. Patel M (2016) Targeting oxidative stress in central nervous system disorders. Trends Pharmacol Sci 37(9):768–778
12. Selivanov VA, Votyakova TV, Zeak JA, Trucco M, Roca J, Cascante M (2009) Bistability of mitochondrial respiration underlies paradoxical reactive oxygen species generation induced by anoxia. PLoS Comput Biol 5(12)
13. Selivanov VA, Votyakova TV, Pivtoraiko VN, Zeak J, Sukhomlin T, Trucco M, Roca J, Cascante M (2011) Reactive oxygen species production by forward and reverse electron fluxes in the mitochondrial respiratory chain. PLoS Comput Biol 7(3):e1001115
14. Selivanov VA, Cascante M, Friedman M, Schumaker MF, Trucco M, Votyakova TV (2012) Multistationary and oscillatory modes of free radicals generation by the mitochondrial respiratory chain revealed by a bifurcation analysis. PLoS Comput Biol 8(9)
15. Petzold L (1981) An efficient numerical method for highly oscillatory ordinary differential equations. SIAM J Numer Anal 18

(3):455–479. https://doi.org/10.1137/0718030
16. Marsden J, McCracken M (1976) Hopf bifurcation and its applications. Springer-Verlag, New York
17. Jones AJY, Blaza JN, Varghese F, Hirst J (2017) Respiratory complex I in *Bos taurus* and *Paracoccus denitrificans* pumps four protons across the membrane for every NADH oxidized. J Biol Chem 292(12):4987–4995
18. Cocco T, Lorusso M, Di Paola M, Minuto M, Papa S (1992) Characteristics of energy-linked proton translocation in liposome reconstituted bovine cytochrome bc1 complex: influence of the protonmotive force on the H+/e− stoichiometry. Eur J Biochem 209 (1):475–481
19. Berg J, Liu J, Svahn E, Ferguson-Miller S, Brzezinski P (1861) Structural changes at the surface of cytochrome c oxidase alter the proton pumping stoichiometry. Biochim Biophys Acta Bioenerg 2020(2):148116
20. Turina P, Samoray D, Gräber P (2003) H+/ATP ratio of proton transport-coupled ATP synthesis and hydrolysis catalysed by CFOF1-liposomes. EMBO J 22(3):418–426
21. Sazanov LA (2015) A giant molecular proton pump: structure and mechanism of respiratory complex I. Nat Rev Mol Cell Biol 16 (6):375–388
22. Crofts AR, Rose S (2007) Marcus treatment of endergonic reactions: a commentary. Biochim Biophys Acta Bioenerg 1767(10):1228–1232
23. Berrisford JM, Sazanov LA (2009) Structural basis for the mechanism of respiratory complex I. J Biol Chem 284(43):29773–29783
24. Sazanov LA, Hinchliffe P (2006) Structure of the hydrophilic domain of respiratory complex I from Thermus thermophilus. Science 311 (5766):1430–1436
25. Moser CC, Farid TA, Chobot SE, Dutton PL (2006) Electron tunneling chains of mitochondria. Biochim Biophys Acta Bioenerg 1757 (9–10):1096–1109
26. Ohnishi T, King TE, Salerno JC, Blum H, Bowyer JR, Maida T (1981) Thermodynamic and electron paramagnetic resonance characterization of flavin in succinate dehydrogenase. J Biol Chem 256(11):5577–5582
27. Ohnishi T, Salerno JC, Winter DB (1976) Thermodynamic and EPR characteristics of two ferredoxin type iron sulfur centers in the succinate ubiquinone reductase segment of the respiratory chain. J Biol Chem 251 (7):2094–2104
28. Ohnishi T, Lim J, Winter DB, King TE (1976) Thermodynamic and EPR characteristics of a HiPIP type iron sulfur center in the succinate dehydrogenase of the respiratory chain. J Biol Chem 251(7):2105–2109
29. Yu L, Xu JX, Haley PE, Yu CA (1987) Properties of bovine heart mitochondrial cytochrome b560. J Biol Chem 262(3):1137–1143
30. Ohnishi T, Ragan CI, Hatefi Y (1985) EPR studies of iron-sulfur clusters in isolated subunits and subfractions of NADH-ubiquinone oxidoreductase. J Biol Chem 260 (5):2782–2788
31. Mitchell P (1976) Possible molecular mechanisms of the protonmotive function of cytochrome systems. J Theor Biol 62(2):327–367
32. Ohnishi T, Trumpower BL (1980) Differential effects of antimycin on ubisemiquinone bound in different environments in isolated succinate·cytochrome c reductase complex. J Biol Chem 255(8):3278–3284
33. Chandel NS, Maltepe E, Goldwasser E, Mathieu CE, Simon MC, Schumacker PT (1998) Mitochondrial reactive oxygen species trigger hypoxia-induced transcription. Proc Natl Acad Sci U S A 95(20):11715–11720
34. Bell EL, Klimova TA, Eisenbart J, Schumacker PT, Chandel NS (2007) Mitochondrial reactive oxygen species trigger hypoxia-inducible factor-dependent extension of the replicative life span during hypoxia. Mol Cell Biol 27 (16):5737–5745
35. Heiske M, Letellier T, Klipp E (2017) Comprehensive mathematical model of oxidative phosphorylation valid for physiological and pathological conditions. FEBS J 284 (17):2802–2828
36. Smith AC, Eyassu F, Mazat JP, Robinson AJ (2017) MitoCore: a curated constraint-based model for simulating human central metabolism. BMC Syst Biol 11(1):114
37. Ohnishi T (1998) Iron-sulfur clusters/semiquinones in Complex I. Biochim Biophys Acta Bioenerg 1364(2):186–206
38. Rich PR, Maréchal A (2010) The mitochondrial respiratory chain. Essays Biochem 47:1–23
39. Calvetti D, Cheng Y, Somersalo E (2016) Uncertainty quantification in flux balance analysis of spatially lumped and distributed models of neuron–astrocyte metabolism. J Math Biol 73(6–7):1823–1849

Chapter 7

Integrated Multiomics, Bioinformatics, and Computational Modeling Approaches to Central Metabolism in Organs

Sonia Cortassa, Pierre Villon, Steven J. Sollott, and Miguel A. Aon

Abstract

Data-driven research led by computational systems biology methods, encompassing bioinformatics of multiomics datasets and mathematical modeling, are critical for discovery. Herein, we describe a multiomics (metabolomics–fluxomics) approach as applied to heart function in diabetes. The methodology presented has general applicability and enables the quantification of the fluxome or set of metabolic fluxes from cytoplasmic and mitochondrial compartments in central catabolic pathways of glucose and fatty acids. Additionally, we present, for the first time, a general method to reduce the dimension of detailed kinetic, and in general stoichiometric models of metabolic networks at the steady state, to facilitate their optimization and avoid numerical problems. Representative results illustrate the powerful mechanistic insights that can be gained from this integrative and quantitative methodology.

Key words Metabolomics, Fluxomics, Glucose and fatty acids catabolism, Heart, Diabetes, Kinetic modeling

1 Introduction

Central catabolism constitutes the universal biochemical backbone from unicellular to multicellular organisms, prokaryotes to eukaryotes, supplying energy, redox intermediates, and key carbon precursor metabolites that, ultimately, lead to the synthesis of new cells, the maintenance of cellular structures, and the resistance to stressful conditions, among other functions. Systems biology computational methods, encompassing bioinformatic analysis and mathematical models, are critical for interpreting the mechanisms underlying the integrated, metabolic dynamics associated with organs function, within and between them [1, 2]. At the organ level, heart function is a particularly challenging case because of its constantly changing dynamics. To analyze heart performance from a metabolic standpoint, herein we combine experimental and computational approaches comprising metabolomics data,

Sonia Cortassa and Miguel A. Aon (eds.), *Computational Systems Biology in Medicine and Biotechnology: Methods and Protocols*, Methods in Molecular Biology, vol. 2399, https://doi.org/10.1007/978-1-0716-1831-8_7,

bioinformatics and kinetic modeling. Using this methodological approach, we determine the fluxome or set of metabolic fluxes giving rise to the heart's metabolome from central catabolic pathways of glucose and fatty acids degradation [2].

Specifically, the methodology presented herein represents an experimental-computational approach to integrate multiomics data (metabolomics–fluxomics) with bioinformatic analysis and computational modeling of central catabolism. A primary goal of our procedure is to enable the translation of high throughput metabolites profile (metabolome) into the set of metabolic fluxes (fluxome) from which the metabolome emerged, as quantitatively described by a computational kinetic model [3]. Results presented here illustrate the powerful mechanistic insights that can be gained from this quantitative and integrative process.

Our approach combines an analytical platform comprising several integrated quantitative methodologies with a detailed computational kinetic model. Overall, the main advantage of our methodology of fluxomic determination over other metabolic flux balance approaches [4, 5] relies on the integration with the computational model that encompasses regulatory information (e.g., effectors, feedbacks), which is crucial for nonlinear dynamic systems. Since this analytical procedure, spanning from metabolome to fluxome, is widely applicable, virtually to any kind of kinetic model, we also present a general method for reducing the dimension of detailed kinetic models of metabolic networks at the steady state, to facilitate their optimization. Heart function is an ideal prototypic system to test the usefulness of the approach proposed herein. Since the heart is constantly changing in response to energy demand, the choice of kinetic modeling is justified because it accounts for time-dependent behavior.

2 Materials

Herein, we focus on new methodological developments while briefly recapitulating some of the methods that have been described in previous volumes of this series [6, 7].

2.1 Metabolite Profiling and Bioinformatic Analyses

1. Biological material. Its collection for metabolomics analysis should come from a physiological steady state or pseudo–steady state. Steady state refers to a condition characterized by constancy in the levels of intermediary metabolites and in the fluxes occurring through the metabolic network. It is an idealized condition for a system that, like the heart, is undergoing permanent changes in activity and organs that are known to be entrained by circadian rhythms. However, a pseudo–steady state is feasible if we know that the average behavior is sustained (*see* **Note 1**).

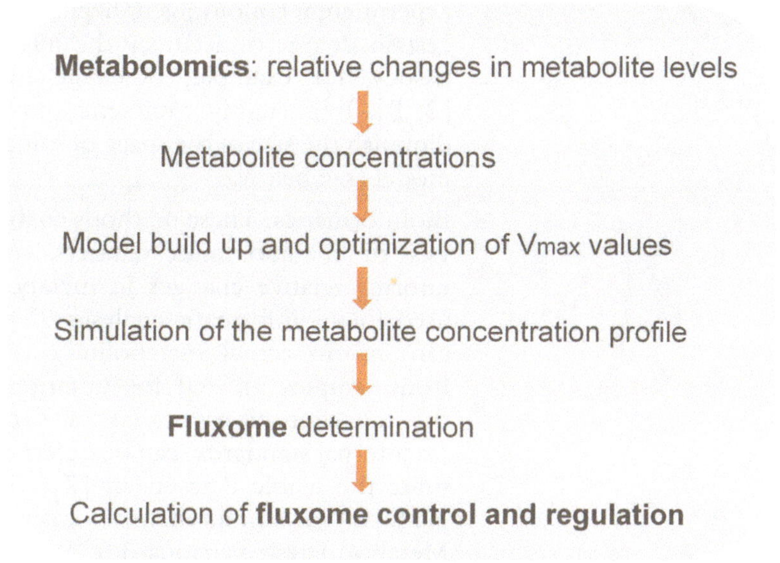

Fig. 1 Scheme of the workflow leading from metabolite profiling to the fluxome and its control and regulation. (Reproduced from Cortassa, Caceres, Bell, O'Rourke, Paolocci, and Aon (2015) *Biophys. J.* 108, 163–172)

2. Experimental data. Model parameters in each of the kinetic expressions, usually correspond to in vitro conditions, which can be obtained from the specialized literature and databases, such as BRENDA (https://www.brenda-enzymes.org). *Metabolomics data*, targeted or untargeted, represent the first step in our work-flow diagram leading to the fluxome determination (Fig. 1). *Targeted* metabolite profiling can inform actual metabolite concentrations and it would be the ideal data source for the analysis described here. However, *untargeted* metabolomics, which informs relative changes in metabolite levels, could be used for the purpose of this fluxomics method if the mass spectrometry (MS) signal can be calibrated. Sample analysis for metabolomics data can be outsourced. Empirically, a minimal size of at least six samples of the biological material per treatment, genotype, or physiological condition, is recommended. In Cortassa et al. [3], the authors used internal standards for calibration with enzymatic determinations (available Sigma kits) and those values used for calibration of the area under the MS peaks. Quantification according to this procedure assumes that the area under the peak is proportional to the mass of the molecular ion impacting the MS detector [7].
3. Experimental values of reference. The present method utilizes experimentally determined metabolite concentration as input for the model equations. However, and given the biological variability of flux values under different physiological conditions, we need "reference fluxes" that apply to the actual

experimental conditions utilized, to be able to calculate with a certain degree of accuracy the internal fluxes within cells or tissues, for example, exchange of substrates or products [2, 8]. Importantly, "reference fluxes" are also necessary to diminish the size of the space of solutions of the computational model (see below).

4. Bioinformatics. These methods enable an initial unbiased analysis of metabolomics datasets. *Untargeted* metabolomics informs relative changes in metabolite levels, and these are used for bioinformatic analysis. *Targeted* metabolite profiling can inform actual metabolite concentrations. Alternatively, from samples utilized for untargeted metabolites profiling, the concentration of a few representative metabolites (as internal standards) can be determined and utilized to quantitate the whole dataset (*see* [7]). The integrated web-based platform for comprehensive analysis of metabolomics data MetaboAnalyst, versions 3.0 [9] and 4.0 [10] can be used. Individual, univariate, metabolites analysis with Prism 8.0 (GraphPad Software, San Diego, CA, USA) can be also utilized as complementary information for specific applications.

2.2 Computational Tools

1. To build up and work with the mathematical model, Matlab®, Wolfram's Mathematica, or any piece of software designed to solve systems of ordinary differential equations by numerical integration, is required. Some useful computational modeling packages have been developed for software such as Matlab. Among them the graphical package MatCont that enables simulation of time-dependent behavior, calculation of dynamic stability [11–13], and parameter sensitivity of the model [14].
2. Elementary flux modes (EFM) analysis that requires the software Metatool (e.g., version 5.1 for Matlab [15], is freely available for academic use (http://pinguin.biologie.uni-jena.de/bioinformatik/networks/metatool/metatool5.1/meta tool5.1.html). Metatool enables the computation of structural properties of biochemical reaction networks while informing about the system's capacity, as designed, to achieve a steady state or if there are missing components.

3 Methods

3.1 Metabolomics Analysis

The relative level of metabolites and their sense of variation (i.e., up- or downmodulation) can be determined using univariate (ANOVA), and multivariate (principal component [PCA] and partial least square discriminant [PLSD] analyses, volcano plots, heat maps, clustering, correlation matrix, pattern search) statistical techniques. A main goal is to assess the set of metabolites responsible for

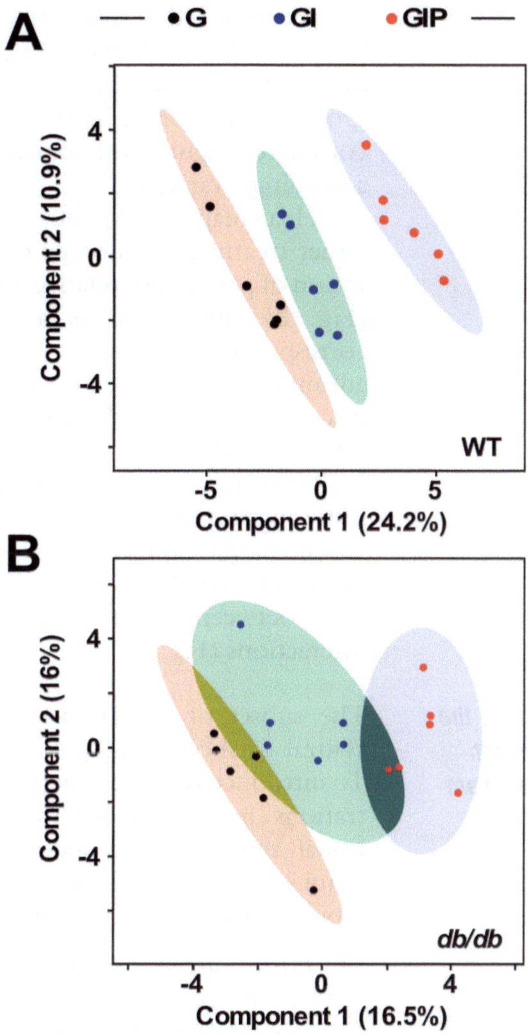

Fig. 2 Partial least square discriminant (PLSD) analysis of WT and diabetic heart metabolomes. PLSD analysis is a cross-validated multivariate supervised clustering/classification method from MetaboAnalyst that we used to determine the extent of separation afforded by a subset of 43 metabolites that exhibited significant changes in response to different treatments (G, GI, GIP) within each group (WT, *db/db*). (Reproduced from Cortassa, Caceres, Tocchetti, Bernier, de Cabo, Paolocci, Sollott, and Aon (2020) *J. Physiol.* 598.7, 1393–1415)

the separation between experimental groups (e.g., genotype, disease, treatment) as can be judged by PCA or PLSDA (Fig. 2).

Comprehensive metabolomics analysis was performed with MetaboAnalyst versions 3.0 [9] and 4.0 [10]. Metabolites were normalized using the autoscaling function of MetaboAnalyst preceded by detection and removal of outliers, which were excluded from the statistics when above or below 1.5 times the interquartile range comprised between the 75 and 25% percentiles, respectively [16]. Combining PLSD and two-way ANOVA analyses, we determined the metabolites responsible for the separation between groups as indicated in the main text. Metabolic pathways were considered significantly enriched at log $p \geq 1.3$ ($p \leq 0.05$), and accordingly ranked.

Heat maps of the set of significantly changed metabolites by either treatment, genotype or their interaction, allows the visualization of patterns of relative abundance or depletion of metabolites (Fig. 3), and serve as a guide for their mapping into the corresponding pathways (Fig. 4). Figure 4 also shows a profound advantage of untargeted vs. targeted metabolomics, that is, the discovery of new pathways related, in this case, to redox modulation of cardiac contractility in diabetes, specifically, methionine cycle and transsulfuration routes leading to major redox couples buffering systems such as glutathione, cysteine, taurine. The correlation matrix captures the existence of clusters of positively and negatively correlated metabolites, offering an overview of interrelations between pathways and novel insights into their functional interactions (Fig. 5).

3.2 Computing the Fluxome Through Central Metabolism

The subset of significantly changed metabolites mapping onto central catabolism (Fig. 4) is subjected to further quantification. To introduce realistic experimental constraints, metabolite concentrations (in molar units) are needed to parameterize the computational model. More specifically, metabolite concentrations are required to calculate the fluxes through the metabolic network as inputs of the rate expressions from the kinetic model (*see* [3]).

1. *Linear optimization of Vmax values.* Vmax optimization is performed from metabolomics data to accurately represent model behavior under the specific conditions of the experimental design. The method involves inserting metabolites concentration into the rate expressions from the kinetic model [7] to solve and optimize the model's $V_{\max}$ values at the steady state for all metabolic steps taken into account. The solution involves finding a minimum (or maximum) of the objective function z corresponding to the following box problem, that is, an optimization problem with bounded solutions:

$$\boldsymbol{S}_{\mathbf{t}} \cdot \mathbf{D}_{\mathbf{r}}\mathbf{v} \cdot \boldsymbol{v}_{\mathbf{max}} = \boldsymbol{b}_{\mathbf{t}} \quad (1)$$

With $\boldsymbol{S_t}$ representing the stoichiometry matrix, that is, the matrix of all corresponding stoichiometric coefficients to the chemical reactions of the metabolic network, organized as m rows (metabolites) and n columns (reactions); $\mathbf{D_r v}$ is the matrix of the derivatives of the rate expressions with respect to the $V_{\max}$ of each reaction in the network; $\boldsymbol{v}_{\mathbf{max}}$ is the vector of maximal rates of each reaction step, and $\boldsymbol{b_t}$ are fluxes of demand

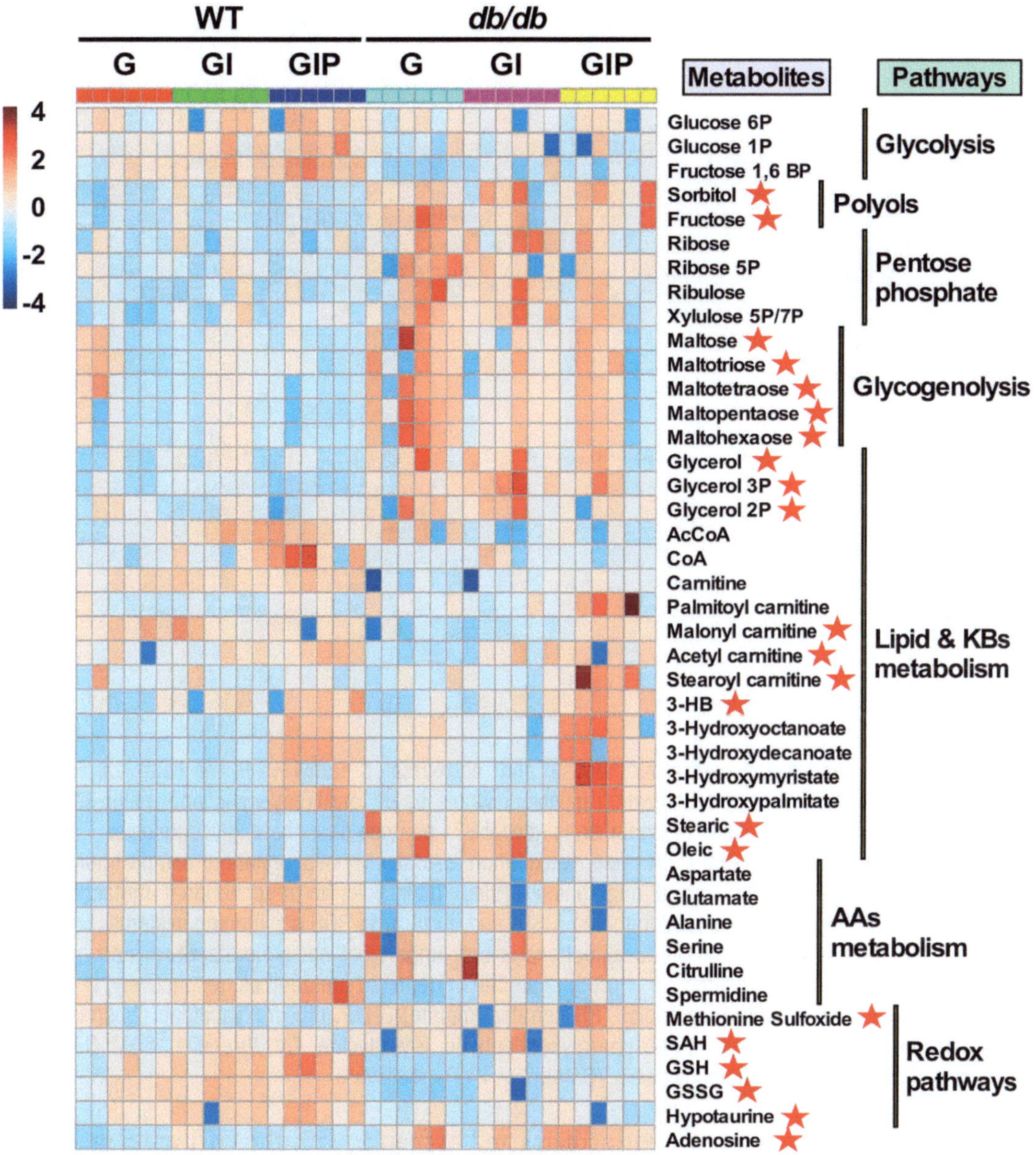

Fig. 3 Heat map of significantly changed metabolites in the heart metabolomes from WT and diabetic mice. The relative abundance of intermediates from different pathways in WT and *db*/*db* hearts is displayed under the different experimental conditions assayed, as described in the text. Depicted is the heat map of the normalized levels of 43 metabolites mainly responsible for the separation between groups (WT, *db*/*db*) and treatments (G, GI, GIP) (*see* Fig. 2) from central glucose and FA degradation pathways, along with redox-related pathways such as methionine cycle and transsulfuration routes. The heat map was constructed using the web-based resource MetaboAnalyst 3.0. Graded pseudocolors brown and blue correspond to metabolites accumulation or depletion, respectively, according to the scale placed at the left of the plot. AA, amino acid; GSH, reduced glutathione; GSSG, oxidized glutathione; 3-HB, 3-hydroxybutyrate; KB, ketone body; SAH, S-adenosylhomocysteine. (Reproduced from Cortassa, Caceres, Tocchetti, Bernier, de Cabo, Paolocci, Sollott, and Aon (2020) *J. Physiol.* 598.7, 1393–1415)

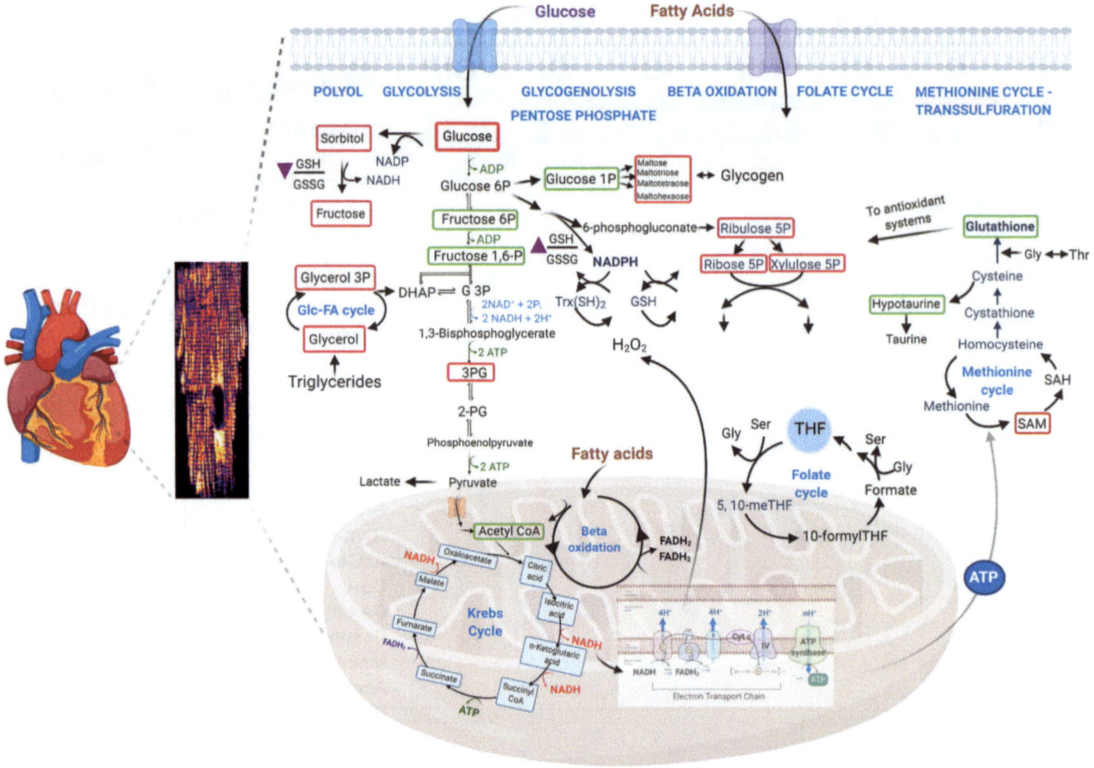

Fig. 4 Metabolites mapping in central metabolism and redox-related pathways in the heart. Depicted is a schematic view of the levels of organization involved, whole heart, cardiomyocyte, and major subcellular pathways from central metabolism, in cytoplasmic (glycolysis, pentose phosphate, glycogenolysis, glucose–-fatty acid cycle) and mitochondrial (TCA cycle, ß-oxidation, oxidative phosphorylation) compartments. Additionally, displayed are the folate and methionine cycles, their links to mitochondrial metabolism and transsulfuration pathways leading to glutathione (GSH) and, in turn, to ROS scavenging systems, both in the cytoplasm and mitochondrial matrix (i.e., mitochondria only import but not generate GSH). Red and green rectangles correspond to significantly ($p < 0.05$) abundant or depleted metabolites, respectively, in the heart of diabetic over WT mice. Key to symbols: THF, tetrahydrofolate; DHAP, dihydroxyacetone phosphate; G3P, glyceraldehyde 3 phosphate; 3PG, 3 phospho-glyceraldehyde; Gly, glycine; Thr, threonine; SAM, S-adenosyl methionine; SAH, S-adenosyl homocysteine. (Created at BioRender.com)

of intermediates and exchange with the environment, such as the transport of a carbon substrate or the efflux of a metabolic product. A description of the maximal error associated with the $V_{\max}$ [17] and metabolites determinations are included (*see* **Notes 2** and **3**, respectively). For a more general treatment of the fluxomics calculation procedure *see* **Note 4**.

Equation 1 is constrained by $\boldsymbol{v}_{\mathbf{min}} \leq \boldsymbol{v}_{\mathbf{max}} \leq \boldsymbol{v}_{\mathbf{M}}$ with $v_{\min}$ and v_{M} being the vectors of minimal and maximal possible values of $v_{\max}$, respectively, and satisfies the objective function, $z(v_{\max}) = c^T.v_{\max}$, with c being the vector of costs. Usually, the components of the vector $v_{\min}$ are zeros.

2. The vector $v_{\max}$ is the set of parameters to be optimized because $V_{\max}$ values are most sensitive to changes in gene

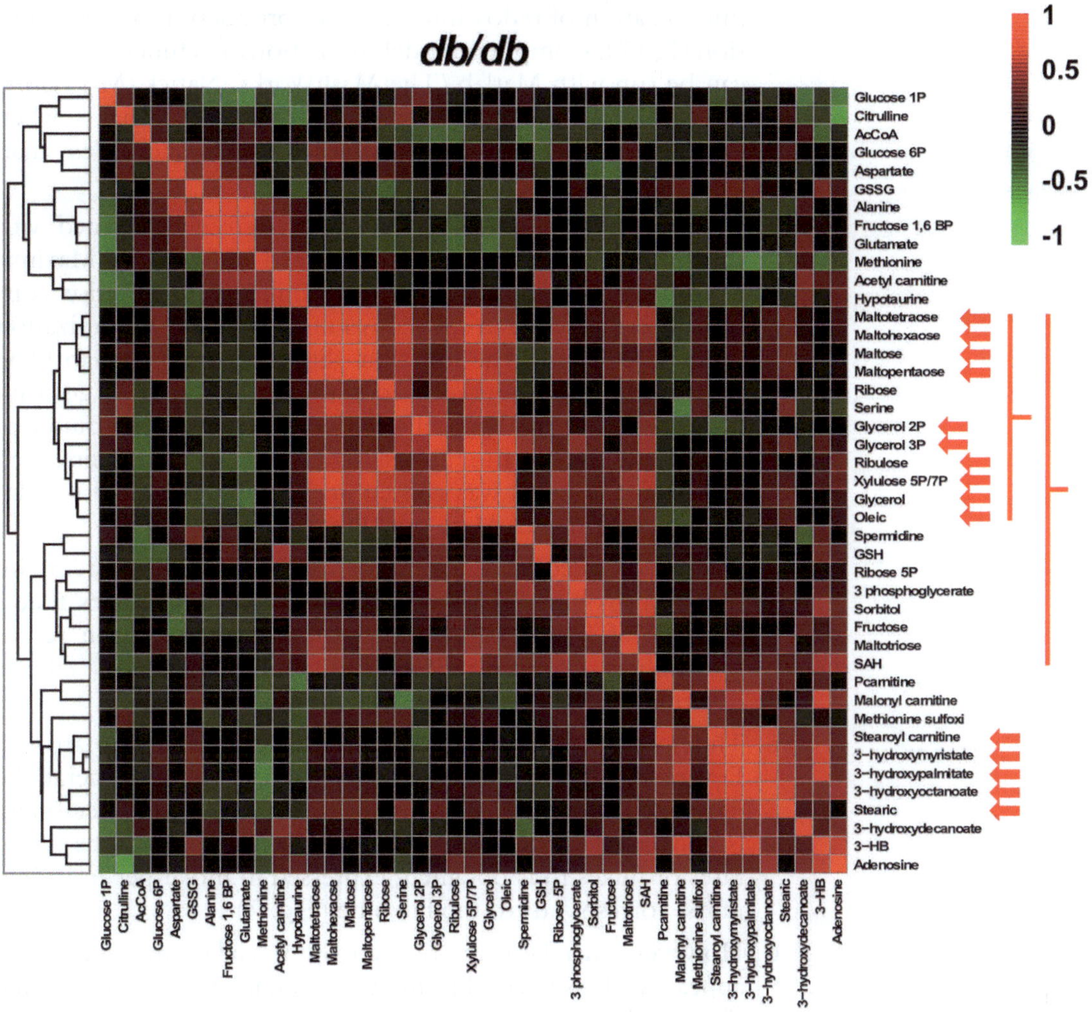

Fig. 5 Correlation matrix of significantly changed metabolites in the heart from diabetic mice. Correlation matrix of the 43 metabolites responsible for treatment separations under G, GI, and GIP conditions in hearts from diabetic mice was obtained using MetaboAnalyst 3.0. The type (positive or negative) and strength (color intensity) of correlation are coded red and green, respectively, according to the bar on the right, and normalized between 1 and −1. Red arrows on the left denote strong positively correlated metabolites. Red on the diagonal correspond to correlation 1 for each metabolite with respect to itself. GSH, reduced glutathione; GSSG, oxidized glutathione; 3-HB, 3-hydroxybutyrate; SAH, S-adenosylhomocysteine. (Reproduced (partially, only bottom panel) from Cortassa, Caceres, Tocchetti, Bernier, de Cabo, Paolocci, Sollott, and Aon (2020) *J. Physiol.* 598.7, 1393–1415)

expression, activity modulation by signaling, and posttranslational modifications [3, 18]. Since all rate expressions depend linearly on V_{max}, linear algebraic methods can be used (*see* **Note 5**). Optimization is performed with an algorithm, for example, implemented in the linprog optimization method in Matlab. Different objective functions may be used as optimization criteria, such as maximization of ATP synthesis fluxes or minimization of redox intermediates production or consumption [7, 17]. Computer model simulations as a function of time can be run with Matlab (The MathWorks, Natick, MA, USA) to corroborate the metabolites' steady state level, that is, when the derivative of the state variables of the model is less than 1×10^{-10}.

3. Linear optimization of the algebraic system is performed with the "Simplex" algorithm (implemented in MatLab). Maximization of ATP synthesis fluxes or minimization of redox consumption were used as constraints for the optimization procedure. Firstly, the function "linprog" from Matlab (The Mathworks, Inc.) was utilized with the "Simplex" algorithm for optimization of V_{max} values. Secondly, within the volume of possible solutions, the highest value of V_{max} for the first step (glucose transport) was chosen and computer model simulations were run with MatCont (Fig. 6).
4. Instead of calculating a unique solution for the system, we obtain a "volume" of possible solutions. This implies that a "range" of fluxes will be compatible with the given set of metabolites concentrations and the reference fluxes determined.
5. The optimized vector v_{max}, when multiplied by the matrix S_t. D_rv, renders the vector of the fluxes occurring in the metabolic network under study, that is, the rates at which each metabolite is converted or transported across compartments. The set of rates or fluxes thus obtained, expressed in molar per unit time units, corresponds to the fluxome (Fig. 7).
6. However, when large matrices are optimized numeric problems appear, such as ill conditioning or degeneration. To overcome this drawback, we describe a procedure to decrease the dimensions of the system to improve our chances to get an accurate and stable solution. In our specific example of central metabolism, a biologically representative and broadly expressed biochemical backbone among prokaryote and eukaryotes, there are processes that occur at high speed (large rate constants) while others occur at much slower pace (small rate constants), the latter being rate-controlling under certain conditions.

3.3 Reduction in the Dimension of the Algebraic Problem for Optimizing $\mathbf{v}_{max}$

1. A more general formulation of the dimension reduction is presented in **Note 4**, in which Table 2 displays the equivalence in terms of central metabolism, the significant biological network that we are using to illustrate this procedure.
2. We consider a metabolic network with n number of processes comprising m metabolites, where $n > m$. In our illustration of the method's implementation, m corresponds to 33 variables,

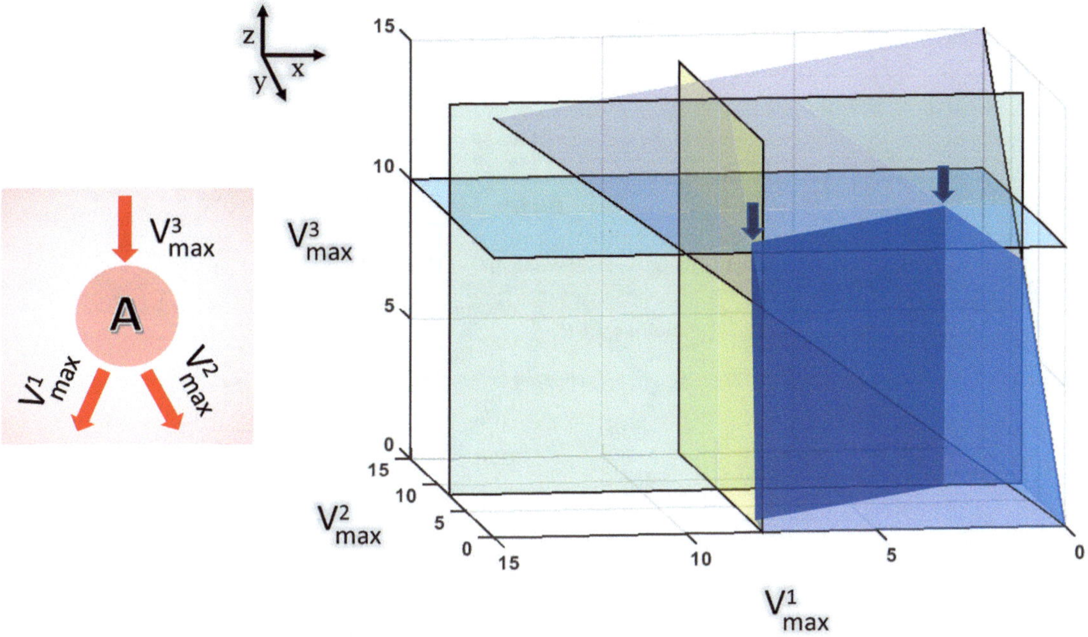

Fig. 6 Solutions space in a hypothetical branched metabolic network with three unknown fluxes. The network displayed on the left comprises three fluxes (unknowns) and a single metabolite M thus underdetermined, meaning that its solution is not unique but presents a solutions space in 2D, that is, a surface of solutions (purple plane). The optimization renders a solution space represented by the volume of the blue "box." According to the procedure proposed to find the V_{max} values, the solutions space chosen (identified by dark arrows) fulfil two conditions: (a) $V_{max} > 0$ for all enzymes in the network; and (b) belongs to the solution space of the network (bright blue pentagon contained between the absolute V_{max} values of which boundaries are given by the yellow, green, and teal surfaces corresponding to $V^1{}_{max}$, $V^2{}_{max}$, and $V^3{}_{max}$, respectively)

and n to 37 rates. We will redefine the product of the S_t times $D_r v$ from Eq. 1 as

$$S_t.D_r v = S_d$$

3. The dimension of the problem is reduced as follows:
   (a) We consider the input required v_{max}^{opt} for the optimization procedure to be given by S_d, b_t, the boundaries of the vector $\boldsymbol{v}_{\mathbf{max}}$ and the costs vector, $\boldsymbol{c}$. The outputs are the optimized maximal rates vector and the optimized objective function

$$z^{opt} = c^T \cdot v_{max}^{opt}$$

   (b) This output will be achieved by finding the matrix F and vector y_{ref}, introduced in **Note 4**, where the general

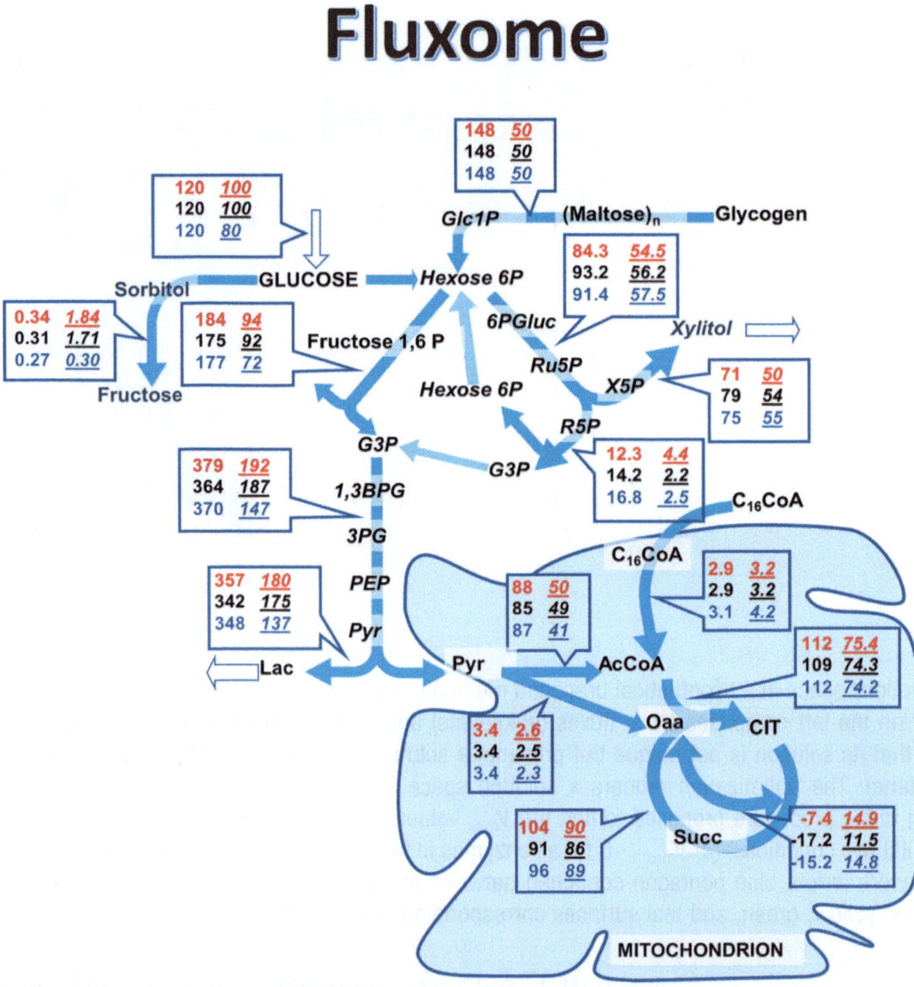

Fig. 7 Fluxome of central metabolism from glucose and fatty acids in the mouse heart. Depicted are the fluxes next to their respective steps of central catabolism of glucose and fatty acids from WT and diabetic mice hearts. The fluxes are expressed in μM s^{-1} (equivalent to nmol s^{-1} ml^{-1} intracellular water). The fluxome was calculated from the experimental data obtained after metabolite profiling of mice hearts perfused as described elsewhere [2], following the workflow diagram shown in Fig. 1, and the integrated analytical procedures described in this chapter. The fluxes displayed correspond to those producing the metabolite concentrations that reproduce the experimentally obtained by metabolomics analysis. Boxed are the flux values corresponding to WT and diabetic hearts (left and right, underlined, number columns, respectively) next to their respective steps in the network. Within each box, the treatment given—glucose, glucose + isoproterenol, or glucose + isoproterenol + palmitate—is denoted by a distinct color (red, black, and blue, respectively). (Reproduced from Cortassa, Caceres, Tocchetti, Bernier, de Cabo, Paolocci, Sollott, and Aon (2020) *J. Physiol.* 598.7, 1393–1415)

method is developed more stepwise, by applying QR decomposition.

(c) $[\mathbf{Q}\ \mathbf{R}] = \mathrm{qr}(S_d^{\mathbf{T}})$ and extracting $Q = [Q_1\ Q_2]$ and $R = \begin{bmatrix} r \\ 0 \end{bmatrix}$

(d) Being Q_1 a matrix with m (33) columns, Q_2 (n – m: 37–33 = 4) columns and r a triangular matrix of dimension m (33) (see the general treatment in **Note 4**).

$$S'_{\mathrm{d}} = \begin{bmatrix} -F \\ F \end{bmatrix}, \; b'_{\mathrm{t}} = \begin{bmatrix} v_{\max}^{\mathrm{ref}} - v_{\min} \\ v_{\mathrm{M}} - v_{\max}^{\mathrm{ref}} \end{bmatrix}, \text{ and } c' = F^T c \qquad (2)$$

(e) Now the dimension of the optimization problem is 4 (= $n - m$) and is represented by the following expression

$$x^{\mathrm{opt}} = \mathrm{linprog}\left(c', S'_{\mathrm{d}}, b'_{\mathrm{t}}\right) \qquad (3)$$

(f) From where we can find the solution

$$v_{\max}^{\mathrm{opt}} = F.x^{\mathrm{opt}} + y_{\mathrm{ref}}, \; z^{\mathrm{opt}} = c^T \cdot v_{\max}^{\mathrm{opt}} \qquad (4)$$

3.4 Representative Results

1. In the present work, we describe a methodology as applied to the determination of the heart fluxome considering glucose catabolism in addition to mitochondrial metabolism to render central metabolism (Fig. 7). We use integrated high throughput -omics data-driven research, metabolomics–fluxomics, that has enabled systems biology approaches to address human health and disease, including metabolic disorders (obesity, diabetes), cancer, aging, health- and lifespan [2, 3, 8, 19].
2. Langendorf-perfused hearts from wild-type (WT) mice under euglycemic conditions, exhibit a palmitate-elicited declining trend in cardiac performance as measured by decreased ISO-evoked enhancement of LV function [2]. Hearts of diabetic and WT (control) mice were subjected to perfusion in the presence of glucose alone (G), glucose + isoproterenol (GI) and Glc + isoproterenol + palmitate (GIP).
3. Snap-frozen Langendorff-perfused hearts were subjected to metabolite profiling at Metabolon, Inc. (Research Park Triangle, NC) (*see* **Note 1**) PLSD analysis of the heart's metabolome revealed that 43 metabolites (out of 261) are significantly involved in the separation of the metabolite profiles from G-, GI-, and GIP-treated hearts (Figs. 2 and 3).
4. When the model can simulate the experimental input of metabolites profile at steady state (Table 1), we can also use it to determine the rate or flux (in μM s^{-1}) through each individual step of the metabolic network. The set of metabolic fluxes determined in this way corresponds to the fluxome (Fig. 7) [2].
5. The fluxome underpinning of central catabolism metabolites' profile (Fig. 4) depicted in Fig. 7, and the corroboration of the computational model's ability to simulate the experimentally determined metabolite's concentration (Table 1) has been reported [2].

Table 1
Comparison of experimental and model-simulated metabolite concentrations obtained from hearts of diabetic (*db/db*) mice

	db/db—G		*db/db*—GI		*db/db*—GIP	
Metabolite	**Exptl.**[a]	**Model**	**Exptl.**	**Model**	**Exptl.**	**Model**
Glucose(ex)	30		30		30	
Glucose(in)	5.5	5.37	5.42	4.94	0.32	0.337
G1P	0.135 ± 0.015	0.195	0.153 ± 0.008	0.149	0.146 ± 0.016	0.180
F6P + G6P	3.71 ± 0.48	3.23	3.75 ± 0.26	2.48	3.55 ± 0.31	2.98
F1,6bP	0.049 ± 0.012	0.060	0.015 ± 0.004	0.035	0.011 ± 0.002	0.012
G3P	<0.01	4.9×10^{-3}	<0.01	6.8×10^{-3}	<0.01	3.4×10^{-3}
1,3 DPG	<0.01	1.7×10^{-4}	<0.01	2.6×10^{-4}	<0.01	1.2×10^{-4}
3-PG	0.504 ± 0.022	0.146	0.64 ± 0.07	0.356	0.62 ± 0.05	0.087
PEP	<0.01	1.8×10^{-3}	<0.01	7.4×10^{-4}	<0.01	9.2×10^{-4}
Pyruvate	<0.01	4.8×10^{-3}	<0.01	4.96×10^{-3}	<0.01	3.9×10^{-3}
Ru5P + X5P	4.2 ± 0.5	2.06	4.4 ± 0.4	2.20	3.7 ± 0.4	2.41
R5P	0.48 ± 0.04	0.87	0.38 ± 0.03	0.93	0.35 ± 0.04	1.02
Xylitol	0.9 ± 0.12	0.429	0.81 ± 0.07	0.473	0.804 ± 0.06	0.550
Sorbitol	1.2 ± 0.1	1.51	1.1 ± 0.3	1.36	1.5 ± 0.2	1.26
Fructose	4.0 ± 0.4	2.57	2.5 ± 0.5	2.24	3.2 ± 0.5	2.28
AcCoA[b]	1.94 ± 0.74	1.42	1.74 ± 0.79	1.40	1.56 ± 0.88	1.41
Succ	0.12 ± 0.02	0.102	0.11 ± 0.05	0.09	0.102 ± 0.023	0.10
FUM	1.86 ± 1.01	1.59	1.69 ± 0.53	1.52	1.55 ± 0.55	1.57
ISO	<0.01	0.014	< 0.01	0.014	<0.01	0.014
aKG	<0.01	0.004	< 0.01	0.0037	<0.01	0.0039

[a]Experimental and model-simulated concentrations are expressed in mmol/liter, and experimental values are displayed as mean ± SEM ($n = 6$)
[b]AcCoA concentration is given in μM units
Key to symbols: G, 30 mM glucose; GI, G + 10 nM ISO; GIP, G + I + 0.4 mM Palm

6. Quantitatively speaking, the fluxome reflects the flux distribution in, for example, cytoplasmic and mitochondrial compartments of glucose and lipids degradation through central metabolism. Hearts of diabetic and WT (control) mice are subjected to three treatments, G, GI, and GIP (*see* **step 2** above). Figure 7 illustrates the results of optimizing the fluxes through the glucose and fatty acids central catabolic network (33 metabolites and 37 reaction steps). The solution shown corresponds to one of the vertices of the "volume of solutions" of the optimization problem (solution dimension = 4) (Fig. 6). To reduce the solution space, more reference fluxes

should be determined, for example, the release of xylitol to the extracellular medium. The metabolomics profiles from hearts of diabetic and WT mice subjected to Langendorf perfusion and the three treatments showed important differential properties of the catabolic networks that reflected more the disease phenotype than the condition in which the hearts were perfused (Fig. 7). The computational model of the hearts from control and diabetic mice were adjusted with their own metabolomics data. A more extended version of the catabolic network displayed in Fig. 7 has been recently published [1].

4 Notes

1. Previously, we emphasized the need to perform the analysis of metabolites under defined and controlled experimental conditions, accompanied by hypothesis-driven experimental design and appropriate challenge (e.g., adrenergic stimulation as it happens during exercise). These are crucial for the success of the present procedure. The rapid sampling can be achieved by methods that assure a very fast stop of all metabolic reactions such as snap freezing in liquid nitrogen. In early work treatment with concentrated perchloric acid or an equivalent intervention was considered accepted practice. However, later on it was demonstrated that, in yeast, the concentration of ATP could drop to a small fraction of its initial concentration in a few seconds before perchloric acid has been able to fully stop metabolic fluxes [20]. Consequently, all interventions requiring a waiting period before metabolic fluxes have completely ceased are, in principle, not appropriate for fluxomics analysis, because the concentration of intermediary metabolites will not reflect the actual concentration in the tissue due to artifacts related to improper sample collection. Also, and as a caveat, the accuracy of the latter procedure may also be affected by the stability of certain metabolites (e.g., NAD^+, during sample processing for MS or LC-MS/MS analysis [21].
2. Estimation of the maximal error associated with the calculation of the Vmax values. There are several sources of error associated with the estimation of the fluxes through this method. Errors associated with the variability in experimental determinations of metabolites that will affect the matrix that contains the derivatives of rate expressions with respect to the Vmax.

   The matrix S_t in Eq. 1 can be made square by appending it with a matrix, size $p \times n$ where is $p = n - m$ and contains 1's and 0's. Obtaining a nonsingular square matrix is required in order to be able to invert it. b_t, the vector of transport and demand processes, for example, biosynthesis also appended with a vector containing p elements that correspond to the measured flux rates. $\mathbf{D_r v}$ is the diagonal matrix of the derivatives of the rate expressions with respect to the *V*max of each rate.

In addition to variability in the experimental values of the transport or demand processes taken into account in vector b_t, sources of variability in the determination of metabolites or in parameter values in each rate expression, such as the affinities of enzymes for their respective substrates, or effectors, are reflected in the variability of $\mathbf{D_r V}$.

$$\boldsymbol{S_t}.\Delta\mathbf{D_r v}.\Delta\boldsymbol{v}_{\mathbf{max}} = \Delta\boldsymbol{b}_{\mathbf{t}} \tag{5}$$

Inspired by the treatment of Savinell and Palsson [17], the matrix norm can be obtained:

$$\|\mathrm{S_t}\|.\|\mathbf{D_r v}\|.\|\boldsymbol{v}_{\mathbf{max}}\| \geq \|\boldsymbol{b}_{\mathbf{t}}\| \tag{6}$$

which, when applied to Eq. 5, can be rearranged as

$$\|\Delta\boldsymbol{v}_{\mathbf{max}}\| \leq \|\mathrm{S_t}^{-1}\|.\|\Delta\mathbf{D_r v}^{-1}\|.\|\Delta\boldsymbol{b}_{\mathbf{t}}\| \tag{7}$$

which lead to an expression for the maximum relative error in the determination of V_{max} as:

$$\frac{\|\Delta\boldsymbol{v}_{\mathbf{max}}\|}{\|\boldsymbol{v}_{\mathbf{max}}\|} \leq \|\mathrm{S_t}\|.\|\mathrm{S_t}^{-1}\|.\frac{\|\Delta\mathbf{D_r v}^{-1}\|}{\|\mathbf{D_r v}^{-1}\|}.\frac{\|\Delta\boldsymbol{b}_{\mathbf{t}}\|}{\|\boldsymbol{b}_{\mathbf{t}}\|} \tag{8}$$

3. Estimation of the error associated with the calculation of the V_{max} values due to variability in metabolites. In the particular case that is being considered in this manuscript, we have estimated the error arising from the variability in metabolites by computing the values of Vmax that would be obtained if all the metabolites values were $M_{\mathrm{av}} - \Delta M$ or $M_{\mathrm{av}} + \Delta M$, where M_{av} is the experimentally determined mean concentration value and ΔM is the standard error of the mean.
4. General formal development of the dimension reduction of the $\boldsymbol{v}_{\mathbf{max}}$ optimization problem. For the sake of a general treatment, we will use more unspecific names for the variables but will redefine them to the equivalent in our problem in Table 2.

   The problem of finding the set of V_{max} that satisfies the levels of intermediates determined by targeted metabolomics and the fluxes used as reference to perform the calculation, boils down to finding the vector x which has a lower dimension than n,

$$A.y = -b \qquad \Leftrightarrow \qquad y = F.x + y_{\mathrm{ref}}$$

$$z' = (c')^T x$$

with $x \in R^{(n\ -\ m)}$ which in our case $n - m = 4$. Thus, an optimization in a space of dimension 4 is less likely to show ill

Table 2
Equivalence between symbols used in the general treatment and fluxome calculation

Notation in the general treatment	Specific notation for fluxome calculation
$\mathbf{A}$	$\mathbf{S_d}$
$\mathbf{b}$	$\mathbf{b_t}$
$\mathbf{y}$	$\mathbf{v_{max}}$
$\mathbf{y_{ref}}$	v_{max}^{ref}
$\mathbf{y^{opt}}$	v_{max}^{opt}
$\underline{y}$	v_{min}
$\overline{y}$	v_M

conditioning and other sources of instability than one in a 37 (n) dimensional space. Then the new objective function is defined as with the constraint

$$A'x \leq b'.$$

The *objective function* describes the condition, for example, "minimizing the rate of substrate consumption" or "maximizing the energy (ATP) output," of the metabolic network. The *cost function* is a weighted stoichiometric vector containing positive or negative values depending upon the process's role in the objective function (e.g., producing, or consuming ATP in the "maximizing the energy output" objective).

The solution of the optimization performed with the linear programing function

$$x^{opt} = \text{linprog}(c', A', b')$$

is given by

$$y^{opt} = F.x^{opt} + y_{ref}$$

Due to the introduction of function $f(x)$, we can write the inequality

$$\underline{y} \leq y \leq \overline{y}$$

as follows:

$$\underline{y} - y_{ref} \leq Fx \leq \overline{y} - y_{ref}$$

which may be rearranged,
$(-F)x \leq y_{ref} - \underline{y}$ and $Fx \leq \overline{y} - y_{ref}$
and, in turn, rewritten as the following inequality:

$$\begin{bmatrix} -F \\ F \end{bmatrix} .x \leq \begin{bmatrix} y_{ref} - \underline{y} \\ \overline{y} - y_{ref} \end{bmatrix}$$

Thus, we can identify now A' and b'.

$$A' = \begin{bmatrix} -F \\ F \end{bmatrix}$$

$$b' = \begin{bmatrix} y_{ref} - \underline{y} \\ \overline{y} - y_{ref} \end{bmatrix}$$

The objective function $z(v)$ becomes

$$c^T y = c^T Fx + c^T y_{\text{ref}}$$

$$z(y) = c^T.y$$

which can be rewritten as follows:

$$c^T.y = c^T.(F.x + y_{\text{ref}})$$

$$z = c^T.F.x + c^T.y_{\text{ref}}$$

Finally, the objective function can be expressed as

$$z' = z - c^T y_{\text{ref}}, \quad \text{and} \quad c' = F^T c.$$

For calculating F and $\mathbf{y_{ref}}$ the QR decomposition of the transpose of matrix **A** can be performed as follows:

$$A^T = [Q_1 \quad Q_2]\begin{bmatrix} r \\ 0 \end{bmatrix} = QR$$

where Q_1 is a 37 × 33 matrix, Q_2 is a 37 × 4 matrix, and r is a triangular matrix of range 33 with

$$Q^T.Q = \underset{n}{I}$$

It can be demonstrated that

$$F = Q_2$$
$$y_{\text{ref}} = -Q_1 r^{-T} \text{b}$$

5. Linearity of rate expressions with respect to V_{max}. The optimization procedure used in these calculations requires linearity with respect to the variable that is "unknown," in our case this is the V_{max}, that is, the rate equations are in all cases linear with respect to V_{max}. Nevertheless, the rate equations in our kinetic models exhibit nonlinear relationships with respect to the metabolites (substrates, products, effectors) in each rate expression participating in the network.

5 Conclusion

This chapter contributes an integrated methodological approach that combines experimental measurements, computational modeling, and bioinformatic tools that, as a showcase, is applied to heart function in diabetes. This approach is widely applicable, as it has been shown in liver function to investigate health and lifespan [8].

Quantitation of the fluxome from central metabolism established the causal relationship, and its mechanistic underpinnings, between myocardial redox improvement and enhanced cardiac contractility in type 2 diabetes [2, 22, 23]. Moreover, key new insights emerged from applying this methodology such as, among others, the discovery of the potential, hitherto unknown, involvement of the methionine cycle and transsulfuration pathways in redox regulation of diabetic heart function (Fig. 4) [2].

Finally, the present methodology allows the optimization and parameterization of the computational model, which can be used to calculate the main rate-controlling and regulatory steps, and as a predictive tool of genetic and pharmacological interventions potentially leading to the development of new therapeutic strategies.

Acknowledements

This work was supported by the Intramural Research Program of the National Institute on Aging, National Institutes of Health.

Author Contributions: Writing first draft and figure creations: S.C. and M.A.A.; Editing: S.C., M.A.A., S.J.S., and P.V.

References

1. Cortassa S, Aon MA, Sollott SJ (2019) Control and regulation of substrate selection in cytoplasmic and mitochondrial catabolic networks. A systems biology analysis. Front Physiol 10: 201. https://doi.org/10.3389/fphys.2019.00201
2. Cortassa S, Caceres V, Tocchetti CG, Bernier M, de Cabo R, Paolocci N, Sollott SJ, Aon MA (2020) Metabolic remodelling of glucose, fatty acid and redox pathways in the heart of type 2 diabetic mice. J Physiol 598(7): 1393–1415. https://doi.org/10.1113/JP276824
3. Cortassa S, Caceres V, Bell LN, O'Rourke B, Paolocci N, Aon MA (2015) From metabolomics to fluxomics: a computational procedure to translate metabolite profiles into metabolic fluxes. Biophys J 108(1):163–172. https://doi.org/10.1016/j.bpj.2014.11.1857
4. Edwards JS, Ibarra RU, Palsson BO (2001) In silico predictions of Escherichia coli metabolic capabilities are consistent with experimental data. Nat Biotechnol 19(2):125–130. https://doi.org/10.1038/84379
5. Winter G, Kromer JO (2013) Fluxomics - connecting 'omics analysis and phenotypes. Environ Microbiol 15(7):1901–1916. https://doi.org/10.1111/1462-2920.12064
6. Cortassa S, Aon MA (2012) Computational modeling of mitochondrial function. Methods Mol Biol 810:311–326. https://doi.org/10.1007/978-1-61779-382-0_19
7. Cortassa S, Sollott SJ, Aon MA (2018) Computational modeling of mitochondrial function from a systems biology perspective. Methods Mol Biol 1782:249–265. https://doi.org/10.1007/978-1-4939-7831-1_14

8. Mitchell SJ, Bernier M, Aon MA, Cortassa S, Kim EY, Fang EF, Palacios HH, Ali A, Navas-Enamorado I, Di Francesco A, Kaiser TA, Waltz TB, Zhang N, Ellis JL, Elliott PJ, Frederick DW, Bohr VA, Schmidt MS, Brenner C, Sinclair DA, Sauve AA, Baur JA, de Cabo R (2018) Nicotinamide improves aspects of healthspan, but not lifespan, in mice. Cell Metab 27(3):667–676. e664. https://doi.org/10.1016/j.cmet.2018.02.001
9. Xia J, Wishart DS (2016) Using MetaboAnalyst 3.0 for comprehensive metabolomics data analysis. Curr Protoc Bioinformatics 55: 14.10.11–14.10.91. https://doi.org/10.1002/cpbi.11
10. Chong J, Soufan O, Li C, Caraus I, Li S, Bourque G, Wishart DS, Xia J (2018) MetaboAnalyst 4.0: towards more transparent and integrative metabolomics analysis. Nucleic Acids Res 46(W1):W486–W494. https://doi.org/10.1093/nar/gky310
11. Aon MA, Cortassa S (1997) Dynamic biological organization: fundamentals as applied to cellular systems, 1st edn. Chapman & Hall, London
12. Kembro JM, Cortassa S, Lloyd D, Sollott SJ, Aon MA (2018) Mitochondrial chaotic dynamics: redox-energetic behavior at the edge of stability. Sci Rep 8(1):15422. https://doi.org/10.1038/s41598-018-33582-w
13. Kurz FT, Kembro JM, Flesia AG, Armoundas AA, Cortassa S, Aon MA, Lloyd D (2017) Network dynamics: quantitative analysis of complex behavior in metabolism, organelles, and cells, from experiments to models and back. Wiley Interdiscip Rev Syst Biol Med 9(1). https://doi.org/10.1002/wsbm.1352
14. Dhooge A, Govaerts W, Kuznetsov YA, Meijer HGE, Sautois B (2008) New features of the software MatCont for bifurcation analysis of dynamical systems. Math Comput Model Dyn Syst 14(2):147–175
15. Schuster S, von Kamp A, Pachkov M (2007) Understanding the roadmap of metabolism by pathway analysis. Methods Mol Biol 358: 199–226. https://doi.org/10.1007/978-1-59745-244-1_12
16. Aitken M, Broadhurst B, Hladky S (2009) Mathematics for biological scientists. CRC Press, New York
17. Savinell JM, Palsson BO (1992) Optimal selection of metabolic fluxes for in vivo measurement. I. Development of mathematical methods. J Theor Biol 155(2):201–214. https://doi.org/10.1016/s0022-5193(05)80595-8
18. Cortassa S, Aon MA, Iglesias AA, Aon JC, Lloyd D (2012) An introduction to metabolic and cellular engineering, 2nd edn. World Scientific Publishers, Singapore
19. Aon MA, Bernier M, Mitchell SJ, Di Germanio C, Mattison JA, Ehrlich MR, Colman RJ, Anderson RM, de Cabo R (2020) Untangling determinants of enhanced health and lifespan through a multi-omics approach in mice. Cell Metab 32(1):100–116. e104. https://doi.org/10.1016/j.cmet.2020.04.018
20. de Koning W, van Dam K (1992) A method for the determination of changes of glycolytic metabolites in yeast on a subsecond time scale using extraction at neutral pH. Anal Biochem 204(1):118–123. https://doi.org/10.1016/0003-2697(92)90149-2
21. Demarest TG, Truong GTD, Lovett J, Mohanty JG, Mattison JA, Mattson MP, Ferrucci L, Bohr VA, Moaddel R (2019) Assessment of NAD(+)metabolism in human cell cultures, erythrocytes, cerebrospinal fluid and primate skeletal muscle. Anal Biochem 572:1–8. https://doi.org/10.1016/j.ab.2019.02.019
22. Bhatt NM, Aon MA, Tocchetti CG, Shen X, Dey S, Ramirez-Correa G, O'Rourke B, Gao WD, Cortassa S (2015) Restoring redox balance enhances contractility in heart trabeculae from type 2 diabetic rats exposed to high glucose. Am J Physiol Heart Circ Physiol 308(4): H291–H302. https://doi.org/10.1152/ajpheart.00378.2014
23. Tocchetti CG, Caceres V, Stanley BA, Xie C, Shi S, Watson WH, O'Rourke B, Spadari-Bratfisch RC, Cortassa S, Akar FG, Paolocci N, Aon MA (2012) GSH or palmitate preserves mitochondrial energetic/redox balance, preventing mechanical dysfunction in metabolically challenged myocytes/hearts from type 2 diabetic mice. Diabetes 61(12): 3094–3105. https://doi.org/10.2337/db12-0072

Part III

Systems Biology of Aging and Longevity

Chapter 8

Understanding the Human Aging Proteome Using Epidemiological Models

Ceereena Ubaida-Mohien, Ruin Moaddel, Zenobia Moore, Pei-Lun Kuo, Ravi Tharakan, Toshiko Tanaka, and Luigi Ferrucci

Abstract

Human aging is a complex multifactorial process associated with a decline of physical and cognitive function and high susceptibility to chronic diseases, influenced by genetic, epigenetic, environmental, and demographic factors. This chapter will provide an overview on the use of epidemiological models with proteomics data as a method that can be used to identify factors that modulate the aging process in humans. This is demonstrated with proteomics data from human plasma and skeletal muscle, where the combination with epidemiological models identified a set of mitochondrial, spliceosome, and senescence proteins as well as the role of energetic pathways such as glycolysis, and electron transport pathways that regulate the aging process.

Key words Aging, BLSA, GESTALT, Proteomics, Epidemiology, SOMAscan, Data model, Plasma, Skeletal muscle, TMT

1 Introduction

Human aging is a complex multifactorial process associated with a decline of physical and cognitive function and high susceptibility to chronic diseases, influenced by genetic, epigenetic, environmental, and demographic factors. In order to identify factors that modulate the aging process, a global view of contributing factors is necessary. Several epidemiological models of aging have been developed to examine the relationship between aging outcomes and relevant exposures measured throughout the lifespan. For example, many of the contributors to age-related health outcomes significantly change with age, and often vary by gender, race, physical activity, and/or other phenotypic measures. It is crucial to control for major confounding factors, which may be addressed directly by a matched study design or through the use of multivariable models, such as multivariable regression models, or Cox proportional hazards

Sonia Cortassa and Miguel A. Aon (eds.), *Computational Systems Biology in Medicine and Biotechnology: Methods and Protocols*, Methods in Molecular Biology, vol. 2399, https://doi.org/10.1007/978-1-0716-1831-8_8,

models. Basic concepts of epidemiology are often ignored in molecular studies, especially in high-throughput -omics aging studies that aim to understand the aging process (*see* **Note 1**). The integration of the -omics aging data with the epidemiological models will help improve the understanding of age-related health and lifespan of older adults.

Herein we present the advantages of the integration of epidemiological models with proteomics data from human biological samples, such as plasma and skeletal muscle, and describe a protocol for performing these studies. Two case studies are also presented where mass spectrometry based Tandem Mass Tag (TMT) protein quantification and SOMAscan proteomics methods are applied to samples from the Baltimore Longitudinal Study of Aging (BLSA) and Genetic and Epigenetic Signatures of Translational Aging Laboratory Testing (GESTALT) studies. The analyses of the proteome in human plasma and skeletal muscle identified factors that modulate the aging process in humans, including the identification of a set of mitochondrial, spliceosome, and senescence proteins as well as the role of energetic pathways such as glycolysis, and electron transport pathways that regulate the aging process.

2 Materials

All analyses presented in this chapter are conducted on proteomics data measured in the BLSA and GESTALT studies. In both studies, a cross-sectional proteomic analysis is carried out in muscle and plasma samples, where the correlation between protein abundance and age is investigated. The description of the study design is detailed in Subheading 2.3.

2.1 Modeling Methods Used in Epidemiology

In cross-sectional epidemiological studies, the goal is to characterize the correlation between variables of interest. The associations between protein abundance and chronological age may be pursued using linear regression models where a linear relationship between age and protein abundance is assumed. For nonlinear relationships, multiple approaches can be used including age categories rather than a single continuous variable of age, polynomial regression models, or spline regression. Further, there is growing interest in using various machine learning methods to select a set of variables (or proteins) that predict the outcome of interest (age). In the plasma case study, we present an example using elastic net regression to select a subset of plasma proteins that can accurately predict age.

2.2 Adjustment for Confounding Factors and Outcomes

An important consideration in epidemiological models is to account for confounding factors when examining the association between variables of interest. Confounders, variables that are associated with both exposure (age) and outcome (protein) but are not mediators of the relationship between exposure and outcome, may alter the observed association of interest if not taken into account. One method to address a confounder is the inclusion of a variable that represents the confounder as an independent variable in regression models. Depending on the research question, covariates, or variables that are associated with the outcome but not the dependent variable are also included in the model to improve the precision of the point estimates.

2.3 Sample Collection

1. Participant details and cohort details (BLSA and GESTALT). The BLSA is a continuously enrolled cohort of community dwelling adults, the goals of the study include characterizing physiological and functional trajectories with aging and identifying factors that affect those trajectories. Started in 1958, the study evaluates contributors to healthy aging in persons 20 years old and older [1]. The BLSA follows participants at intervals from one to 4 years, depending on their age (annual visits for participants older than 80 years, every 2 years for participants between ages 60 and 79 years, and every 4 years for participants age 60 and younger).
2. The GESTALT study is a closed cohort of 100 community dwelling adults, begun in April 2015 with the goal of discovering novel molecular biomarkers of aging in different cell types for the identification of new phenotypes that are highly age sensitive and can be potentially applied in epidemiological studies of aging.
3. For both BLSA and GESTALT, participants 20 years or older are recruited from the DC/Baltimore metropolitan area, at enrollment participants are considered healthy based on stringent criteria, including the absence of any chronic disease (with the exception of controlled hypertension), cognitive impairment or impairment of physical function.
4. For the GESTALT plasma study, baseline samples were run in the SOMAscan Assay. For the BLSA plasma study, samples collected at times when all healthy criteria were still met were selected. The studies have similar protocols for clinical and functional assessments as well as biochemical measurements. Sample collection and research testing are conducted by trained study staff. The study protocol for both studies was reviewed and approved by the Internal Review Board of the National Institute for Environmental Health Sciences (NIEHS, NIH, IRB), and all participants provided written informed consent.
5. GESTALT skeletal muscle study is conducted on 58 participants. A 6-mm Bergstrom biopsy needle was inserted through the skin and fascia incision into the muscle, and muscle tissue samples were obtained using a standard method. Biopsy

specimens cut into small sections were snap frozen in liquid nitrogen and subsequently stored at <−80 °C until used for analysis.

2.4 Phenotypic Information of the Sample

1. Information about demographics such as race and age were assessed by self-report (Table 1 for plasma and Table 2 for skeletal muscle).
2. Body mass index (BMI) is the ratio of weight in kg to square of height in meters were objectively assessed during a standard medical exam.
3. White blood cell count measured as part of the standard CBC using SYSMEX SE-2100 (Sysmex, Kobe, Japan). Plasma creatinine was measured using an enzymatic method (Ortho Clinical Diagnostics, Raritan, NJ, USA).
4. The level of physical activity for a participant was determined using an interview-administered standardized questionnaire. Total participation time in moderate to vigorous physical activity per week was calculated by multiplying the frequency by amount of time performed for each activity, summing all of the activities, then dividing by two to derive minutes of moderate to vigorous physical activity per week, the following categories were used: <30 min per week of high intensity physical activity was considered "not active" and coded as 0; high-intensity physical activity > = 30 and < 75 min was considered "moderately active" and coded as 1, high-intensity physical activity > = 75 and < 150 min was considered "active" and coded as 2, and high-intensity physical activity > = 150 min was considered to "highly active" and coded as 3. An ordinal variable from 0 to 3 was used in the analysis.

3 Methods

3.1 Sample Preparation for SOMAscan Based Plasma Analysis [2]

1. Proteomic profiles of 240 plasma samples with 1322 Slow Offrate Modified Aptamer (SOMAmers) were assessed using the 1.3 K SOMAscan Assay at the Trans-NIH Center for Human Immunology and Autoimmunity, and Inflammation (CHI), National Institute of Allergy and Infectious Disease, National Institutes of Health (Bethesda, MD, USA) [3].
2. Each 1.3 K SOMAscan plate holds 96 samples that include buffer wells, quality control and calibrator samples provided by SOMAlogic, and an additional bridging sample that allows for normalization across plates.
3. Each plate, therefore, holds 80 test samples, and the 240 BLSA and GESTALT samples were run across three plates. The samples were randomized by age, sex, and study (BLSA or GESTALT) across the three plates.

Table 1

Clinical and demographic characteristics of BLSA and GESTALT participants with plasma proteomic data. Reproduced from [2]. Information about years of education were assessed by self-report. Waist circumference, BMI (ratio of weight in kg to square of height in meters), and blood pressure were objectively assessed during a standard medical exam. Grip strength was measured three times on each of the right and left hand. The highest average grip strength was used. Usual gait speed was measured in two trials of a 6-m walk; the faster time between the two trials was used in the analysis. Blood tests were performed at a Clinical Laboratory Improvement Amendments certified clinical laboratory at Harbor Hospital, home of the National Institute of Aging (NIA) intramural research program clinical unit. White blood cell count was measured as part of the standard CBC using SYSMEX SE-2100 (Sysmex, Kobe, Japan). Total cholesterol, creatinine with enzymatic methods, HDL and LDL with dextran magnetic, triglycerides with colorimetric methods, glucose with glucose oxidase using the Vitros system (Ortho Clinical Diagnostics, Raritan, NJ, USA). Serum inflammatory markers IL6 (R&D System, Minneapolis, MN, USA) and CRP (Alpco, Salem, NH, USA) were measured with enzyme-linked immunosorbent assay (ELISA)

	Age 20–35 years ($n = 48$)	Age 35–50 years ($n = 48$)	Age 50–65 years ($n = 48$)	Age 65–80 years ($n = 48$)	Age 80+ years ($n = 48$)	p
White Blood Cell Count	5.6 ± 1.4	5.3 ± 1.4	5.2 ± 1.4	5.6 ± 1.5	5.4 ± 1.4	0.898
IL-6 (pg/mL)	3.1 ± 1.8	3.2 ± 1.1	4.1 ± 3.5	3.7 ± 1.7	4.4 ± 4.3	0.011
CRP (μg/mL)	1.8 ± 2.6	1.7 ± 1.7	2.0 ± 2.9	2.8 ± 4.1	2.2 ± 2.1	0.125
Creatinine (mg/dL)	0.9 ± 0.2	0.9 ± 0.2	0.9 ± 0.2	0.9 ± 0.2	0.9 ± 0.2	0.240
Triglyceride	91.5 ± 50.1	92.6 ± 72.4	90.4 ± 46.5	97.4 ± 42.2	97.9 ± 47.0	0.459
HDL-C	60.8 ± 17.0	61.8 ± 18.3	66.2 ± 16.4	61.3 ± 18.5	65.5 ± 15.3	0.250
LDL-C	99.8 ± 32.9	100.7 ± 28.8	106.3 ± 24.9	113.1 ± 31.0	105.2 ± 28.5	0.085
Total cholesterol	178.8 ± 34.5	180.9 ± 31.1	190.6 ± 30.5	193.8 ± 37.8	190.3 ± 34.7	0.020
Glucose (mg/dL)	84.4 ± 7.1	84.8 ± 8.6	88.7 ± 8.5	90.6 ± 10.7	88.7 ± 6.9	<0.001
BMI (kg/m^2)	25.8 ± 5.5	26.1 ± 4.4	27.3 ± 4.5	27.1 ± 4.0	25.5 ± 3.2	0.878
Grip strength right (kg)	39.8 ± 11.5	38.5 ± 12.8	36.5 ± 12.4	30.9 ± 9.6	26.8 ± 8.8	<0.001
Usual gait speed (m/s)	1.3 ± 0.2	1.3 ± 0.2	1.3 ± 0.2	1.2 ± 0.2	1.1 ± 0.2	<0.001
Years of education	16.4 ± 2.2	17.5 ± 2.7	17.2 ± 2.4	17.1 ± 2.5	16.6 ± 3.2	0.973

Table 2

Baseline characteristics of the GESTALT skeletal muscle participants. Reproduced from [5]. Participants are classified into 5 different age groups. Gender: *M* is Male, *F* is Female; the number of participants is indicated. Age is indicated in years as mean and standard deviation (SD ±) for each age group. Race: number of participants is shown on the left and race is shown in italics; *C* is Caucasian, *AA* is African American, and *A* is Asian. Body Mass Index (BMI) is expressed as mean and SD (±) for each group. *P*-value is calculated by 1-way ANOVA with Kruskal-Wallis test. Race is analyzed by chi-square test

Age group	20–34 ($n = 13$)	35–49 ($n = 11$)	50–64 ($n = 12$)	65–79 ($n = 12$)	80+ ($n = 10$)	*P*-value –	R^2 –
Age (year)	27.2 ± 3.3	41.3 ± 4.5	57.1 ± 4.7	70.3 ± 2.3	82.4 ± 2.4	–	–
Gender	*M*8, *F*5	*M*7, *F*4	*M*7, *F*5	*M*8, *F*4	*M*6, *F*4	–	–
Education (year)	16 ± 3	14 ± 3	14 ± 2	16 ± 2	17 ± 2	0.3305	–
Race	9*C*, 2*AA*, 2*A*	5*C*, 6*AA*	8*C*, 4*AA*	10*C*,1*AA*,1*A*	9*C*,1*AA*	0.0958	–
*BMI, kg/m²	25.9 ± 2.8	26.4 ± 2.6	26.6 ± 3.2	26.4 ± 2.4	25.2 ± 3.9	0.3458	0.007
Height (cm)	172 ± 11	177 ± 10	169 ± 4	172 ± 11	172 ± 6	0.3985	–
*Weight (kg)	76 ± 10	81 ± 9	77 ± 12	75 ± 13	73 ± 16	1.74E-05	0.34
*Waist circumference (cm)	82 ± 7	87 ± 7	90 ± 11	92 ± 11	92 ± 13	6.32E-06	0.39
*KEIS (left) ±	192 ± 31	208 ± 55	200 ± 71	165 ± 62	130 ± 42	4.29E-07	0.40
*KEIS (right) ±	194 ± 38	220 ± 65	194 ± 78	169 ± 53	147 ± 57	2.41E-07	0.41
†Physical activity	1.8 ± 1.4	1.8 ± 1.3	2 ± 1.1	2.3 ± 1	1.5 ± 1.1	0.5145	–

**P*-value calculated from linear regression model, gender adjusted

±Knee Extension Isokinetic Strength (KEIS) (30°/s; Nm)

†Physical activity is calculated from self-report involvement in weight circuit, vigorous exercise, brisk walking and casual walking and summed as high-intensity physical activity hours per week. This is further categorized into 0 (not active), 1 (moderately active), 2 (active), and 3 (highly active) and expressed as mean of categorical variables (0,1,2,3) ± SD

4. The SOMAmer reagents are aliquoted into three groups based on their expected abundance in plasma.
5. Plasma samples are diluted into three concentrations of 0.005%, 1%, and 40% to measure high, medium, and low abundance proteins, respectively. The diluted plasma samples are then incubated with their respective SOMAmers.
6. The SOMAscan assays are conducted semiautomatically with Tecan Freedom Evo 200 *High Throughput System* (HTS).
7. Data normalization was conducted in three stages. First, hybridization control normalization removes individual sample variance on the basis of signaling differences between microarray or Agilent scanner. Second, median signal normalization removes inter-sample differences within a plate due to technical

differences such as pipetting variation. Finally, calibration normalization removes variance across assay runs. Further, there is an additional inter-plate normalization process that utilizes CHI calibrator that allows normalization across all experiments conducted at CHI laboratory [4].

8. The 1322 SOMAmer Reagents, 12 hybridization controls and 4 viral proteins (HPV type 16, HPV type 18, isolate BEN, isolate LW123) and 5 SOMAmers that were reported to be nonspecific (P05186; ALPL, P09871; C1S, Q14126; DSG2, Q93038; TNFRSF25, Q9NQC3; RTN4) were removed leaving 1301 SOMAmer Reagents in the final analysis.
9. The protein data reported are SOMAmer Reagent abundance in relative fluorescence units (RFU).

3.2 Sample Preparation for Mass Spectrometry (MS) Based Skeletal Muscle Analysis [5]

1. Muscle tissue (~8 mg) from 58 participants was pulverized in liquid nitrogen and mixed with the lysis buffer containing protease inhibitor cocktail (8 M Urea, 2 M Thiourea, 4% CHAPS, 1% Triton X-100, 50 mM Tris, pH 8.5) [5].
2. Determined protein concentration using a commercially available 2-D quant kit (GE Healthcare Life Sciences). Sample quality was confirmed using NuPAGE protein gels stained with fluorescent Sypro Ruby protein stain (Thermo Fisher).
3. ~300 μg of muscle tissue lysate was precipitated with Methanol/Chloroform (2:1) extraction protocol to remove lipids and detergents [5, 6].
4. Proteins were resuspended in concentrated urea buffer (8 M Urea, 2 M Thiourea, 150 mM NaCl) and reduced with 50 mM DTT for 1 h at 36 °C. The solution is then alkylated with 100 mM iodoacetamide for 1 h at 36 °C in the dark. The concentrated urea was diluted 12 fold with 50 mM ammonium bicarbonate buffer [5].
5. Proteins were digested for 18 h at 36 °C using trypsin–LysC mixture in 1:50 (w/w) enzyme to protein ratio (https://doi.org/10.7554/eLife.49874). The digests were desalted on 10 × 4.0 mm C18 cartridge using Agilent 1260 Bio-inert HPLC system with the fraction collector. Purified peptides were speed vacuum–dried and stored at −80 °C until analysis.
6. Tandem Mass Tags (TMT) based quantitative proteomics was carried out on the purified peptides from the skeletal muscle tissue (*see* **Note 2**) [5]. 200 femtomoles of bacterial beta-galactosidase digest was spiked into each sample prior to TMT labeling to control for labeling efficiency and overall instrument performance. Each TMT labeling reaction contained six labels to be multiplexed in a single MS run.

7. The TMT-labeled peptides were separated using a linear gradient from 5% B to 50% B over 100 min with 10 mM ammonium formate pH 10 as mobile phase A and 10 mM ammonium formate and 90% ACN (pH 10) as mobile phase B using a 3.9 mm × 5 mm XBridge BEH Shield RP18 XP VanGuard cartridge and 4.6 mm × 250 mm XBridge Peptide BEH C18 column [5].
8. 99 fractions were collected during each LC run at 1 min interval. Three individual high-pH fractions were concatenated into 33 combined fractions with the 33 min interval between each fraction (fractions 1, 34, 67 = combined fraction 1, fractions 2, 35, 68 = combined fraction two, and so on). Combined fractions were speed vacuum–dried, desalted, and stored at −80 °C until final LC-MS/MS analysis [5].
9. Purified peptide fractions from skeletal muscle tissues were analyzed using UltiMate 3000 Nano LC Systems coupled to the Q Exactive HF mass spectrometer on a 35 cm capillary column (3 μm C18 silica) with 150 μm ID at 650 nl/min with a linear gradient using from 5% to 35% B over 205 min, with mobile phases A and B consisting of 0.1% formic acid in water and 0.1% formic acid in acetonitrile, respectively [5]. Q Exactive HF mass spectrometer was set with the heated capillary temperature + 280 °C and spray voltage set to 2.5 kV. Full MS1 spectra were acquired from 300 to 1500 *m/z* at 120,000 resolution and 50 ms maximum accumulation time with automatic gain control [AGC] set to 3×10^6 [5]. Dd-MS2 spectra were acquired using dynamic *m/z* range with fixed first mass of 100 *m/z*. MS/MS spectra were resolved to 30,000 with 155 ms of maximum accumulation time and AGC target set to 2×10^5. Twelve most abundant ions were selected for fragmentation using 30% normalized high collision energy. A dynamic exclusion time of 40s was used to discriminate against the previously analyzed ions.

3.3 Bioinformatics Analysis of the SOMAscan Plasma Data

1. Protein RFU values were natural log transformed and outliers outside 4SD were removed. Association of each protein with chronological age was assessed using linear regression using the R function lm(). Potential confounders including sex, study (BLSA or GESTALT), plate ID, and race (white, black, other) were accounted for by including them in the regression model. A Bonferroni corrected *p*-value of 3.84×10^{-5} (0.05/1301) was considered significant for the analysis of 1301 SOMAmer Reagents.
2. To construct a proteomic age predictor, a penalized regression model was implemented using the R package glmnet. First, the 240 samples were split into training and validation sample. The training set of 120 participants were selected using a random

sampling method choosing 24 subjects from each of the 15-year age strata (20–35, 35–50, 50–65, 65–80, 80+ years). The remaining 120 subjects were used as a validation sample. In the training dataset, chronological age was regressed on 1301 log-transformed protein abundances. The alpha value was set to 0.5 (for elastic net regression), and a lambda of 0.8767859 was selected using a tenfold cross-validation on the training set using the cv.glmnet function. The resulting age-prediction model from the penalized regression was applied to the validation data.

3.4 Bioinformatics Analysis of the MS Skeletal Muscle Data

1. The MS raw files were converted to searchable files with ion list and mass (.mgf files) using MSConvert, ProteoWizard 3.0.6002. MGF files were searched with Mascot 2.4.1 and X! Tandem CYCLONE (2010.12.01.1) using the SwissProt Human sequences from Uniprot (Version Year 2015, 20,200 sequences, appended with 115 contaminants) database. The search engines were set with fixed medication and variable modifications (*see* **Note 3**).
2. The peptide and protein data were extracted from the search files by Scaffold Q+ analysis system (Scaffold Q+ 4.4.6, Proteome Software, http://www.proteomesoftware.com/), and the result files were combined. Peptide and protein probability were calculated, False Discovery Rate (FDR) is measured by using a decoy database. Proteins were filtered at the threshold of 0.01 peptide FDR and 0.1 protein FDR.
3. The reporter ion intensity from the proteins are measured for each sample from the peptides quantified and the protein data is log2 transformed. Relative protein abundance was estimated by median of all peptides for a protein combined. Protein sample loading effects from sample preparations were corrected by median polishing, that consists in subtracting the channel median from the relative abundance estimate across all channels to have a median zero. The TMT normalization is implemented in R using Limma library and the methods from [7, 8].
4. Linear mixed regression model was used to examine age effects, and the model was adjusted for physical activity, gender, race, bmi, type I and type II myosin fiber ratio and TMT mass spectrometry experiments. Protein significance from the regression model was determined with *p*-values derived from lmerTest. The regression model was implemented using R 3.3.4 (R Development Core Team, 2016) with lme4 v1.1. library.
5. From the linear model outcome any protein which has a positive age beta is considered as upregulated in the muscle data and any protein with negative age beta is considered as down

regulated in the muscle data. If a protein is present in 50% of the samples or at least in three samples in each age group, and the *p*-value is significant ($p < 0.05$), the proteins were considered age-associated.

6. Annotation of the proteins were performed by manual literature curation and combining information from Uniprot (https://www.uniprot.org/), Gene Ontology (http://geneontology.org/), and PANTHER (http://www.pantherdb.org/) database. Further bioinformatics analysis was performed using R programming language (3.4.0) and the free libraries available on Bioconductor such as DEXSeq, ggplot2, pheatmap, pie3D, and limma.

3.5 Plasma Proteome Data Interpretation and Data Visualization

1. From the linear regression analysis of plasma proteome, 217 proteins (20 negatively associated, 197 positively associated) were significantly associated with chronological age. These results are displayed in a volcano plot with the beta from the linear regression model on the *x*-axis and -log10 of the *p*-value on the *y*-axis (Fig. 1a).
2. The association of the most significant protein positively associated (GDF15) and negatively (CTSV) associated with age is displayed as a scatter plot (Fig. 1b, c).
3. The correlation between the predicted age based on the elastic net regression and observed chronological age is displayed as a scatter plot in the validation set of 120 individuals (Fig. 1d).

3.6 Skeletal Muscle Proteome Data Interpretation and Data Visualization

1. All the proteins identified in the data were annotated, so the significant proteins enrichment analysis would be fairly simple. This was shown by a pie chart diagram of all the categories of proteins (Fig. 2a).
2. The skeletal muscle analysis using mixed linear model identified 1265 proteins that were differentially regulated with aging; 904 proteins up-regulated with age, and 361 proteins down-regulated. A volcano plot is constructed using R plot function to show all the differentially expressed proteins (Fig. 5a).
3. Some proteins have multiple functions, and the same protein is involved in multiple pathways, so an accurate protein annotation is key for useful data interpretation. Finding the most relevant protein function based on previous research of the same tissue and understanding the protein function using multiple databases (PANTHER, Uniprot, Gene Ontology, Proteomics DB) will help data interpretation.
4. With an extended and elaborate protein annotation, a pie chart of differentially expressed proteins were generated using R package pie3D. All proteins were categorized into "Hallmarks of Aging" [9] and some additional categories (*see* **Note 4**). This

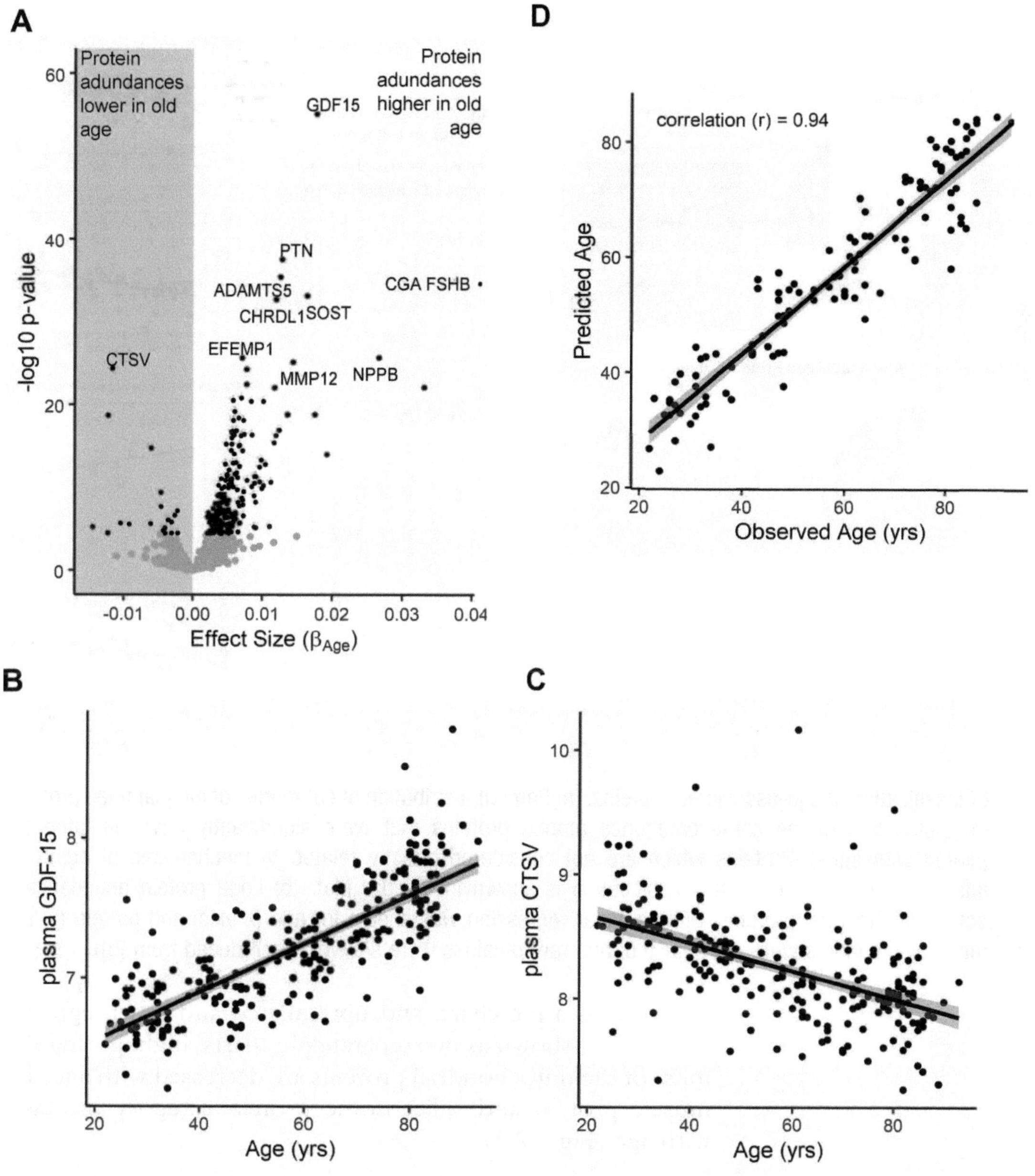

Fig. 1 Plasma proteome data interpretation and data visualization. (**a**) Volcano plot displaying the associations of 1301 plasma proteins with chronological age. Protein abundances were log transformed and association with chronological age was tested using linear regression model adjusting for sex, race, study (BLSA/GESTALT) and batch. The figure displays beta estimates (effect size) from the linear regression model and significance expressed as the -$\log_{10}$(P-value). Scatterplot displays the linear association of GDF15 (**b**) and CTSV (**c**) with chronological age. (**d**) Using elastic net regression model, 76 proteins were selected to create a proteomic predictor of chronological age. The correlation between predicted age and observed age was 0.94. (Reproduced from [2])

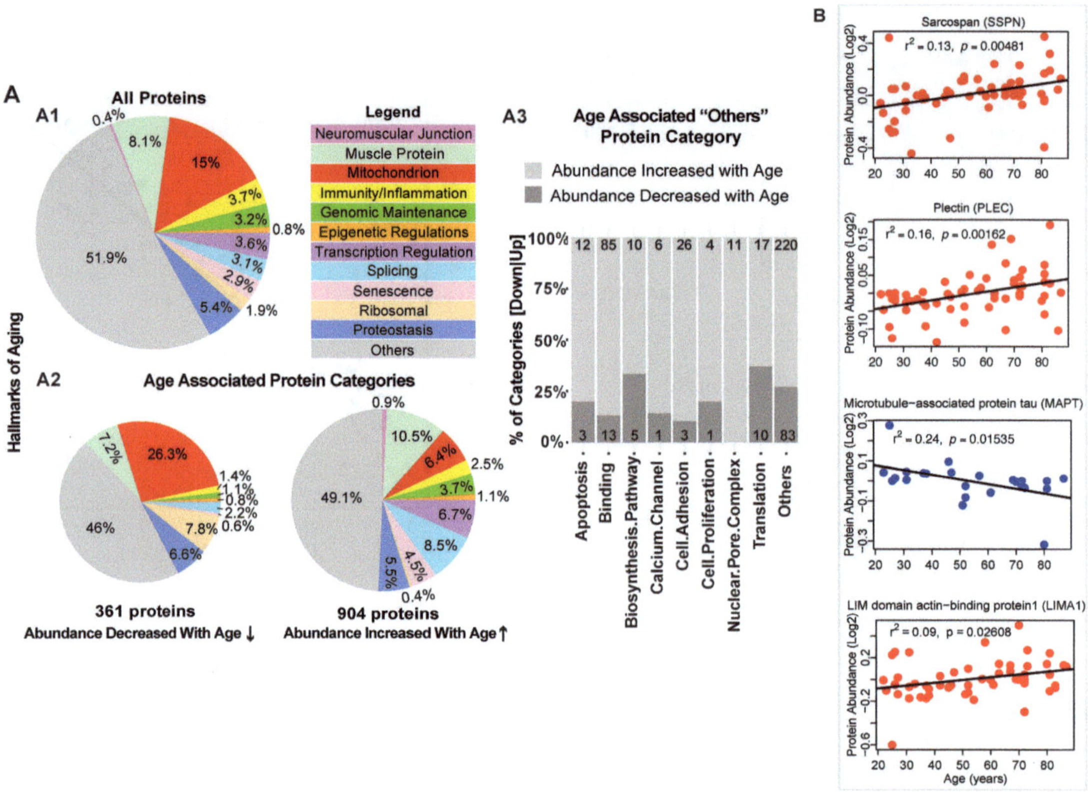

Fig. 2 Classification of age-associated proteins. (**a**) Percent distribution of categories of all quantified proteins, percent distribution of the same categories among proteins that were significantly downregulated and upregulated with aging. Proteins which are not considered directly related to mechanisms of aging are annotated as others and their subclassification is shown in the bar plot. (**b**) Log2 protein abundance of contractile, architectural proteins. Simple linear regression was shown for age (x-axis) and protein (y-axis) correlation, confounders were not adjusted, and raw *p*-values were shown. (Reproduced from [5])

is shown as a pie chart, and upregulated and downregulated proteins are shown as two separate pie charts, highlighting that most of the mitochondrial proteins are decreased with age, and muscle protein and spliceosome protein category increased with age (Fig. 2A2).

5. Since the annotation showed mitochondrial proteins enrichment in the dataset, an additional validation for the mitochondrial proteins are performed using MitoCarta2.0 database (https://www.broadinstitute.org/mitocarta), which is an inventory for mitochondrial proteins with experimental and theoretical evidence for mitochondrial localization.

6. A set of functional proteins are annotated as glycolysis or TCA, and a barplot is created for each protein to show the log2 protein abundance change with each year of age. Then the barplot is overlayed on a manually created pathway using wiki pathways (https://www.wikipathways.org/) as reference. The

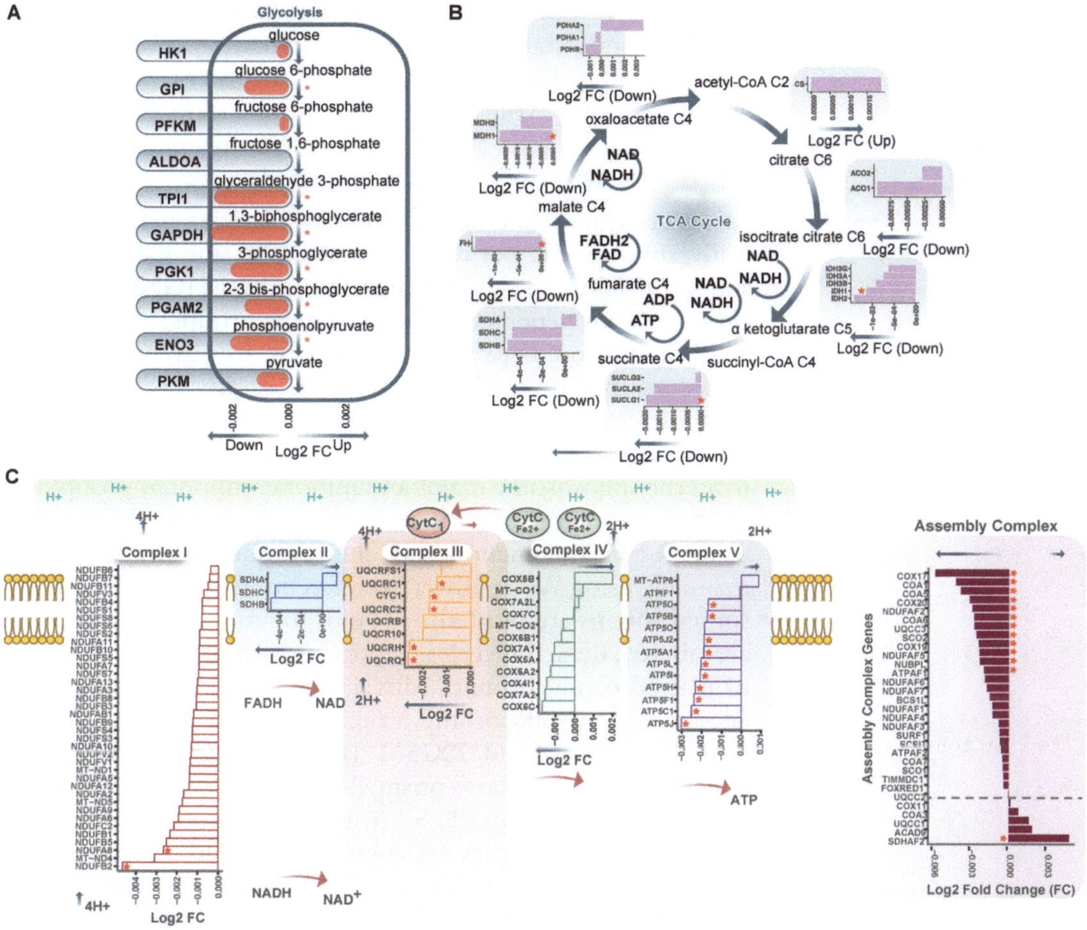

Fig. 3 Dysregulation of Bioenergetic Pathway. (**a**) Proteins quantified from glycolysis and TCA cycle are shown. Of the 26 glycolysis proteins quantified, six are significantly underrepresented with age. (**b**) Of the TCA cycle gene products shown, four are significantly decreased with aging. A red asterisk indicates genes significantly changed with age ($p < 0.05$). (**c**) Respiratory Chain Complex I–V and Aging. Electron Transport Chain protein quantification is shown. Proteins quantified from Complex I, Complex II, Complex III, Complex IV, and Complex V, and Assembly complex proteins are represented. Age-associated proteins are marked by a red asterisk (*). Log2 fold ratios of the gene are on x-axis; arrows pointing to left shows underrepresented proteins and arrows pointing to the right are overrepresented proteins. (Reproduced from [5])

pathway map help to understand the proteins identified for each pathway and the protein expression pattern across ages for each protein in the pathway (Fig. 3a, b).

7. The major energetics pathway for skeletal muscle is the electron transport chain (ETC), the proteins from five ETC complexes are plotted using R barplot function. Unlike other figures, the ETC pathway figure shows all proteins quantified for this pathway, significant proteins represented with * (Fig. 3c). While many complex proteins have low abundance and are undetectable with MS proteomics methods, we would advise to also

consider the nonsignificant proteins in the visualization in order to understand the complete regulation of the proteins in the pathway. For example, proteins that do not change with age in a pathway would suggest a temporal state of the protein and therefore implies that certain proteins in a biological pathway do not act in a linear fashion (Fig. 3).

8. In order to understand proteins in a complex pathway, especially spliceosome pathway complex, the annotated spliceosome protein data is reannotated using HGNC database (https://www.genenames.org/) to gene groups, and the hierarchical relation between gene groups are categorized. The gene groups are mapped to protein groups using Uniprot database (https://www.uniprot.org/), and recategorized the different spliceosome proteins to U1, U2, U4/U6, and U5, from the spliceosome major and minor component groups and plotted the abundance of each component and the proteins using R barplot function. The spliceosome pathway is created manually using Wiki pathway as reference (Fig. 4a). All spliceosome proteins are plotted by R plot function and major genes are labeled, depicting a clear representation of the upregulation of all spliceosome proteins with age (Fig. 4b). The average of all age-associated spliceosome proteins within each age group is plotted by GraphPad PRISM 10.1 (https://www.graphpad.com/scientific-software/prism/) (Fig. 4c). Effect of age (1-year difference) on the 57 reannotated proteins of the spliceosome major complex are color coded based on spliceosome domains. Inset (left) is a legend for the complex domains and inset (right) shows that PRPF8 protein is robustly overrepresented with age (Fig. 4d). R barplot function is used to generate the plot, labeled protein log2 abundance as y-axis, each participant as *X*-axis, and the category of the spliceosome is represented on the barplot annotation (Fig. 4d).

3.7 Integrations of Epidemiological Models and Proteomic Analysis Results

1. Analysis of plasma proteome reveal that the protein with the strongest association with age was the GDF15, a member of the transforming growth factor-b cytokine superfamily. GDF15 has been shown to have important roles in cellular response to stress signals in cardiovascular diseases and has been associated with cardiovascular disease mortality [10]. More recently, GDF15 has been identified as a senescence-associated secretory phenotype (SASP), support the role of cell senescence in aging [11]. The other plasma proteins identified represented proteins in the blood coagulation, chemokine and inflammatory pathways. Use of machine learning technique resulted in a highly accurate proteomic predictor of age based on data from 76 proteins. These analyses show that the plasma proteome capture changes that occur with age.

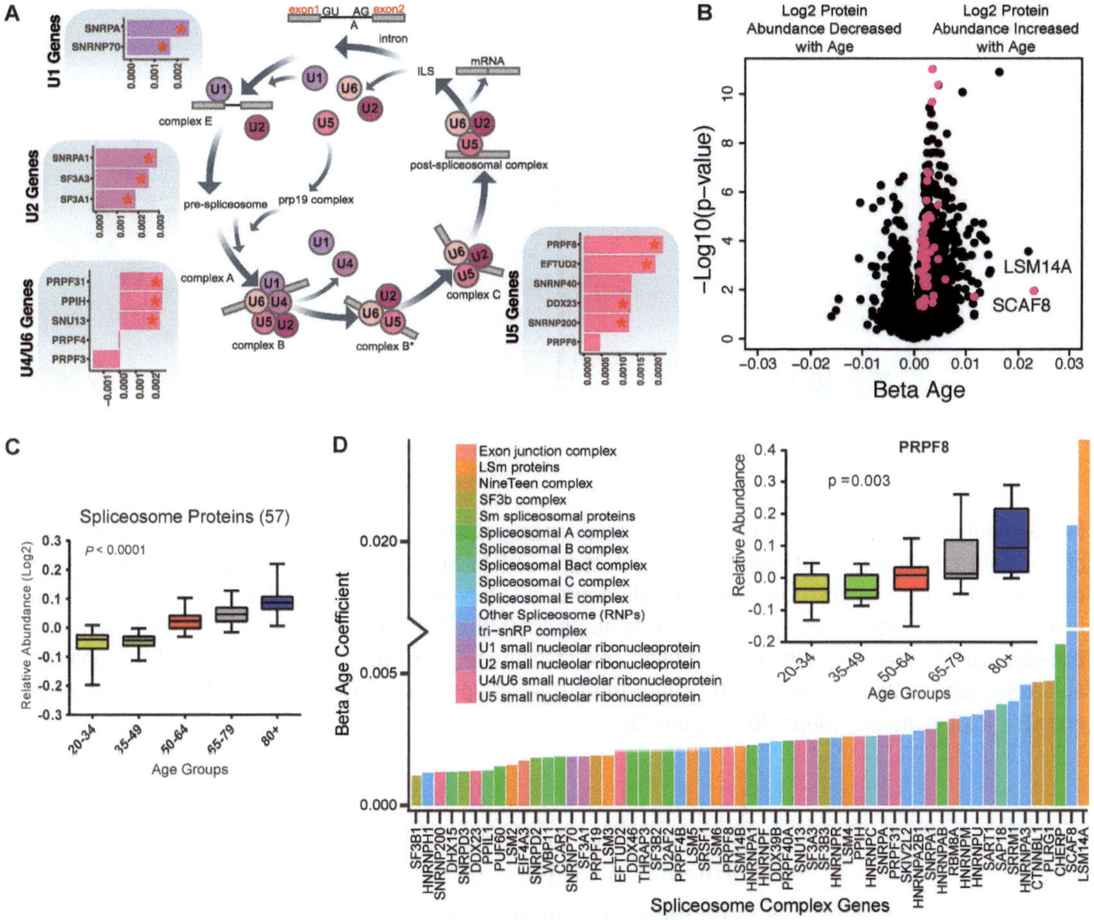

Fig. 4 Implications of proteins that modulate transcription and splicing. (**a**) Spliceosome major complex pathway protein expression abundance and dysregulation. KEGG major spliceosome complex pathway representation and spliceosome complex proteins quantified (associated with splicing RNAs U1, U2, U4/U6, and U5) as plotted in the side square boxes. (**b**) The log2 abundance expression of 57 spliceosome complex proteins associated with age ($p < 0.05$) are depicted as magenta circles, while all other quantified proteins are black circles. All snRNPs and spliceosome regulatory proteins are upregulated with age. (**c**) The average of all age-associated spliceosome proteins within each age group reveals an upregulation of spliceosome proteins with age. (**d**) Effect of age (1-year difference) on the 57 reannotated proteins of the spliceosome major complex and color coded based on spliceosome domains. Inset (left) is a legend for the complex domains and inset (right) shows that PRPF8 protein is robustly overrepresented with age. (Reproduced from [5])

2. Integrating epidemiological models in proteomic analysis helps to explore the differences of protein expression profiles across the lifespan (20–87 years). A heatmap of the 1265 age-associated proteins were shown from the mixed linear model (Fig. 5a). Hierarchical clustering of protein expression suggested that the strongest difference was between young (20–34) and old (80+), which also shows the linear relation of the protein and age with 15 years of age interval. The

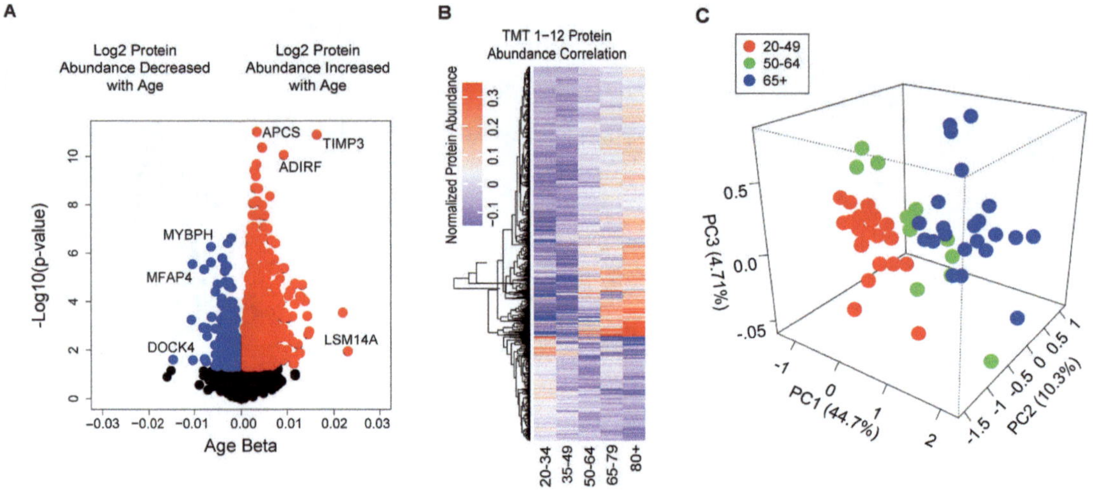

Fig. 5 Effect of age on protein expression levels. (**a**) In this volcano plot the x-axis represents the size and sign of the beta coefficient of the specific protein regressed to age (adjusted for covariates) and the y-axis represents the relative -log10 *p*-value. Each dot represents a protein, and all significant proteins are indicated in blue and red (age-associated 1265 proteins, $p < 0.05$). (**b**) The heatmap of the 1265 significantly age-associated proteins reveals changing expression profiles across aging. (**c**) PLS analysis of age-associated proteins were classified into three age groups: 20–49 (young), 50–64 years (middle age), and 65+ (old) years old. (Reproduced from [5])

separation of protein expression among the participants between three age groups (20–49, 50–64, and 65+) was confirmed by principal least square analysis (Fig. 5b, c).

3. The most notable decrease, 70% of the mitochondrial proteins, were observed in the skeletal muscle data. Most of these mitochondrial proteins are the energy resource for the skeletal muscle, and the reduction of the mitochondrial protein with age suggests that changes in muscle with aging are characterized by a profound change in energy metabolism, in particular, oxidative phosphorylation. An evident mitochondrial protein impairment across the age can be shown here because of the meticulous data analysis methodology using a linear model adjusted for gender, race, BMI, physical activity, muscle fiber type, and so on (Fig. 6).
4. While mitochondrial proteins were decreased in the data, a set of cytoplasmic proteins were increased with age, notably spliceosome complex proteins. The positive association of spliceosome complex with age is a novel finding here given the fact that the skeletal muscle data is adjusted for participants' physical activity. A positive association of spliceosome complex proteins are reported with increased physical activity.

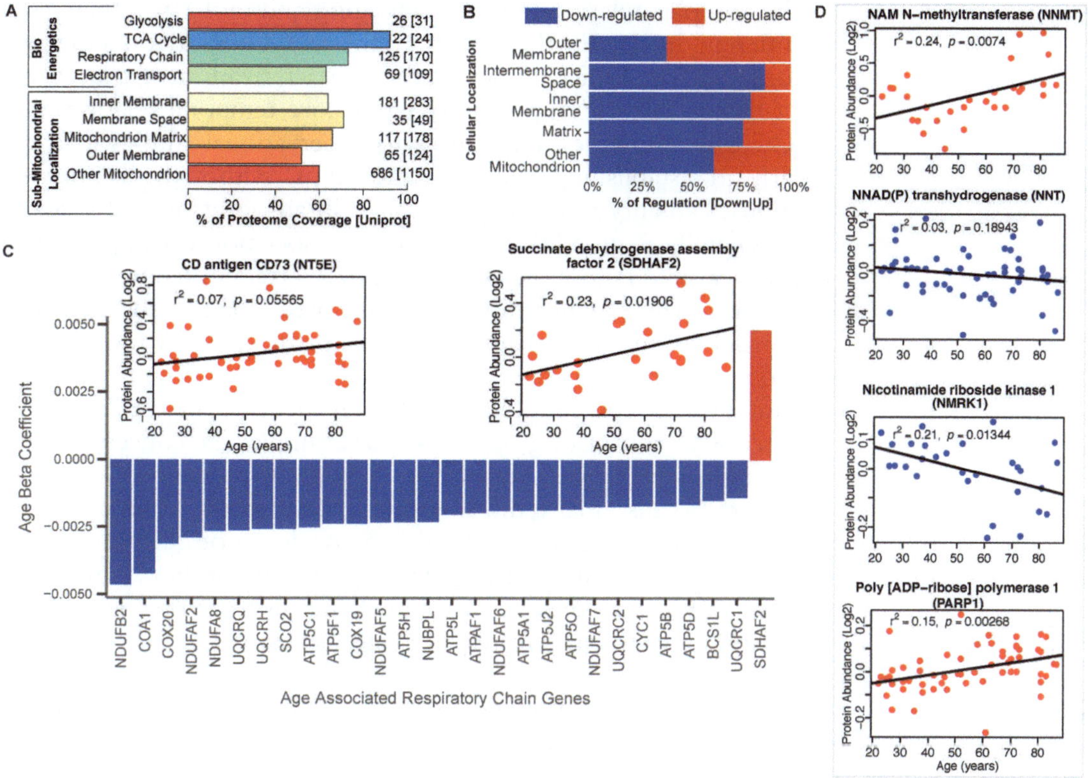

Fig. 6 Functional decline of mitochondrial proteins with age. (**a**) Percent coverage within categories of skeletal muscle proteins compared to the Uniprot database. The top section shows various energetics categories, while the *z*-axis indicates the number of proteins identified for each protein category and in parenthesis the number of proteins reported in Uniprot for the same category. (**b**) Subcellular location of age-associated mitochondrial proteins based on up- or downregulation. Of note, most of the mitochondrial proteins are downregulated. (**c**) Age-dependent decline of respiratory and electron transport chain proteins. All mitochondrial proteins in the respiratory and electron transport chain that are significantly associated with age are downregulated ($p < 0.05$) except SDHAF2. The inset panel reports data on the proteins that are significantly upregulated with aging, SDHAF2 (mitochondrial) and the membrane protein CD73. (**d**) Simple linear regression was shown for some of the NMT mitochondrial proteins, age is (*x*-axis) and protein (*y*-axis) correlation, confounders were not adjusted, and raw *p*-values were shown. (Reproduced from [5])

3.8 Advantages and Limitations of the Epidemiological Models in Proteomic Analysis

In this case study we presented examples of proteomic analyses of age in muscle and plasma in the BLSA and GESTALT studies using two platforms, MS based proteomics and SOMAscan, with the goal of identifying important proteomic biomarkers of age, and to understand molecular pathways underlying aging. The greatest advantage of conducting proteomics within an observational study is the larger sample size and depth of clinical data available to explore complex relationships. Both the BLSA and GESTALT studies have a wide range of demographic and clinical data. In addition, there are multiple levels of -omics data that have been collected, including genetics, epigenetics, gene expression, and metabolomics, allowing for layering of these data to understand the complex relationship between aging and molecular changes at different levels. Further, both BLSA and GESTALT studies are longitudinal studies, allowing the characterization of trajectories of health and the opportunity to explore temporal relationships. There are, however, limitations to observational studies, particularly in establishing causal relationships. Although establishing robust causality is almost impossible in an observational study, it is generally assumed that longitudinal association where predictors are measured before the occurrence of a certain outcome, or longitudinal analyses where a change in a predictor is associated with parallel change in an outcome, approximate causality more than a cross-sectional association. Overall, the issue of how to establish causation in epidemiology is strongly debated in the literature, often with contrasting opinions. For example, the concept of necessary and sufficient causes does not apply to many disease-risk factors because many diseases often occur in the absence of risk factors, and the presence of a risk factor does not guarantee that a certain disease will eventually develop. In spite of these limitations, observational studies are an important first step in identifying important candidates to test in clinical trials.

3.9 Conclusion

Integrating epidemiological models to proteomics analysis is a powerful strategy for discovering aging biomarkers. This methodology allows the integration of phenotypic data with biological data providing critically detailed information on the molecular and biological processes that change with age, using study design and analytic methods that account for potential biases such as confounding by race, gender, BMI, physical activity, fiber types, or other assay/technical errors. The case study presented here clearly shows the molecular and biological pattern of protein markers in healthy aging muscle and plasma, thus highlighting how epidemiological models may be applied to proteomics to gain biological insights. The underlying biological mechanism for these aging markers can be further validated through bioinformatics and experimental methods and thus facilitate new therapeutic targets and interventions for healthy aging.

4 Notes

1. Epidemiology, in broad terms, is focused on the determinants of health and disease in populations. Epidemiologists use a range of different study designs and analytic techniques to address specific questions on the relationships between exposures and outcomes. However, the concepts and tools of epidemiology are applicable across disciplines. Epidemiological methods focused on study design and measurement, as well

as choice of statistical model may be implemented in molecular studies to discern and address not only confounding but also information bias and selection bias. Further, techniques for the description of mediation and effect modification may be used to generate mechanistic hypotheses and the extensive discourse on causation in epidemiology may help to contextualize molecular observations.

2. TMT is a relative quantification method for the accurate quantification of peptides and proteins. Here we used TMT6+ where each sample peptides were labeled with unique isobaric tandem mass tags (i.e., 126, 127, 128, 129, 130, and 131) and six samples are mixed (multiplexed). Using this method, the peptides from different samples to be identified by their relative abundance with greater ease and accuracy than other MS methods.
3. The search engine was set with the following search parameters: TMT6plex lysine and N-terminus as fixed modifications and variable modifications of carbamidomethyl cysteine, deamidation of asparagine and glutamate, carbamylation of lysine and N-terminus, and oxidized methionine. A peptide mass tolerance of 20 ppm and 0.08 Da, respectively, and two missed cleavages were allowed for precursor and fragment ions in agreement with the known mass accuracy of the instrument.
4. Protein annotations are categorized into some of the nine tentative hallmarks that represent common denominators of aging in different organisms, specifically on mammalian aging. These hallmarks comprise genomic instability, telomere attrition, epigenetic alterations, loss of proteostasis, deregulated nutrient-sensing, mitochondrial dysfunction, cellular senescence, stem cell exhaustion, and altered intercellular communication. In addition to some of these categories, we included other categories relevant to skeletal muscle, such as muscle protein, neuromuscular junction, immunity, splicing, and ribosomes.

Acknowledgments

This work was supported by the Intramural Research Program of the National Institute on Aging, National Institutes of Health.

Author Contributions: Supervision: L.F; writing first draft and figure creations: CU-M, T.T, R.T., and R.M.; Editing: C.U-M, T.T, A. Z.M., P.Q, R.T, R.M., and L.F.

References

1. Kuo PL, Schrack JA, Shardell MD, Levine M, Moore AZ, An Y, Elango P, Karikkineth A, Tanaka T, de Cabo R, Zukley LM, AlGhatrif M, Chia CW, Simonsick EM, Egan JM, Resnick SM, Ferrucci L (2020) A roadmap to build a phenotypic metric of ageing: insights from the Baltimore longitudinal study of aging. J Intern Med 287(4):373–394. https://doi.org/10.1111/joim.13024
2. Tanaka T, Biancotto A, Moaddel R, Moore AZ, Gonzalez-Freire M, Aon MA, Candia J, Zhang P, Cheung F, Fantoni G, CHI consortium, Semba RD, Ferrucci L (2018) Plasma proteomic signature of age in healthy humans. Aging Cell 17(5):e12799. https://doi.org/10.1111/acel.12799
3. Rohloff JC, Gelinas AD, Jarvis TC, Ochsner UA, Schneider DJ, Gold L, Janjic N (2014) Nucleic acid ligands with protein-like side chains: modified aptamers and their use as diagnostic and therapeutic agents. Mol Ther Nucleic Acids 3:e201. https://doi.org/10.1038/mtna.2014.49
4. Candia J, Cheung F, Kotliarov Y, Fantoni G, Sellers B, Griesman T, Huang J, Stuccio S, Zingone A, Ryan BM, Tsang JS, Biancotto A (2017) Assessment of variability in the SOMAscan assay. Sci Rep 7(1):14248. https://doi.org/10.1038/s41598-017-14755-5
5. Ubaida-Mohien C, Lyashkov A, Gonzalez-Freire M, Tharakan R, Shardell M, Moaddel R, Semba RD, Chia CW, Gorospe M, Sen R, Ferrucci L (2019) Discovery proteomics in aging human skeletal muscle finds change in spliceosome, immunity, proteostasis and mitochondria. elife 8:e49874. https://doi.org/10.7554/eLife.49874
6. Bligh EG, Dyer WJ (1959) A rapid method of total lipid extraction and purification. Can J Biochem Physiol 37(8):911–917. https://doi.org/10.1139/o59-099
7. Kammers K, Cole RN, Tiengwe C, Ruczinski I (2015) Detecting significant changes in protein abundance. EuPA Open Proteom 7:11–19. https://doi.org/10.1016/j.euprot.2015.02.002
8. Herbrich SM, Cole RN, West KP Jr, Schulze K, Yager JD, Groopman JD, Christian P, Wu L, O'Meally RN, May DH, McIntosh MW, Ruczinski I (2013) Statistical inference from multiple iTRAQ experiments without using common reference standards. J Proteome Res 12(2):594–604. https://doi.org/10.1021/pr300624g
9. Lopez-Otin C, Blasco MA, Partridge L, Serrano M, Kroemer G (2013) The hallmarks of aging. Cell 153(6):1194–1217. https://doi.org/10.1016/j.cell.2013.05.039
10. Xie S, Lu L, Liu L (2019) Growth differentiation factor-15 and the risk of cardiovascular diseases and all-cause mortality: a meta-analysis of prospective studies. Clin Cardiol 42(5):513–523. https://doi.org/10.1002/clc.23159
11. Basisty N, Kale A, Jeon OH, Kuehnemann C, Payne T, Rao C, Holtz A, Shah S, Sharma V, Ferrucci L, Campisi J, Schilling B (2020) A proteomic atlas of senescence-associated secretomes for aging biomarker development. PLoS Biol 18(1):e3000599. https://doi.org/10.1371/journal.pbio.3000599

Chapter 9

Unraveling Pathways of Health and Lifespan with Integrated Multiomics Approaches

Miguel A. Aon, Michel Bernier, and Rafael de Cabo

Abstract

Distinct and shared pathways of health and lifespan can be untangled following a concerted approach led by experimental design and a rigorous analytical strategy where the confounding effects of diet and feeding regimens can be dissected. In this chapter, we use integrated analysis of multiomics (transcriptomics–metabolomics) data in liver from mice to gain insight into pathways associated with improved health and survival. We identify a unique metabolic hub involving glycine–serine–threonine metabolism at the core of lifespan, and a pattern of shared pathways related to improved health.

Key words Integrated pathway analysis, Gene and functional ontologies, Topology-based pathway network analysis, Healthy aging, Computational systems biology, Diet, Time-restricted feeding

1 Introduction

Systems biology is right at the transition between information and knowledge which, essentially, consists in the search for "meaning." A way for distilling knowledge (meaning) from comprehensive "omics" datasets derived/collected from, for example, cells and organs, is to have a connection map of their elements.

Complex systems are organized in *networks*, which are wiring maps of *nodes* and *links* (or *edges*) of their components [1]. The network components determine their nature, which can be very diverse. For example, in an *actor network*, actors are nodes and links are movies where they were coacting like a *researchers' collaboration network* where nodes are researchers and links are papers coauthored. Other examples are, for example, the Internet, power grid, the web (WWW), and protein–protein interactions [1]. *Network Science* is a new discipline, and we refer the reader to the book by Albert-László Barabási [1] as its main source.

Miguel A. Aon and Michel Bernier contributed equally to this work.

Sonia Cortassa and Miguel A. Aon (eds.), *Computational Systems Biology in Medicine and Biotechnology: Methods and Protocols*, Methods in Molecular Biology, vol. 2399, https://doi.org/10.1007/978-1-0716-1831-8_9,

Metabolism can be represented as networks in the form of graphs, in which nodes are *metabolites* and links are *chemical reactions catalyzed by enzymes*. Enzymes are proteins, thus remitting us to a genetic ontology, whereas metabolites are part of pathways, usually known as metabolic pathways, that represent a functional ontology, that is, the coexpression and activity of genes (enzymatic proteins) and metabolites. As a caveat, the framework provided by networks is topological, that is, how elements are wired between them, but remains mute about the dynamics of the network elements. For this, we need a different approach as shown in Cortassa et al., from this book.

The accessibility to high-throughput "omics" and performant, user-friendly, web-based software, have facilitated the utilization of multiomics approaches and the integration of data/information (e.g., transcripts/genes, metabolites) with different ontological origins.

In the present work, we will be dealing with bipartite networks, genes-metabolites, where a gene and a metabolite are linked if they belong to the same pathway. The integration of transcripts (genes) and metabolites data will result in metabolic pathways exhibiting specific functions, for example, energetic, redox, signaling, epigenetic, and transport. The overall analysis will lead to several pathways, whose pattern will suggest new, and hopefully, insightful, and testable hypotheses about the biological problem under study.

The case study for the present work corresponds to studies performed at the National Institute on Aging/NIH with the aim of investigating the role played by diet composition and chronic feeding paradigms on health and lifespan in mice (*Mus musculus*). Mice were fed ad libitum (AL), 30% caloric restriction (CR), and single-meal feeding (MF), the latter accounting for differences in energy density and caloric intake consumed by mice subjected to AL from two very distinct diets, a natural ingredient chow (NIC) and a purified ingredient diet (PID) (Table 1) [2]. Figure 1a shows a scheme of the experimental design, and Fig. 2 an overview of the analytical strategy.

The results showed mean lifespan extensions of 11% and 28% by MF and CR, respectively, and the increased survival effect was independent of diet type [2] (Fig. 1b, c). The duration of eating varied dramatically depending on diet type, NIC and PID (natural ingredients vs rich in fat and refined sugar, respectively) and feeding protocol (AL, MF, CR), thus resulting in differences in the length of the fasting period. Unlike in mice under AL, the MF and CR groups had extended periods of daily fasting associated with high-amplitude metabolic rhythms, consistent with high metabolic flexibility [2] (Fig. 1b). Overall, the data obtained suggested that extended periods of fasting, independent of diet composition or total caloric intake, might be an effective intervention to enhance health span and longevity [2].

Table 1
Diets composition

Purified ingredient diet (PID)		Natural ingredient chow (NIC)	
Component	**(% by weight)**	**Component**	**(% by weight)**
Lactalbumin	15	Ground wheat	34.3
Sucrose	28.5	Ground corn	22.0
Corn starch	30	Soybean hulls	10.9
Dextrin	5	Soybean meal	8.2
Corn oil	10	Fish meal	5.3
Cellulose	5	Sucrose	3.9
		Soybean oil	3.1
Vitamin mix	1	Alfalfa meal	2.9
Calcium phosphate, dibasic	2.16	Dried whey	2.9
Calcium carbonate	0.37	Brewer's yeast	1.94
Potassium citrate, monohydrate	2.08		
Sodium chloride	0.54	Limestone	1.26
Magnesium oxide	0.20	Dicalcium phosphate	0.9675
Ferric citrate	0.12	Iodized salt	0.5805
Manganous carbonate	0.013	Mineral mix	0.387
Zinc carbonate	0.006	Folate 2%	0.039
Cupric carbonate	0.002	DL-methionine	0.1258
Chromium potassium sulfate, dodecahydrate	0.002	Potassium carbonate	0.72
Potassium iodate	0.0004	Vitamin C	0.39
Sodium selenite, pentahydrate	0.00004	Vitamin mix	0.0968

2 Materials

1. Diets composition. Table 1 displays the composition of PID and NIC diets. The natural ingredient chow (NIC) is composed of agricultural byproducts, such as ground wheat, ground corn, soybean hulls, alfalfa and soybean meals, dried whey, brewer's yeast, fish meal as a protein source, low amount of sucrose (3.9%, w/w), and soybean oil (3.1%, w/w) and is supplemented with minerals and vitamins (Table 1). Thus, NIC constitutes a high fiber diet containing complex carbohydrates, with fats from soybean oil. In contrast, the purified ingredient diet (PID) consists of lactalbumin as a protein source, with carbohydrates provided by corn starch, dextrin, and 28.5% sucrose (w/w), and corn oil for fat (10%). Fiber is provided

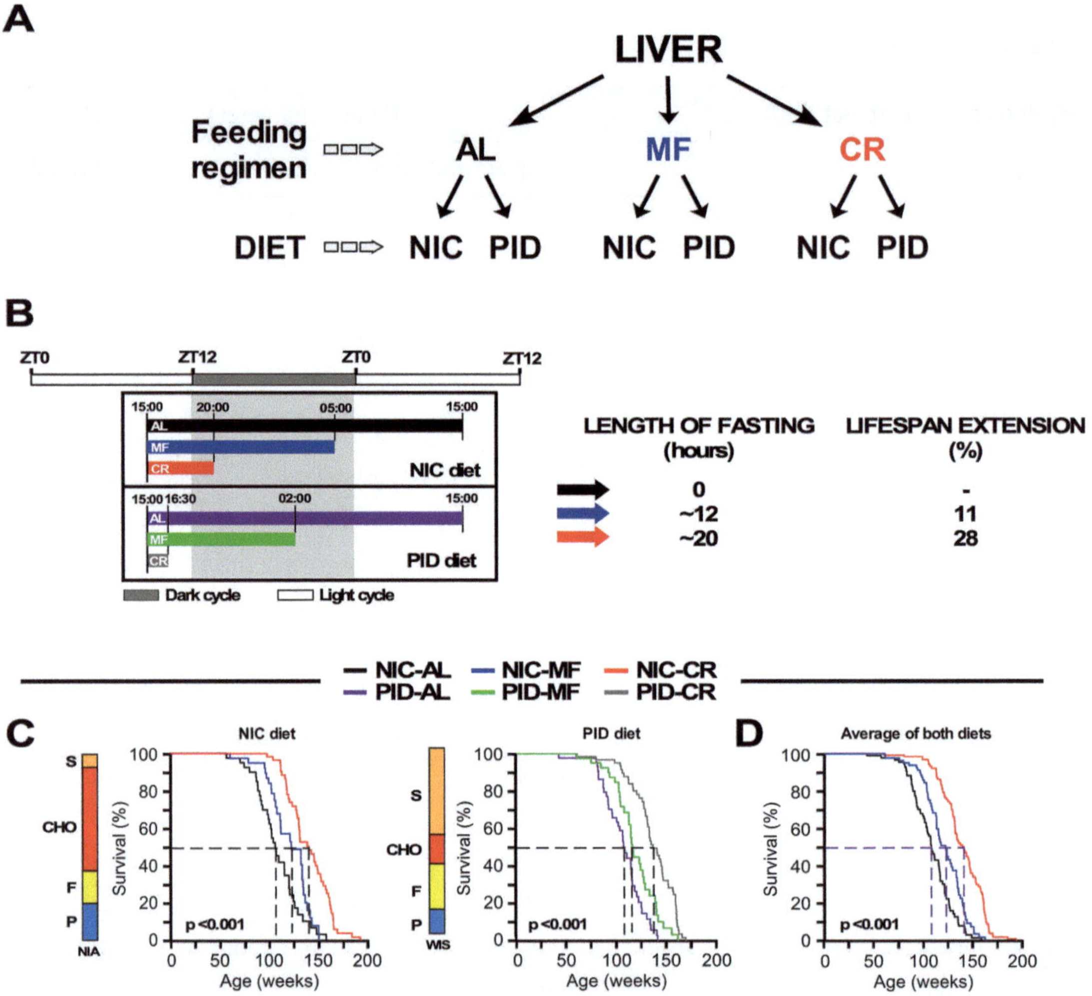

Fig. 1 Overview of experimental design and key findings previously reported. (**a**) Scheme of the experimental design showing the groups and the two main factors studied, feeding regime (AL, MF, CR, see text Section *i* "*The power of experimental...*" for details) and diet (Table 1, and Section ii, 2.1. "*Diets composition*"). (**b**) Schematic showing the estimated fasting time for each of the two diets, and three feeding paradigms across the study with respect to the zeitgeber time, which is defined as any external clue, such as the 12:12-h light/dark cycle that synchronizes common biological rhythms in an organism [2]. AL mice had constant access to food and so were not subjected to a daily fasting time. (**c**) Kaplan-Meier survival curves for mice fed either NIC diet (left panel) or PID diet (middle panel) ad libitum (AL), meal-fed (MF), or maintained on 30% calorie restriction (CR). Stacked bars depict the relative composition of the NIC and PID diets, expressed as % kcal. P, protein; F, fat; CHO, carbohydrates other than sucrose (S) [2]. (Reproduced, (panels B and C), from Mitchell et al. (2019) *Cell Metabolism* 29, 221–228)

by cellulose and the diet was also supplemented with minerals and vitamins (Table 1). Therefore, the two diet types are using nutrients from different sources (natural vs. purified) differing not only in their relative amounts of fat and carbohydrate but also in micro- and macronutrient content. NIC and PID diets differ in their phytoestrogen content from soy that is subject to seasonal and geographical variations in yield and isoflavone content which is absent from the latter diet. Sucrose is a predominant carbohydrate source in PID, and because sucrose is composed of 50% fructose, this could help explain the weight gain and development of insulin resistance and dyslipidemia seen in PID-fed nonhuman primates fed [3]. In addition to phytoestrogen or fructose content, variance in measured phenotypes between NIC- and PID-fed animals may also be related to the types of fat ingested (*see* **Note 1**).

2. Transcriptomics and metabolomics. Snap freeze in liquid nitrogen livers from 24-month-old mice fed either with NIC or PIC diet for 20 months under AL, MF, or CR. Process total liver samples for transcriptomics and metabolomics analyses.
3. Bioinformatic analysis of metabolomics data. Analyze metabolite profiles using MetaboAnalyst versions 3.0 and 4.0 [4, 5], an integrated web-based platform for comprehensive analysis of metabolomic data (https://www.metaboanalyst.ca/faces/home.xhtml). Utilize univariate and multivariate built-in analytical methods from modules of the web-based platform of MetaboAnalyst, as specified (see below). Apply pairwise comparisons, multivariate (principal component analysis, PCA; partial least square discriminant analysis, PLSDA), clustering (heat map, correlation matrix), and pattern finding statistical analyses to the analysis of metabolite profiles.
4. Multiomics transcriptomics–metabolomics. Utilize the Joint Pathway Analysis (JPA) module from MetaboAnalyst 3.0 and 4.0 that enables the combination of transcriptomics and metabolomics data for functional enrichment analysis and pathway topology analysis [4, 5]. Utilize topological analysis based on *degree centrality* and *betweenness centrality* metrics [5] in addition to enrichment analysis that evaluates whether the observed genes or metabolites in a pathway appear more frequently than expected by random chance.

3 Methods

3.1 Sample Preparation for Liver Transcriptomics

1. Extract RNA from livers of mice using the TRIzol reagent (Invitrogen, Carlsbad, CA) according to standard protocols.
2. Determine total RNA quantity and quality using, for example, the Agilent Bioanalyzer RNA 6000 Chip (Agilent, Santa Clara, CA).
3. Label five hundred ng total RNA according to the manufacturer's instructions, for example, the Illumina® TotalPrep™ RNA amplification kit (Illumina, San Diego, CA).

4. Hybridize a total of 750 ng biotinylated aRNA to the Illumina Mouse Ref-8 v2 BeadChip overnight.
5. Following post hybridization rinses, incubate arrays with streptavidin-conjugated Cy3, and scan at a resolution of 0.53 μm using an Illumina iScan scanner.
6. Extract hybridization intensity data from the scanned images using Illumina Bead Studio Genome Studio software, V2011.1.
7. Subject rRaw data to *Z*-normalization as described [6, 7].
8. Perform principal component analysis (PCA) on the normalized *Z*-scores of all the detectable probes in the samples.
9. Select significant genes by the *z*-test <0.05, false discovery rate < 0.30, as well as *z*-ratio > 1.5 in both directions and ANOVA *p* value <0.05.
10. Calculate the pairwise distances between samples considering each microarray as a point in a high-dimensional space by treating each probe as a variable. For parametric analysis of gene set enrichment (PAGE), test the expression data using the PAGE method as previously described [8]. Briefly, compute an aggregated *Z* score for each pathway under each pair of conditions:

$$= \frac{\frac{Z-\text{score pathway}}{n\text{pathway}} Z - \text{ratio genes in the pathway} - Z - \text{ratio genes on the array}}{\sigma\text{sample}}$$

where *n*pathway is the number of genes in the specific pathway and *σ*sample is the standard deviation of *Z*-ratio on the comparison sample arrays. Compute for each *Z* (pathway), a *P* value to the total *Z-ratio*.

3.2 Sample Preparation for Liver Metabolomics

Metabolomics analysis on mouse liver extracts was performed by the West Coast Metabolomics Center at UC Davis. A detailed account of the procedure employed to prepare the samples for analysis is described in Mitchell et al. (2016) [9].

3.3 Pathways of Lifespan

1. Since the results indicated that MF and CR interventions prolonged the lifespan of mice, irrespective of diet (Fig. 1c), we used the CR–AL and MF–AL ratios of transcripts or metabolites to assess which of them were significantly up- or down-modulated in each feeding regimen. For statistical significance, define a threshold for transcripts (genes) *Z* ratio ≥ 1.5 in either direction, false discovery rate < 0.3, $p < 0.05$, and for metabolites, fold change $\geq$1.2-fold and $\leq$ 0.8-fold) (*see* **Note 2**).
2. Pairwise comparisons of CR–AL and MF–AL identified 1926 and 1032 unique gene transcripts, respectively, whereas similar comparisons for the metabolome data identified a total of

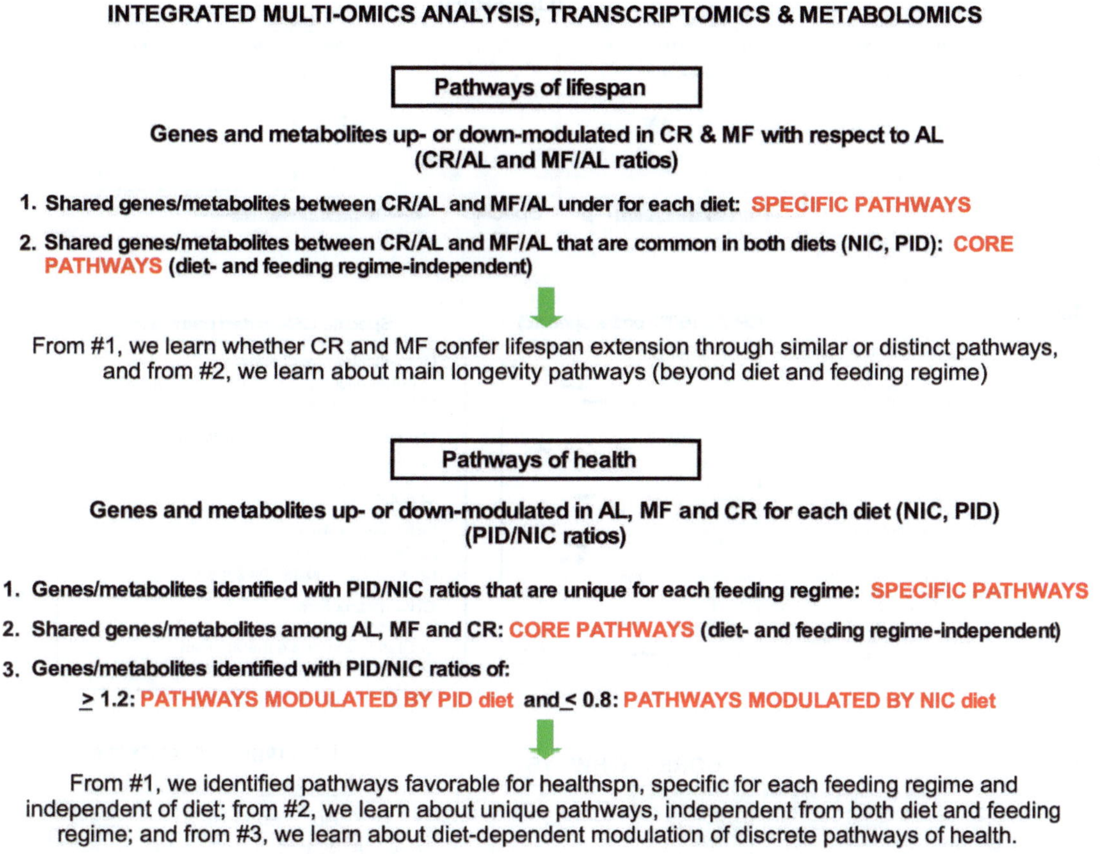

Fig. 2 Overview flow diagram of the analytical scheme employed in the multiomics analysis. The figure describes how pathways of lifespan or health span, shared and specific, were detected. Explicit are the rationale and outcome of each step performed in the analysis which is described in detail in the main text

155 metabolites, of which CR–AL and MF–AL comparisons had 39 and 43 unique, significantly changed metabolites, respectively (Fig. 3); *see also* Figs. 1a and 2, which shows an overview and rationale of the analytical strategy (*see* **Note 3**).

3. Venn diagrams were then used to identify *shared* or *unique* genes or metabolites that were significantly altered by MF or CR (independent of diet) (Fig. 3). In each feeding regimen, two criteria to discover pathway enrichment among experimental groups can be employed: Pathways were defined as “specific” when independent of diet (PID or NIC) but responsive to either MF or CR, and “core” when independent of both diet and feeding regime (Figs. 2 and 3) (*see* **Note 4**).

4. Essentially, the overlapping parts of the Venn diagrams correspond to “core” genes or metabolites. Subsets of shared transcripts/metabolites exhibiting significant up- or downmodulation with respect to AL were as follows: in CR, 700 genes and 19 metabolites, and in MF, 69 genes and

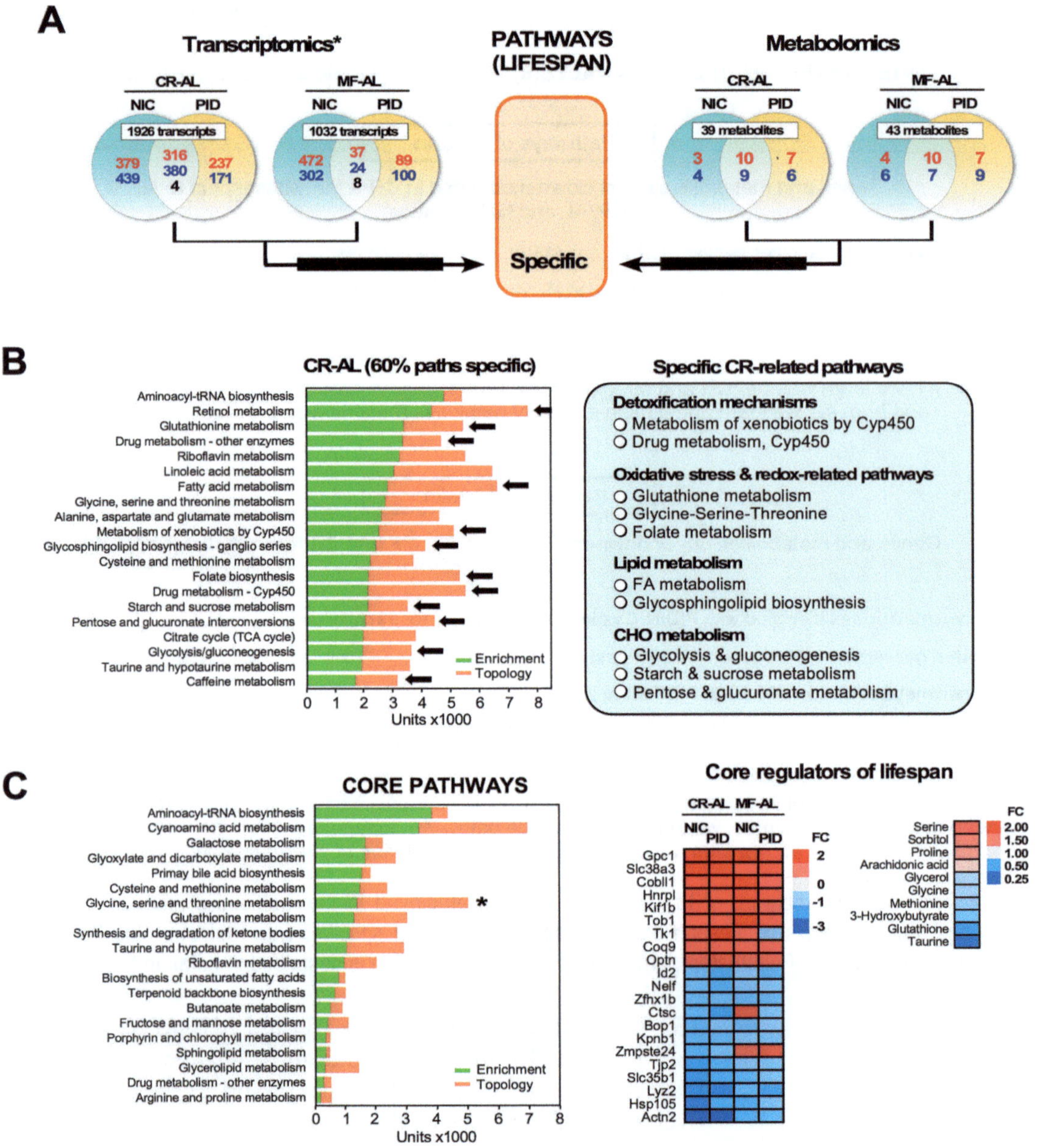

Fig. 3 Multiomics analysis of liver extracts: Specific pathways of lifespan. (**a**) Venn diagrams depict the distribution of transcripts (left panels) and metabolites (right panels) in the liver of mice on NIC or PID diet in response to the indicated pairwise comparisons (CR-AL and MF-AL). Shared elements constitute "specific" attributes associated with lifespan within each pairwise comparison. Upregulation (red font), downregulation (blue font), and reciprocal regulation (black font) of significantly impacted elements (transcripts and metabolites) are depicted. (**b**) Multiomics analysis using transcriptomics and metabolomics data was performed according to the analytical scheme shown in (**a**). The Joint Pathway Analysis from MetaboAnalyst 3.0 was used to calculate the bar chart, which is a combination of enrichment *p* values (green bars) and topology analysis (orange bars) of the pathways denoted on the *y*-axis. The pathways impacted by CR are shown as an example, where black arrows denote biosynthetic and metabolic pathways specifically influenced by CR when compared to AL-fed controls. The number/scale in *x*-axis is arbitrary and is calculated by scaling enrichment and topology to the same range (0–1), then summed up and multiplied by 1000. (**c**) Heatmap visualization of 21 core transcripts (left) and 10 core metabolites (right) shared regardless of diet or feeding regimen (*see* Tables 2 and 3). Upregulation (red font) and downregulation (blue font). FC, fold change. (Reproduced from Aon, Bernier et al. (2020) *Cell Metabolism* 32, 1–17)

17 metabolites (Fig. 3a). Utilize these CR and MF gene/metabolite subsets as input for multiomics analysis.

5. Utilize the Joint Pathway Analysis (JPA) module from MetaboAnalyst 3.0 or 4.0 to perform a multiomics approach for integrated gene- and metabolite-based pathway analysis. JPA allows for functional enrichment and pathway topology analyses from a combination, in this case, of transcriptomics and metabolomics data (*see* **Note 5**).
6. Seek for factors collectively responding to CR and MF that differ from AL but are independent of diet and of feeding paradigm to discern which pathways are contributing to the shared outcome of improved longevity (Fig. 2). A "core" subset of 21 genes (Table 2) and 10 metabolites (Table 3) were shared between CR and MF (Fig. 3b). This analytical strategy led to the discovery of glycine–serine–threonine metabolism as a major metabolic hub of lifespan in liver, and a source of main chemical donors of posttranslational and epigenetic modification such as methylation, acetylation, phosphorylation, and redox (Fig. 4) (*see* **Note 6**).

3.4 Pathways of Health Span

1. To investigate the role of feeding regimen in health, identify the NIC- and PID-responsive transcriptome and metabolome for each feeding regimen (AL, MF, and CR) rather than using AL as a reference, and then look for signatures that are either specific to each regimen or shared among regimens (Fig. 5).
2. To obtain diet-dependent enrichment in transcripts sort differentially abundant species in PID over NIC with *Z* score cutoffs of $\geq$1.5 in both directions and ratio fold-change (FC) (cut-off $\geq$1.2-fold or $\leq$ 0.8-fold), that yield a set of nonredundant annotated genes. Conduct a similar analysis with the metabolome. For metabolites, perform this step with the MetaboAnalyst software using the module "Statistical Analysis" (Fig. 5a) (*see* **Note 7**).
3. From these lists of genes/metabolites, conduct a separate analysis of each of the three regimens, ranking the top 20 pathways based on enrichment and topology (Fig. 5b, left two panels). Perform this step with the JPA module from MetaboAnalyst 3.0. In this approach, the analysis is based on enrichment and topological metrics and therefore does not account for the quantitative fold-changes between both diets. Thus, the results shown in Fig. 5b, left two panels, are, de facto, diet-independent (see next subsection, and **Note 8**).
4. Six pathways were identified as being common (magenta arrows) among feeding regimens, including metabolism of xenobiotics and drugs by cytochrome P450, glutathione metabolism, linoleic acid metabolism, TCA cycle (aka Krebs

Table 2
List of differentially expressed mRNA genes associated with the pro longevity effects of CR and MF regardless of the diet type

Accession	Gene symbol	Name	FC (CR-AL) NIA	FC (CR-AL) WIS	FC (MF-AL) NIA	FC (MF-AL) WIS
NM_016696	Gpcl	Glypican 1	1.85	1.56	1.89	1.46
NM_023805.2	Slc38a3	Solute carrier family 38, member 3	1.85	1.63	1.35	1.43
NM_177025	Cobll1	Cordon-bleu WH2 repeat protein like 1	1.45	1.45	1.66	1.21
NM_177301.3	Hnrpl	Heterogeneous nuclear ribonucleoprotein L	1.44	1.38	1.51	1.38
NM_207682.1	Kif1b	Kinesin family member 1B	1.39	1.31	1.39	1.21
NM_009427.1	Tob1	Transducer of ERBB-2.1	1.33	1.32	1.78	1.36
NM_009387.1	Tk1	Thymidine kinase 1	1.32	1.73	1.21	−1.23
NM_026452	Coq9*	Coenzyme Q9	1.27	1.26	1.23	1.24
NM_181848.3	Optn	Optineurin	1.26	1.41	1.24	1.26
NM_010496.2	Id2	Inhibitor of DNA binding 2	−1.46	−1.77	−1.22	−1.66
NM_020276.2	Nelf	Nasal embryonic LHRH factor	−1.47	−1.70	−1.30	−1.29
NM_172700.1	Zeb2*	Zinc finger E-box-binding homeobox 2	−1.55	−1.47	−1.57	−1.58
NM_009982.2	Ctsc	Cathepsin C	−1.56	−1.92	1.40	−1.34
NM_013481.1	Bop1	Block of proliferation 1	−1.58	−1.26	−1.50	−1.20
NM_008379.2	Kpnb1	Karyopherin subunit beta 1	−1.60	−1.55	−1.26	−1.38
NM_172700.1	Zmpste24	Zinc metallopeptidase STE24	−1.65	−1.37	1.28	1.40
NM_011597	Tjp2	Tight junction protein 2	−1.98	−1.66	−1.24	−1.34
NM_016752.1	Slc35b1*	Solute carrier family 35, member B1	−2.04	−1.71	−1.62	−1.36
NM_017372.2	Lyzs	Lysozyme	−2.16	−2.29	−1.56	−1.95
NM_013559.1	Hsp105*	Heat shock protein 105/110-kDa	−2.43	−1.76	−1.75	−1.38
NM_145368.2	Acnat2*	Acyl-coenzyme A amino acid N-acyltransferase 2	−3.04	−3.44	−1.77	−1.65

These significant, differentially expressed genes were selected based on the following criteria: Zratio >1.5 in both directions, false discovery rate < 0.30, ANOVA p value <0.05, and fold change (FC) > 1.2 in both directions. A heatmap was generated and can be found in Fig. 3b, that is, "Core regulators of lifespan." *Asterisks denote that these genes have aliases which were originally used in the microarray analysis: Coq9, 2310005O14Rik; Zeb2, Zfhx1b; Slc35b1, Ugalt2; Hsp105, Hsph1; Acnat2, C730036D15Rik

cycle), and fatty acid (FA) elongation. Conservation of these pathways is consistent with their critical role in hepatic function (Fig. 5b, two right panels).

These common pathways were not equivalently ranked among the feeding regimens, suggesting that there may be

Table 3
Two-way ANOVA dissection of the effect of feeding regimen (treatment) or diet on lifespan according to the % effect on the metabolite's variance in mouse liver and serum

Metabolite	Liver		
	Treatment	Diet	Interaction
Glucose	NS	NS	NS
Glucose 6P	NS	** (18%)	NS
Glucose 1P	NS	** (24%)	NS
Ribose	* (18%)	NS	NS
Maltose	NS	**** (39.5%)	NS
Maltotriose	NS	** (21%)	NS
Sorbitol	* (15%)	* (9%)	NS
Fructose	* (12%)	NS	NS
Lactate	* (14%)	NS	NS
Pyruvate	* (17%)	NS	NS
Citrate	** (28%)	NS	NS
Malate	NS	NS	NS
Asparagine	* (22%)	NS	NS
Valine	NS	**** (35%)	NS
Isoleucine	NS	** (20%)	NS
Leucine	NS	** (17%)	NS
Aspartate	* (16%)	NS	* (20%)
Serine	** (20%)	*** (20%)	NS
Alanine	NS	NS	NS
Tryptophan	** (28%)	NS	NS
Tyrosine	NS	* (15%)	NS
Threonine	NS	** (24%)	NS
Glutamate	NS	NS	NS
Glutamine	NS	NS	NS
Phenylalanine	NS	** (19%)	NS
Proline	** (20%)	** (16%)	NS
Glycine	* (9%)	** (26%)	NS
Lysine	NS	NS	NS
Fumarate	NS	NS	NS
Ornithine	*** (30%)	NS	** (30%)
Urea	* (20%)	NS	NS
N-acetylglutamate	**** (54%)	NS	NS
Methionine	* (16%)	*** (27%)	NS
Cysteine	NS	NS	NS
Nicotinamide	NS	NS	NS
Glycerol	**** (45%)	* (8%)	NS
Oleate	** (31%)	NS	NS
Palmitic	*** (36%)	NS	* (14%)
Palmitoleic	* (22%)	NS	NS
Linoleic	** (30%)	NS	NS
Stearic	NS	NS	NS
Arachidonic acid	** (12%)	**** (51%)	NS
Myristic	NS	NS	NS
Squalene	**** (56%)	NS	NS
Cholesterol	NS	NS	** (23%)

3-HB	* (17%)	** (16%)	NS
Taurine	*** (33%)	*** (22%)	NS
Glutathione	* (17%)	* (13%)	NS
AMP	** (31%)	NS	NS
IMP	**** (47%)	NS	NS
Hypoxanthine	** (26%)	NS	NS
Adenosine	* (18%)	NS	NS
Metabolite	**Serum**		
	Treatment	**Diet**	**Interaction**
Glucose	NS	* (16%)	NS
Ribose	* (17%)	NS	NS
Sorbitol	** (26%)	NS	NS
Fructose	* (15%)	NS	NS
Lactate	NS	NS	NS
Citrate	* (20%)	NS	NS
Malate	** (27.5%)	NS	NS
Isocitrate	* (21%)	NS	NS
Asparagine	*** (35%)	NS	NS
Valine	**** (57%)	* (5%)	S* (8%)
Isoleucine	**** (48%)	**** (20%)	NS
Leucine	**** (56%)	NS	NS
Aspartate	NS	* (10%)	L* (20%)
Serine	* (18%)	NS	NS
Alanine	** (15%)	**** (43%)	NS
Tryptophan	NS	* (19%)	NS
Tyrosine	*** (41%)	NS	NS
Threonine	*** (39%)	NS	NS
Glutamate	** (32%)	NS	NS
Glutamine	NS	NS	NS
Phenylalanine	*** (38%)	NS	NS
Proline	* (18%)	NS	NS
Glycine	** (24%)	NS	NS
Lysine	**** (32%)	* (7%)	S** (20%)
Fumarate	** (27%)	NS	NS
Ornithine	* (16%)	NS	L** (30%) S** (18%)
Urea	NS	NS	NS
N-acetyl Glutamate	*** (40%)	NS	NS
Citrulline	NS	**** (38%)	NS
Methionine Sulfoxide	**** (62%)	* (5%)	NS
Glycerol	NS	* (11%)	NS
Oleate	* (19%)	NS	NS
Palmitic	*** (29%)	*** (20%)	L* (14%)

Palmitoleic	NS	NS	NS
Linoleic	NS	NS	NS
Stearic	*** (27%)	** (15%)	NS
Arachidonic	* (12%)	**** (49%)	NS
Myristic	*** (20%)	**** (34%)	S** (16%)
Cholesterol	NS	NS	L** (23%)
3-HB	* (14%)	** (19%)	NS
Taurine	* (18%)	NS	NS
Adenosine	** (28%)	NS	NS

Metabolites influenced by treatment (AL, MF, CR) alone, diet alone, or by "diet x treatment" interaction according to two-way ANOVA analysis. Metabolites influenced by the interaction between treatment and diet are depicted in red. * $p < 0.05$; ** $p < 0.01$; *** $p < 0.001$; **** $p < 0.0001$. *NS* not significant

differences in how these pathways are engaged as a function of feeding behavior [10]. A separate set of pathways, indicated by the black arrows, were specific to one regimen or another. For example, folate metabolism was detected for both MF and CR but appeared more prominently in the CR ranking (Fig. 5b).

5. Then, use Venn diagrams to identify the genes and metabolites that are common among the three feeding regimens (Fig. 5b, two right panels) independent of diet composition (Fig. 5a and Table 4). A subset of 41 transcripts and 14 metabolites was found and employed for multiomics analysis using JPA. Four of the six shared pathways (xenobiotics and drug metabolism by cytochrome P450, glutathione metabolism, and ω6-linoleic acid metabolism) rank at the top of the "core pathways" according to combined enrichment and network topology metrics (Fig. 5b).
6. At this point, a metabolic interpretation of the JPA results becomes paramount. Particularly, FA elongation and TCA cycle, in addition to the aforementioned pathways (Fig. 5b), closely interrelate with steroid hormone biosynthesis (via cytochrome P450), taurine–hypotaurine and cysteine–methionine metabolism (via glutathione), and lipid metabolism (via ω3-linolenic and ω6-arachidonic acids, and glycerophospholipids) (Fig. 6). Cytochrome P450-related xenobiotic and drug metabolism exhibits high *centrality* values (0.49 and 0.86) but low *betweenness centrality* (0.012 and 0.018). On the contrary, glutathione, TCA cycle, ω6-linoleic acid and FA elongation display high values of both topological metrics (e.g., glutathione 0.39 and 0.78; TCA cycle 0.63 and 1.38; FA elongation 0.60 and 1.36, respectively) meaning that, unlike xenobiotics-drug metabolism, which constitutes central but isolated networks, the other pathways exhibit both centrality and high traffic properties suggesting that they connect diverse networks (Figs. 4 and 6) (*see* **Note 8**).

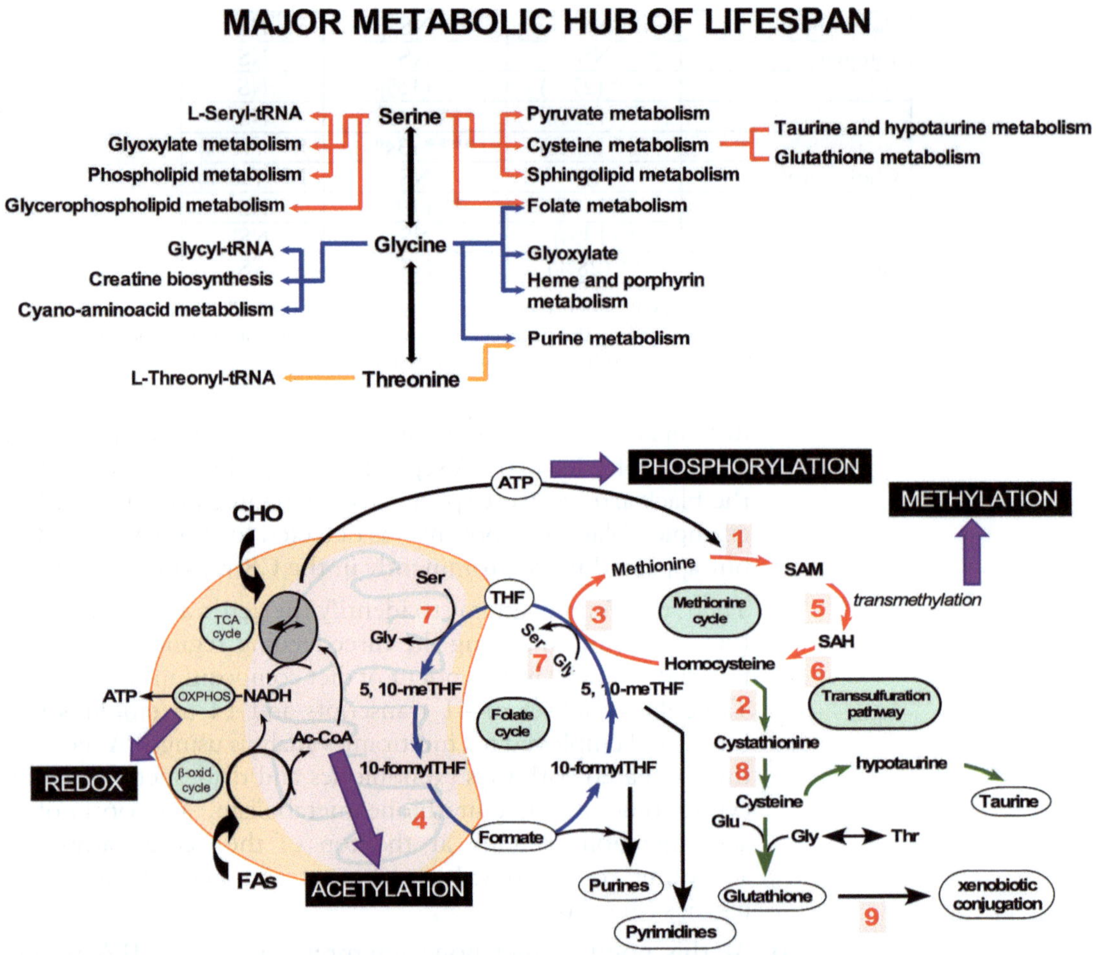

Fig. 4 Glycine (Gly)–serine (Ser)–threonine (Thr) metabolism as a major metabolic hub in lifespan. Top, scheme depicting the hub nature of Gly, Ser, and Thr metabolic network as it relates to multiple pathways. Bottom, integration of the Gly-Ser-Thr metabolism with folate, methionine, and transsulfuration pathways leading to the biosynthesis of nucleotides, transmethylation reactions, and glutathione generation. This metabolic hub is also a source of chemical donors (acetyl-CoA, S-adenosyl methionine, ATP, H_2O_2) of posttranslational and epigenetic modifications (acetylation, methylation, phosphorylation, redox) leading to changes in enzyme activity, metabolic fluxes, and gene expression. Enzyme-catalyzed reactions: (1) methionine adenosyltransferase (MAT, Mat); (2) cystathionine beta-synthase (CBS, Cbs); (3) methionine synthase (MS, Mtr); (4) methylenetetrahydrofolate reductase (MTHFR, Mthfr); (5) S-adenosyl-L-methionine-dependent methyltransferase (Mtase); (6) S-adenosylhomocysteine hydrolase (SAHH, Sahh); (7) serine hydroxymethyltransferase (SHMT, Shmt); (8) cystathionine g-lyase (CTH, Cth); (9) glutathione S-transferase (GST, Gst). (Reproduced (with partially modified bottom panel) from Aon, Bernier et al. (2020) *Cell Metabolism* 32, 1–17)

3.5 The Impact of Diet on Health Preservation

1. To investigate the impact of diet (PID, NIC) within each feeding paradigm (AL, MF and CR), utilize, as input for the multiomics JPA, the respective genes and metabolites whose fold-change from the ratio PID/NIC was upmodulated (threshold >1.2) thus influenced by the PID diet (Fig. 5a,

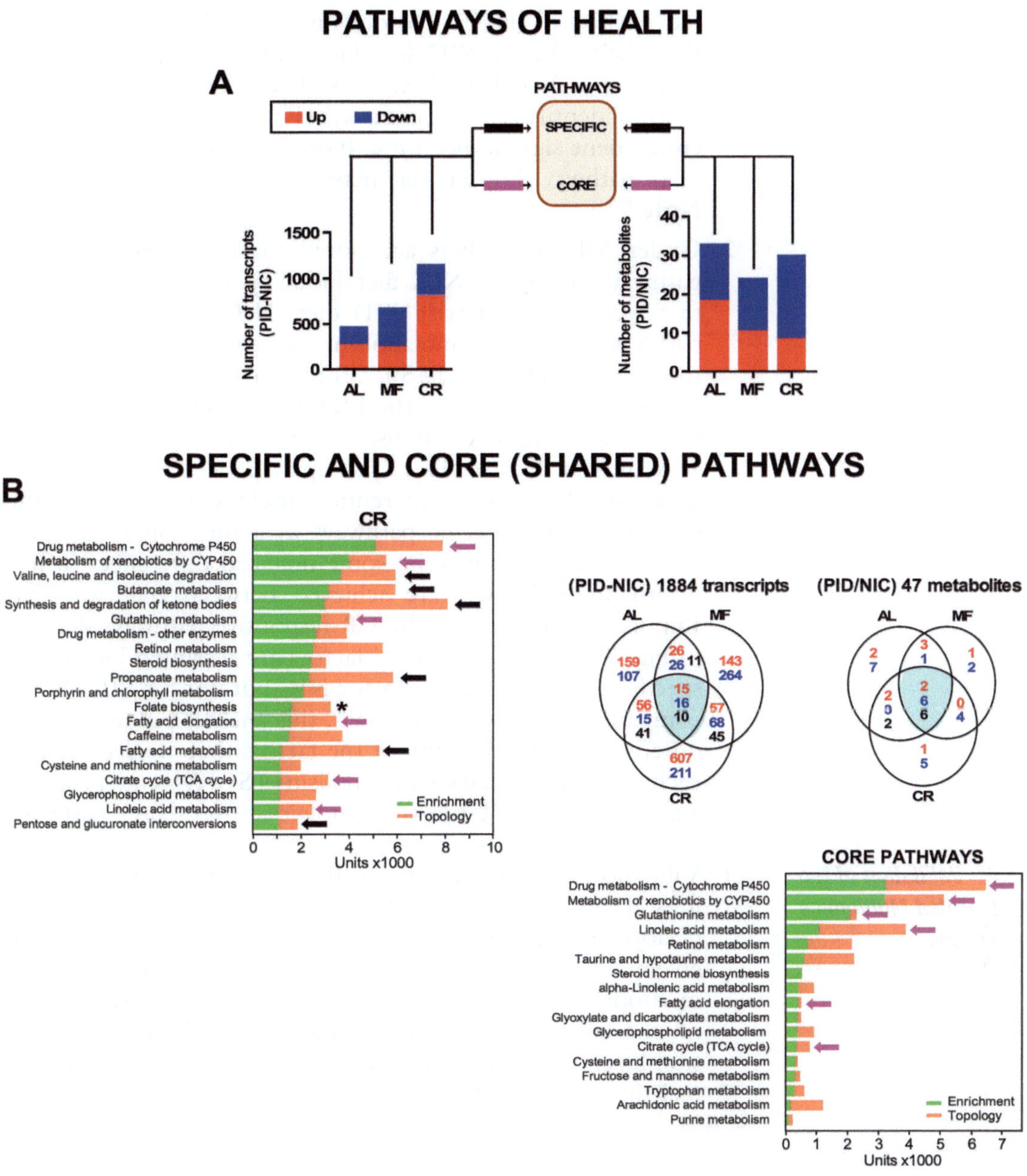

Fig. 5 Multiomics analysis of liver extracts: *Specific* and *Core* pathways of health preservation in response to feeding regimes. (**a**) Bar graphs show the number of up- (red) and downmodulated (blue) liver genes and metabolites in PID over NIC diet from mice fed with AL, MF, and CR. The shared genes/metabolites complying with the cutoff threshold (fold change >1.2 or < 0.8) and possessing valid identification (ID) were utilized as input for multiomics analysis shown in (**b**). (**b**) Multiomics analysis using transcriptomics and metabolomics data was performed according to the analytical scheme shown in panel A (*see also* Fig. 2). JPA analysis from MetaboAnalyst 3.0 was used to calculate composite bars comprising enrichment (green) and topology (orange) of top pathways for each feeding paradigm. (**b**, top left) Displayed, as an example, are the pathways influenced by the feeding regime corresponding to 30% calorie restriction (CR) where black arrows denote biosynthetic and metabolic pathways specific for CR while magenta arrows indicate pathways that are common to all feeding regimes tested in this study, that is, AL, MF, and CR [10]. Asterisk (*) denotes the

red portion of the bar) or downmodulated (threshold <0.8) thus elicited by the NIC diet (Fig. 5a, blue portion of the bar). Conduct JPA with both subsets. The following criteria were used to identify relevant pathways (Fig. 7, green boxes): *y*-axis, enrichment significance ($p < 0.05$ [$-\log(p) > 1.3$]) and *x*-axis, pathway impact for network topology (> 0.5) (*see* **Note 9**).

2. Under MF, both diets accounted for numerous significant pathways, with the NIC diet inducing changes in the liver that were not seen with PID diet. NIC diet was linked to pathways such as "metabolism of xenobiotics and drugs through cytochrome P450," the "ω6 PUFA linoleic," and "NAD salvage," whereas the PID diet promoted mostly pathways from central catabolism, that is, carbohydrate, amino acids and TCA cycle (Fig. 7, compare panel E with B). These data show that there is differential enrichment of factors within the shared and specific pathways as a function of diet and suggest that there may be some nuanced differences in how health and longevity are achieved between the two diets. Importantly, looking at the CR-responsive pathways depicted in Fig. 7b, several additional factors are recruited beyond the specific and shared pathways for NIC but not for PID (Fig. 7, compare panel F with C). Again, this analysis points to subtle differences in the identity of the factors that are recruited by each diet to the specific—metabolism of SCFAs (e.g., propionic and butyric) and PUFAs (i.e., linoleic) —and shared pathways.

3.6 Validation of the Integrated Multiomics Analyses

1. Validation of the multiomics analyses leading to the main findings of the work is a very important step. On the one hand, we performed an assessment of the presence of proteins with enzymatic activity, suggested by the pathway analysis, using methods independent from those utilized for generating the microarray data (and other -omics, if applicable). The expression level of key genes and enzymes was investigated utilizing real-time PCR (qRT-PCR) and immunoblotting, respectively.

Fig. 5 (continued) relevance of folate biosynthesis in MF and CR groups. (**b**, top right) Three-way Venn diagram depicting the distribution of common elements regardless of the feeding regimen (CR, MF, or AL). Highlighted are shared 41 out of 1884 transcripts and 14 out of 47 metabolites (*see* Table 4). These shared elements constitute common attributes regardless of diet type and feeding regimen, which determine *Core* pathways as described next. Upregulation (red font), downregulation (blue font), and reciprocal regulation (black font) of significantly impacted transcripts/metabolites are depicted. (**b**, bottom right) Top 17 *Core* pathways calculated by JPA with similar bar coding described above in the legend to (**b**). Magenta arrow denotes common pathways independent of the feeding regime, as shown in (B, top left). (Reproduced from Aon, Bernier et al. (2020) *Cell Metabolism* 32, 1–17)

Table 4
List of CORE transcripts and metabolites in the liver associated with the effects of AL, CR and MF regardless of the diet type

			Fold change (WIS-NIA)		
Accession	Gene symbol	Name	AL	MF	CR
NM_007468.2	Apoa4	Apolipoprotein A-IV	4.11	3.81	4.34
NM_016696	Gpcl	Glypican 1	1.78	1.39	1.50
NM_199472.1	MGC68323	Glyceraldehyde-3-phosphate dehydrogenase pseudogene	1.56	1.25	1.70
NM_025703.2	Tceal8	Transcription elongation factor A-like 8	1.55	1.56	1.39
NM_019699	Fads2	Fatty acid desaturase 2	1.52	1.95	1.32
NM_016895.2	Ak2	Adenylate kinase 2	1.51	1.22	1.24
NM_008439.2	Khk	Ketohexokinase	1.42	1.46	1.43
NM_153173	Histlh4h	Histone cluster 1, H4h	1.38	1.51	1.30
NM_009929.2	Col18al	Collagen, type XV111, α1	1.36	1.20	1.43
NM_026395.1	Rerl	RER1 retention in ER 1 homolog	1.36	1.26	1.29
NM_0100002.1	Cyp2c38	Cytochrome P450, family 2, subfamily c, polypeptide 38	1.35	−1.36	1.35
NM_030693.1	Atf5	Activating transcription factor 5	1.33	1.50	1.42
XM_109657.3	Fnip1*	Folliculin interacting protein	1.31	−1.32	−1.20
NM_029362.2	Chmp4b*	Charged multivesicular body protein 4B	1.30	1.44	−1.43
NM_028710.1	Arsg*	Arylsulfatase G	1.29	1.53	1.29
NM_172265.1	Eif2b5	Eukaryotic translation initiation factor 2B, subunit 5	1.27	1.39	−1.21
NM_025633	Metapl1*	Methionyl aminopeptidase type 1D (mitochondrial)	1.22	1.34	1.26
NM_010756.3	Mafg	v-maf musculoaponeurotic fibrosarcoma oncogene family, protein G	1.22	−1.29	1.38
NM_001001892.1	H2-Kl	Histocompatibility 2, K1, K region	1.20	1.32	1.25
NM_019828.1	Trpc4ap	Transient receptor potential cation channel, subfamily C, member 4 associated protein	1.20	1.37	1.24
NM_029017	Mrpl47*	Mitochondrial ribosomal protein L47	−1.21	−1.40	−1.33
NM_028868.1	Cxxcl	CXXC finger 1 (PHD domain)	−1.23	−1.33	1.29
NM_007496	Atbfl*	Zinc finger homeobox 3	−1.24	−1.27	1.24
NM_001039198.1	Zfhx2	Zinc finger homeobox 2	−1.25	−1.26	1.38
NM_025782	Ttc39b*	Tetratricopeptide repeat domain 39B	−1.26	−1.69	−1.59
NM_001013785.1	Akrlc19	Aldo–keto reductase family 1, member C19	−1.29	−1.67	−1.37
NM_019422	Elovl1	Elongation of very long chain fatty acids-like 1	−1.30	1.34	1.28
NM_024268.1	Zc3h18*	Zinc finger CCCH-type containing 18	−1.32	−1.20	−1.23
NM_009454.2	Ube2e3	Ubiquitin-conjugating enzyme E2E3,	−1.33	−1.26	1.20
NM_027249.2	Tlcd2*	TLC domain containing 2	−1.44	−1.32	−1.33
NM_00998.1	Pcytla	Phosphate cytidylyltransferase 1, choline, alpha isoform	−1.47	−1.42	−1.30
NM_019792.1	Cyp3a25	Cytochrome P450, family 3, subfamily a, polypeptide 25	−1.48	−2.23	−1.90

(continued)

Table 4 (continued)

Accession	Gene symbol	Name	Fold change (WIS-NIA)		
			AL	MF	CR
XM_354627.1	2010305C02Rik	TLC domain containing 2	−1.50	−1.31	−1.26
NM_183278.1	Fam25c*	Family with sequence similarity 25, member C	−1.51	−2.07	−1.28
NM_026764.2	Gstm4	Glutathione *S*-transferase, mu 4	−1.66	−1.37	−1.38
NM_028064.2	Slc39a4	Solute carrier family 39 (zinc transporter), member 4	−1.69	−2.11	−1.95
NM_016865.2	Htatip2	HIV-1 tat interactive protein 2, homolog	−1.88	−1.47	−1.42
NM_007818.2	Cyp3a11	Cytochrome P450, family 3, subfamily a, polypeptide 11	−1.90	−2.28	−2.81
NM_144942.1	Csad	Cysteine sulfinic acid decarboxylase	−2.25	−2.45	−2.28
NM_008182.1	Gsta2	Glutathione *S*-transferase, alpha 2	−3.85	−2.87	−1.78
NM_008181.2	Gsta1	Glutathione *S*-transferase, alpha 1	−6.68	−2.54	−1.45
Metabolite					
		Cysteine	1.850	3.100	0.674
		Glucose 1-phosphate	1.772	1.322	1.293
		Taurine	1.702	1.714	4.682
		Adenosine	1.637	1.310	0.727
		Arachidonic acid	1.566	1.770	1.406
		Lactic acid	1.227	0.750	1.382
		Ornithine	1.222	1.756	0.698
		Threonine	0.825	0.801	0.735
		Citric acid	0.821	2.017	1.251
		Valine	0.807	0.625	0.750
		3-Hydroxybutyric acid	0.733	0.407	0.826
		Proline	0.721	0.806	0.802
		Maltotriose	0.688	0.597	0.588
		Maltose	0.631	0.814	0.586

These significant, differentially expressed transcripts were selected based on the following criteria: Zratio >1.5 in both directions, false discovery rate < 0.3, ANOVA p value <0.05, and fold change (FC) > 1.2 in both directions

For example, in the Gly-Ser-Thr metabolic hub (Fig. 4), we investigated the expression level of genes encoding for enzymes from methionine cycle transsulfuration (*Cbs, Gst2*), cytochrome P450-related detoxification (*Cyp3a11*), and lipid desaturation (*Fads2*) pathways. These measurements were also utilized to cross-validate liver microarray results. Another example was for testing nutrient-sensitive factors linked to energy storage and utilization that could be involved in synchronization of metabolic processes for each of the diets and

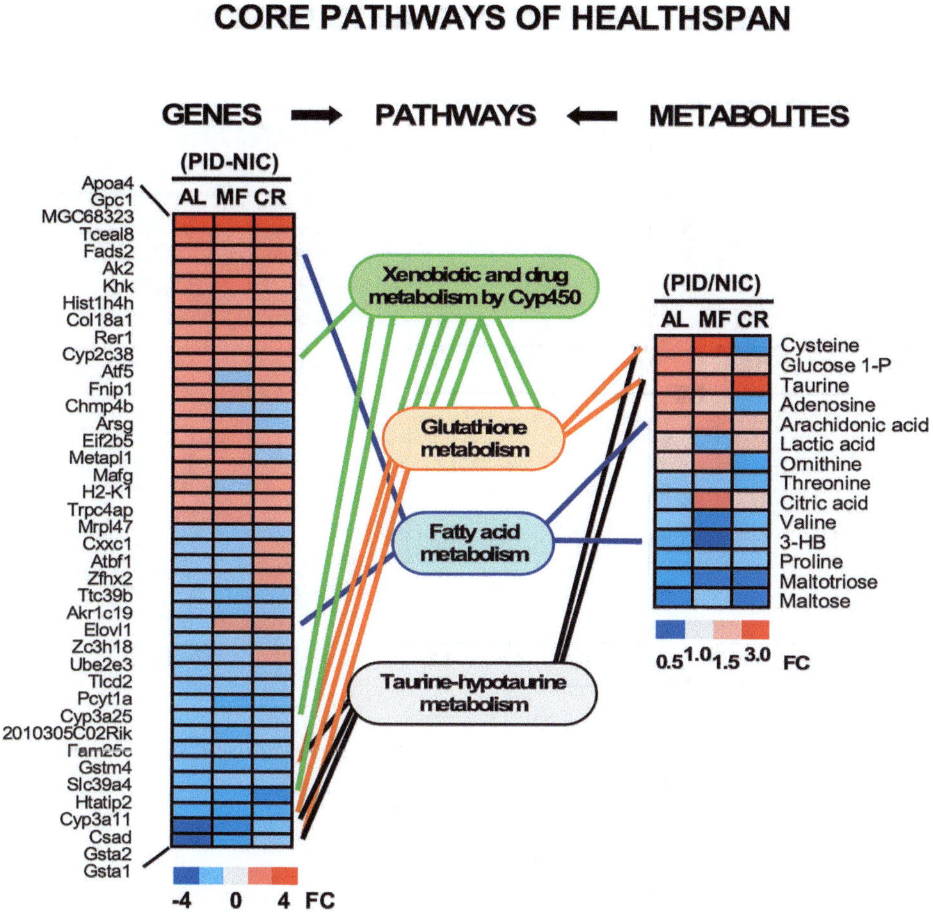

Fig. 6 Core pathways of healthspan: Bipartite networks of genes and metabolites. Heatmaps of shared genes (left) and metabolites (right) derived from Fig. 5 (**b**, top right) (*see* Table 4 for quantitative values). Also displayed are the *links* (genes) between network *nodes* (metabolites) belonging to the same pathways. (Reproduced from Aon, Bernier et al. (2020) *Cell Metabolism* 32, 1–17)

feeding regimens, as suggested by the large-scale reprogramming of metabolism according to the multiomics analyses. We assessed key players, such as AMPK, SIRT1, and NAMPT (nicotinamide phosphoribosyltransferase) [10].

2. For quantitative RT-PCR analysis, total RNA extraction from mouse livers can be performed with the RNeasy mini kit (Qiagen, Waltham, MA) according to the manufacturer's protocol. Quantify RNA concentration with a Nanodrop spectrophotometer (Nanodrop® ND1000, Thermo Scientific), and reverse-transcribe 2 μg of RNA into cDNA using the iScript Advanced cDNA Synthesis Kit for RT-qPCR (Bio-Rad Laboratories, Hercules, CA). Perform quantitative real-time RT-PCR using the iTaq universal SYBR® Green Supermix (Bio-Rad) and incubate the samples at 95 °C for 30 s, followed by 35 cycles

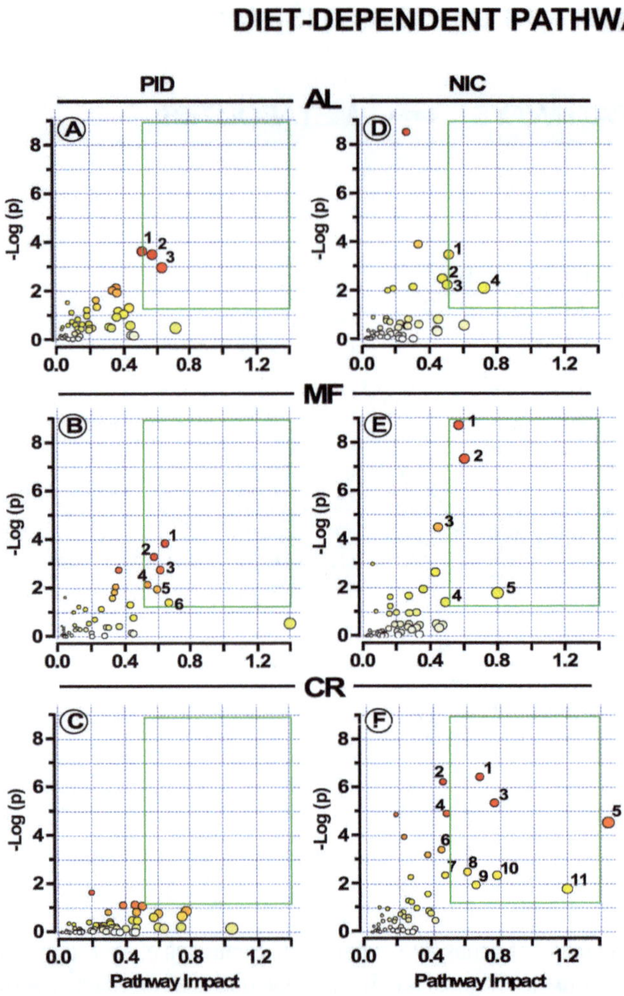

Diet and feeding regimen	Pathways
PID-AL (A)	1. Starch, sucrose metab. 2. Glutamine & glutamate metab. 3. Alanine, aspartate, and glutamate metab.
PID-MF (B)	1. Glycolysis/gluconeogenesis 2. Glutamine & glutamate metab. 3. Galactose metab. 4. Amino sugar and nucleotide sugar metab. 5. TCA cycle 6. Butirosin & neomycin biosynthesis
PID-CR (C)	None
NIC-AL (D)	1. Drug metab. -CyP450 2. Cysteine and methionine metab. 3. Cyanoamino acid metab. 4. Glycine, serine, threonine metab.
NIC-MF (E)	1. Metabolism of xenobiotics by CyP450 2. Drug metab. -Cyp450 3. Linoleic metab. 4. Retinol metab. 5. Phenylalanine, tyrosine, tryptophan metab. 6. Nicotinate & nicotinamide
NIC-CR (F)	1. Drug metab. -Cyp450 2. Valine, isoleucine, leucine degradation 3. Retinol metab. 4. Metabolism of xenobiotics by CyP450 5. Synthesis and degradation of ketone bodies 6. Linoleic metab. 7. Taurine & hypotaurine 8. Butanoate metab. 9. Phenylalanine metab. 10. Propanoate metab. 11. Phenylalanine, tyrosine, tryptophan metab.

Fig. 7 Identification of pathways impacted by diet and feeding regimens. Input for the multiomics JPA consisted of the fold change derived from the PID/NIC ratio of transcripts or metabolites gathered in Fig. 5a, whereby threshold >1.2 indicates upregulation by the PID diet and threshold <0.8 signifies NIC diet-mediated upregulation. The impact of feeding regimen (AL, MF, or CR) within each diet toward pathway enrichment and their network topology is depicted. *y*-axis, enrichment significance; *x*-axis, pathway impact for network topology. Green box highlights pathways significantly impacted as defined by enrichment significance $p < 0.05$ [log (p) > 1.3] and pathway impact >0.5. The NIC diet is linked to pathways such as "metabolism of xenobiotics and drugs through cytochrome P450," the "ω6 PUFA linoleic" and "NAD salvage," whereas the PID diet promoted pathways from central catabolism, that is, carbohydrate, amino acids, and TCA cycle [10]. (Reproduced from Aon, Bernier et al. (2020) *Cell Metabolism* 32, 1–17)

composed of a 3-s period at 95 °C and a 30-s period at 60 °C per cycle with the Applied Biosystems™ QuantStudio™ 6 Flex Real-Time PCR System (Thermo Fisher). Perform the calculation of mRNA expression with the $2 - \Delta\Delta CT$ method using the geometric mean of the housekeeping genes Actb, Gapdh, and Rn18s. The oligonucleotide primers were purchased from

Table 5
List of murine oligonucleotide primers used for validation of microarray analysis

Accession number	Target mRNA	Product length (nt)	Primer orientation	Sequence (5′ - > 3′)
NM_010361.1	Gstt2	114	Forward Reverse	CCGTGGATATACTCAAACAGCAC AGATGGCTGTCCTTTCGGTC
NM_144855.3	Cbs	111	Forward Reverse	GCAGTTCAAACCGATCCACC GCCTGGTCTCGTGATTGGAT
NM_007818.3	Cyp3a11	121	Forward Reverse	ACCTGGGTGCTCCTAGCAAT GCACAGTGCCTAAAAATGGCA
NM_019699.1	Fads2	99	Forward Reverse	GCCCCTTGAGTATGGCAAGA TACATAGGGATGAGCAGCGG
NM_007393.5	Actb	102	Forward Reverse	CACTGTCGAGTCGCGTCC CGCAGCGATATCGTCATCCA
NM_001289726.1	Gapdh	110	Forward Reverse	AAGAGGGATGCTGCCCTTAC ATCCGTTCACACCGACCTTC
NR_003278.3	Rn18s	151	Forward Reverse	GTAACCCGTTGAACCCCATT CCATCCAATCGGTAGTAGCG

IDT (San Jose, CA) and are listed in Table 5. Fold-changes in gene expression were quantified relative to the NIC-AL group. Comparisons between groups were performed using one-way ANOVA with Dunnett's multiple post hoc tests. $n = 4$ per group.

3. For Western blotting (WB) validation, perform protein extraction and immunoprecipitation (IP) of frozen liver tissues. Lyse tissue in radioimmunoprecipitation buffer containing ethylenediaminetetraacetic acid (EDTA) and ethylene glycol tetraacetic acid (EGTA) (Boston BioProducts, Ashland, MA) supplemented with protease inhibitor cocktail (Sigma-Aldrich, St-Louis, MO), phosphatase inhibitor cocktail sets I and II (Calbiochem, San Diego, CA), and protein deacetylase inhibitors [5 μM trichostatin A, 10 mM nicotinamide, and 10 mM

sodium butyrate, all from Sigma-Aldrich]. Following homogenization using TissueLyser II (Qiagen) with bead mill and adapter set, centrifuge samples (18,407 × *g*, 30 min at 4 °C) and determine protein concentration in clarified lysates using the bicinchoninic acid reagent (Pierce BCA Protein Assay Kit, Thermo Fisher Scientific, Waltham, MA). Separate proteins (10–20 μg/well) on 4–15% Criterion TGX precast gels (Bio-Rad) using SDS–polyacrylamide gel electrophoresis under reducing conditions and then electrophoretically transfer onto nitrocellulose membranes (Trans-Blot Turbo Transfer System, Bio-Rad). Perform western blots according to standard methods, which involve a blocking step in phosphate-buffered saline/0.1% Tween 20 (PBS-T) supplemented with 5% nonfat milk and incubation with primary antibodies of interest. All antibodies were detected with horseradish peroxidase-conjugated secondary antibodies (Santa Cruz Biotechnology, Dallas, TX) and visualized by enhanced chemiluminescence (Immobilon Western Chemiluminescent HRP Substrate, Millipore, Billerica, MA). Imaging of the signal was captured with Amersham Imager 600 (GE Healthcare, Piscataway, NJ). Perform quantification of the protein bands by volume densitometry using ImageJ software (National Institutes of Health, Bethesda, MD) and normalization to Ponceau S staining of the membranes.

Carry out IP of liver lysates with anti-acetyl lysine antibody according to the Signal-Seeker Acetyl-Lysine Detection Kit (cat.: BK163; Cytoskeleton, Inc., Denver, CO) for tissue preparation. In short, dilute equal amounts of proteins from mouse liver lysates with buffer mix (1:4 Blast R lysis: Blast R dilution) to a final concentration of 1 mg/ml. Perform a preclearing step using 25 μl of agarose bead suspension for 30 min at 4 °C, after which transfer each sample to a clean tube containing 50 μl of prewashed acetyl-lysine Affinity Bead suspension as provided in the Signal-Seeker Acetyl-Lysine Detection Kit. Also incubate a pair of samples with a 50-μl aliquot of Control IP Bead suspension. After overnight incubation at 4 °C, collect the beads by centrifugation at 5000 × *g* for 1 min at 4 °C and washed 5 times for 5 min each with 1 ml of BlastR-2 Wash Buffer. Add bead elution buffer (50 μl) to each tube followed by a 5-min incubation at room temperature with occasional tapping of the tube. Subsequently, add 2 μl of 2-mercaptoethanol to each sample followed by 5 min of incubation at 70 °C. Spun down samples at 10,000 × *g* for 1 min and use 10 μl/well of supernatant for WB, as above described.

3.7 Conclusions

Integrated multiomics analysis into well-designed experiments is a powerful discovery strategy. From the start, it is important to realize the role played by the experimental design and the underlying questions implicit in the design, which are key for devising the analytical strategy. The ideal experimental design will include, or leave open the possibility, of making available multiomics data. Depending on results, and the biological phenomenon under study, this may enable us to use integrated multiomics analyses to potentiate meaning, understanding, and the generation of new insights as well as testable hypotheses.

4 Notes

1. One of the motivations of the study on mice described in Fig. 1 came from two longitudinal studies of calorie restriction (CR) in nonhuman primates (NHP), carried out at the National Institute on Aging (NIA)/NIH), and at the University of Wisconsin-Madison (WIS). These two independent studies demonstrated differences in longevity outcomes which created some controversy about the efficacy of CR in NHPs and the translatability of the paradigm [3, 11, 12]. Differences in diet composition (Table 1) between the two laboratories: NIC (e.g., NIA diet) and PID (e.g., WIS diet) was investigated further, along with feeding regimens (AL, MF, CR) in a study with laboratory mice [2]. The possible effects on molecular and phenotypic biomarkers of health and lifespan introduced by the variance in diet composition (NIC vs. PID), in addition to the genetic background and feeding paradigms, were addressed (*see* Figs. 3, 5 and 6).
2. The significance threshold election for transcripts is based on the Z ratio, where a 1.5 ratio is a common and safe choice; in the case of metabolites, a 1.2-fold change up or 0.8-fold change down implies values above or below, at least, a 20% experimental error/noise, which is a generous limit.
3. We detected few metabolites as outliers and excluded them from the statistics when above or below 1.5 times the interquartile range comprised between the 75 and 25% percentiles, respectively [13]. Beyond the use of multivariate statistics like Principal Component Analysis (PCA), we further ascertained that the 39 and 43 metabolites were the main ones contributing to the groups' separation by eliminating them from the list of 155 metabolites and repeating PCA to confirm overlapping groups, as expected from the absence of metabolites that contribute to the groups' separation.
4. This is a critical analytical decision that ties directly with experimental findings showing that mean lifespan extension in MF and CR groups was independent of diet type. This enabled us to discover "specific" (independent of diet) and "core"

(independent of both diet and feeding regime) pathways of lifespan (Figs. 2, 3, and 4).

5. After selecting the module *Joint Pathway Analysis* from MetaboAnalyst, the user will have to enter two separate lists of genes and metabolites without numbers associated with them, for example, fold-change or *Z*-ratio. Options for the ID of each gene or metabolite, that is, official gene symbol or name of chemical compound, HMBD, KEGG, will be offered, and these should match the respective IDs of the program's database; otherwise, the program will alert that there was no match for a gene or metabolite query. In that case, try other synonyms until matching is achieved. Regarding the parameter selection for "topology" metrics, for example, degree centrality or betweenness centrality, MetaboAnalyst will guide you. Conveniently, using the same input data, you can choose one or the other topology metric to perform your pathway analysis and compare results (*see* Subheading 3.4, **steps 3** and **6** and **Note 8** for explanation of how topology metrics may provide new insights into pathway analysis).

   Another good feature of this accessible web-based software from McGill University is that it is updated on a regular basis, and at each update the Authors publish a paper in a peer-reviewed journal specifying the new functions along with already present ones in the form of a user's manual, where the function, aim, and capabilities of a module are explained with screenshots and examples; *see* [4, 5] for MetaboAnalyst versions 3.0 and 4.0, respectively. The online Q&A of the software is also very useful.

6. The 10 metabolites at the "core" also fulfilled the "shared" condition of statistical significance for feeding regime alone, diet alone, and "feeding × diet" interaction, by two-way ANOVA (Table 3).

   An important additional feature of MetaboAnalyst is that it enables doing, in addition to JPA, a "gene-centric" or "metabolic-centric" analysis that you can conveniently choose using the same input data. The availability of this feature is based upon the idea of being able to discern whether there is bias introduced by underdetermination of metabolites, which is usually the case.

7. After selecting the module "Statistical Analysis" and uploading your data (csv Excel table) you will be prompted for a "data integrity check"; the software will offer the option of "refilling" missing values, had you this case, by using small values or "Missing value imputation" to choose other methods. In case those missing values are outliers in the dataset we do not choose the "missing values imputation" because it would be

using an arbitrary number instead of a real number that, in fact, was an outlier (*see also* **Note 3**).

8. Besides *enrichment analysis* that evaluates whether the observed genes or metabolites in a pathway appear more frequently than expected by random chance, we utilized topological analysis based on *degree centrality* and *betweenness centrality* metrics [1, 5]. Topology combined with enrichment metrics constitute a powerful method to estimate the relevance of pathways as determined by integrated pathway analysis from both gene (transcriptomics) and functional (metabolomics) ontologies. *Network topology* corresponds to the way pathways (metabolic or signaling) are wired or connected. *Degree* and *betweenness centrality* are two key metrics of network topology. *Degree centrality* refers to how many inputs (indegree) and outputs (outdegree) a node has in a network. The higher the degree, the higher the relevance of the node, whether a metabolite, protein, or transcription factor, because of its potential to elicit both upstream and downstream effects in a network. *Betweenness centrality* measures the number of times a protein or metabolite acts as a bridge along the shortest path between two other metabolites, estimating how connected a pathway is to the rest of the metabolic network, an indication of high metabolic traffic.
9. JPA can be performed with the list of significantly changes genes and metabolites. However, if quantitative data, for example, fold-change, is available for each gene or metabolite (red and blue portions of the bars, respectively, in each feeding regime, that is, left and right panels in Fig. 5a, respectively), then that can be exploited depending on the experimental design and objective of the study. In the present work, the availability of fold-changes was relevant to account for the influence of diet on health span, because that enabled us to separate the effects of genes and metabolites that were significantly increased or decreased by each diet (PID or NIC) (Fig. 5a, b).

Acknowledgments

This work was supported by the Intramural Research Program of the National Institute on Aging, National Institutes of Health. We thank Dr. Sonia Cortassa for critically reading the manuscript.
Author Contributions: Writing first draft and figure creations: M.A.A. and M.B.; Editing: M.A.A., M.B., and R.d.C.

References

1. Barabási A-L (2016) Network science. Cambridge University Press, Cambridge
2. Mitchell SJ, Bernier M, Mattison JA, Aon MA, Kaiser TA, Anson RM, Ikeno Y, Anderson RM, Ingram DK, de Cabo R (2019) Daily fasting improves health and survival in male mice independent of diet composition and calories. Cell Metab 29(1):221–228 e223. https://doi.org/10.1016/j.cmet.2018.08.011
3. Mattison JA, Colman RJ, Beasley TM, Allison DB, Kemnitz JW, Roth GS, Ingram DK, Weindruch R, de Cabo R, Anderson RM (2017) Caloric restriction improves health and survival of rhesus monkeys. Nat Commun 8:14063. https://doi.org/10.1038/ncomms14063
4. Chong J, Wishart DS, Xia J (2019) Using MetaboAnalyst 4.0 for comprehensive and integrative metabolomics data analysis. Curr Protoc Bioinformatics 68(1):e86. https://doi.org/10.1002/cpbi.86
5. Xia J, Wishart DS (2016) Using MetaboAnalyst 3.0 for comprehensive metabolomics data analysis. Curr Protoc Bioinformatics 55:14.10.11–14.10.91. https://doi.org/10.1002/cpbi.11
6. Cheadle C, Cho-Chung YS, Becker KG, Vawter MP (2003) Application of z-score transformation to Affymetrix data. Appl Bioinforma 2(4):209–217
7. Lee JS, Ward WO, Ren H, Vallanat B, Darlington GJ, Han ES, Laguna JC, DeFord JH, Papaconstantinou J, Selman C, Corton JC (2012) Meta-analysis of gene expression in the mouse liver reveals biomarkers associated with inflammation increased early during aging. Mech Ageing Dev 133(7):467–478. https://doi.org/10.1016/j.mad.2012.05.006
8. Kim SY, Volsky DJ (2005) PAGE: parametric analysis of gene set enrichment. BMC Bioinformatics 6:144. https://doi.org/10.1186/1471-2105-6-144
9. Mitchell SJ, Madrigal-Matute J, Scheibye-Knudsen M, Fang E, Aon M, Gonzalez-Reyes JA, Cortassa S, Kaushik S, Gonzalez-Freire M, Patel B, Wahl D, Ali A, Calvo-Rubio M, Buron MI, Guiterrez V, Ward TM, Palacios HH, Cai H, Frederick DW, Hine C, Broeskamp F, Habering L, Dawson J, Beasley TM, Wan J, Ikeno Y, Hubbard G, Becker KG, Zhang Y, Bohr VA, Longo DL, Navas P, Ferrucci L, Sinclair DA, Cohen P, Egan JM, Mitchell JR, Baur JA, Allison DB, Anson RM, Villalba JM, Madeo F, Cuervo AM, Pearson KJ, Ingram DK, Bernier M, de Cabo R (2016) Effects of sex, strain, and energy intake on hallmarks of aging in mice. Cell Metab 23(6):1093–1112. https://doi.org/10.1016/j.cmet.2016.05.027
10. Aon MA, Bernier M, Mitchell SJ, Di Germanio C, Mattison JA, Ehrlich MR, Colman RJ, Anderson RM, de Cabo R (2020) Untangling determinants of enhanced health and lifespan through a multi-omics approach in mice. Cell Metab 32(1):100–116. e104. https://doi.org/10.1016/j.cmet.2020.04.018
11. Colman RJ, Anderson RM, Johnson SC, Kastman EK, Kosmatka KJ, Beasley TM, Allison DB, Cruzen C, Simmons HA, Kemnitz JW, Weindruch R (2009) Caloric restriction delays disease onset and mortality in rhesus monkeys. Science 325(5937):201–204. https://doi.org/10.1126/science.1173635
12. Mattison JA, Roth GS, Beasley TM, Tilmont EM, Handy AM, Herbert RL, Longo DL, Allison DB, Young JE, Bryant M, Barnard D, Ward WF, Qi W, Ingram DK, de Cabo R (2012) Impact of caloric restriction on health and survival in rhesus monkeys from the NIA study. Nature 489(7415):318–321. https://doi.org/10.1038/nature11432
13. Aitken M, Broadhurst B, Hladky S (2010) Mathematics for biological scientists. Garland Science, New York

Part IV

Systems Biology of Disease

Chapter 10

UT-Heart: A Finite Element Model Designed for the Multiscale and Multiphysics Integration of our Knowledge on the Human Heart

Seiryo Sugiura, Jun-Ichi Okada, Takumi Washio, and Toshiaki Hisada

Abstract

To fully understand the health and pathology of the heart, it is necessary to integrate knowledge accumulated at molecular, cellular, tissue, and organ levels. However, it is difficult to comprehend the complex interactions occurring among the building blocks of biological systems across these scales. Recent advances in computational science supported by innovative high-performance computer hardware make it possible to develop a multiscale multiphysics model simulating the heart, in which the behavior of each cell model is controlled by molecular mechanisms and the cell models themselves are arranged to reproduce elaborate tissue structures. Such a simulator could be used as a tool not only in basic science but also in clinical settings. Here, we describe a multiscale multiphysics heart simulator, UT-Heart, which uses unique technologies to realize the abovementioned features. As examples of its applications, models for cardiac resynchronization therapy and surgery for congenital heart disease will be also shown.

Key words Heart simulation, multiscale, multiphysics, Finite-element method, Monte-Carlo simulation, Personalization

1 Introduction

From the second half of the twentieth century, a reductionist or top-down approach has achieved great success in biological research, and we have now gained a plethora of knowledge at molecular and cellular levels. However, because our ultimate goal is to understand how each finding contributes to the health and disease of the human body, the importance of an integrative bottom-up approach using computer simulation is also recognized.

In the case of heart research, pioneering work by Denis Noble 1] opened the door for in silico across-scale integration in cardiac electrophysiology. Inspired by Hodgkin and Huxley's 2] model of nerve impulse conduction, Noble succeeded in modelling the cardiac rhythm from channel activity, thereby clearly demonstrating

Sonia Cortassa and Miguel A. Aon (eds.), *Computational Systems Biology in Medicine and Biotechnology: Methods and Protocols*, Methods in Molecular Biology, vol. 2399, https://doi.org/10.1007/978-1-0716-1831-8_10,

that the rhythm is not driven by an oscillator component, but emerges as a result of the interactions of functional proteins 3]. Following this report, cell models of electrophysiology were applied to diverse cell types such as atrial cells, Purkinje cells, and ventricular cells, from various animal species and under normal and diseased conditions 4–10]. Explosive progress in computer science accelerated these in silico approaches and extended the scale of integration, and now multiscale simulations covering molecular level to tissue and organ level functions are possible. These simulation models are recognized as useful tools for elucidating the mechanisms behind arrhythmias and drug effects 11–13].

Comparing the functional mechanisms of the heart to that of an artificial heart, the simulation of electrophysiology only deals with the performance of the controlling circuit, and the function of the power unit and the behavior of the blood are totally ignored. Of course, the mechanical functions of the heart have been approached in simulation studies that follow a similar history to those on electrophysiology. In 1957, Andrew Huxley proposed a crossbridge model that beautifully reproduced the fundamental properties of skeletal muscle contraction, such as the hyperbolic force-velocity relation based on the stochastic interaction of myosin and actin molecules 14]. As cardiac muscle shares the basic mechanisms of contraction with skeletal muscle, the crossbridge model has been widely used to model the sarcomere dynamics of the heart. However, unlike in the case of electrophysiology, the extension of scale from the sarcomere model to the atrium or ventricle is not straightforward, because accurate description of the large deformation of a realistic heart model by solving the force equilibrium requires a computationally heavy finite element method (FEM). In fact, some studies avoided the use of FEM analysis by assuming a simplified sphere or cylinder model of the ventricle 15]. Furthermore, phenomenological models of muscle contraction are often used in such simplified models 16].

The motion of the blood inside the heart chamber is powered by the active contraction of the heart muscle and is governed by the principles of fluid dynamics. Furthermore, it is important to note that blood also applies force to the heart wall, thereby influencing contractile behavior. Analysis of this complex interaction between the structure and fluid requires the coupling of solid mechanics and fluid dynamics, which is termed fluid–structure interaction analysis, and thus further makes the modeling a challenging task. In fact, only a few studies have used fluid–structure interaction analysis with a strong coupling strategy to accurately analyze the blood flow in heart chambers 17, 18].

Until very recently, simulation studies on cardiac electrophysiology were generally pursued independently to those on cardiac mechanics. However, to fully understand the health and pathology of the heart, these functional aspects governed by distinct physical principles must be integrated to produce a comprehensive view of heart function. Furthermore, description of each functional aspect should be multiscale in nature, covering the phenomena observed at molecular, cellular, and organ levels. In other words, in addition to the dimension of scale, another dimension representing the physical principles needs to be added to the concept of heart simulation. These physical principles relevant to heart simulation include electricity, solid mechanics, and fluid dynamics.

This type of simulation is referred to as multiscale multiphysics simulation, and when applied to the heart it is highly complex and computationally heavy and is only made possible by recent advances in high-performance computing. Efforts toward realizing a truly multiscale multiphysics heart simulation have just begun, and various approaches aiming at diverse applications are currently being attempted. In addition to these functional aspects directly related to the pumping action of the heart, metabolic process and growth of the myocardium should be included in the model, but the time constants of these processes are mostly much longer than those of beat-to-beat activities and are thus hard to integrate into the current scheme of the beating heart simulation.

In this article, we describe the current methods for multiscale multiphysics simulation of the heart. Among the various projects currently ongoing world wide, we pay particular focus to our mul tiscale multiphysics heart simulator "UT-Heart", which was developed at the University of Tokyo. This simulator is based on the finite element method and reproduce the propagation of excitation, contraction, and relaxation of the heart wall, and the accompanying blood flow in the heart chambers, with realistic motion of the valves using the strongly coupled fluid–structure interaction analysis.

First, we present methods for simulating the electrophysiology and mechanics separately, then we introduce the integrated models with some examples of their applications.

2 Methods

2.1 Mesh Generation

Three-dimensional (3D) reconstruction of the heart is based on segmented magnetic resonance imaging (MRI) or multidetector computer tomography (CT) images. Segmentation is usually performed on images taken at end-diastole using commercial software (Virtual Place, Cannon Medical Systems, Tokyo, Japan). From the 3D heart model thus prepared, we make a fine voxel mesh for electrophysiology and a coarse tetrahedral finite element mesh for mechanical analysis. The size of the voxel mesh needs to be small enough (~0.2 mm) to reproduce the proper propagation velocity of the excitation wave using the physiological conductivities of heart tissue. However, when we have to use a larger voxel size because of limited computational resources, conductivities are adjusted to

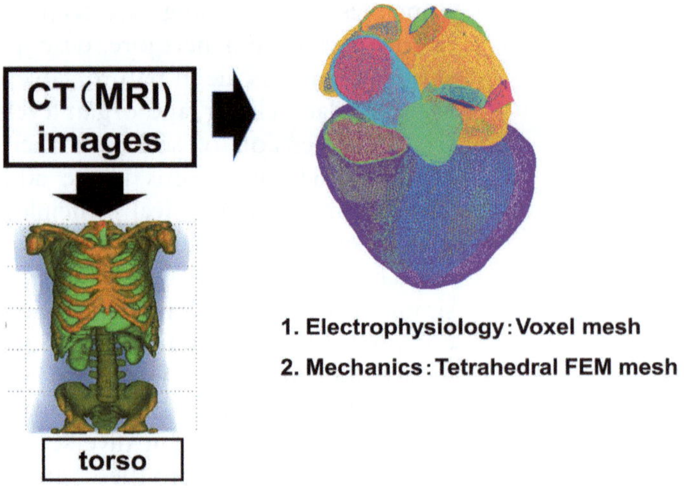

Fig. 1 FEM models of the heart and torso. Left: torso mesh Right: heart mesh

achieve the proper conduction velocity. The conduction system is modeled using one-dimensional elements based on Tawara's monograph 19]. For the analysis of the surface electrocardiogram (ECG), we also make a voxel mesh of the upper body with major organs surrounding the heart (torso) in the same manner (Fig. 1). However, to save computational time, a larger-sized voxel mesh (1.6 mm) is used for the torso model, including the blood in the heart chambers.

On the other hand, because the mechanical analysis is complex and computationally heavy compared with the electrophysiology, we need to use a larger size of the finite element mesh (~2 mm). The blood domain inside the heart chambers is also modeled using a tetrahedral mesh of the same size.

In the walls of the atria and ventricles, myocytes are regularly arranged on a local scale, but change their orientation (fiber orientation) depending on their location and depth in the wall 20]. Because the electrophysiological and mechanical properties of myocytes are anisotropic, fiber orientation has a significant impact on heart function. We therefore map the fiber orientation to meshes using either of two methods we have developed. One of these is the rule-based method, by which data on the local fiber orientation are assigned according to coordinates determined by solving Laplace's equation 21], so that the fiber orientation changes gradually from an endocardial to an epicardial surface (left ventricular [LV] free wall −90° to 60°; interventricular septum −90° to 70°, right ventricular [RV] free wall −60° to 60°).

The other method is a fiber optimization algorithm, in which native fiber structure reported in the literature 20, 22] is self-organized while the beating heart simulation is repeated 23]. This algorithm assumes that the fiber structure of the heart is organized

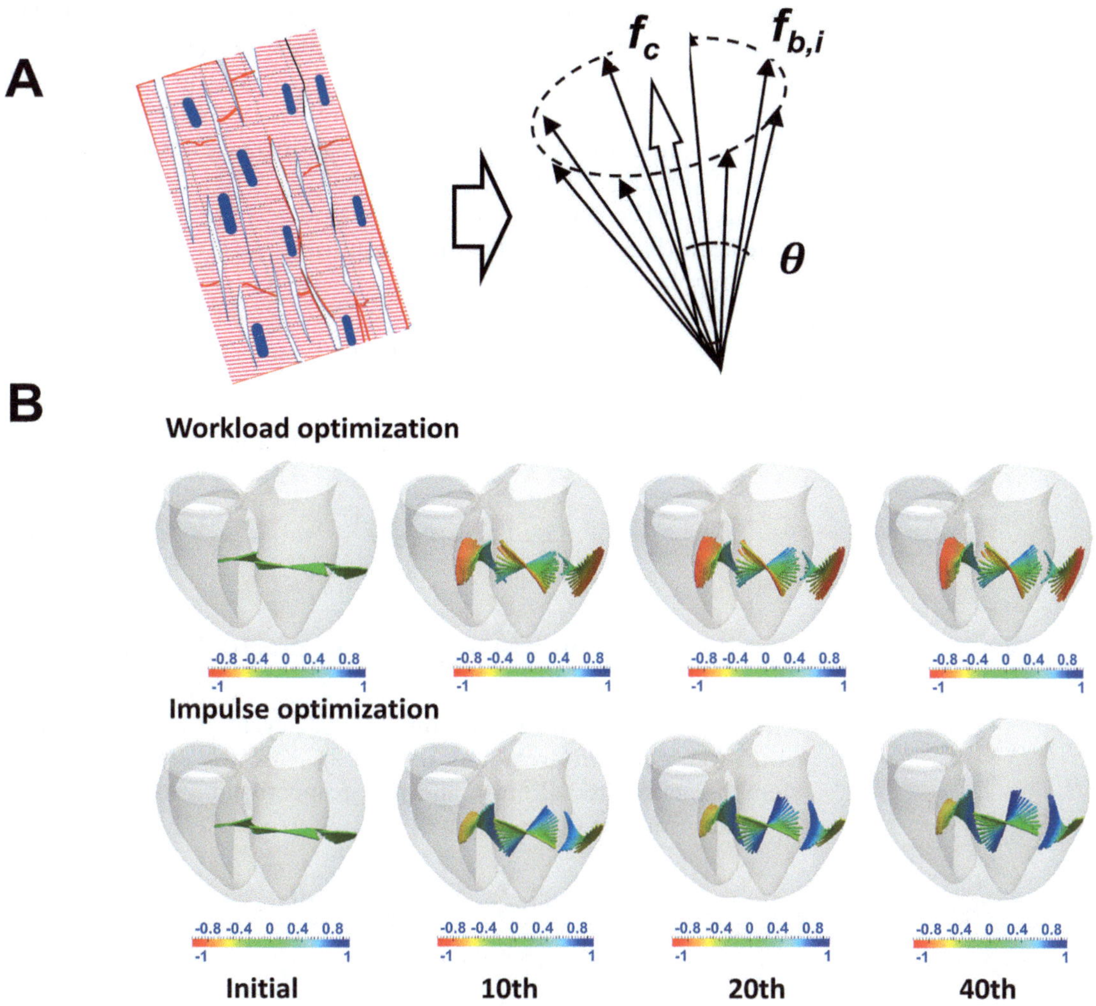

Fig. 2 Self-organization of fiber structure. (**a**) Branching structure of cardiac muscle as an angle sensor. f_c: a central unit vector; $f_{b,\ i}$: unit vector distributed regularly along the base circle of a cone with angle θ. (**b**) Initial fiber orientations (horizontal) and fiber orientations during optimization for workload optimization (top) and impulse optimization (bottom). The number at the bottom indicates the beat number. (From 23] with permission)

to optimize the local parameters, rather than a global parameter like the external work of the ventricles. In other words, tissue remodeling is mediated by biological signals in the microenvironment. We also assume that the branching structures of the cardiomyocytes for the lateral connection serve not only as multidirectional force generators but also as direction sensors. The remodeling process promoted by this sensing mechanism was formulated in the following manner. We modeled the branching structure at each point in the ventricular wall with a central unit vector f_c and n unit vectors $\{f_{b,\ i}\}_{i\ =\ 1,\ \ldots n}$ distributed regularly along the base circle of a cone with angle θ (Fig. 2a). The density ratios of myofibrils in these directions were assumed to be γ_c and γ_b, with $\gamma_c + n\gamma_b = 1$. We hypothesized that each myofibril senses either workload or mechanical impulse during a cardiac cycle as the signal in the microenvironment, and that the number of myocytes aligned along a direction with a larger signal increases whereas those along a direction with a smaller signal decrease. As a result, the direction of the central unit vector is updated for the next cardiac cycle according to the following equations.

With workload (W) as a signal,

$$W_{fc} = -\gamma_c \int_0^{t_c} \dot{\lambda}_{fc} T_{fc} dt \tag{1}$$

$$W_{fb,i} = -\gamma_b \int_0^{t_c} \dot{\lambda}_{fb,i} T_{fb,i} dt, \quad i = 1, \ldots .n \tag{2}$$

$$f^W_{c,updated} = \frac{\tilde{f}^W_c}{\left\| \tilde{f}^W_c \right\|}, \quad \text{where } \tilde{f}^W_c$$

$$= \max\left(0, W_{fc}\right) f_c + \sum_{i=1}^{n} \max\left(0, W_{fb,i}\right) f_{b,i}, \tag{3}$$

where t_c is the duration of a single cardiac cycle, T_{fc} is the contraction force per unit area, and $\dot{\lambda}_{fc}$ and $\dot{\lambda}_{fb,i}$ are the stretches of the elements calculated along f_c and $f_{b,\ i}$, respectively.

With impulse (J) as a signal,

$$J_{fc} = -\gamma_c \int_0^{t_c} T_{fc} dt \tag{4}$$

$$J_{fb,i} = -\gamma_b \int_0^{t_c} T_{fb,i} dt, i = 1, \ldots .n, \tag{5}$$

$$f^J_{c,updated} = \frac{\tilde{f}^J_c}{\left\| \tilde{f}^J_c \right\|}, \text{where } \tilde{f}^J_c$$

$$= \max\left(0, J_{fc}\right) f_c + \sum_{i=1}^{n} \max\left(0, J_{fb,i}\right) f_{b,i}, \tag{6}$$

Starting from the nearly horizontal fiber orientation (10° at the endocardium and -10° at the epicardium), we repeat the simulation of the 3D heart model connected to physiological pre- and afterload. The simulated fiber structure converges fairly rapidly (~10 iterations) and approaches the measured structure reported in the literature (Fig. 2b) 23]. We found that with the impulse as a signal, the optimized fiber structure achieves better agreement with the measured human fiber orientation 24].

Recently, we reported that the local stretch ratio calculated only during the isovolumic contraction phase can be used as a signal for the reorientation of fibers. The final results are similar to those obtained using the above mentioned method, but are achieved with much less computational time 25].

2.2 Electrophysiology

2.2.1 Cell Model of Electrophysiology

As stated above, various types of cell electrophysiology models have been reported. Among these, we use models of ventricular myocytes by ten Tusscher et al. 8] or O'Hara et al. 26], a model for atrial cells by Courtemanche et al. 6], and a model for the conduction system by Stewart et al. 7]. The two ventricular cell models include three cell species having different action potential durations (APDs), that is, endocardial cells, mid-myocardial (M) cells, and epicardial cells, the distributions of which are known to depend on their depth in the wall. In our previous studies, we found that the physiological morphologies of T waves in the surface electrocardiogram can be reproduced by locating M cells in the endocardial side (10–40% from the endocardium) using the ten Tusscher model 27]. In the case of the O'Hara model, we need to locate M cells within 25–75% of the wall thickness from the endocardial side 13, 28]. In either case, the differences in APD are attenuated because of intercellular coupling.

When using the O'Hara model, we replace the equations describing the kinetics of the m gate of the sodium channel with those of ten Tusscher model 8], to reproduce the physiological conduction velocity in myocardial tissue. Similar care was also taken by other researchers 29].

2.2.2 Propagation of Excitation

We simulated the propagation of excitation by solving the bidomain equations described below. These equations are defined in the domain consisting of two subdomains, with their boundaries representing the heart (Ω_H, Γ_H) and the surrounding tissue (torso) and blood in the heart chamber (Ω_C, Γ_C) (Fig. 3a). In the Ω_C, we need to solve only the equation defined in the extracellular space.

In the heart domain,

$$-\nabla\cdot\boldsymbol{\sigma}_i\nabla\emptyset_i = -\beta I_m \text{ on } \Omega_H, \tag{7}$$

$$\vec{n}_H\cdot\boldsymbol{\sigma}_i\nabla\emptyset_i = 0 \; on \; \Gamma_H, \tag{8}$$

$$-\nabla\cdot\boldsymbol{\sigma}_e\nabla\emptyset_e = -\beta I_m \text{ on } \Omega_H, \tag{9}$$

$$\vec{n}_H\cdot\boldsymbol{\sigma}_e\nabla\emptyset_e = J_H \; on \; \Gamma_H, \tag{10}$$

$$I_m = C_m\frac{\partial V_m}{\partial t} + I_{ion}(V_m, S) \text{ on } \Omega_H, \tag{11}$$

where $\emptyset_i$ is the intracellular potential, $\emptyset_e$ is the extracellular potential, V_m is the membrane potential calculated as $V_m = \emptyset_i - \emptyset_e$, β is the surface-to-volume ratio of the tissue, C_m is the membrane capacitance per unit area, and $\vec{n}_H$ is the unit vector normal to the boundary. Conductivity tensors of the intracellular ($\boldsymbol{\sigma}_i$) and extracellular ($\boldsymbol{\sigma}_e$) spaces are anisotropic with respect to the fiber (f), sheet (s), and sheet normal (n) directions.

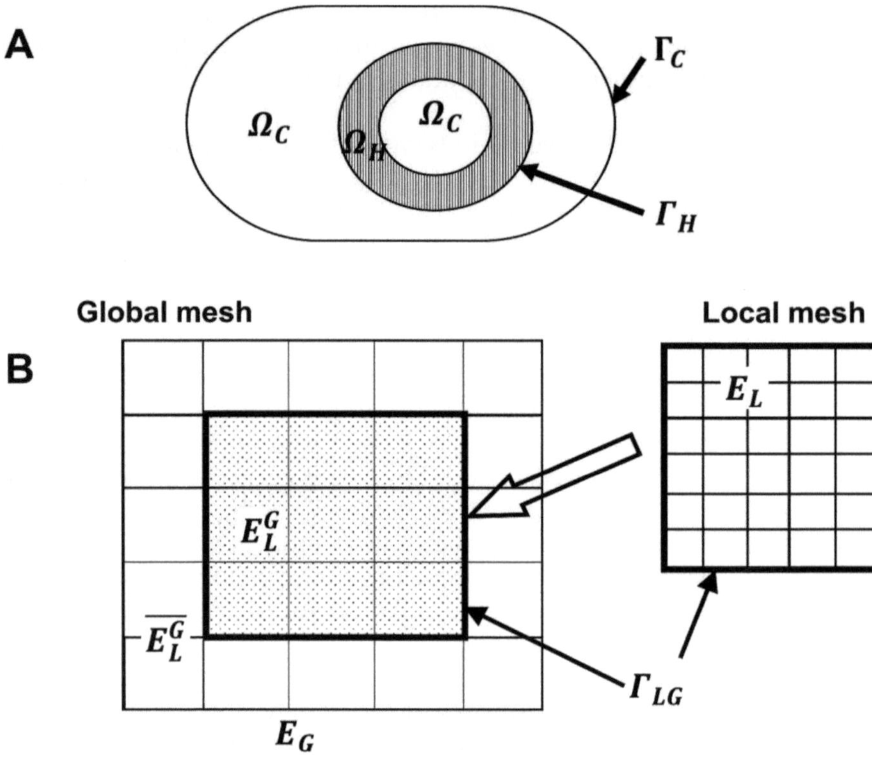

Fig. 3 Parallel multilevel technique for solving the bidomain equation. (**a**) Two-dimensional image of the heart and torso. Ω_H: heart domain; Ω_C: torso domain; Γ_H: boundary of heart domain; Γ_C: boundary of torso domain. (**b**) A composite global mesh (left: Ω^G) and local mesh (right: Ω^G). E^G: the set of nodes in Ω^G; E^L: the set of nodes in Ω^L; Ω_L^G: the subsets of Ω^Gon Ω^L; $\overline{\Omega_L^G}$: the subsets of Ω^Goutside of Ω^L; E_L^G: the subsets of E^Gin Ω_L^G; $\overline{E_L^G}$: the subsets of E^Gin $\overline{\Omega_L^G}$

In the torso domain and on its boundaries,

$$-\nabla\cdot\boldsymbol{\sigma}_C\nabla\emptyset_C = 0 \text{ on } \Omega_C, \tag{12}$$

$$\vec{n}_H\cdot\boldsymbol{\sigma}_C\nabla\emptyset_C = \vec{n}_H\cdot\boldsymbol{\sigma}_e\nabla\emptyset_e = J_H \ on\ \Gamma_H, \tag{13}$$

$$\vec{n}_C\cdot\boldsymbol{\sigma}_C\nabla\emptyset_C = 0 \text{ on } \Gamma_C, \tag{14}$$

where $\emptyset_C$ is the potential on Ω_C and $\boldsymbol{\sigma}_C$ is the conductivity tensor, for which a specific value is assigned to each organ in the torso.

The weak (integral) forms of Eqs. 7)–(14) are given by.

$$\int_{\Omega_H}\nabla\omega_i\cdot\boldsymbol{\sigma}_i\nabla\emptyset_i d\Omega = -\int_{\Omega_H}\omega_i\beta I_m d\Omega, \tag{15}$$

$$\int_{\Omega_H}\nabla\omega_e\cdot\boldsymbol{\sigma}_e\nabla\emptyset_e d\Omega = \int_{\Omega_H}\omega_e\beta I_m d\Omega + \int_{\Gamma_H}\omega_e J_H d\Gamma, \tag{16}$$

$$\int_{\Omega_C}\nabla\omega_C\cdot\boldsymbol{\sigma}_C\nabla\emptyset_C d\Omega = -\int_{\Gamma_H}\omega_C J_H d\Gamma, \tag{17}$$

where ω_i, ω_e, and ω_C are arbitrary test functions. Using the relation (Eq. 13), we can replace (Eqs. 16 and 17) by.

$$\int_{\Omega} \nabla \omega_e \cdot \boldsymbol{\sigma}_e \nabla \emptyset_e d\Omega = -\int_{\Omega_H} \omega_e \beta I_m d\Omega, \tag{18}$$

which can be applied to the whole domain $\Omega = \Omega_H \cap \Omega_C$, where extracellular potentials and the conductivity tensors are combined as.

$$\emptyset_e = \left\{ \begin{matrix} \emptyset_e \ on \ \Omega_H \\ \emptyset_C \ on \ \Omega_C \end{matrix} \right\}, \sigma_e = \left\{ \begin{matrix} \sigma_e \ on \ \Omega_H \\ \sigma_C \ on \ \Omega_C \end{matrix} \right\}.$$

The finite element discretization of (Eqs. 15 and 18) can be described by the matrix representation as.

$$\boldsymbol{K}_i \emptyset_i = -\beta \boldsymbol{I}_m, \tag{19}$$

$$\boldsymbol{K}_e \emptyset_e = \boldsymbol{R}_H^T \beta \boldsymbol{I}_m, \tag{20}$$

where $\boldsymbol{I}_m$ is the transmembrane current per unit area, $\boldsymbol{R}_H$ represents a restriction operator from the whole domain Ω to the subdomain Ω_H on the heart muscle. Sinc $\boldsymbol{R}_H$ is a simple injection (mapping) on Ω_H in this case, $\boldsymbol{R}_H$ and its transpose $\boldsymbol{R}_H^T$ are omitted in the following equations. $\boldsymbol{K}_{i \ or \ e}$ is a matrix representing the conductance of the tissue. By multiplying the potential vector ($\emptyset_i$ or $\emptyset_e$) to rows of $\boldsymbol{K}_{i \ or \ e}$, current vector ($\boldsymbol{I}_m$) is obtained.

Using the relation between the potentials, $V_m = \emptyset_i - \emptyset_e$, and representing $\boldsymbol{R}_H^T \boldsymbol{K}_i$ and $\boldsymbol{R}_H^T \boldsymbol{K}_i \boldsymbol{R}_H$ by $\boldsymbol{K}_i$ for simplicity, (Eqs. 19 and 20) are rewritten as.

$$\boldsymbol{K}_i V_m + \beta \boldsymbol{I}_m + \boldsymbol{K}_i \emptyset_e = 0, \tag{21}$$

$$\boldsymbol{K}_i V_m + (\boldsymbol{K}_i + \boldsymbol{K}_e) \emptyset_e = 0. \tag{22}$$

These systems are integrated along the temporal axis by the explicit scheme shown below.

$$\begin{bmatrix} \frac{\beta C_m}{\Delta t} \boldsymbol{M}_H & \boldsymbol{0} \\ \boldsymbol{K}_i & \boldsymbol{K}_i + \boldsymbol{K}_e \end{bmatrix} \begin{bmatrix} \boldsymbol{V}_m^{t+\Delta t} \\ \emptyset_e^{t+\Delta t} \end{bmatrix} = \begin{bmatrix} \frac{\beta C_m}{\Delta t} \boldsymbol{M}_H - \boldsymbol{K}_i & -\boldsymbol{K}_i \\ \boldsymbol{0} & \boldsymbol{0} \end{bmatrix} \begin{bmatrix} \boldsymbol{V}_m^t \\ \emptyset_e^t \end{bmatrix} - \begin{bmatrix} \beta \boldsymbol{M}_H & \boldsymbol{0} \\ \boldsymbol{0} & \boldsymbol{0} \end{bmatrix} \times \begin{bmatrix} I_{ion}\left(\boldsymbol{V}_m^t, \boldsymbol{S}^t\right) \\ \boldsymbol{0} \end{bmatrix}, \tag{23}$$

where $\boldsymbol{M}_H$ is the lumped matrix on Ω_H and $\boldsymbol{S}$ is the state vector computed by the cell model.

Because a very small time step is required for the computation of the ion currents $I_{ion}\left(\boldsymbol{V}_m^t, \boldsymbol{S}^t\right)$, we adopt a previously reported "inner-outer" time integration scheme 30]. In this scheme, the equation for the intracellular domain (Eq. 21) is integrated with a small timestep in the inner iteration, while the extracellular potential $\emptyset_e$ is fixed. The extracellular potential $\emptyset_e$ is then updated in the outer iteration with a large step.

To further speedup the calculation, we also introduce a composite voxel mesh in Ω_C for spatial discretization. This composite mesh consists of a fine mesh on the local rectangular parallelepiped domain surrounding the heart, and a coarse mesh on the global domain covering the whole torso, because a fine resolution is not required for the extracellular equation outside the heart. We consider the weak form of (Eq. 22).

$$\int_{\Omega} \nabla\omega \cdot \boldsymbol{\sigma} \nabla \emptyset d\Omega = -\int_{\Omega_H} \nabla\omega \cdot \boldsymbol{\sigma}_i \nabla V_m d\Omega, \tag{24}$$

where $\sigma = \sigma_e + \sigma_i$, and $\emptyset = \emptyset_e$ for notational convenience. To discretize this equation on the composite mesh, we applied a Lagrange multiplier method for the constraints at the interface of the local fine mesh and the coarse global mesh, by starting with a variational formulation of the problem. The energy functional for the formulation is described as.

$$\varepsilon(\emptyset) = \int_{\Omega} \frac{1}{2} \nabla\emptyset \cdot \boldsymbol{\sigma} \nabla \emptyset d\Omega + \int_{\Omega_H} \nabla\emptyset \cdot \boldsymbol{\sigma}_i \nabla V_m d\Omega. \tag{25}$$

In the following discussion, we define the domains and nodes as shown in Fig. 3b.

Ω^G: the domain covered by the global mesh.

Ω^L:the domain covered by the local mesh.

E^G: the set of nodes in Ω^G.

E^L: the set of nodes in Ω^L.

Ω_L^G: the subsets of Ω^Gon Ω^L.

$\overline{\Omega_L^G}$: the subsets of Ω^Goutside of Ω^L.

E_L^G: the subsets of E^Gin Ω_L^G.

$\overline{E_L^G}$: the subsets of E^Gin $\overline{\Omega_L^G}$.

We introduce an interpolated function of $\emptyset^L$ for the nodal values $\emptyset^L$ on the local mesh as

$$\emptyset^L = \boldsymbol{N}^L \cdot \emptyset^L = \sum_{i \in \Omega^L} N_i^L \emptyset_i^L, \tag{26}$$

where $\left\{N_i^L\right\}_{\in \Omega^L}$ are the shape functions. Similarly, for the global mesh, $\emptyset^G$ is defined as

$$\emptyset^G = \boldsymbol{N}^G \cdot \emptyset^G = \sum_{i \in \Omega^G} N_i^G \emptyset_i^G. \tag{27}$$

Using these definitions, the energy functional representing electrical energy for a given nodal function $\emptyset = \{\emptyset^L, \ \emptyset^G\}$ on the composite mesh is defined as

$$\begin{aligned}
\varepsilon(\emptyset) &= \int_{\Omega_L} \frac{1}{2} \nabla\emptyset^L \cdot \boldsymbol{\sigma} \nabla\emptyset^L d\Omega + \int_{\overline{\Omega_L^G}} \frac{1}{2} \nabla\emptyset^G \cdot \boldsymbol{\sigma} \nabla\emptyset^G d\Omega \\
&+ \int_{\Omega_H} \frac{1}{2} \nabla\emptyset^L \cdot \boldsymbol{\sigma}_i \nabla \boldsymbol{V}_m^L d\Omega \\
&= \sum_{e^L \in E^L} \int_{e^L} \frac{1}{2} \nabla \boldsymbol{N}^L \emptyset^L \cdot \boldsymbol{\sigma} \nabla \boldsymbol{N}^L \emptyset^L d\Omega \\
&+ \sum_{e^G \in \overline{E_L^G}} \int_{e^G} \frac{1}{2} \nabla \boldsymbol{N}^G \emptyset^G \cdot \boldsymbol{\sigma} \nabla \boldsymbol{N}^G \emptyset^G d\Omega \\
&+ \sum_{e^L \in E_H^L} \int_{e^L} \frac{1}{2} \nabla \boldsymbol{N}^L \emptyset^L \cdot \boldsymbol{\sigma}_i \nabla \boldsymbol{N}^L \boldsymbol{V}_m^L d\Omega \qquad (28)
\end{aligned}$$

where E_H^L are elements in E^L inside Ω_H.

By applying the Lagrange multiplier method to the variational problem (Eq. 28), we obtain

$$\begin{aligned}
&\int_{\Omega_L} \nabla\omega^L \cdot \boldsymbol{\sigma} \nabla\emptyset^L d\Omega + \int_{\overline{\Omega_L^G}} \nabla\omega^G \cdot \boldsymbol{\sigma} \nabla\emptyset^G d\Omega \\
&+ \int_{\Omega_H^L} \frac{1}{2} \nabla\omega^L \cdot \boldsymbol{\sigma}_i \nabla \boldsymbol{V}_m^L d\Omega + \left(\boldsymbol{\omega}^L - \boldsymbol{I}_G^L \boldsymbol{\omega}^G\right)_{\Gamma_{LG}} \cdot \boldsymbol{\lambda} \\
&+ \boldsymbol{\omega}^L \cdot \left(\emptyset^L - \boldsymbol{I}_G^L\right)_{\Gamma_{LG}}, \qquad (29)
\end{aligned}$$

where ω^L and ω^G are arbitrary test functions and $\boldsymbol{I}_G^L$ is an interpolation operator from Ω_G to Ω_L.

To this problem, we impose the following constraint condition on Γ_{LG}.

$$\emptyset^L = \boldsymbol{I}_G^L \emptyset^G. \qquad (30)$$

Finally, we obtain

$$\begin{aligned}
&\sum_{e^L \in E^L} \int_{e^L} \nabla \boldsymbol{N}_i^L \cdot \boldsymbol{\sigma} \nabla N_j^L d\Omega \emptyset^L + \sum_{e^L \in E^L} \int_{e^L} \nabla \boldsymbol{N}_i^L \cdot \boldsymbol{\sigma} \nabla \boldsymbol{N}_j^L d\Omega \boldsymbol{V}^m + \boldsymbol{\lambda} \\
&= \boldsymbol{0} \text{ on } \Omega^L \qquad (31)
\end{aligned}$$

$$\sum\nolimits_{e^G \in \overline{E_L^G}} \int_{e^L} \nabla \boldsymbol{N}_i^G \cdot \boldsymbol{\sigma} \nabla N_j^G d\Omega - {\boldsymbol{I}_G^L}^T \boldsymbol{\lambda} \text{ on } \overline{\Omega_L^G}. \qquad (32)$$

Equation (31) implies that the nodal values of the Lagrange multiplier ($\boldsymbol{\lambda}$) can be interpreted as the electric currents from the local mesh, and that they are distributed by ${\boldsymbol{I}_G^L}^T$ to the global mesh nodes at Γ_{LG} according to (Eq. 32). Therefore, a current balance is ensured at the interface. These equations are solved in an iterative manner.

2.2.3 Personalization of Electrophysiology

In clinical applications, we modify the model parameters to reproduce the surface ECG of each patient. First, we reproduce the morphologies of the QRS waves of a twelve-lead ECG by changing the locations and timings of the earliest activation sites on the endocardial surface. Next, we modify the distributions of APD both transmurally and longitudinally to match the T-waves of the simulated ECG with those of the real ECG. Each procedure is iterated until good agreement is achieved, with goodness of agreement being evaluated by cross-correlation (R_{cc}) calculated according to the following equation:

$$R_{cc} = \frac{\sum_{j=1}^{12} \sum_{i=1}^{N} A(i,j) \times B(i,j)}{\sqrt{\sum_{j=1}^{12} \sum_{i=1}^{N} A(i,j)^2 \times \sum_{l=1}^{12} \sum_{k=1}^{N} B(k,l)^2}}, \tag{33}$$

where $A(i,j)$ and $B(i,j)$ are the ECG values at time point i of the j-th lead of the simulated and real ECG, respectively.

2.3 Mechanics

2.3.1 Sarcomere Model

Because both cardiac and skeletal muscles share a common mechanism of contraction, the concept of the cycling cross bridge model described by Huxley 14] has been adopted in many cardiac sarcomere models. However, most of these models failed to reproduce the high sensitivity of the developed force to changes in cytosolic calcium concentration, such as is observed in cardiac muscle. This distinct property of cardiac muscle is believed to be caused by cooperative interactions among molecules in the sarcomere, and end-to-end interactions of regulatory troponin/tropomyosin (T/T) units along the thin filament are proposed as a potential mechanism for this cooperativity. To faithfully model this phenomenon, spatially distributed models mimicking the physical arrangement of the functional units of a sarcomere, including the crossbridges in the thick filament and T/T units in the thin filament, are proposed. Our spatially distributed sarcomere model composed of a pair of thin filaments and a thick filament is illustrated in Fig. 4. The numbers of myosin head (MH) and T/T units in a half sarcomere were 38 and 32, respectively. In this model, the crossbridge formation in the T/T unit is regulated by calcium binding, and the myosin head (MH) that forms a crossbridge (XB) makes transitions among the nonpermissive (N_{XB}), permissive (P_{XB}), prepower stroke (XB_{PreR}), and postpower stroke (XB_{PostR}) states. The framework of the model and the notations of the states were adopted from the work by Rice et al. 31], but in our model P_{XB} was assumed to be in an attached state, thus contributing to the force generation. The end-to-end interaction of T/T units is modeled by introducing factors γ^n into the transitions from N_{XB} to P_{XB} and γ^{-n} into transitions from the two binding states P_{XB} and XB_{PreR}

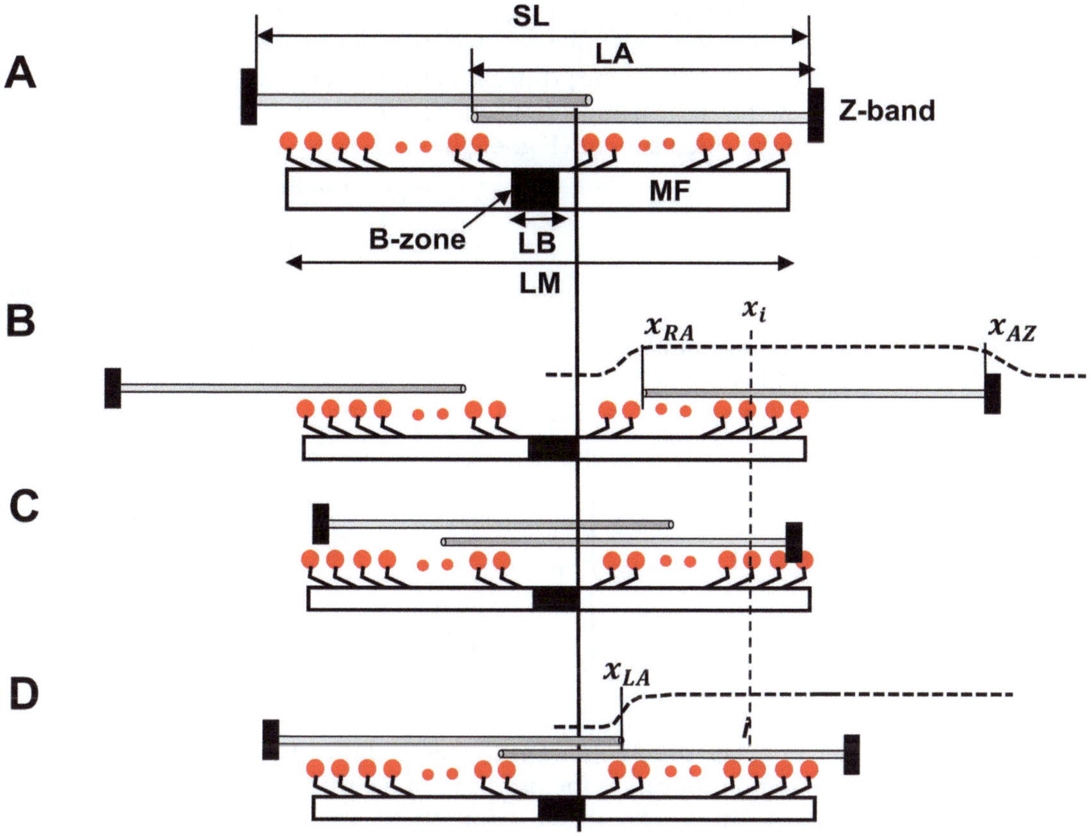

Fig. 4 Sarcomere model. (**a**) Schematic representation of the sarcomere structure. (**b**) Relative position of filaments in the single overlapping state (SL > 2LA + LB). (**c**) State of no overlapping at the MF ends (SL < LM). (**d**) The double overlapping state (LM < SL < 2LA − 2 LB). *MF* thick filament, *MH* myosin head, *B-zone* bare zone, *AF* thin filament, *SL* sarcomere length, *LA* thin filament length, *LM* thick filament length, *LB* bare zone length. *xLA* position of the end of the left thin filament

to N_{XB}, where n takes the value 0, 1, or 2, depending on the states of the neighboring MHs in the following manner:

$$n = \begin{cases} 0 \; if \; both \; neighboring \; MHs \; are \; in \; N_{XB} \\ 1 \; if \; one \; of \; the \; two \; neighboring \; MHs \; is \; in \; a \; binding \; state \\ 2 \; if \; both \; neighboring \; MHs \; are \; in \; binding \; states \end{cases} . \tag{34}$$

We set γ to 80 to reproduce the force-pCa relation with high cooperativity that is reported in the literature 32, 33]. Furthermore, we introduced a length-dependence into the crossbridge kinetics by modifying the rate constant.

$$\overline{k}_{np0}(SL, i) = \chi_{LA}(SL, i)\chi_{RA}(SL, i)k_{np0} \tag{35}$$

$$\overline{k}_{np1}(SL, i) = \chi_{LA}(SL, i)\chi_{RA}(SL, i)k_{np1}. \tag{36}$$

The factors $\chi_{RA}(SL, im)$ and $\chi_{LA}(SL, im)$ are defined for each T/T unit as the function of its position (x_i) and the filament overlap determined by the positions of the free end (x_{RA}) and Z-band (x_{AZ}) of the right-hand side filament and the free end (x_{LA}) of the left-hand side filament (Fig. 4a, b).

$$x_{AZ} = (SL - LB)/2 \tag{37}$$

$$x_{LA} = LA - x_{AZ} - LB \tag{38}$$

$$x_{RA} = x_{AZ} - LA \tag{39}$$

SL: sarcomere length, LA: length of actin filament, LB: length of bare zone.

$\chi_{RA}(SL, i)$ is defined so as to attenuate the rate constant of cross-bridges in the nonoverlapping region ($x_i \leq x_{RA}$) (Fig. 4b).

$$\chi_{RA}(SL, i) = \begin{cases} \exp\left(-\frac{(x_{RA} - x_i)^2}{a_R^2}\right), & x_i \leq x_{RA} \\ 1, & x_{RA} < x_i < x_{AZ} \\ \exp\left(-\frac{(x_i - x_{AZ})^2}{a_R^2}\right), & x_i \geq x_{AZ} \end{cases} \tag{40}$$

The third condition applies to the case where a nonoverlapping region appears at the right end of the thick filament (MF in Fig. 4c).

By $\chi_{LA}(SL, i)$, we assume that the cross-bridge formation is inhibited in the double overlapping region of the thin filament (Fig. 4d, SL < 2LA – LB).

$$\chi_{LA}(SL, i) = \begin{cases} \exp\left(-\frac{(x_{LA} - x_i)^2}{a_L^2}\right), & x_i \leq x_{LA} \\ 1, & x_i \geq x_{LA} \end{cases} \tag{41}$$

The state transition of each MH at each time step was calculated by Monte Carlo (MC) simulation.

2.3.2 Heart Mechanics

For the heart mechanics, the fluid–structure interaction problem is solved and the equations for the muscle part are given by

$$\int_{\Omega_s} \left\{ \delta\dot{\boldsymbol{u}}\cdot\rho_s\boldsymbol{u} + \delta\dot{\boldsymbol{Z}} : \left(\boldsymbol{\Pi} + 2p_s J\boldsymbol{F}^{-1}\right)^T \right\} d\Omega_s = \int_{\Gamma_{fs}} \delta\dot{\boldsymbol{u}}\cdot\boldsymbol{\tau}_f d\Gamma_{fs} \tag{42}$$

$$\int_{\Omega_s} \delta p_s \left(2(J-1) - \frac{p_s}{K_s} \right) d\Omega_s = 0. \tag{43}$$

Ω_s: heart and vessel wall domains in the reference configuration.

Γ_{fs}: the blood–muscle interface in the current configuration.

$\boldsymbol{u}(\boldsymbol{X}, t) = \boldsymbol{x}(\boldsymbol{X}, t) - \boldsymbol{X}$: displacement of the material point $X \in \Omega_s$ at time t.

ρ_s: density of heart tissue.

$\boldsymbol{F} = \frac{\partial \boldsymbol{x}}{\partial \boldsymbol{X}}$: deformation gradient tensor.

$\boldsymbol{Z} = \frac{\partial \boldsymbol{u}}{\partial \boldsymbol{X}}$: displacement gradient tensor.

$J =$ det ($\boldsymbol{F}$): Jacobian.

$\boldsymbol{\tau}_f$: traction force vector of the blood on the internal surface of the wall.

p_s: hydrostatic pressure.

K_s: modulus of volume elasticity.

$\boldsymbol{\Pi}$: first Piola–Kirchhoff stress tensor.

The first equation is the momentum equation and the second gives the incompressibility constraint.

$\boldsymbol{\Pi}$ is composed of active ($\boldsymbol{\Pi}_{act}$), passive ($\boldsymbol{\Pi}_{pas}$), and viscous ($\boldsymbol{\Pi}_{vis}$) parts.

$$\boldsymbol{\Pi} = \boldsymbol{\Pi}_{act} + \boldsymbol{\Pi}_{pas} + \boldsymbol{\Pi}_{vis} \tag{44}$$

The active part is related to the contraction force per unit area (T) calculated by the sarcomere model according to the following equation:

$$\frac{T}{\lambda} \boldsymbol{f} \bigotimes \boldsymbol{f} \cdot \boldsymbol{F}^T, \tag{45}$$

where $\boldsymbol{f}$ is a unit vector in the fiber direction and λ is the stretch ratio along $\boldsymbol{f}$.

The passive stress tensor is defined using the deformational potential function W as

$$\boldsymbol{\Pi}_{act} = \frac{\partial W^T}{\partial \boldsymbol{Z}}. \tag{46}$$

$$W = c_1 \left(\widetilde{I}_1 - 3 \right) + c_u \frac{1}{2} \left(e^{Q_u} - 1 \right) + W_{sar}. \tag{47}$$

Here, $\widetilde{I}_1$ is the reduced invariant defined as

$$\widetilde{I}_1 = det(\boldsymbol{C})^{-\frac{1}{3}} Tr(\boldsymbol{C}), \tag{48}$$

where C is the right Cauchy–Green deformation tensor:

$$\boldsymbol{C} = \boldsymbol{F}^T \boldsymbol{F},$$

and Q_u is the quadratic form of the Green–Lagrange strain tensor given by

$$Q_u = b_{ff} E_{ff}^2 + b_{ss} E_{ss}^2 + b_{nn} E_{nn}^2 + 2 b_{fs} E_{fs}^2 + 2 b_{fn} E_{fn}^2 + 2 b_{sn}. \tag{49}$$

E_{ff}, E_{ss}, E_{nn}, E_{fs}, and E_{fn} are components of $\boldsymbol{E} = \frac{1}{2}(\boldsymbol{C} - 1)$ defined in the fiber (f), sheet (s), and sheet normal (n) coordinates.

Parameter values need to be adjusted to reproduce the diastolic pressure–volume relationship of the subject estimated by the Klotz formula 34].

For the viscous part, we formulated the Newtonian viscosity as:

$$\boldsymbol{\Pi}_{vis} = 2\mu_s J \boldsymbol{F}^{-1} \boldsymbol{D}_s,$$

where $\boldsymbol{D}_s$ is the deformation velocity tensor

$$\boldsymbol{D}_s = \frac{1}{2}\left(\frac{\partial \dot{\boldsymbol{u}}}{\partial t} + \frac{\partial \dot{\boldsymbol{u}}^T}{\partial t}\right).$$

To describe the behavior of the blood in the chamber, we adopted the Navier-Stokes equation assuming that the blood is incompressible and Newtonian. The governing equations are:

$$\int_{\Omega_f} \left\{ \boldsymbol{\delta v} \cdot \rho_f \boldsymbol{\alpha}_x + 2\mu_f \boldsymbol{\delta D}_f : \boldsymbol{D}_f - p_f \nabla_x \cdot \boldsymbol{\delta v} \right\} d\Omega_f$$

$$= \int_{\Gamma_f} \boldsymbol{\delta v} \cdot \boldsymbol{t}_f d\Gamma_f + \int_{\Gamma_{fs}} \boldsymbol{\delta v} \cdot \boldsymbol{\tau}_s d\Gamma_{fs}$$

$$\int_{\Omega_f} \delta p_f \nabla_x \cdot \boldsymbol{v} d\Omega_f = 0, \tag{50}$$

where χ denotes the arbitrary Lagrangian–Eulerian (ALE) coordinate system on the fluid domain Ω_f, and $\boldsymbol{\tau}_s$ is the traction force vector applied by the muscle at the fluid–structure interface Γ_{fs}. The acceleration of the fluid in the ALE coordinate $\boldsymbol{\alpha}_x$ includes the artificial convective term:

$$\boldsymbol{\alpha}_x(x, t) = \frac{\partial \boldsymbol{v}(x, t)}{\partial t} + (\boldsymbol{c} \cdot \nabla)\boldsymbol{v} \text{ on } \Omega_f, \tag{51}$$

where $\boldsymbol{c} = \boldsymbol{v} - \hat{\boldsymbol{v}}$ is the relative velocity observed from the ALE coordinate system moving at a velocity $\hat{\boldsymbol{v}}$. $\boldsymbol{D}_f$ is the deformation velocity tensor defined as

$$\boldsymbol{D}_f = \frac{1}{2}\left(\frac{\partial \boldsymbol{v}}{\partial \boldsymbol{x}} + \frac{\partial \boldsymbol{v}^T}{\partial \boldsymbol{x}}\right) \text{ on } \Omega_f. \tag{52}$$

Γ_f corresponds to the inlet and outlet boundaries of the heart chambers, where the traction force $\boldsymbol{t}_f$ is determined through the interaction with the circulatory system. The second equation of (Eq. 50) gives the incompressibility constraint of the fluid.

2.3.3 Circulatory Model

The finite element method (FEM) heart model is coupled with the lumped parameter models of the systemic and pulmonary circulation (Fig. 5) in a similar manner to that of Kerckhoffs et al. 35]. In this scheme, R1a is often called as characteristic impedance, representing the resistance of proximal aorta, R2a is systemic resistance representing the rest of the resistance on the arterial side including

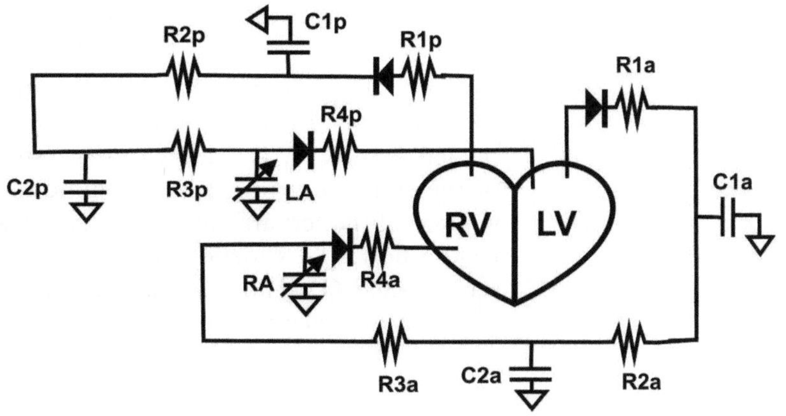

Fig. 5 Model of systemic and pulmonary circulations. R_{ia} ($i = 1$ to 4): resistance in systemic circulation; C_{ia} ($i = 1$ to 4): capacitance in systemic circulation; R_{ip} ($i = 1$ to 4): resistance in pulmonary circulation; C_{ip} ($i = 1$ to 4): capacitance in pulmonary circulation; *LV* left ventricle, *RV* right ventricle, *LA* left atrium, *RA* right atrium

capillaries, R3a is venous resistance and R4a is filling resistance to associated with the tricuspid valve, C1a is arterial capacitance and C2a is venous capacitance. Resistances and capacitances in the pulmonary circulation were named in a similar manner. This model is described by a set of equations governing the relation between flow (Q) and pressure (P) at each segment as follows.

Systemic circulation:

$$P_{as} = V_{as}/C_{1a} \tag{53}$$

$$P_{vs} = V_{vs}/C_{2a} \tag{54}$$

$$Q_{ao} = \begin{cases} \dfrac{P_{LV} - P_{as}}{R_{1a}} & (P_{LV} > P_{as}) \\ 0 & (P_{LV} < P_{as}) \end{cases} \tag{55}$$

$$Q_{as} = (P_{as} - P_{vs})/R_{2a} \tag{56}$$

$$Q_{vs} = (P_{VS} - P_{RA})/R_{3a} \tag{57}$$

$$Q_{mitral} = \begin{cases} \dfrac{P_{LA} - P_{LV}}{R_{4a}} & (P_{LA} > P_{LV}) \\ 0 & (P_{LA} < P_{LV}) \end{cases} \tag{58}$$

$$\frac{dV_{LA}}{dt} = Q_{vp} - Q_{mitral} \tag{59}$$

$$\frac{dV_{LV}}{dt} = Q_{mitral} - Q_{ao} \tag{60}$$

$$\frac{dV_{as}}{dt} = Q_{ao} - Q_{as} \tag{61}$$

$$\frac{dV_{vs}}{dt} = Q_{as} - Q_{vs}, \tag{62}$$

where P_{as} is pressure of the systemic arteries; P_{VS} is pressure of the systemic veins; P_{LV} is pressure in the left ventricle; P_{LA} is pressure in the left atrium; V_{as} is volume of the systemic arteries; V_{vs} is volume of the systemic veins; V_{LV} is volume of the left ventricle; V_{LA} is volume of the left atrium; Q_{mitral} is mitral flow; Q_{VP} is pulmonary venous flow; Q_{ao} is aortic flow; Q_{as} is flow going out of the systemic arteries; and Q_{vs} is flow going out of the systemic veins.

Pulmonary circulation:

$$P_{ap} = V_{ap}/C_{1p} \tag{63}$$

$$P_{vp} = V_{vp}/C_{2p} \tag{64}$$

$$Q_{pa} = \begin{cases} \dfrac{P_{RV} - P_{ap}}{R_{1p}} & (P_{RV} > P_{ap}) \\ 0 & (P_{RV} < P_{ap}) \end{cases} \tag{65}$$

$$Q_{ap} = (P_{ap} - P_{vp})/R_{2p} \tag{66}$$

$$Q_{vp} = (P_{vp} - P_{LA})/R_{3p} \tag{67}$$

$$Q_{tricus} = \begin{cases} \dfrac{P_{RA} - P_{RV}}{R_{4p}} & (P_{RA} > P_{RV}) \\ 0 & (P_{RA} < P_{RV}) \end{cases} \tag{68}$$

$$\frac{dV_{RA}}{dt} = Q_{vs} - Q_{tricus} \tag{69}$$

$$\frac{dV_{RV}}{dt} = Q_{tricus} - Q_{pa} \tag{70}$$

$$\frac{dV_{ap}}{dt} = Q_{pa} - Q_{ap} \tag{71}$$

$$\frac{dV_{vp}}{dt} = Q_{ap} - Q_{vp}, \tag{72}$$

where P_{ap} is pressure of the pulmonary arteries; P_{vp} is pressure of the pulmonary veins; P_{RV} is pressure in the right ventricle; P_{RA} is pressure in the right atrium; V_{ap} is volume of the pulmonary arteries; V_{vp} is volume of the pulmonary veins; V_{RV} is volume of the right ventricle; V_{RA} is volume of the right atrium; Q_{tricu} is tricuspid flow; Q_{vs} is systemic venous flow; Q_{pa} is pulmonary arterial flow; Q_{ap} is flow going out of the pulmonary arteries; and Q_{vp} is flow going out of the pulmonary veins.

Atria can be modeled by either the lumped parameter model or the FEM model with realistic morphology. In the former approach, we adopted the time-varying elastance model proposed by Kaye et al. 36], in which the instantaneous right or left atrial pressure

$[P(t)]$ is related to the instantaneous volume of the chamber according to the following equations:

$$P(t) = P_{ed}(V(t)) + e(t)[P_{es}(V(t)) - P_{ed}(V(t))] \quad (73)$$

$$P_{ed}(V(t)) = \beta[\exp(\alpha(V(t) - V_0) - 1] \quad (74)$$

$$P_{es}(V(t)) = E_{es}[\mathrm{V(t)} - V_0], \quad (75)$$

where $P_{ed}(V(t))$ is the end-diastolic pressure with scaling factors α and β; $P_{es}(V(t))$ is the end-systolic pressure with end-systolic elastance (E_{es}) and the volume axis intercept (V_0). Time-varying elastance ($e(t)$) was also adopted from this report, and was used for the right and left atria.

$$e(t) = \begin{cases} 0.5 \cdot \left[sin\left(\frac{\pi}{T_{max}} t - \frac{\pi}{2}\right) + 1 \right] & \left(t < \frac{3}{2} T_{max}\right) \\ 0.5 \cdot exp\left[\frac{-\left(t - \frac{3}{2} T_{max}\right)}{\tau_a} \right] & \left(t > \frac{3}{2} T_{max}\right) \end{cases}, \quad (76)$$

where T_{max} is the time to maximum elastance and τ_a is the time constant of relaxation. We also previously used an FEM model of atria 37], but in this previous study we did not simulate the propagation of excitation in the atria, so that the atria tissue contracted simultaneously.

2.3.4 Personalization of the Circulatory Model

We estimated the parameter values of the circuit (Fig. 5) in the following manner. For the systemic circulation, we calculated the total resistance ($R = R_{1a} + R_{2a} + R_{3a}$) as R = (mean arterial pressure − mean right atrial pressure)/(cardiac output). Then, the total resistance was subdivided into R_{1a} (=5%), R_{2a} (=93%), and R_{3a} (2%) according to the literature 38–40]. We estimated the time constant of the arterial pressure decay during diastole (τ) using the exponential function

$$P_d = P_{es} \times \exp(t_d/\tau), \quad (77)$$

where P_d is diastolic pressure, P_{es} is end-systolic pressure, and t_d is the diastolic time interval. By dividing the τ value by R_{2a}, we could obtain C_{1a}. C_{2a} was assumed to be 40 times C_{1a}, and R_{4a}, which is the filling resistance of the tricuspid valve (R_{4a}), was set at 0.0025 mmHg/ml/s 36]. Parameter values for the pulmonary circulation were estimated similarly. Finally, using the parameter values thus estimated as the initial condition, fine tuning was made using a simpler model in which the FE model of the ventricles was replaced with time-varying elastance models of the right and left ventricles 35]. Use of the simple system that ran much faster than the FEM simulation enabled efficient tuning of the parameter values.

2.4 Integrated Model

To show the usefulness of the multiscale multiphysics heart simulation, we show two example applications.

2.4.1 Prediction of the Therapeutic Effect of Cardiac Resynchronization Therapy (CRT)

CRT is a pacing therapy for the dyssynchronous failing heart using a pair of ventricular leads. Although its effectiveness has been confirmed by clinical trials, a significant number of patients who are indicated for the treatment by the current guidelines fail to show a benefit from CRT (nonresponders). We therefore tested the ability of heart simulation using patient-specific models to predict the outcome of CRT 41–43].

Using clinical data collected before the treatment, we created patient-specific models of dyssynchronous failing hearts, and, to these heart models, we performed simulations of biventricular pacing. As shown in Fig. 6a, the simulated ECGs well-reproduced the real EEGs measured before and after the treatment. From the hemodynamics simulations (Fig. 6b), we retrieved multiple functional indices that are used in clinical settings to find that the maximum value of the time derivative of the left ventricular pressure (max dP/dt) can predict the clinical outcome of CRT. Besides the clinical researches seeking for the biomarkers, such simulation studies can help optimize patient selection, determining who is likely to benefit from CRT.

2.4.2 In Silico Surgery

Surgical correction of the structural anomaly is the main therapeutic approach for congenital heart disease (CHD), but variations in morphology and function among affected individuals may hamper accumulation of the experience that surgeons require to improve their expertise. Multiscale multiphysics heart simulation can help in the design of efficient surgical strategies by facilitating an understanding of anomalous geometry and function in CHD patients. We have previously shown the feasibility of in silico heart surgery, which was capable of predicting postoperative cardiac function in a complex CHD case 37]. We simulated a case of double outlet right ventricle (DORV), a condition in which both the aorta and pulmonary artery originate from the right ventricle. The patient's circulation was supported by the shunt flow through the atrial and ventricular septal defects. Immediately after birth, a pulmonary artery banding operation was performed as palliative therapy, but at the age of two, surgical repair for the restoration of physiological circulation was attempted, with creation of an intracardiac tunnel (conduit) connecting the ventricular septal defect to the aortic root, closure of the septal defects, and pulmonary artery debanding. We created a patient-specific heart model of this patient in the preoperative state, then, the morphology of this heart model was modified to reproduce the surgical procedure. As shown in Fig. 7, the heart simulation successfully reproduced the pathophysiology of the patient and the cure by surgical treatment. We also point out

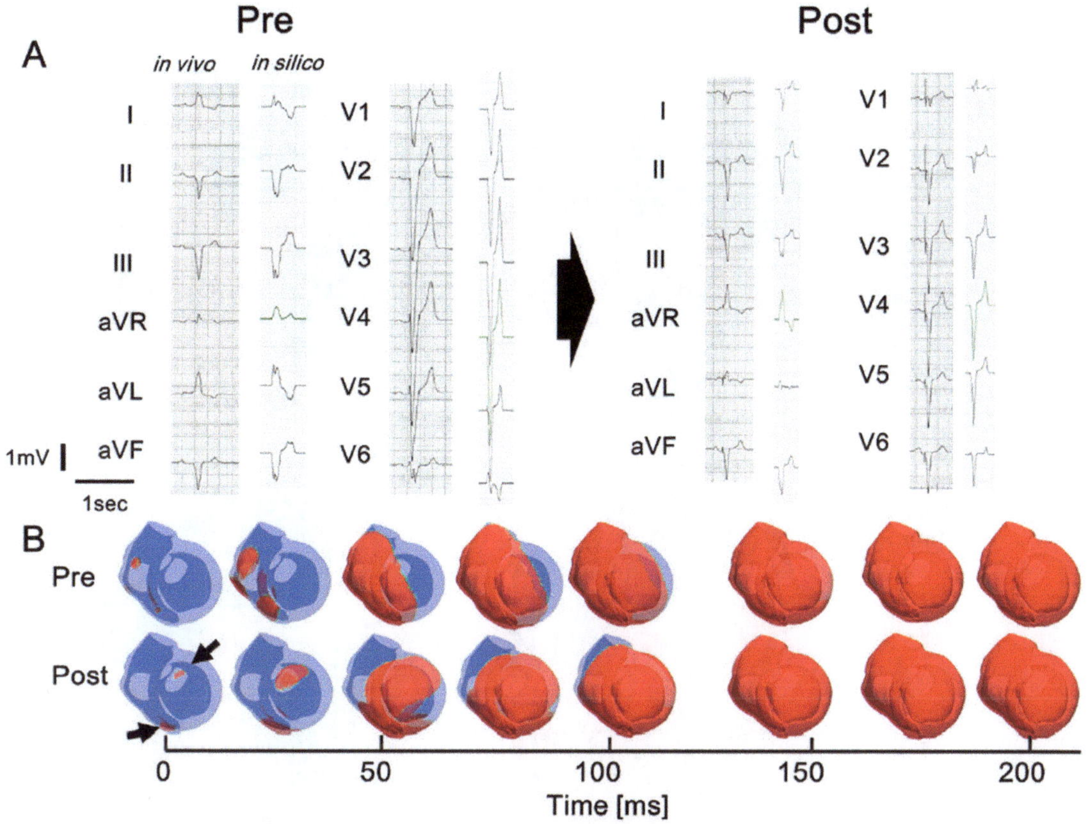

Fig. 6 Simulated effects of CRT. (**a**) ECG before (left) and after (right) CRT. In each panel, ECGs are compared between the simulation (in silico, right column) and clinical record (in vivo, left column). (**b**) Time-lapse images of the propagation of excitation and contraction before (Pre: top row) and after (Post: bottom row) CRT. Numbers at the bottom indicate the time after the onset of excitation in milliseconds. Arrows indicate the pacing sites. (From 42])

that, in these simulation models, realistic behaviors of heart valves were also reproduced by our fluid–structure interaction analysis technique, the details of which can be seen in our previous publication 37] (Fig. 8). Simulations were also used to compare the predicted outcomes resulting from different surgical procedures.

3 Notes

1. If the electrophysiology of the heart is the only target of analysis, monodomain equations can be used.
2. Because MC simulation of sarcomere dynamics is computationally heavy, we also use an ordinary differential equation (ODE) model, in which the results of MC simulations are approximated by the ODEs 44].

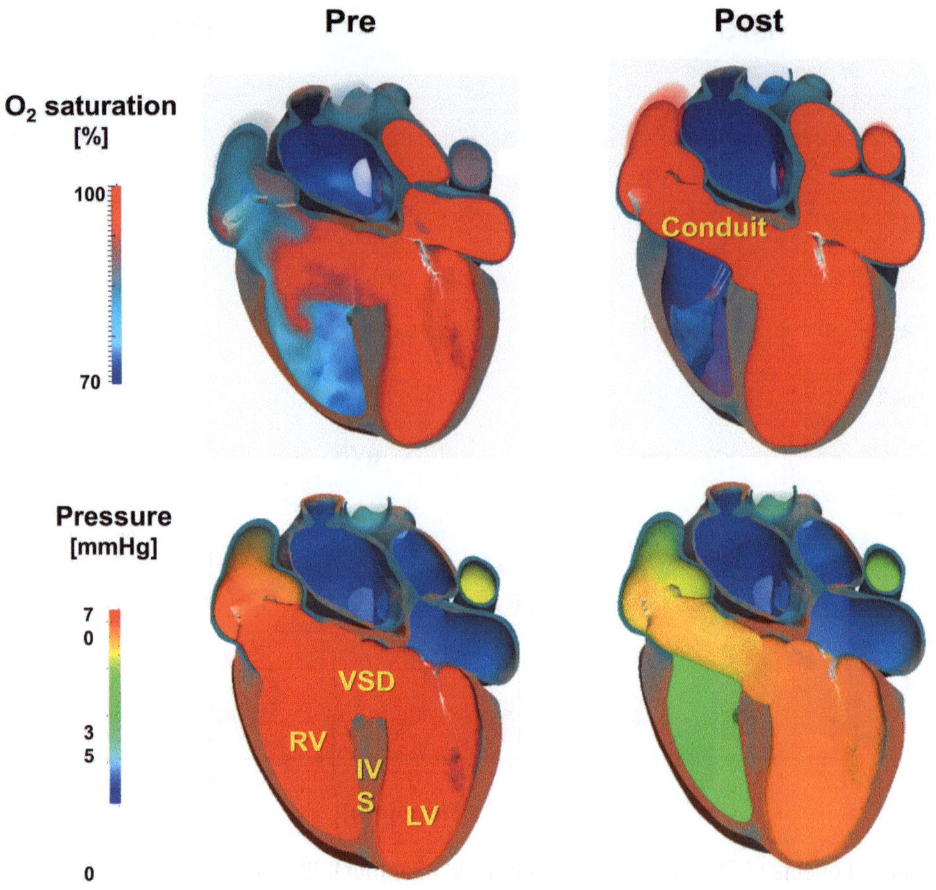

Fig. 7 Simulation of congenital heart disease. O_2 saturation (top row) and systolic blood pressure (bottom) in the patient-specific models of congenital heart disease are compared before (Pre) and after (Post) surgery. *RV* right ventricle, *LV* left ventricle, *VSD* ventricular septal defect, *IVS* interventricular septum. (From 37] with permission)

3. The complexity of the model can be reduced by simplifying or ignoring a part. For example, analysis of electrophysiology can be performed with the static heart model, ignoring the mechanics. However, we must be careful of the influence of tissue stretch on the activation of certain ion channels and conduction velocity.

4 Conclusion

In this study, we demonstrate our approaches to develop a multi-scale multiphysics heart simulator, UT-Heart, and show some example clinical applications. Heart diseases are becoming a world-wide health problem and novel treatment measures are vigorously sought in the fields of clinical medicine and basic sciences. Because

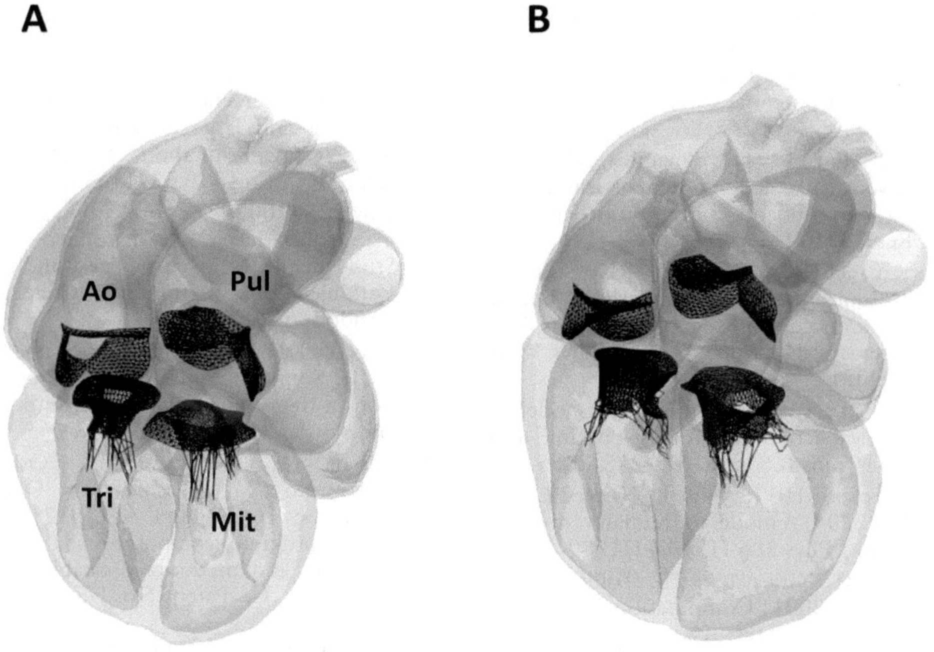

Fig. 8 Simulation of valves. Images showing the motion of four valves in the heart. (**a**) During systole, aortic (Ao) and pulmonary (Pul) valves are open, while tricuspid (Tri) and mitral (Mit) valves are closed. (**b**) During diastole, Tri and Mit valves are open, while Ao and Pul valves are closed

of its ability to simulate unrealizable experiments, a multiscale multiphysics heart simulator can contribute to such efforts in many ways. As a tool in basic science, a multiscale heart simulator can visualize the behavior of a single key molecule in the live organ. As a tool for translational research, the heart simulator allows for the testing of new drugs and devices in the human heart without any ethical concerns. However, to fulfill all these objectives, we need to further improve the power of the simulator in respect to its multiscale and multiphysics nature. We are continuing research into the model with support by MEXT in the form of the "Program for Promoting Researches on the Supercomputer Fugaku."

Acknowledgments

We thank Edanz Group (https://en-author-services.edanzgroup.com/ac) for editing a draft of the manuscript.

References

1. Noble D (1960) Cardiac action and pacemaker potentials based on the Hodgkin-Huxley equations. Nature 188:495–497
2. Hodgkin AL, Huxley AF (1952) A quantitative description of membrane current and its application to conduction and excitation in nerve. J Physiol 117:500–544. https://doi.org/10.1113/jphysiol.1952.sp004764
3. Noble D (2006) The rhythm section: the heartbeat and other rhythm. In: The music of

life, biology beyond the genome. Oxford University Press, New York, pp 55–73
4. Beeler GW, Reuter H (1977) Reconstruction of the action potential of ventricular myocardial fibers. J Physiol 268:177–210
5. Luo C, Rudy Y (1994) A dynamic model of the cardiac ventricular action potential - simulatons of ionic currents and concentration changes. Circ Res 74:1071–1097
6. Courtemanche M, Ramirez RJ, Nattel S (1998) Ionic mechanisms underlying human atrial action potential properties: insights from a mathematical model. Am J Phys 275: H301–H321
7. Stewart P et al (2009) Mathematical models of the electrical action potential of Purkinje fibre cells. Phil Trans R Soc A 367:2225–2255
8. Ten Tusscher KHWJ, Noble D, Noble PJ, Panfilov AV (2004) A model for human ventricular tissue. Am J Phys 286:H1573–H1589
9. Winslow RL, Greenstein JL, Tomaselli GF, O'Rouke B (2001) Computational models of the failing myocyte: relating altered gene expression to cellular function. Phil Trans R Soc A 359:1187–1200
10. Grandi E, Pasqualini FS, Bers DM (2010) A novel computational model of the human ventricular action potential and Ca transient. J Mol Cell Cardiol 48:112–121. https://doi.org/10.1016/j.yjmcc.2009.09.019
11. Vigmond E et al (2009) Towards predictive modelling of the electrophysiology of the heart. [review]. Exp Physiol 94:563–577
12. Trayanova NA (2011) Whole heart modeling: Applications to cardiac electrophysiology and electromechanics. Circ Res 108:113–128. https://doi.org/10.1161/CIRCRESAHA.110.223610
13. Okada J-I et al (2015) Screening system for drug-induced arrhythmogenic risk combining a patch clamp and heart simulator. Sci Adv 1: e1400142
14. Huxley AF (1957) Muscle structure and theories of contraction. Prog Biophys Biophys Chem 7:255–318
15. Beyar R, Sideman S (1984) A computer study of the left ventricular performance based on fiber structure, sarcomere dynamics, and transmural electrical propagation velocity. Circ Res 55:358–375
16. Negroni JA, Lascano EC (1996) A cardiac muscle model relating sarcomere dynamics to calcium kinetics. J Mol Cell Cardiol 28: 915–929
17. Watanabe H, Sugiura S, Kafuku H, Hisada T (2004) Multiphysics simulation of left ventricular filling dynamics using fluid-structure interaction finite element method. Biophys J 87:2074–2085
18. Zhang Q, Hisada T (2001) Analysis of fluid-structure interaction problem with structural buckling and large domain change by ALE finite element method. Comput Methods Appl Mech Eng 190:6341–6357
19. Tawara S (2000) The condcution system of the mammalian heart An Anatomico-histological Study of the Atrioventricular Bundle and the Purkinje Fibers. Imperial College Press, London, p 256
20. Streeter DD Jr, Spotnitz HM, Patel DP, Ross JR Jr, Sonnenblick ED (1969) Fiber orientation in the canine left venticle during diastole and systole. Circ Res 24:339–347
21. Hisada T, Kurokawa H, Oshida M, Yamamoto M, Washio T, Okada J-I, Watanabe H, Sugiura S (2012) Modeling device, program, computer-readable recording medium, and method of establishing correspondence, US Patent No. US 8,095,321 B2, Jan 10, 2012
22. Helm P, Winslow R, McVeigh E DTMRI data sets [Internet]. 2004 [cited Dec. 1, 2014]. Available from: https://gforge.icm.jhu.edu/gf/project/dtmridata_setshttps://gforge.icm.jhu.edu/gf/project/dtmridata_sets
23. Washio T et al (2015) Ventricular fiber optimization utilizing the branching structure. Int J Numer Meth Biomed Eng 32:e02753. https://doi.org/10.1002/cnm.2753
24. Lombaert H et al (2012) Human atlas of the cardiac fiber architecture: study on a healthy population. IEEE Trans Med Imag 31: 1436–1447
25. Washio T, Sugiura S, Okada J-I, Hisada T (2020) Using systolic local mechanical load to predict fiber orientation in ventricles. Front Physiol 11:467. https://doi.org/10.3389/fphys.2020.00467
26. O'Hara T, Virag L, Varro A, Rudy Y (2011) Simulation of the undiseased human cardiac ventricular action potential: model formulation and experimental validation. PLoS Comput Biol 7:e1002061
27. Okada J et al (2011) Transmural and apicobasal gradients in repolarization contribute to T-wave genesis in human surface ECG. Am J Phys 301:H200–H208
28. Okada J-I et al (2018) Arrhythmic hazard map for a 3D whole-ventricles model under multiple ion channel block. Brit J Pharmacol 175: 3435–3452
29. Sanchez-Alonso JL et al (2016) Microdomain-specific modulation of L-type calcium channels

leads to triggered ventricular arrhythmia in heart failure. Circ Res 119:944–955
30. Vigmond EJ, Aguel F, Trayanova NA (2002) Computational techniques for solving the bidomain equation. IEEE Trans Biomed Eng 49:1260–1269
31. Rice JJ, Stolovitzky G, Tu Y, de Tombe PP (2003) Ising model of cardiac thin filament activation with nearest-neighbor cooperative interactions. Biophys J 84:897–909
32. van der Velden J, de Jong JW, Owen VJ, Burton PBJ, Stienen GJM (2000) Effect of protein kinase a on calcium sensitivity of force and its sarcomere length dependence in human cardiomyocytes. Cardiovasc Res 46:487–495
33. Konhilas JP, Irving TC, de Tombe PP (2002) Length-dependent activation in three striated muscle types of the rat. J Physiol 544:225–236
34. Klotz S, Dickstein ML, Burkhoff D (2007) A computational method of prediction of the enddiastolic pressure–volume relationship by single beat. Nat Protoc 2:2152–2158
35. Kerckhoffs RCP et al (2006) Coupling of a 3D finite element model of cardiac ventricular mechanics to lumped systems models of the systemic and pulmonic circulation. Ann Biomed Eng 35:1–18
36. Kaye D et al (2014) Effects of an internal shunt on rest and exercise hemodynamics: results of a computer simulation in heart failure. J Cardiac Fail 20:212–221
37. Kariya T et al (2020) Personalized perioperative multi-scale, multi-physics heart simulation of double outlet right ventricle. Ann Biomed Eng 48:1740–1750
38. O'Rourke MF, Taylor MG (1967) Input impedance of the systemic circulation. Circ Res 20:365–380
39. Westerhof N, Elzinga G, Sipkema P (1971) An artificial arterial system for pumping hearts. J Appl Physiol 36:123–127
40. Nichols WW, Pepine CJ, Geiser EA, Conti R (1980) Vascular load defined by the aortic input impedance spectrum. Fed Proc 39: 196–201
41. Panthee N et al (2016) Tailor-made heart simulation predicts the effect of cardiac resynchronization therapy in a canine model of heart failure. Med Image Anal 31:46–62
42. Okada J-I et al (2017) Multi-scale, tailor-made heart simulation can predict the effect of cardiac resynchronization therapy. J Mol Cell Cardiol 108:17–23
43. Isotani A et al (2020) Patient-specific heart simulation can identify non-responders to cardiac resynchronization therapy. Heart Vessels 35:1135–1147. https://doi.org/10.1007/s00380-020-01577-1
44. Washio T, Okada J, Sugiura S, Hisada T (2011) Approximation for cooperative interactions of a spatially-detailed cardiac sarcomere model. Cell Mol Bioeng 5:113–126. https://doi.org/10.1007/s12195-011-0219-2

Chapter 11

Multiscale Modeling of the Mitochondrial Origin of Cardiac Reentrant and Fibrillatory Arrhythmias

Soroosh Solhjoo, Seulhee Kim, Gernot Plank, Brian O'Rourke, and Lufang Zhou

Abstract

While mitochondrial dysfunction has been implicated in the pathogenesis of cardiac arrhythmias, how the abnormality occurring at the organelle level escalates to influence the rhythm of the heart remains incompletely understood. This is due, in part, to the complexity of the interactions formed by cardiac electrical, mechanical, and metabolic subsystems at various spatiotemporal scales that is difficult to fully comprehend solely with experiments. Computational models have emerged as a powerful tool to explore complicated and highly dynamic biological systems such as the heart, alone or in combination with experimental measurements. Here, we describe a strategy of integrating computer simulations with optical mapping of cardiomyocyte monolayers to examine how regional mitochondrial dysfunction elicits abnormal electrical activity, such as rebound and spiral waves, leading to reentry and fibrillation in cardiac tissue. We anticipate that this advanced modeling technology will enable new insights into the mechanisms by which changes in subcellular organelles can impact organ function.

Key words Cardiac arrhythmia, Computational modeling, Mitochondrial dysfunction, Optical mapping, Neonatal rat ventricular myocyte, Action potential

1 Introduction

Cardiovascular disease (CVD) is a leading cause of death in the world; in 2017 alone, CVD accounted for ~17.8 million (95% CI, 17.5–18.0 million) deaths [1]. A large proportion of these deaths occur as a consequence of sudden cardiac death resulting from cardiac arrhythmias [2, 3]. Cardiac arrhythmias refer to conditions in which the heart's rhythm is disrupted, or the electrical activity is

The original version of this chapter was revised. The correction to this chapter is available at https://doi.org/10.1007/978-1-0716-1831-8_18

Supplementary Information The online version of this chapter (https://doi.org/10.1007/978-1-0716-1831-8_11) contains supplementary material, which is available to authorized users.

Sonia Cortassa and Miguel A. Aon (eds.), *Computational Systems Biology in Medicine and Biotechnology: Methods and Protocols*, Methods in Molecular Biology, vol. 2399, https://doi.org/10.1007/978-1-0716-1831-8_11,

abnormal. Regardless of the site of occurrence, cardiac arrhythmias can be attributed to abnormality in either impulse initiation or electrical propagation [4, 5]. Abnormal electrical propagation has been associated with conduction block or formation of reentrant waves (reentry), which occurs when a single propagating electrical wave traveling through the heart excites a region of the heart more than once. Reentrant arrhythmias can occur via several mechanisms (for a review see reference [6]): (1) compromised intercellular electrical coupling caused by dysfunction of gap junctions [7]; (2) regional electrical uncoupling caused by anatomical barriers such as scar tissue formed during ischemic infraction [8]; and (3) dynamic functional block due to heterogeneity of intrinsic electrophysiological restitution properties [9]. More recent work has added to this list "metabolic sinks" which are associated with failure of mitochondrial energetics, involving myocardial tissue regions [10, 11], although the precise underlying tissue/(sub)-cellular/molecular mechanisms remain incompletely understood.

Mitochondria lie at the crossroad of cellular metabolic and signaling pathways, and play a pivotal role in regulating key processes maintaining cellular functions and health [12]. The loss of mitochondrial function has emerged as a key contributor to the generation of arrhythmias [10, 13, 14]. Mitochondrial dysfunction can cause excess reactive oxygen species (ROS) generation and profound dissipation of mitochondrial membrane potential ($\Delta\Psi_m$) leading to reduced ATP production, which modulate a variety of redox- and energy-sensitive ion channels/transporters involved in ionic and action potential modulation, such as sarcoplasmic reticulum ryanodine receptors and Ca^{2+} ATPase [15–18], Ca^{2+}/calmodulin-dependent protein kinase II [19–21], cell membrane Na^+ channels [22], and sarcolemmal ATP-sensitive potassium channels (K_{ATP}) [23]. For instance, previous studies have shown that K_{ATP} channels are rapidly activated upon $\Delta\Psi_m$ depolarization and energy depletion, causing shortening of action potential duration and decrease of action potential amplitude in a cardiomyocyte [23, 24]. While the effect of mitochondrial dysfunction on cellular electrophysiology has been extensively examined [18, 21–23], how the disruption of organelle function escalates to influence the rhythm of the whole heart remains elusive. In intact hearts, the electrical, mechanical, and energetic subsystems interact at various spatiotemporal scales, constituting complex networks that are difficult to fully comprehend solely by experiments.

The present chapter describes a combined approach of experimental work with multiscale mathematical modeling to unravel the mechanisms underlying the origin of arrhythmias in heart tissue. Together, computer simulations and optical mapping experiments help us to explore mechanisms linking mitochondrial dysfunction (in particular dissipation of mitochondrial membrane potential, $\Delta\Psi_m$) and incidence of arrhythmias in cardiac tissue. Among the mechanisms investigated are the roles of regional mitochondrial depolarization at the cardiac tissue level, the activation of sarcolemmal K_{ATP} channels, and the resultant formation of metabolic current sinks, in which the large background K^+ conductance locks the sarcolemmal membrane potential close to the equilibrium membrane potential of K^+ (E_k), rendering cardiac cells unexcitable. Under this condition, the cardiomyocyte is unable to generate an action potential due to $\Delta\Psi_m$ depolarization and lack of ATP eliciting the opening of sarcolemmal K_{ATP} channels [14, 25]. This *metabolic sink* mechanism is distinct from (but could be occurring in parallel with) blocks caused according to existing paradigms of electrical dysfunction in the heart (*see* 1–3 above).

Herein, we specifically describe the general method of integrating a previously developed cardiomyocyte electrophysiology model incorporating excitation–contraction coupling, mitochondrial energetics, and ROS-induced ROS-release (ECME-RIRR) [23] into a two-dimensional (2D) finite element model of ventricular tissue. Then, we illustrate how to use this tissue model to simulate the effect of regional mitochondrial depolarization on the propagation of electrical activity leading to the triggering of arrhythmias characterized by wave rebound, reentry, and fibrillation. Finally, we show how the regional $\Delta\Psi_m$ depolarization induced by chemical uncoupling of mitochondrial oxidative phosphorylation can elicit arrhythmias in a monolayer of neonatal rat ventricular myocytes (NRVMs), lending validation to the simulation findings, while providing key mechanistic insights into a complex nonlinear dynamic phenomenon such as the generation of arrhythmias in the heart.

2 Materials

1. Integrated excitation–contraction coupling, mitochondrial energetics and ROS-induced ROS release (ECME-RIRR) model of a cardiomyocyte, previously described [23] (Fig. 1).
2. A 2D finite element model of ventricular tissue, in which the spread of electrical activity is described by a monodomain equation [26], given by

$$\beta C_m \frac{\partial V_m}{\partial t} + \beta I_{ion}(V_m, \eta) = \nabla\cdot(\sigma_m \nabla V_m) + I_{tr}$$

   where β is the membrane surface to volume ratio, C_m is the membrane capacitance, V_m is the transmembrane voltage, I_{ion} is the density of the total ionic current which is a function of V_m and a set of state variables η, σ_m is the monodomain conductivity tensor, and I_{tr} is a transmembrane stimulus current.
3. A custom-tailored ordinary differential equation (ODE) integration technique, named temporal multiscale decoupling (TMSD), developed previously for systems with high stiffness [27].

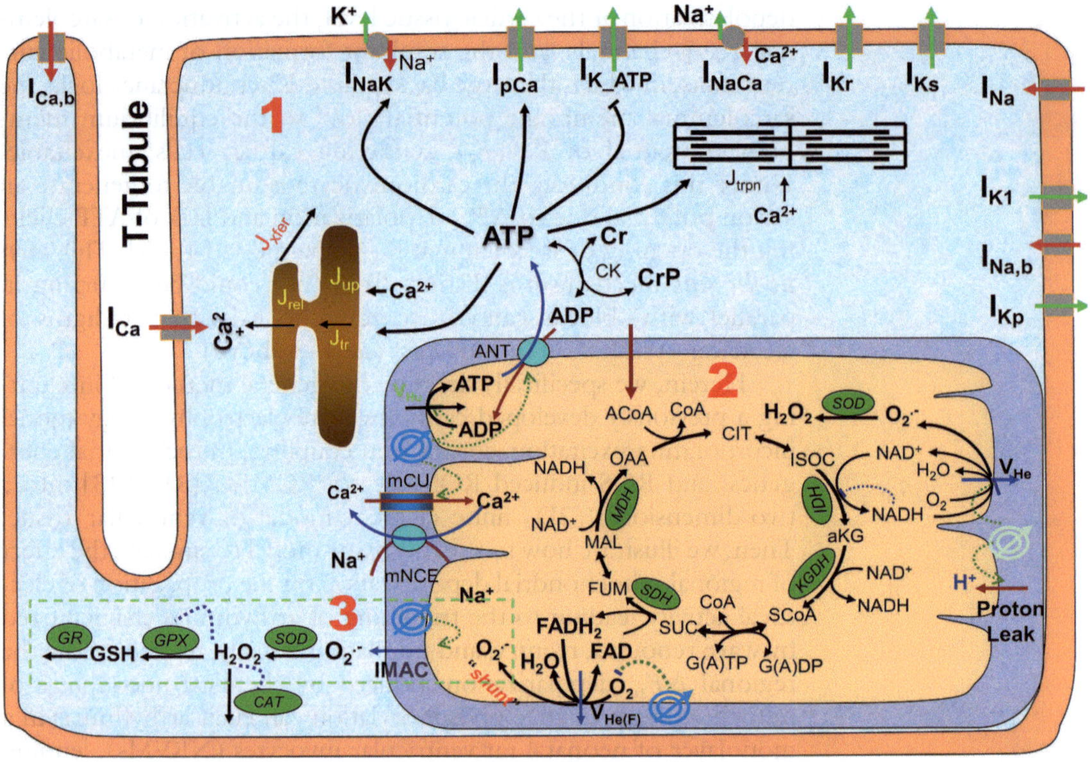

Fig. 1 The scheme of the ECME-RIRR cardiomyocyte model, which consists of three modules. Module 1: Electrophysiological module describing the major ion channels underlying ionic and action potential dynamics. Module 2: Mitochondrial energetics module accounting for the tricarboxylic acid cycle, oxidative phosphorylation, and inner membrane channels and transporters. Module 3: RIRR module describing ROS production (from the electron transport chain), transport across mitochondrial membrane, and scavenging (e.g., by the superoxide dismutase and glutathione peroxidase enzymes). For details *see* Ref. 23

4. A programing language such as C++, MATLAB, or Python.
5. A simulation package such as the Cardiac Arrhythmia Research Package (CARP) developed by Vigmond et al. [28], which is built on top of the message passing interface-based library PETSc [29].
6. Fibronectin-coated coverslips ($r = 2.1$ cm).
7. Culture medium: medium 199 supplemented with 10% heat-inactivated fetal bovine serum. Starting on the second day of culture, the serum level is lowered to 2%.
8. Tyrode's solution: 135 mM NaCl, 5.4 mM KCl, 1.8 mM $CaCl_2$, 1 mM $MgCl_2$, 0.33 mM NaH_2PO_4, 5 mM HEPES, and 5 mM glucose.
9. An optical mapping system: a highly sensitive fluorescence imaging system, such as a photodiode array, to detect the changes in the sarcolemmal membrane potential and the propagation of electrical activity in monolayer cultures of cardiomyocytes.

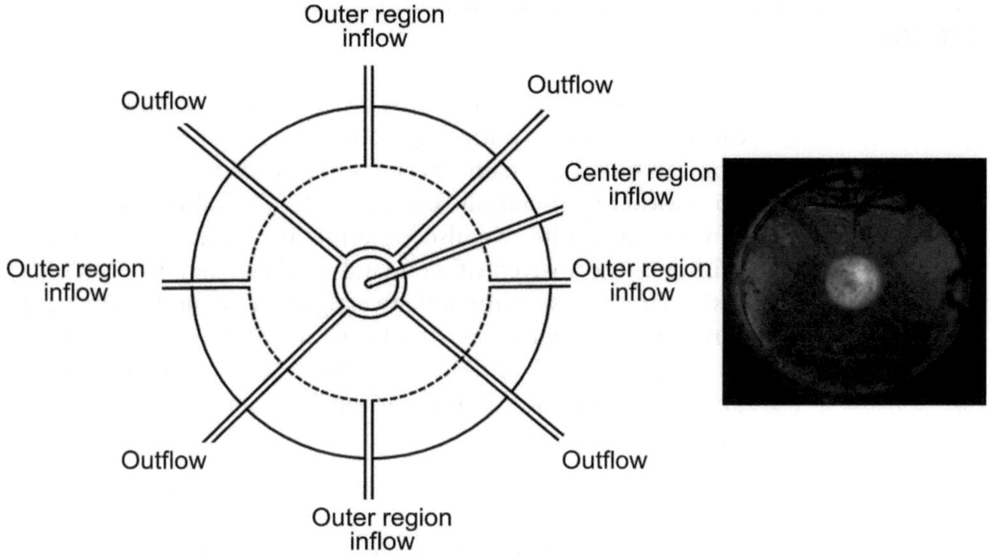

Fig. 2 The scheme of the customized local perfusion system. *Left:* The local perfusion system divides the chamber of the optical mapping setup into two sections: one outer region superfused with normal Tyrode's solution, and a center region superfused with Tyrode's solution supplemented with a chemical mitochondrial uncoupler to induce a metabolic sink. Normal Tyrode's solution enters the chamber from the outer edges of the chamber and is suctioned out from the borders of the center region. Mitochondrial uncoupler-supplemented Tyrode's solution enters from the center of the chamber and is suctioned out at the borders of the metabolic sink. The solutions were heated to 37 °C prior to entering the chamber. A pair of electrodes at the edge of the lid are used to apply voltage pulses that propagate through the monolayer. The dashed line shows the extent of the chamber of the optical mapping system where the monolayer is placed. *Right:* An example fluorescent image showing the effect of local perfusion with FCCP on mitochondrial inner membrane potential. An increase in TMRM emission signal in the dequenching mode in the center region of the monolayer of cardiomyocytes indicates depolarization of mitochondria in that region. Optical mapping can confirm formation of a metabolic sink in the region with depolarized mitochondria

10. A customized local perfusion system. For details see reference [30] (Fig. 2).
11. A fluorescent dye sensitive to sarcolemmal membrane potential, such as 4-(2-(6-(dibutylamino)-2-naphthalenyl) ethenyl)-1-(3-sulfopropyl)pyridinium hydroxide inner salt (di-4-ANEPPS).
12. A chemical mitochondrial uncoupler, such as carbonyl cyanide-*p*-trifluoromethoxyphenylhydrazone (FCCP).
13. An indicator for mitochondrial inner membrane potential, such as the potentiometric fluorescent dye tetramethylrhodamine methyl ester (TMRM).
14. A chemical blocker of sarcolemmal KATP channels, such as glibenclamide.

3 Methods

Methodologically, we describe a computational model of the myocardial syncytium to investigate whether regional $\Delta\Psi_m$ depolarization can initiate a chain of events that leads to reentry, through the formation of a metabolic current sink. To evaluate the goodness of simulation results, we also describe an experimental model comprising a monolayer of neonatal cardiomyocytes (NRVMs) in a dish, for the experimental investigation of how $\Delta\Psi_m$ instability can induce arrhythmias. Iteration between modeling simulations and experimental findings is applied to explore the mechanisms linking mitochondrial dysfunction with cardiac arrhythmias at the cardiac tissue level.

1. 2D tissue model development, numeric solution, and simulation strategy.
   (a) Describe cellular ionic and metabolic dynamics using the ECME-RIRR cardiomyocyte model. Use the same model parameters described previously [23] unless indicated otherwise.
   (b) Incorporate the ECME-RIRR cell model into a 2D finite element model of ventricular tissue (5×5 cm^2), which is composed of elements of size 200×200 μm^2. Each element represents an ensemble of about 20 cells (assuming a length of 100 μm and a diameter of 20 μm), that are homogenized into a continuum. Thus, the total number of cells is 1,250,000 (Fig. 3).
   (c) Describe the spread of electrical activity in the tissue using the monodomain equation. Implement no-flux conditions on membrane voltage (V_m) at all model boundaries.
   (d) Discretize the monodomain equation (a partial differential equation, PDE) at 200 μm spatial resolution (*see* **Note 1**).
   (e) Use a forward Euler method to solve the PDE, and the TMSD approach to integrate the ODEs of the ECME-RIRR model. Use different time steps for the PDE and ODEs (*see* **Note 2**).
   (f) Perform simulations using CARP.
2. Simulate the formation of a metabolic current sink induced by regional $\Delta\Psi_m$ depolarization.
   (a) Pace the 2D tissue model at 1 Hz (S1) from the lower left corner of the sheet (Fig. 3).
   (b) Induce regional mitochondrial depolarization in cells within a central circular region ($r = 1$ cm) by increasing

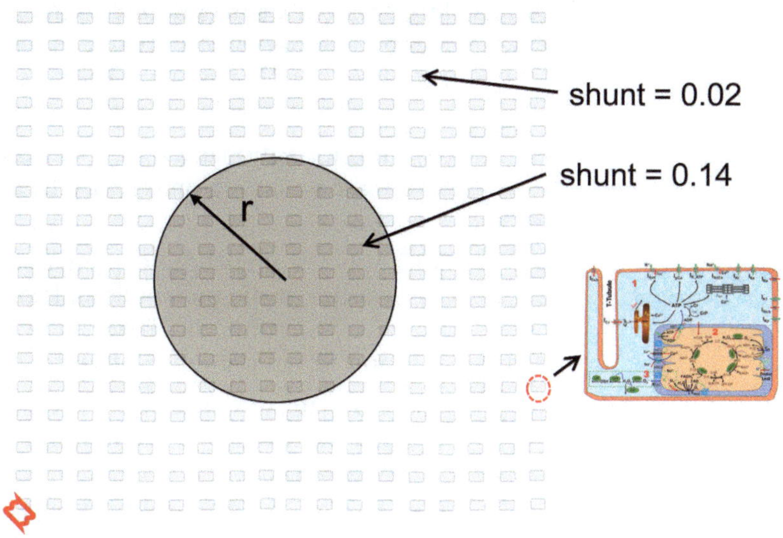

Fig. 3 The scheme of the two-dimensional monodomain myocardial tissue model. This tissue sheet (5 × 5 cm^2) consists of ~63,000 cells, with each cell described by the ECME-RIRR model. To induce regional mitochondrial depolarization, the level of ROS production (i.e., parameter *shunt* in the ECME-RIRR) is increased from 2% to 14%, in cells within the central region of the tissue sheet. Tissue is paced by a pulse train stimulus (S1, 1 Hz) at the lower left corner. For details *see* Ref. 25

the fraction of ROS production (i.e., parameter *shunt* in the ECME-RIRR model) from 0.02 to 0.14 in the mitochondria of those cardiomyocytes (Fig. 3).

(c) Plot the simulated $\Delta\Psi_m$ in the tissue to confirm the formation of regional $\Delta\Psi_m$ depolarization.

(d) Simulate and plot the electrical wave propagation triggered by S1 stimulus, with varying K_{ATP} channel density (σ_{KATP}, 0–3.8/μm^2).

(e) Plot V_m and availability of sodium channels (jNa) of a representative cardiomyocyte in the central region.

(f) Analyze simulation results to determine the effects of σ_{KATP} on the amplitude and duration of the V_m, as well as on jNa, in the central region where $\Delta\Psi_m$ is depolarized.

(g) Analyze the effect of metabolic sink and σ_{KATP} on electrical wave propagation, including wavelength, wavefront, and refractory period in the sink zone.

(h) Change the size of the central region, that is, $r = 0.5$ or 2 cm, and repeat **steps 1–7** to examine its effect on regional mitochondrial depolarization-induced metabolic sink and, consequently, on electrical wave propagation.

3. Determine the susceptibility of the metabolic sink substrate to reentry and fibrillation evoked by a second stimulus (S2).
   (a) Pace the 2D tissue model at 1 Hz (S1) from the lower left corner of the sheet. Set $\sigma_{\text{KATP}} = 3.8/\mu\text{m}^2$.
   (b) Induce regional $\Delta\Psi_{\text{m}}$ depolarization in cells within a central circular region ($r = 1$ cm) as described in Subheading 3.2.
   (c) Apply a single pulse premature stimulus (S2) near the border of the metabolic sink at various coupling intervals (i.e., the time difference between the applications of S1 and S2) in the range of 10 to 300 ms.
   (d) Simulate the electrical wave propagation and determine the occurrence of reentry and fibrillation (*see* example in Video S1).
   (e) Analyze the effect of S1–S2 coupling interval on the incidence of reentry and the duration of fibrillatory activity.
   (f) Change the size of the central region, that is, $r = 0.5$ or 2 cm, and repeat **steps 1–5** to determine its effect on the propensity of metabolic sink-induced fibrillation.
4. Investigate the effect of $\Delta\Psi_{\text{m}}$ recovery in the metabolic sink on electrical wave propagation.
   (a) Induce regional $\Delta\Psi_{\text{m}}$ depolarization in the cells within a central circular region as described in Subheading 3.2, with various zone sizes (i.e., $r = 0.5$, 1, or 2 cm).
   (b) Pace the 2D tissue model at 1 Hz (S1) from the lower left corner of the sheet, at the time point so that the wavefront reaches the edge of the central region when $\Delta\Psi_{\text{m}}$ in those cells is repolarizing.
   (c) Simulate the electrical wave propagation.
   (d) Analyze the effect of sink size on the tendency for abnormal electrical activity (e.g., rebound, spiral wave, and turbulence).
5. Analyze the effect of the timing of metabolic sink recovery (relative to S1) on the induction of spontaneous arrhythmias.
   (a) Induce regional $\Delta\Psi_{\text{m}}$ depolarization in the cells within a central circular region as described in Subheading 3.2 ($r = 2$ cm).
   (b) Pace the 2D tissue at 1 Hz (S1) from the lower left corner of the sheet at different time points so that the wavefront reaches the edge of the central region when $\Delta\Psi_{\text{m}}$ in those cardiomyocytes is repolarizing to different extents (e.g., 10%, 30%, 60%, or 90%).
   (c) Simulate electrical wave propagation.

(d) Analyze the effect of the lag between electrical stimulation and mitochondrial $\Delta\Psi_m$ recovery on electrical activity (e.g., type and duration of arrhythmias), and dissect the spatiotemporal determinants of the aberrant electrical behavior elicited by $\Delta\Psi_m$ changes.

6. Prepare NRVM monolayer cultures.
   (a) Place plastic or glass coverslips in a 6-well culture dish, and superfuse them with fibronectin (25 μg/mL) (*see* **Note 3**).
   (b) After the coverslips are coated with fibronectin (30 min to an hour), wash them with PBS and remove the media.
   (c) Isolate ventricular cardiomyocytes from the 2-day-old Sprague-Dawley rats, as previously described [25, 31].
   (d) Suspend the isolated cardiomyocytes in the culture medium.
   (e) Plate 850,000 cardiomyocytes on each coverslip and incubate them at 37 °C with CO_2 5% (*see* **Note 4**).
   (f) Culture the cardiomyocytes for 5–7 days. Renew the culture medium every day.
7. Induction of metabolic sink through regional $\Delta\Psi_m$ depolarization.
   (a) After 5–7 days of culture, examine the monolayer under the microscope to make sure it is confluent with no gaps between the cells (*see* **Note 5**).
   (b) Transfer the monolayer to a small petri dish and superfuse it with Tyrode's solution at 37 °C (*see* **Note 6**).
   (c) Load the monolayer with TMRM (2 μmol/L) for 2 h in 37 °C incubator (dequenching mode) [32].
   (d) Fill the chamber of the optical mapping setup with Tyrode's solution at 37 °C and place the monolayer in the chamber. Cover the chamber with the local perfusion lid.
   (e) Start by perfusing the monolayer with normal Tyrode's in both the outer and the center regions.
   (f) Start imaging the monolayer from above by means of a camera (MicroMax 1300Y cooled CCD, Princeton Instruments) to record the changes in TMRM signal.
   (g) To induce mitochondrial uncoupling, switch the central perfusion medium to Tyrode's solution supplemented with FCCP (1 μmol/L) for 30 min (*see* **Note** 7). Switch back to normal Tyrode's solution to wash FCCP out and repolarize mitochondria.
   (h) Process images to confirm the formation of the region with depolarized mitochondria in the center of the monolayer (*see* example in Video S2).

8. Optical mapping of the action potential propagation.
   (a) Stain a confluent monolayer with di-4-ANEPPS (5 μmol/L, a fluorescent indicator of plasma membrane) for 15 min and then wash it with Tyrode's solution (*see* **Note 8**).
   (b) Fill the chamber of the optical mapping setup with Tyrode's solution at 37 °C and place the monolayer in the chamber (*see* **Note 9**).
   (c) Cover the chamber with the local perfusion lid.
   (d) Perfuse the monolayer with normal Tyrode's solution in both the outer and the center regions.
   (e) Using the electrodes incorporated in the lid, start pacing the monolayer by applying a train of voltage pulses at 1 Hz at the edge of the monolayer.
   (f) Start imaging the monolayer using the photodiode array to map the sarcolemmal electrical activity.
   (g) To induce mitochondrial uncoupling as explained in Subheading 3.7, switch the central perfusion medium to Tyrode's solution supplemented with FCCP. Switch back to normal Tyrode's solution to wash FCCP out and repolarize mitochondria.
   (h) Process the data to study the effect of regional mitochondrial depolarization on action potential characteristics, propagation of the voltage wave and its characteristics such as wavelength and conduction velocity, and the incidence of reentrant waves and arrhythmic behavior (*see* example in Video S3) while the mitochondria depolarize and as they recover during the FCCP washout (*see* **Note 10**).
   (i) Repeat this section while including chemical blockers of sarcolemmal K_{ATP} channels, such as glibenclamide (10 μmol/L), to assess the role of this channel in scaling the mitochondrial dysfunction to cellular electrical malfunction and arrhythmogenicity.

4 Notes

1. It is important to use a spatial resolution well below the spatial extent of the wavefront in the range of 200–700 μm, to avoid discretization artifacts leading to conduction slowing.
2. The ECME-RIRR model contains both fast (in the submillisecond range, such as formulations describing the intrinsically fast dynamics of the calcium-induced calcium release process including L-type calcium channel and the ryanodine receptor,

as well as the calcium dynamics in the dyadic space) and slow (in the hundreds of millisecond range, such as mitochondrial tricarboxylic acid cycle and oxidative phosphorylation) responses; thus, different time steps are used to integrate the PDE and ODEs. Specifically, the PDE is integrated using a time step of 20 μs, and the set of ODEs is split into groups of variables that operate at similar time scales so that appropriate time steps can be chosen for each group. This numeric scheme leads to a substantial reduction in execution time. For details please *see* Ref. 27.

3. If using plastic coverslips, prior to coating with fibronectin, they should be treated with UV light.
4. When plating the cardiomyocytes on the fibronectin-coated coverslips, the cells tend to gather in the center of the tissue culture dish/well. You can distribute the cells evenly by shaking the culture dish slightly.
5. At this point, the cells throughout the monolayer should be beating synchronously at a spontaneous rate of ~1 Hz.
6. Be careful not to damage and scratch the monolayer while handling the coverslip with forceps.
7. To adjust the pumps speed to control the size of sink area prior to the experiment, you can use a colored dye.
8. di-4-ANEPPS is very susceptible to bleaching and all the steps of the preparation for this experiment should be performed in the dark. During the experiment, the excitation light should be limited to the minimum intensity needed and recording should be done in short periods lowering the chance for bleaching.
9. We use custom-made optics to detect the small potential-dependent fluorescence changes of di-4-ANEPPS (~10% per 100 mV). To produce a low-attenuation filter for recording the di-4-ANEPPS emission, we coated a coverslip with a red dye. The coverslip bearing the monolayer is placed directly on top of the red filter.
10. Fluorescent emission signal from di-4-ANEPPS was transferred to a computer after digitization and analyzed using software developed in LabView and MATLAB to determine the changes in the sarcolemmal membrane potential. Subsequently, the software is used to measure other parameters such as conduction velocity and to produce videos of the propagation of the voltage wave through the monolayer.

Acknowledgments

This work was supported by National Institute of Health (NIH) 5T32HL007227, American Heart Association 14POST20000018, and Defense Health Agency HU00011920029 (to S.S); NIH R01HL137259 (to B. O'R.); and NIH R01s HL121206 and HL128044 (to L.Z.).

References

1. Virani SS, Alonso A, Benjamin EJ, Bittencourt MS, Callaway CW, Carson AP, Chamberlain AM, Chang AR, Cheng S, Delling FN, Djousse L, Elkind MSV, Ferguson JF, Fornage M, Khan SS, Kissela BM, Knutson KL, Kwan TW, Lackland DT, Lewis TT, Lichtman JH, Longenecker CT, Loop MS, Lutsey PL, Martin SS, Matsushita K, Moran AE, Mussolino ME, Perak AM, Rosamond WD, Roth GA, Sampson UKA, Satou GM, Schroeder EB, Shah SH, Shay CM, Spartano NL, Stokes A, Tirschwell DL, VanWagner LB, Tsao CW, American Heart Association Council on E, Prevention Statistics C, Stroke Statistics S (2020) Heart disease and stroke Statistics-2020 update: a report from the American Heart Association. Circulation 141(9): e139–e596. https://doi.org/10.1161/CIR.0000000000000757
2. Cohn JN (1996) Prognosis in congestive heart failure. J Card Fail 2(4 Suppl):S225–S229
3. Cohn JN, Archibald DG, Ziesche S, Franciosa JA, Harston WE, Tristani FE, Dunkman WB, Jacobs W, Francis GS, Flohr KH et al (1986) Effect of vasodilator therapy on mortality in chronic congestive heart failure. Results of a veterans administration cooperative study. N Engl J Med 314(24):1547–1552. https://doi.org/10.1056/NEJM198606123142404
4. Antzelevitch C, Burashnikov A (2011) Overview of basic mechanisms of cardiac arrhythmia. Card Electrophysiol Clin 3(1):23–45. https://doi.org/10.1016/j.ccep.2010.10.012
5. Tse G (2016) Mechanisms of cardiac arrhythmias. J Arrhythm 32(2):75–81. https://doi.org/10.1016/j.joa.2015.11.003
6. Kleber AG, Rudy Y (2004) Basic mechanisms of cardiac impulse propagation and associated arrhythmias. Physiol Rev 84(2):431–488. https://doi.org/10.1152/physrev.00025.2003
7. Jongsma HJ, Wilders R (2000) Gap junctions in cardiovascular disease. Circulat Res 86 (12):1193–1197. https://doi.org/10.1161/01.RES.86.12.1193
8. Siebermair J, Kholmovski EG, Marrouche N (2017) Assessment of left atrial fibrosis by late gadolinium enhancement magnetic resonance imaging: methodology and clinical implications. JACC Clin Electrophysiol 3 (8):791–802. https://doi.org/10.1016/j.jacep.2017.07.004
9. Ciaccio EJ, Coromilas J, Wit AL, Peters NS, Garan H (2018) Source-sink mismatch causing functional conduction block in re-entrant ventricular tachycardia. JACC Clin Electrophysiol 4(1):1–16. https://doi.org/10.1016/j.jacep.2017.08.019
10. Akar FG, Aon MA, Tomaselli GF, O'Rourke B (2005) The mitochondrial origin of postischemic arrhythmias. J Clin Invest 115 (12):3527–3535
11. Aon MA, Cortassa S, Akar FG, Brown DA, Zhou L, O'Rourke B (2009) From mitochondrial dynamics to arrhythmias. Int J Biochem Cell Biol 41(10):1940–1948. https://doi.org/10.1016/j.biocel.2009.02.016
12. Aon MA, Camara AKS (2015) Mitochondria: hubs of cellular signaling, energetics and redox balance. A rich, vibrant, and diverse landscape of mitochondrial research. Front Physiol 6:94. https://doi.org/10.3389/fphys.2015.00094
13. Song J, Yang R, Yang J, Zhou L (2018) Mitochondrial dysfunction-associated Arrhythmogenic substrates in diabetes mellitus. Front Physiol 9:1670. https://doi.org/10.3389/fphys.2018.01670
14. Solhjoo S, O'Rourke B (2015) Mitochondrial instability during regional ischemia-reperfusion underlies arrhythmias in monolayers of cardiomyocytes. J Mol Cell Cardiol 78:90–99. https://doi.org/10.1016/j.yjmcc.2014.09.024
15. Zhou L, Aon MA, Liu T, O'Rourke B (2011) Dynamic modulation of ca(2+) sparks by mitochondrial oscillations in isolated Guinea pig cardiomyocytes under oxidative stress. J Mol

Cell Cardiol 51(5):632–639. https://doi.org/10.1016/j.yjmcc.2011.05.007

16. Barrington PL, Meier CF Jr, Weglicki WB (1988) Abnormal electrical activity induced by H2O2 in isolated canine myocytes. Basic Life Sci 49:927–932
17. Horackova M, Ponka P, Byczko Z (2000) The antioxidant effects of a novel iron chelator salicylaldehyde isonicotinoyl hydrazone in the prevention of H(2)O(2) injury in adult cardiomyocytes. Cardiovasc Res 47(3):529–536
18. Xie LH, Chen F, Karagueuzian HS, Weiss JN (2009) Oxidative-stress-induced afterdepolarizations and calmodulin kinase II signaling. Circ Res 104(1):79–86. https://doi.org/10.1161/CIRCRESAHA.108.183475
19. Erickson JR, He BJ, Grumbach IM, Anderson ME (2011) CaMKII in the cardiovascular system: sensing redox states. Physiol Rev 91 (3):889–915. https://doi.org/10.1152/physrev.00018.2010
20. Erickson JR, Joiner ML, Guan X, Kutschke W, Yang J, Oddis CV, Bartlett RK, Lowe JS, O'Donnell SE, Aykin-Burns N, Zimmerman MC, Zimmerman K, Ham AJ, Weiss RM, Spitz DR, Shea MA, Colbran RJ, Mohler PJ, Anderson ME (2008) A dynamic pathway for calcium-independent activation of CaMKII by methionine oxidation. Cell 133(3):462–474. https://doi.org/10.1016/j.cell.2008.02.048
21. Yang R, Ernst P, Song J, Liu XM, Huke S, Wang S, Zhang JJ, Zhou L (2018) Mitochondrial-mediated oxidative ca(2+)/calmodulin-dependent kinase II activation induces early afterdepolarizations in Guinea pig cardiomyocytes: an in silico study. J Am Heart Assoc 7(15):e008939. https://doi.org/10.1161/JAHA.118.008939
22. Liu M, Liu H, Dudley SC Jr (2010) Reactive oxygen species originating from mitochondria regulate the cardiac sodium channel. Circ Res 107(8):967–974. https://doi.org/10.1161/CIRCRESAHA.110.220673
23. Zhou L, Cortassa S, Wei AC, Aon MA, Winslow RL, O'Rourke B (2009) Modeling cardiac action potential shortening driven by oxidative stress-induced mitochondrial oscillations in Guinea pig cardiomyocytes. Biophys J 97 (7):1843–1852
24. Aon MA, Cortassa S, Marban E, O'Rourke B (2003) Synchronized whole cell oscillations in mitochondrial metabolism triggered by a local release of reactive oxygen species in cardiac myocytes. J Biol Chem 278 (45):44735–44744. https://doi.org/10.1074/jbc.M302673200
25. Zhou L, Solhjoo S, Millare B, Plank G, Abraham MR, Cortassa S, Trayanova N, O'Rourke B (2014) Effects of regional mitochondrial depolarization on electrical propagation: implications for arrhythmogenesis. Circ Arrhythm Electrophysiol 7(1):143–151. https://doi.org/10.1161/CIRCEP.113.000600
26. Niederer SA, Kerfoot E, Benson AP, Bernabeu MO, Bernus O, Bradley C, Cherry EM, Clayton R, Fenton FH, Garny A, Heidenreich E, Land S, Maleckar M, Pathmanathan P, Plank G, Rodriguez JF, Roy I, Sachse FB, Seemann G, Skavhaug O, Smith NP (2011) Verification of cardiac tissue electrophysiology simulators using an N-version benchmark. Philos Trans A Math Phys Eng Sci 369(1954):4331–4351. https://doi.org/10.1098/rsta.2011.0139
27. Plank G, Zhou L, Greenstein JL, Cortassa S, Winslow RL, O'Rourke B, Trayanova NA (2008) From mitochondrial ion channels to arrhythmias in the heart: computational techniques to bridge the spatio-temporal scales. Philos Transact A Math Phys Eng Sci 366 (1879):3381–3409. https://doi.org/10.1098/rsta.2008.0112
28. Vigmond EJ, Hughes M, Plank G, Leon LJ (2003) Computational tools for modeling electrical activity in cardiac tissue. J Electrocardiol 36(Suppl):69–74. https://doi.org/10.1016/j.jelectrocard.2003.09.017
29. Balay S, Abhyankar S, Adams M, Brown J, Brune P, Buschelman K, Dalcin L, Dener A, Eijkhout V, Gropp W, Karpeyev D, Kaushik D, Knepley M, MAY D, Curfman McInnes L, Mills R, Munson T, Rupp K, Sanan P, Smith B, Zampini S, Zhang H, Zhang H, MAY D (2019) PETSc users manual. Argonne National Laboratory, Lemont
30. Lin JW, Garber L, Qi YR, Chang MG, Cysyk J, Tung L (2008) Region [corrected] of slowed conduction acts as core for spiral wave reentry in cardiac cell monolayers. Am J Physiol Heart Circ Physiol 294(1):H58–H65. https://doi.org/10.1152/ajpheart.00631.2007
31. Li Q, Ni RR, Hong H, Goh KY, Rossi M, Fast VG, Zhou L (2017) Electrophysiological properties and viability of neonatal rat ventricular myocyte cultures with inducible ChR2 expression. Sci Rep 7(1):1531. https://doi.org/10.1038/s41598-017-01723-2
32. Davidson SM, Yellon D, Duchen MR (2007) Assessing mitochondrial potential, calcium, and redox state in isolated mammalian cells using confocal microscopy. Methods Mol Biol 372:421–430. https://doi.org/10.1007/978-1-59745-365-3_30

Chapter 12

Automated Quantification and Network Analysis of Redox Dynamics in Neuronal Mitochondria

Felix T. Kurz and Michael O. Breckwoldt

Abstract

Mitochondria are complex organelles with multifaceted roles in cell biology, acting as signaling hubs that implicate them in cellular physiology and pathology. Mitochondria are both the target and the origin of multiple signaling events, including redox processes and calcium signaling which are important for organellar function and homeostasis. One way to interrogate mitochondrial function is by live cell imaging. Elaborated approaches perform imaging of single mitochondrial dynamics in living cells and animals. Imaging mitochondrial signaling and function can be challenging due to the sheer number of mitochondria, and the speed, propagation, and potential short half-life of signals. Moreover, mitochondria are organized in functionally coupled interorganellar networks. Therefore, advanced analysis and postprocessing tools are needed to enable automated analysis to fully quantitate mitochondrial signaling events and decipher their complex spatiotemporal connectedness. Herein, we present a protocol for recording and automating analyses of signaling in neuronal mitochondrial networks.

Key words Mitochondria, Redox potential, Grx1-roGFP2, Fluorescence microscopy, Computational wavelet analysis, Mitochondrial cluster

1 Introduction

Mitochondria are the "powerhouse" of the cell to which they provide the vast majority of ATP. Neurons downregulate glycolysis and depend on mitochondrial oxidative phosphorylation (OXPHOS) [1, 2]. Related to ATP generation, mitochondrial function includes β-oxidation of fatty acids, calcium-buffering, and the control of apoptosis and necrosis [3, 4]. Neurons harbor extraordinary long processes and mitochondria are actively transported anterogradely toward the synapse and retrogradely toward the cell body [5]. "Dysfunctional" or "aged" mitochondria are degraded in a process called "mitophagy" [6]. Given the multifaceted functions of mitochondria, it is not surprising that their dysfunction is implicated in various diseases of neurological, cardiovascular, neoplastic, and inflammatory origin [7–10].

Sonia Cortassa and Miguel A. Aon (eds.), *Computational Systems Biology in Medicine and Biotechnology: Methods and Protocols*, Methods in Molecular Biology, vol. 2399, https://doi.org/10.1007/978-1-0716-1831-8_12,

Mitochondria function as signaling hubs for redox- and calcium-mediated signaling, as well as for orchestrating apoptosis and necrosis [11–14]. Redox signals originate from the electron transport chain, located at the inner mitochondrial membrane. The free radical superoxide ($O_2 \cdot ^-$) is generated mostly at complex I and III from the respiratory chain [15, 16], and can be quickly scavenged by highly expressed mitochondrial matrix antioxidant reactive oxygen species (ROS) scavengers, such as manganese superoxide dismutase, glutathione peroxidases, and peroxiredoxins [17]. ROS escaping antioxidant systems can, in excess, cause membrane lipids peroxidation or other molecular damage (e.g., mitochondrial DNA), but at nondamaging concentration levels, hydrogen peroxide (H_2O_2) can also be an autocrine and paracrine signaling molecule that can be sensed by immune cells or other cell types [13, 18, 19].

The exploration of redox signaling is a fast-growing field [20, 21]. Redox mediators, including ROS derived from mitochondria, play an important role in cellular signaling [22], however their assessment has been challenging due to their short half-life. A mutated form (ro-GFP) of the redox-sensitive green fluorescent protein (GFP), developed by the groups of Tsien and Remington [23, 24], harbors two engineered cysteine residues into the beta sheet backbone of GFP close to the chromophore, rendering GFP redox-sensitive. Its usefulness to measure the redox potential has been abundantly demonstrated in different compartments (e.g., mitochondria, endoplasmic reticulum, cytoplasm) and systems (plant cells, cell lines, neurons) [25–27]. Grx1-roGFP2, a second generation redox sensor with improved kinetics [28], comprises the fusion of roGFP2 with the human glutaredoxin 1 (Grx1). This modification accelerates the redox relay by an order of magnitude and is specific for sensing the glutathione redox potential (E_{GSH}) within a dynamic range of ~2.6 [28] up to ~5 [29].

The protocol described herein applies to the automated quantitation of the mitochondrial glutathione redox potential (E_{GSH}), using Grx1-roGFP2 in vitro and ex vivo preparations. Our procedure assesses redox signals and their connectedness throughout entire mitochondrial networks, utilizing image segmentation complemented by mathematical analysis. This protocol can also be easily adapted to other imaging probes (sensing, e.g., calcium or pH) and experimental systems (e.g., plant cells, ex vivo preparations, in vivo imaging) (*see* **Note 1**).

2 Materials

2.1 Experimental Agents

1. The plasmid pLPCX mito-Grx1-roGFP2 is available from Addgene [28].

2. Human Embryonic Kidney (HEK-293 cells) are available for example from ATCC®.
3. Chemicals for measuring dose responses of the sensor, H_2O_2 (Sigma); dithiothreitol (DTT; Sigma or VWR).
4. Dulbecco's Modified Eagle Medium (DMEM, Invitrogen) and Ringer's Lactate solution [30] for cell culture and ex vivo preparations.
5. Widefield (e.g., BX-51 Olympus) or confocal microscope setups (e.g., FV-1000 Olympus).
6. For widefield microscopy polychrome V polychromator system (Till Photonics) as light source equipped with cooled CCD camera (Sensicam, pco imaging) controlled by TillVision software (Till Photonics).

2.2 Image Analysis Software

1. Mitochondrial properties can be analyzed using the open-source software Fiji (http://fiji.sc) [31] and Matlab v7.14.0.0739 (R2012a).

3 Methods

3.1 Experimental Methods

3.1.1 Imaging Mitochondrial Redox Dynamics in Cell Culture In Vitro

1. Culture appropriate cell line, for example, Hek 293 cells using standard cell culture conditions, for example, Dulbecco's Modified Eagle Medium (DMEM, Invitrogen) substituted with 10% fetal bovine serum and 1% Normocin (Normocin™, Invivogen) or penicillin/streptomycin.
2. Grow cells on sterile, poly-L-lysine coated glass cover slides using cloning cylinders to reduce media volumes needed for transfection.
3. Transfect cells with 500 ng plasmid DNA of the construct, mixed with 0.5 μl lipofectamine (Invitrogen) in 500 μl PBS. Incubate for 10 min at room temperature.
4. Add 500 μl of the mixture on top of the cells for 2–4 h in the incubator.
5. Gently remove glass cylinders without disrupting the adherent cells and add 1 ml DMEM. Let cells grow for 1–3 days to express the construct. Expression levels improve over time and is optimal 2–3 days post transfection. >30% of cells should express the fluorescent protein.
6. Transfer the glass cover slips with transfected HEK-293 cells to a heated flow chamber (33–35 °C), continuously perfused with carbogen-bubbled normal Ringer. Treatment with, for example, H_2O_2, DTT, can be administered through a perfusion system.

7. Perform widefield (e.g., BX-51 Olympus) or confocal microscopy (e.g., FV-1000 Olympus) using 408/488 nm excitation laser lines.
8. For the analysis (segmentation and quantification) of redox signal *see* Subheading 3.1.2.

3.1.2 Imaging Neuronal Mitochondrial Redox Dynamics in Ex Vivo Preparations

1. The transgenic mouse line Thy1-mito-Grx1-roGFP2 expresses the redox sensor Grx1-roGFP2 in neuronal mitochondria [32] and is used in this protocol. Grx1-roGFP2 reports the glutathione redox potential (E_{GSH}). Imaging can be performed in neurons of the central and peripheral nervous system. The protocol describes imaging approaches of the peripheral nerve (triangularis sterni explant). The imaging approach is also amenable to in vivo imaging, for example, spinal cord [33] or cerebral cortex [32, 34].
2. Prepare explants of the triangularis sterni muscle as previously described [35]. In brief, euthanize mice in deep anesthesia (e.g., with isoflurane or ketamine–xylazine) and remove the rib cage (with the attached triangularis sterni muscle and its innervating intercostal nerves). Isolate the rib cage by paravertebral cuts and pin explant in a Sylgard-coated dish using insect pins and maintain temperature on a heated stage (32–35 °C) in normal Ringer solution, bubbled with carbogen gas (95% O_2, 5% CO_2). Recordings can be performed in proximal or distal intercostal axons or at the neuromuscular junction (NMJ).
3. Record the mitochondrial glutathione redox potential in motor axons and NMJs by wide-field microscopy (e.g., a BX51 Olympus), using an appropriate objective (e.g., ×20/0.5 N.A. or ×100/1.0 N.A. dipping-cone water immersion objective), a filter wheel with shutter, a dichroic filter (D/F 500 DCXR ET 525/36), a Polychrome V polychromator system (Till Photonics), and a cooled CCD camera (Sensicam, pco imaging) controlled by TillVision software.
4. Acquire images at rates of 1 Hz with exposure times of 150 ms for 408 nm excitation and 30 ms for 488 nm excitation.
5. To measure the physiological signals of the mitochondrial glutathione redox potential (E_{GSH}) in axons and NMJs, record time-lapse movies in intercostal nerves for 5–10 min at 1 Hz. (Fig. 1).
6. Measure the dose response of the sensor proteins, if required, at the end of the experiment. Incubate triangularis sterni explants with exogenous H_2O_2 (Sigma; concentration of 6.25, 12.5, 25, 50, 100, 200, 400, 800, 1000 μM H_2O_2 diluted in normal Ringer solution) or dithiothreitol (DTT; Sigma or VWR; concentration of 500 μM diluted in normal Ringer solution) for 5 min.

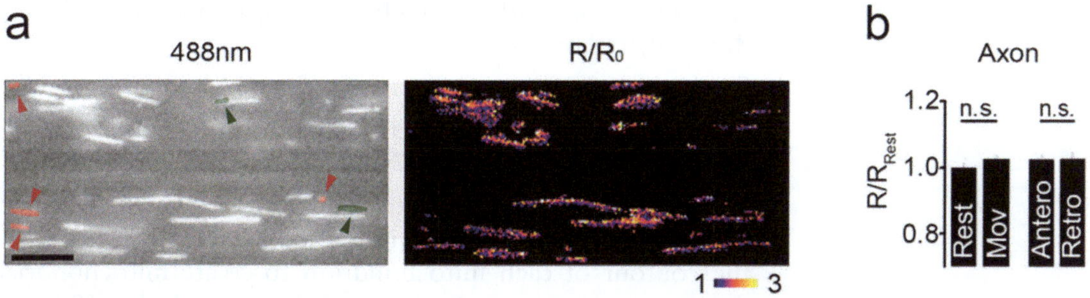

Fig. 1 The glutathione potential is tightly regulated and independent of organelle location or movement. Illustration of mitochondrial redox levels in the intercostal nerve measured in triangularis sterni explants from Thy1-Grx1-roGFP2 mice. Panel shows two parallel-running axons in the proximal intercostal nerve (**a**). The redox level is almost completely reduced and homogenous within the mitochondrial population. There is no apparent difference between resting and moving mitochondria. Also, mitochondria that are anterogradely (red overlay in 488 nm image) or retrogradely (green overlay) transported show no difference in their redox potential. Quantification of redox levels in (**b**) shows different populations of axonal mitochondria (normalized to resting mitochondria; $n = 88$ mitochondria, 3 explants). Scale bar is 5 μm

3.2 Analytical Methods

3.2.1 Image Analysis

1. Select single mitochondria or clusters of mitochondria and background as regions of interest. Measure mean intensity values in the 408 nm and 488 nm channel and subtract background. Divide the two channels (408/488) to indicate the sensor's state of oxidation. This ratio can be normalized by dividing by the mean ratio measured after reduction with DTT (R/R_{DTT}) if DTT was used in the same experiment. Otherwise, the experimental results can be shown either as R/R_0, with R_0 being the ratio at the time before the mitochondrial signal. Export values from Fiji and perform calculations, for example, in Excel (Microsoft).
2. For the generation of pseudocolor images, "threshold" the 488 nm channel and "binarize" the image to serve as a segmentation mask. This mask is used to segment both channels. The resulting images can be divided (408/488 nm) and "normalized" to the DTT measurements (R/R_{DTT}) or the ratio before the signaling event. The spectral pseudocolor look-up table "fire" (Fiji) can be used to display images.
3. For each recorded video, correct mitochondrial translational drift within the imaging plane with the image stabilizer plugin for ImageJ (Version 1.49 s) [36].

3.2.2 Extract Individual Mitochondrial Fluorescence Traces

1. Calculate the average projection of all images in each recorded video.
2. In a raster graphics editor program (e.g., Adobe Photoshop CS6 v 13.0), manually draw the contour of each single mitochondrion and axon border in the average projection image to create a mask image with mitochondria and axon borders.

Assign the space within mitochondrial contours and axon borders as cytoplasm.

3. Save the mask image as a ternary grid template.
4. Allocate numerical identifiers to each mitochondrion, for example, using the function *bwlabel* in Matlab).
5. At each time point, average the mitochondrial intensity signal for all mitochondrial pixels within each mitochondrion and on the contour of each mitochondrion to create mitochondrial fluorescence traces at both the 408 nm and the 488 nm channel.
6. For Grx1-roGFP2 optical sensor traces, determine mitochondrial intensity traces as the ratio of the 408 nm and the 488 nm channel.

3.2.3 Mitochondrial Signal Events

1. Determine the onset t_0 of mitochondrial events as deviations of average mitochondrial signal intensity of more than 10% relative to the mitochondrial intensity baseline.
2. Determine the end of the mitochondrial event, that is, the return to baseline, as t_{end}.
3. Determine the maximum mitochondrial intensity at time $t_0 + \Delta t_{up}$, and its intensity difference to baseline, or amplitude, as ΔA.
4. Determine the decay time Δt_{down} as $\Delta t_{down} = t_{end} - t_0 - \Delta t_{up}$.
5. Determine the rise of the redox event as the slope of a linear polynomial fitted to subsequent time-points in the interval $[t_0 + 0.1\ \Delta t_{up}, t_0 + 0.9\ \Delta t_{up}]$.
6. Determine the decay of the mitochondrial event as the decay rate a in the function $f(t) = f_0 + b\exp(-a \times t)$ fitted to time-points in the interval $[t_{end} - 0.9\ \Delta t_{down}, t_{end} - 0.1\ \Delta t_{down}]$.
7. For multiple events within one mitochondrial intensity trace, determine the frequency of subsequent events as the inverse of the difference of the peak time-points.

3.2.4 Mitochondrial Intensity Trace Wavelet Analysis

Since mitochondrial signal traces possess time-varying frequencies of mitochondrial events, it is helpful to use wavelet analysis to allocate signal frequency content at specific time-points. This procedure can also be used in the analysis of mitochondrial signal oscillations in cardiac myocytes during oxidative stress [37–40]. For instance, wavelet analysis for mitochondrial Grx1-roGFP2 traces enables detection of the dynamic frequency of mitochondrial redox signaling events (Fig. 2).

1. Normalize the mitochondrial intensity signal trace by its standard deviation and pad the number of time-points with zero to the next higher power of 2. The zero-padding will make the following calculations more efficient.

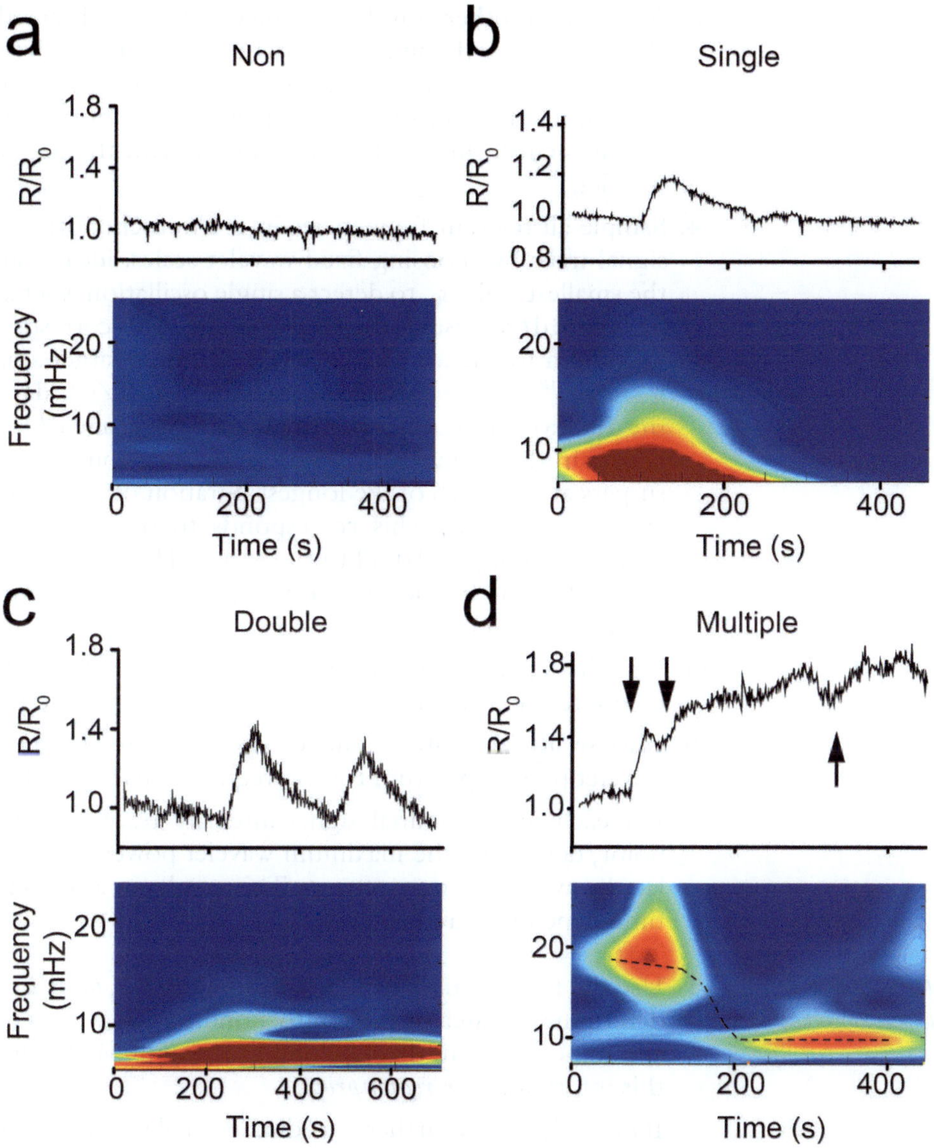

Fig. 2 Mitochondrial redox event detection. Typical mitochondrial signal traces using mito-Grx1-roGFP2 are shown for no event (**a**), a single event (**b**), two events (**c**), and multiple events (**d**), as well as their associated absolute squared wavelet transforms (lower panels). Nonevents do not contain any relevant frequency content, whereas single events produce a wavelet transform smeared around the inverted signal length, corresponding to approximately 5–10 mHz in (**b**). Additional events produce additional frequency content, for example, approximately 8 mHz for the doublet event in (**c**), and approximately 10–20 mHz for the multievent in (**d**). (Adapted from Supplementary Fig. 6 in [42], with permission from Ref. 42. Copyright 2016)

2. Use Matlab's built-in wavelet toolbox or an equivalent wavelet software package for other computational programs/platforms to apply the wavelet transform to each mitochondrial signal intensity trace.

3. Adapt the mother wavelet parameters to the observed signal changes during the mitochondrial events. The Morlet wavelet is recommended for the continuous analysis of mitochondrial oscillations due to its higher frequency resolution. Alternative wavelet transforms are the Paul wavelet and the Mexican hat wavelet.
4. Sample all relevant frequencies in the mitochondrial intensity signal trace by choosing fixed wavelet scales, ideally such that the smallest scale, s_0, to detect a single oscillation is set as $s_0 = 4$ dt, where dt represents the sampling rate. This corresponds to a maximum frequency of $f_{max} = s_0^{-1}$. Choose larger scales s_k as $s_k = s_0\, 2^{kdk}$ with $k = 0,1,\ldots,K$, and $K = \log_2(T/s_0)/dk$, T being the duration of the mitochondrial signal trace, and $dk = 0.1$. To constrain the largest scale, one can exclude long periods that surpass at least 10% of the longest duration of a mitochondrial event ($t_{end} - t_0$). This corresponds to setting a minimum frequency of $f_{min} = 10/(11\,(t_{end} - t_0))$. The focus on frequencies within the frequency interval $[f_{min}, f_{max}]$ allows more efficient computation.
5. Use the squared absolute value of the wavelet transform to compute the wavelet power spectrum for every time-point.
6. Choose an adequate frequency resolution to interpolate the wavelet power spectrum to; we frequently use 0.1 mHz.
7. For each mitochondrial signal intensity trace, at each time-point, determine the maximum wavelet power in the interpolated wavelet power spectrum. This results in a mitochondrial time-dependent frequency.

3.2.5 Mitochondrial Morphological Properties

1. Use the ternary mask from Subheading 3.2.2, **step 2** and **3** to determine the area of each mitochondrion, as well as its major and minor axis length. A helpful function in Matlab to extract this information is *regionprops*.
2. If needed, extract further two-dimensional mitochondrial morphological information such as the mitochondrial eccentricity and perimeter.
3. Determine the mitochondrial shape factor as the ratio of mitochondrial major axis length and mitochondrial minor axis length.

3.2.6 Mitochondrial Clusters

A combined mitochondrial morphological and signal event analysis allows linking morphological and functional information within the mitochondrial network [41]. One should, however, differentiate between a local neighborhood of mitochondria, that is, mitochondria that are in close proximity to each other, and mitochondrial event clusters, that is, clusters of mitochondria that show signal events and whose distance to a cluster mitochondrion is less than twice the length of a mitochondrion (e.g., ~4 μm for cardiac mitochondria; highly variable for axonal mitochondria).

1. Determine a local neighborhood for each mitochondrion as the area of cytoplasm within a radius of 4 μm around the mitochondrial center-point. This corresponds roughly to twice the length of one cardiac muscle cell mitochondrion, see also [42].
2. Assign each mitochondrion m_n within a local neighborhood of mitochondrion m as a nearest neighbor of mitochondrion m if a straight line through the center-points of m_n and m does not contain a cut through the area of another mitochondrion.
3. Determine mitochondrial event clusters as the area spanned by all mitochondria that exhibit events, although performing a morphological closing procedure using a disk with 4 μm diameter. The morphological closing procedure in a binary image first dilates and then erodes with a structuring element (here: a disk) to result in areas of clustered elements. A useful function is *imclose* in Matlab, that performs this procedure on binary images. With such a procedure, mitochondrial events can be grouped either as cluster events (i.e., happening within a mitochondrial cluster) or as isolated events (in mitochondria not belonging to any cluster) (*see* also Fig. 3).

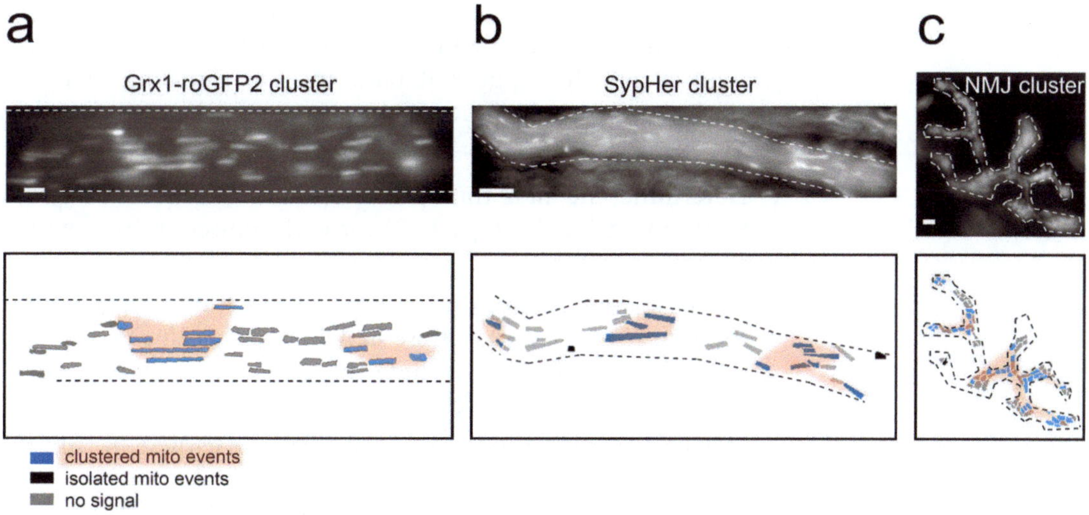

Fig. 3 Morphological clustering of mitochondrial signals. Spatial clusters of mitochondria are shown with redox signaling events in an axon (**a**), pH signaling events in an axon (**b**) and pH signaling events in the neuromuscular junction (**c**). Signaling mitochondria are depicted in blue and their associated clusters in light orange. Signaling mitochondria that are not part of a cluster are depicted in black, and nonsignaling mitochondria in gray. The axon and neuromuscular junction borders are shown in dashed lines. Scale bars are 2 μm. (Adapted from Supplementary Fig. 8 in [42], with permission from Ref. 42. Copyright 2016)

4. Determine the density of a mitochondrial event cluster as the ratio of sum of all mitochondrial areas within the cluster and the cluster area. If one wishes to compare this density to the density of mitochondria in axons, one must only consider the axon area spanned by all mitochondria within the axon. Axon mitochondrial density is then determined as the ratio of the sum of all mitochondrial areas within the axon and the axon area.

3.2.7 Mitochondrial Signal Propagation

Isochronal maps allow visualizing the propagation of a mitochondrial signal within the mitochondrial network of an axon or a myocyte. Starting from the mitochondrion with the first signal event, later events within the mitochondrial network follow a color code that can show if a signal propagates homogeneously throughout the network, if mitochondrial signal events appear at random, if local clusters of mitochondria contain propagation of events that appear simultaneously across clusters, or if a specific number of mitochondria need to show simultaneous signaling events before a signal propagates through the network in analogy with the synchronization of mitochondrial oscillations in cardiac myocytes [43]. The isochronal map of signaling mitochondria after nerve crush injury in triangularis sterni explant axons (Fig. 4a), imaged with mito-Grx1-roGFP2 fluorescence, is shown in Fig. 4b, c: mitochondrial signaling events propagate from mitochondria proximal to the crush site on the left to distal mitochondria on the right in a mostly homogeneous manner. Small islands of mitochondria with late signaling events can be identified in the middle third of the axon and likely correspond to mitochondrial signaling events that appear at random. The signaling dynamics provide insights into the functional properties of the mitochondrial network.

1. Determine the first time-points t_0 and $t_1 = t_0 + \Delta t_{up}$ (*see* Subheading 3.2.3, **steps 1–3**) for each mitochondrion within a signal recording.
2. For each mitochondrion, within the interval $[t_0, t_1]$, exclude the 10% of time-points whose signal intensity value is closest to that of t_0 and likewise exclude the 10% of time-points whose signal intensity value is closest to that of t_1.
3. Of the remaining time-points, calculate the earliest, t_I, and the latest time-point, t_E.
4. Use $t_S = (t_I + t_E)/2$, that is, the arithmetic mean of t_I and t_E, as the reference point for each mitochondrial signal.
5. Among all mitochondrial reference points, determine the earliest reference point as the initial signal event.

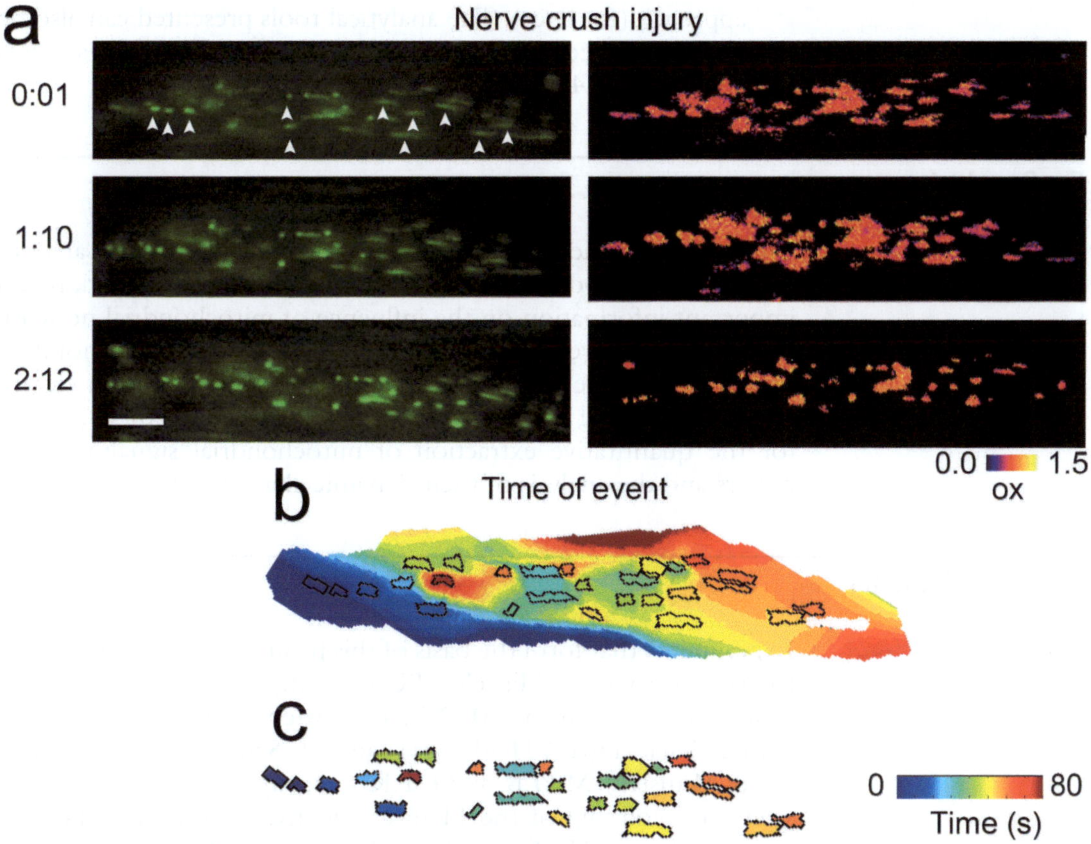

Fig. 4 Isochronal analysis of signaling mitochondria after nerve crush injury. (**a**) Triangularis sterni explant axons after crush injury imaged with mito-Grx1-roGFP2 fluorescence. Starting from the crush site on the left, mitochondria in locations more distal (right) from the crush site become increasingly rounder and oxidized. Time points are indicated in min:s. (**b**) Isochronal analysis identifies the first signaling event for a mitochondrion in the upper right corner, and a propagation of subsequent mitochondrial signaling events toward a signaling cluster in the lower left corner. White mitochondria show no event. (**c**) After a nerve crush injury, mitochondrial Grx1-roGFP2 signaling events in axonal mitochondria propagate from left to right and show clustered oxidation. (Adapted from Fig. 5 in [42], with permission from Ref. 42. Copyright 2016)

6. Assign all reference points a color based on a linear color scale starting at the initial signal event to create an isochronal map over all reference points in the mitochondrial network with color interpolation.

4 Notes

1. Other fluorescent probes can be used to measure other functional parameters, such as mitochondrial membrane potential (TMRM [44]), pH (SypHer [45]) or calcium levels (e.g., with genetically encoded calcium indicators (GECIs [46]), using an appropriate expression system (for genetic sensors) or loading

approach (for dyes). The analytical tools presented can also be easily adapted to other fluorescent probes or systems applications [47].

5 Conclusions

Imaging of individual mitochondrial signaling events and analysis of the spatiotemporal relation of the mitochondrial network reveal important information on the influence of mitochondrial network function, structure, and organization on individual mitochondrial behavior and vice versa. The presented protocol, which details advanced imaging procedures and image analysis methods, allows for the quantitative extraction of mitochondrial signaling parameters and the analysis of their dynamic characteristics.

Acknowledgments

Experiments that form the basis of this protocol were performed in the laboratory of T. Misgeld (TU Munich) and Martin Kerschensteiner (LMU Munich). M.O.B. acknowledges helpful discussions with T. Dick (DKFZ Heidelberg) and M. Schwarzländer (University of Münster). M.O.B. and F.T.K. were supported by a physician-scientist fellowship of the Medical Faculty, University of Heidelberg and by the Hoffmann-Klose Foundation (University of Heidelberg). F.T.K. was supported by the Deutsche Forschungsgemeinschaft (DFG, German Research Foundation, KU 3555/1-1) and a research grant from Heidelberg University Hospital.

Author Contributions: F.T.K. and M.O.B. conceived the study. M.O.B. performed microscopy experiments. F.T.K. provided analytical tools. M.O.B. and F.T.K. performed image analysis of the data. F.T.K. and M.O.B. wrote the manuscript.

References

1. Choi HB, Gordon GRJ, Zhou N, Tai C, Rungta RL, Martinez J et al (2012) Metabolic communication between astrocytes and neurons via bicarbonate-responsive soluble adenylyl cyclase. Neuron 75:1094–1104
2. Herrero-Mendez A, Almeida A, Fernández E, Maestre C, Moncada S, Bolaños JP (2009) The bioenergetic and antioxidant status of neurons is controlled by continuous degradation of a key glycolytic enzyme by APC/C–Cdh1. Nat Cell Biol 11:747–752
3. Hoppins S, Nunnari J (2012) Mitochondrial dynamics and apoptosis-the ER connection. Science 337:1052–1054
4. Vaseva AV, Marchenko ND, Ji K, Tsirka SE, Holzmann S, Moll UM (2012) p53 opens the mitochondrial permeability transition pore to trigger necrosis. Cell 149:1536–1548
5. MacAskill AF, Kittler JT (2010) Control of mitochondrial transport and localization in neurons. Trends Cell Biol 20(2):102–112
6. Youle RJ, Narendra DP (2011) Mechanisms of mitophagy. Nat Rev Mol Cell Biol 12:9–14

7. Wallace DC (2012) Mitochondria and cancer. Nat Rev Cancer 12:685–698
8. Lin MT, Beal MF (2006) Mitochondrial dysfunction and oxidative stress in neurodegenerative diseases. Nature 443:787–795
9. Nunnari J, Suomalainen A (2012) Mitochondria: in sickness and in health. Cell 148:1145–1159
10. Corrado M, Scorrano L, Campello S (2012) Mitochondrial dynamics in cancer and neurodegenerative and neuroinflammatory diseases. Int J Cell Biol 2012:729290
11. Kurz FT, Kembro JM, Flesia AG, Armoundas AA, Cortassa S, Aon MA et al (2017) Network dynamics: quantitative analysis of complex behavior in metabolism, organelles, and cells, from experiments to models and back. Wiley Interdiscip Rev Syst Biol Med 9(1)
12. Hamanaka RB, Chandel NS (2010) Mitochondrial reactive oxygen species regulate cellular signaling and dictate biological outcomes. Trends Biochem Sci 35:505–513
13. Al-Mehdi AB, Pastukh VM, Swiger BM, Reed DJ, Patel MR, Bardwell GC et al (2012) Perinuclear mitochondrial clustering creates an oxidant-rich nuclear domain required for hypoxia-induced transcription. Sci Sign 5:ra47
14. Kurz FT, Aon MA, O'Rourke B, Armoundas AA (2018) Assessing spatiotemporal and functional Organization of Mitochondrial Networks. In: 1st (ed) Mitochondrial Bioenergetics. Humana Press, NY, New York, NY, pp 383–402
15. Murphy MP (2008) How mitochondria produce reactive oxygen species. Biochem J 417 (1):1–13
16. Hirst J (2013) Mitochondrial complex I. Annu Rev Biochem 82:551–575
17. Ibrahim W, Lee US, Yen HC, St Clair DK, Chow CK (2000) Antioxidant and oxidative status in tissues of manganese superoxide dismutase transgenic mice. Free Radic Biol Med 28:397–402. https://doi.org/10.1016/S0891-5849(99)00253-1
18. Niethammer P, Grabher C, Look AT, Mitchison TJ (2009) A tissue-scale gradient of hydrogen peroxide mediates rapid wound detection in zebrafish. Nature 459:996–999
19. Weismann D, Hartvigsen K, Lauer N, Bennett KL, Scholl HPN, Issa PC et al (2011) Complement factor H binds malondialdehyde epitopes and protects from oxidative stress. Nature 478:76–81
20. Schwarzländer M, Dick TP, Meyer AJ, Morgan B (2016) Dissecting redox biology using fluorescent protein sensors. Antioxid Redox Signal 24(13):680–712
21. Breckwoldt MO, Wittmann C, Misgeld T, Kerschensteiner M, Grabher C (2015) Redox imaging using genetically encoded redox indicators in zebrafish and mice. Biol Chem 396:511–522.0294
22. Kurz CT, Aon MA, O'Rourke B, Armoundas AA (2017) Functional implications of cardiac mitochondria clustering, in: mitochondrial dynamics in cardiovascular medicine. Springer, Cham, Cham, pp 1–24
23. Hanson GT, Aggeler R, Oglesbee D, Cannon M, Capaldi RA, Tsien RY et al (2004) Investigating mitochondrial redox potential with redox-sensitive green fluorescent protein indicators. J Biol Chem 279:13044–13053
24. Dooley CT, Dore TM, Hanson GT, Jackson WC, Remington SJ, Tsien RY (2004) Imaging dynamic redox changes in mammalian cells with green fluorescent protein indicators. J Biol Chem 279:22284–22293
25. Schwarzländer M, Fricker MD, Sweetlove LJ (2009) Monitoring the in vivo redox state of plant mitochondria: effect of respiratory inhibitors, abiotic stress and assessment of recovery from oxidative challenge. Biochim Biophys Acta 1787:468–475
26. Guzman JN, Sanchez-Padilla J, Wokosin D, Kondapalli J, Ilijic E, Schumacker PT et al (2010) Oxidant stress evoked by pacemaking in dopaminergic neurons is attenuated by DJ-1. Nature 468:696–700
27. van Lith M, Tiwari S, Pediani J, Milligan G, Bulleid NJ (2011) Real-time monitoring of redox changes in the mammalian endoplasmic reticulum. J Cell Sci 124:2349–2356
28. Gutscher M, Pauleau AL, Marty L, Brach T, Wabnitz GH, Samstag Y et al (2008) Real-time imaging of the intracellular glutathione redox potential. Nat Methods 5:553–559
29. Albrecht SC, Barata AG, Großhans J, Teleman AA, Dick TP (2011) In vivo mapping of hydrogen peroxide and oxidized glutathione reveals chemical and regional specificity of redox homeostasis. Cell Metab 14(6):819–829
30. Singh S, Kerndt CC, Davis D, Ringer's Lactate (2020) StatPearls. StatPearls Publishing, Treasure Island (FL)
31. Schindelin J, Arganda-Carreras I, Frise E, Kaynig V, Longair M, Pietzsch T et al (2012) Fiji: an open-source platform for biological-image analysis. Nat Methods 9:676–682
32. Breckwoldt MO, Pfister FMJ, Bradley PM, Marinković P, Williams PR, Brill MS et al (2014) Multiparametric optical analysis of mitochondrial redox signals during neuronal

physiology and pathology in vivo. Nat Med 20:555–560
33. Misgeld T, Nikić I, Kerschensteiner M (2007) In vivo imaging of single axons in the mouse spinal cord. Nat Protoc 2:263–268
34. Drew PJ, Shih AY, Driscoll JD, Knutsen PM, Blinder P, Davalos D et al (2010) Chronic optical access through a polished and reinforced thinned skull. Nat Methods 7:981–984
35. Kerschensteiner M, Reuter MS, Lichtman JW, Misgeld T (2008) Ex vivo imaging of motor axon dynamics in murine triangularis sterni explants. Nat Protoc 3:1645–1653
36. Li K (2008) The image stabilizer plugin for ImageJ. www.cs.cmu.edu/~kangli/code/Image_Stabilizer.html (02/17/2022)
37. Kurz FT, Derungs T, Aon MA, O'Rourke B, Armoundas AA (2015) Mitochondrial networks in cardiac myocytes reveal dynamic coupling behavior. Biophys J 108:1922–1933
38. Kurz FT, Aon MA, O'Rourke B, Armoundas AA (2010) Spatio-temporal oscillations of individual mitochondria in cardiac myocytes reveal modulation of synchronized mitochondrial clusters. Proc Natl Acad Sci 107:14315–14320
39. Kurz FT, Aon MA, O'Rourke B, Armoundas AA (2010) Wavelet analysis reveals heterogeneous time-dependent oscillations of individual mitochondria. Am J Physiol Heart Circ Physiol 299(5):H1736–H1740
40. Vetter L, Cortassa S, O'Rourke B, Armoundas AA, Bedja D, Jende JME et al (2020) Diabetes increases the vulnerability of the cardiac mitochondrial network to criticality. Front Physiol 11:175
41. Kurz FT, Aon MA, O'Rourke B, Armoundas AA (2014) Cardiac mitochondria exhibit dynamic functional clustering. Front Physiol 5:599
42. Breckwoldt MO, Armoundas AA, Aon MA, Bendszus M, O'Rourke B, Schwarzländer M et al (2016) Mitochondrial redox and pH signaling occurs in axonal and synaptic organelle clusters. Sci Rep 6:23251–23212
43. Aon MA, Cortassa S, Marbán E, O'Rourke B (2003) Synchronized whole cell oscillations in mitochondrial metabolism triggered by a local release of reactive oxygen species in cardiac myocytes. J Biol Chem 278:44735–44744
44. Chazotte B (2011) Labeling mitochondria with TMRM or TMRE. Cold Spring Harb Protoc:895–897
45. Poburko D, Santo-Domingo J, Demaurex N (2011) Dynamic regulation of the mitochondrial proton gradient during cytosolic calcium elevations. J Biol Chem 286:11672–11684
46. Akerboom J, Carreras Calderón N, Tian L, Wabnig S, Prigge M, Tolö J et al (2013) Genetically encoded calcium indicators for multicolor neural activity imaging and combination with optogenetics. Front. Mol. Neurosci 6:2
47. Schwarzländer M, Logan DC, Johnston IG, Jones NS, Meyer AJ, Fricker MD et al (2012) Pulsing of membrane potential in individual mitochondria: a stress-induced mechanism to regulate respiratory Bioenergetics in Arabidopsis. Plant Cell 24:1188–1201

Part V

Systems Biology of Rhythms, Morphogenesis, and Complex Dynamics

Chapter 13

Computational Approaches and Tools as Applied to the Study of Rhythms and Chaos in Biology

Ana Georgina Flesia, Paula Sofia Nieto, Miguel A. Aon, and Jackelyn Melissa Kembro

Abstract

The temporal dynamics in biological systems displays a wide range of behaviors, from periodic oscillations, as in rhythms, bursts, long-range (fractal) correlations, chaotic dynamics up to brown and white noise. Herein, we propose a comprehensive analytical strategy for identifying, representing, and analyzing biological time series, focusing on two strongly linked dynamics: periodic (oscillatory) rhythms and chaos. Understanding the underlying temporal dynamics of a system is of fundamental importance; however, it presents methodological challenges due to intrinsic characteristics, among them the presence of noise or trends, and distinct dynamics at different time scales given by molecular, dcellular, organ, and organism levels of organization. For example, in locomotion circadian and ultradian rhythms coexist with fractal dynamics at faster time scales. We propose and describe the use of a combined approach employing different analytical methodologies to synergize their strengths and mitigate their weaknesses. Specifically, we describe advantages and caveats to consider for applying probability distribution, autocorrelation analysis, phase space reconstruction, Lyapunov exponent estimation as well as different analyses such as harmonic, namely, power spectrum; continuous wavelet transforms; synchrosqueezing transform; and wavelet coherence. Computational harmonic analysis is proposed as an analytical framework for using different types of wavelet analyses. We show that when the correct wavelet analysis is applied, the complexity in the statistical properties, including temporal scales, present in time series of signals, can be unveiled and modeled. Our chapter showcase two specific examples where an in-depth analysis of rhythms and chaos is performed: (1) locomotor and food intake rhythms over a 42-day period of mice subjected to different feeding regimes; and (2) chaotic calcium dynamics in a computational model of mitochondrial function.

Key words Biological clocks, Circadian and ultradian rhythms, Wavelet, Synchrosqueezing, Wavelet coherence, Power spectrum analysis, Phase space reconstruction, Lyapunov exponent

Sonia Cortassa and Miguel A. Aon (eds.), *Computational Systems Biology in Medicine and Biotechnology: Methods and Protocols*, Methods in Molecular Biology, vol. 2399, https://doi.org/10.1007/978-1-0716-1831-8_13,

1 Introduction

1.1 Acknowledging the Importance of Time-Dependent Fluctuations in Complex Biological Systems

In physiology, the idea of "constancy of the internal environment" has become an all-pervading concept ever since its introduction by Claude Bernard in 1865 (*see* [1]). Homeostasis refers to the notion that the relative constancy over time of the physicochemical properties of an organism is kept by regulation (Commission for Thermal Physiology of the International Union of Physiological Sciences, homeostasis). The strong biophysical implications of this conceptual framework have profoundly influenced the way temporal variability is recognized in Biology and Medicine. Chronobiologists have challenged the idea of homeostasis [1–4] given that it seemingly does not account for the broad range of dynamic modes exhibited by living systems. For example, it is widely considered that most, if not all, life forms can generate rhythmicity endogenously [5–7]. Attempts to reconcile obvious theoretical contradictions between rhythmicity and homeostasis have been proposed [5, 8]. In this context, the concept of homeodynamics (instead of homeostasis) conceptualizes the capability of a biological system to switch between monotonic states (*see* fixed points, Box 1), and other possible states, such as periodic (*see* limit cycle, Box 1 and Fig. 6d) and chaotic behaviors (*see* strange attractors, Box 1 and Fig. 6e) [1]. Modern views visualize living systems as dynamic evolving entities [1, 9] in response to the interacting genome and environment and the emerging phenome, that is, the ensemble of, for example, physiological, molecular, cellular, traits exhibited under specific conditions given by challenges such as caloric restriction, the alternation of light and dark phases, genetic mutations, or inherited epigenetic changes. Moreover, each component of a biological system can be viewed as part of a network of interacting parts (i.e., ecological webs, metabolic maps, clustered cardiac mitochondria) from which dynamical properties arise (*see* in-depth analysis in [1–3].

The entrenched logic associated with homeostasis demands the underlying assumption that fluctuations over time in biological variables, for example, heart rate, blood pressure, metabolites, movement, are due to random, stochastic effects. As a result, mean values are extensively used to characterize time series (experimental data values of, for example, membrane potential, movement, respiration, recorded over time). Among other frequent examples are estimating the mean time an ion channel remains open for patch clamp recordings, or an animal locomotion is active. Moreover, researchers often assume that the time series collected have no temporal patterns or correlation structure, which implies that the data points result from independent, additive rather than multiplicative, variables (i.e., white noise), usually normally distributed. Mathematically, independence means that the correlation between time points is zero (Subheading 2.1.5); thus, the system has no memory, that is past and future events are unrelated, which is, biologically speaking, profoundly counterintuitive.

In this context, a less restrictive hypothesis enabling the use of mean values from a time series, would be to describe a biological time series as a trajectory of a stationary ergodic process: a stochastic process is said to be ergodic if its statistical properties can be deduced from a single, sufficiently long, random sample of the process. Thus, the mean of the *process* can be estimated with the mean of the *time series.* Having said that, time series from living systems are most often *not* independent of each other and do not exhibit the characteristics of a stationary ergodic process. On the other hand, correlation patterns in time series from living systems indicate "memory," that is, time points in the "past" influence the value of time points in the "future," not only in the short term but in the long term as well (*see* Subheading 2.1.5). Although this statement may not hold for every biological system (especially isolated simplified ones), one should not a priori assume that fluctuations in time series correspond to white noise but rather properly evaluate whether that is the case.

1.2 Clocks, Chaos, and a Wide Range of Dynamic Regimes

The broad range of dynamic modes that can be exhibited by living systems are displayed in Tables 1, 2, and 3. Periodic oscillations are the most familiar since they have been observed at every level of organization from molecules to organisms [1–3]. Other dynamic regimes are rhythms, bursts, long-range (fractal) correlations (*see* pink noise in Subheading 2.1.7), chaotic dynamics, white noise (i.e., completely random, temporally independent fluctuations over time; *see* Subheadings 2.1.5 and 2.1.7) and brown noise (i.e., temporal integration of white noise; *see* Subheading 2.1.7). In many cases, distinct dynamics at different temporal scales can coexist (*see* example of locomotion Subheading 1.2.1) or be dynamically linked as in the case of periodicity and chaos, through the "route to chaos" (*see* example Subheading 1.2.2).

We illustrate the use and analysis of a wide range of tools available, as applied to two case studies corresponding to significant biological examples: circadian oscillations and intracellular calcium (Ca^{2+}) dynamics. We highlight how these two systems are involved in the staging and modulation of biological temporal dynamics, and the importance of temporal scales. These concepts will be revisited in Subheading 2.2.

1.2.1 Biological Circadian and Ultradian Rhythms

Descriptions of cyclical behavior in plants and animals date from a long time ago, being Linnaeus's "flower clock," a beautiful, long-standing, example of early scientists' fascination with biological rhythmicity. From seasonal collective bird migration to daily chromatin remodeling in mammals, many biological rhythms were described and characterized according to their periodicity as ultradian (period < 24 h), infradian (period > 24 h) and circadian (period near 24 h), being the latter the most pervasive in nature, extensively studied and the focus of this section [10].

Two possible scenarios have been proposed to explain the origin of circadian rhythmicity: First, they generated by endogenous timing mechanisms or clocks, and second, by an external zeitgeber given by an exogenous periodic cycle or perturbation, such as the light–dark cycle. Many elegant experiments demonstrated that distinguishing between self-sustained vs. driven biological rhythmicity is possible under constant environmental (free-running) conditions [4]. Technical developments made possible to conclusively demonstrate that most living systems have developed endogenous, innate, persistent, and temperature-compensated circadian clocks, responsible for generating circadian rhythms. These circadian clocks can also be entrained by environmental periodic cues, suggesting that their evolutionary development is linked to the organisms' ability to anticipate such changes for survival [11, 12].

Early approaches, led by analogy between biological rhythms and mechanical clocks, produced models based on negative feedback loops [13–16]. Specifically, conceptualization around stable limit cycles (*see* Box 1 for definition of limit cycle) helped to explain the resetting of a circadian rhythm's phase after a perturbation [17]. These earlier models appeared even before the discovery of the molecular underpinnings of sustained circadian rhythmicity.

Genetics enabled findings such as the cell-autonomous nature of circadian clocks and their dependence on the interaction between genes and proteins in Transcriptional Translational feedback loops (TTFL) [18–21]. TTFL is a general functional mechanism found in all organisms studied that are able to display circadian rhythmicity, despite the existing diversity in the identity of specific genes and proteins. The core TTFL consists of a set of transcription factors acting as positive elements, since they induce the expression of downstream transcription factors, some of them acting as negative elements of the TTFL, that is, repressing their own expression and closing the molecular loop after circa 24 h [22–24]. This common genetic core design, found from bacteria to humans, has been conceptualized as a negative feedback loop, that Goodwin-Griffith based-models had previously shown to be at the origin of periodic rhythmicity in the form of limit cycle dynamics [25–30]. Additional evidence has shown that the core feedback loop is strengthened by additional positive and negative secondary loops, conferring robustness to the molecular clock, and are key for the fine tuning of period, amplitude and phase of the circadian molecular rhythms [4, 31] (*see* definitions in Subheading 1.3 and Fig. 1). The current view of the circadian molecular clock has several layers of regulation, and new layers are continuously found and described [32].

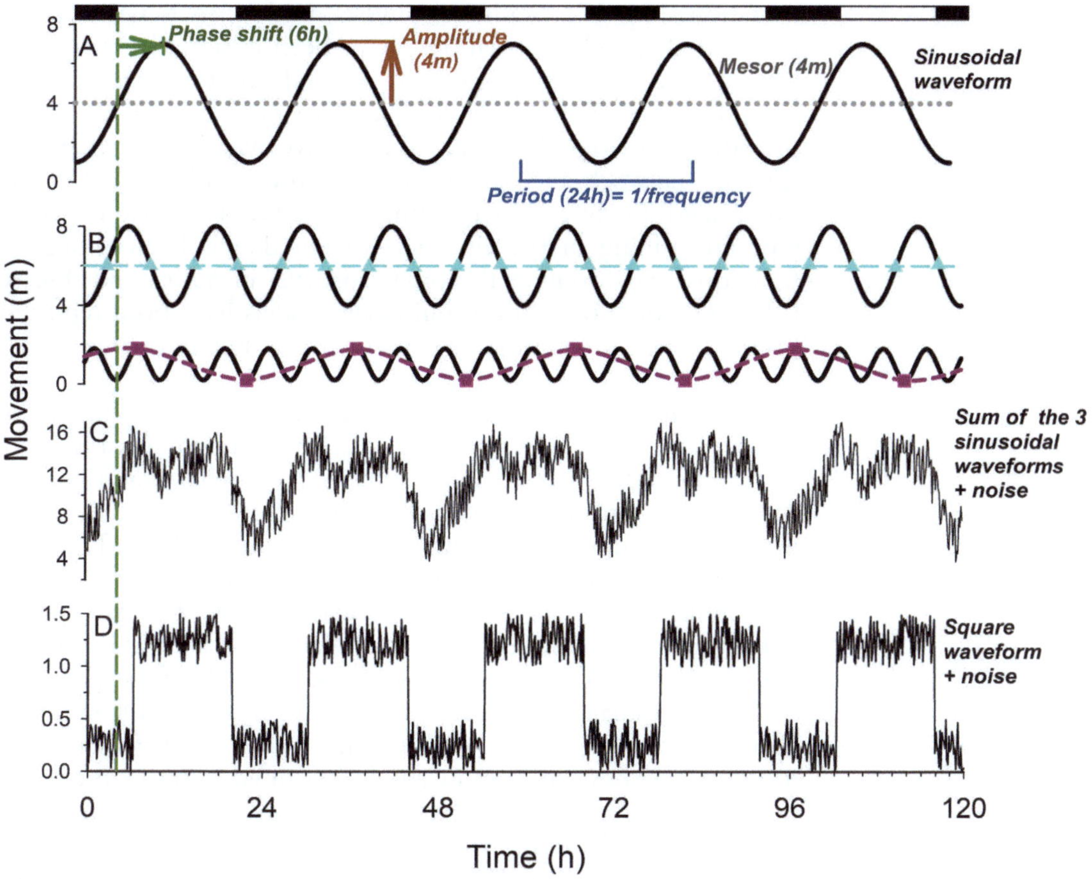

Fig. 1 Examples of different waveforms and their characterization. (**a** and **b**) Sinusoidal waveforms with roughly 24, 12, and 6 h periods. (**c**) Sum of the three black waves represented in (**a** and **b**), plus random uniform noise (range 0 y 0.5). (**d**) Square wave plus random uniform noise. In panel (**a**), the gray dotted line indicates the mean (mesor) value of the time series. The brown arrow indicates the amplitude, and the blue bracket the period. The green broken line represents the initiation of the 14 h light period of the circadian day–night cycle, the associated green arrow indicates the 6 h phase shift of the sinusoidal wave. In panel "**b**" cyan triangles show sampling points at 6 h intervals starting at 9 AM, if this were the case no oscillations would be observed (discontinuous cyan line). The purple squares represent sampling at 18 h intervals, in this case a spurious wave is observed (discontinuous purple line)

Although clock genes and proteins are examples about how genes can directly control behavior, the links between genes and behavior are not always straightforward since the circadian system comprises different levels of organization, structured as a multilayered network of cellular clocks acting in coordinated fashion and integrating quantitative information to produce different behaviors, as exemplified the function of the mammalian suprachiasmatic nucleus (SCN) [33] and the internal synchrony among tissues [32]. The SCN example underscores the role of robust network function for synchronizing heterogeneous single cell circadian oscillators, which reduces the impact of single gene mutations [34–37]. Statistical physics models, such as the Kuramoto model, and its variations, have been critical for hypothesis testing concerning the role of network architecture in determining the properties of the SCN compared to other tissues as well as providing a biological example of collective dynamics generalizable through network and synchronization theory [35–37]. On the other hand, the internal synchronization studies revealed the complex fine tuning of the circadian system that influences organs' functional synchrony at the organism level in health and disease [38, 39]. Some theoretical multiscale models have been reported [40] but their integration to experiments is still an open research field.

Circadian rhythms have been found to coexist with other dynamic modes. In Subheading 2.2.1 we present a case study about the effect of caloric restriction on behavioral rhythms. Another interesting example is locomotor activity (Fig. 3), which not only presents circadian and ultradian rhythms [30, 41] but also a temporal architecture obeying long-range (fractal) correlations over multiple time scales from seconds to hours. This association between different dynamic regimens has been shown in diverse species, including humans [42, 43], rodents [41] and quail [30], with the circadian system seemingly playing a central role in fractal regulatory networks [44]. The SCN modulates both circadian and ultradian rhythms, at least in mammals, in addition to the long-range (fractal) correlations patterns [41]. Moreover, it has been proposed that the underlying control network is also fractal. Specifically, Hu [45] showed that, in vivo, SCN-neural activity exhibits fractal dynamic patterns, virtually identical in mice and rats, and similar to those in motor activity for time scales ranging from minutes up to 10 h. Interestingly, ultradian calcium rhythms with periods of 0.5–4.0 h were also shown in the SCN, as well as the subparaventricular zone and paraventricular nucleus [46]. These results are indications of the importance of considering potential coexisting dynamical patterns in biological time series.

1.2.2 Calcium Dynamics as an Example of the Diversity of Possible Dynamic States

Calcium is one of the most important cellular cations, its dynamics underlying many biological phenomena in (patho)physiology such as muscle contraction, and calcium waves in oocytes fertilization, and development [47]. Ca^{2+} periodicities in eukaryotic cells are an interesting example of biological rhythms spanning from milliseconds to minutes. Usually, these rhythms are displayed in response to diverse cellular stimuli, thus representing an example of stimuli-driven rhythms. From muscle contraction, neurotransmitter release, neurite growth, activation of gene expression to cell growth and death, the ubiquitous involvement of Ca^{2+} signaling, in both excitable and nonexcitable cells, highlights its importance in the regulation of living systems' dynamics. The spatial and temporal encryption of information involved in Ca^{2+} dynamics is exquisitely diverse and precise: from local and fast elementary events in cellular microdomains [48] to global and long-lasting oscillations or waves widespread to the whole cell and other cells [49–51]. How global oscillations, well described with deterministic mathematical models, emerge from the intrinsically stochastic (i.e., random) and aperiodic Ca^{2+} elementary events are still a matter of debate. However, a recent mathematical model, based on a nucleation mechanism, proposes a unifying theory [52]. Nevertheless, from elementary events to global waves, all these responses are possible because the cytoplasmic Ca^{2+} concentration ($[Ca^{2+}]$) is several orders of magnitude smaller than those found within some organelles (i.e., the endoplasmic reticulum (ER), mitochondria, lysosomes) or in the extracellular space.

Ca^{2+} oscillations have been described in a variety of cell types [38–40, 46]. Typically, the oscillations are produced after an external signal triggers the intracellular rise in 1,4,5-trisphosphate (IP3), which activates the Ca^{2+} ionic channels sensitive to IP3 (IP3 receptors) expressed in the ER membranes. The activation of IP3 receptors produces an initial increase of cytoplasmic $[Ca^{2+}]$ released from the ER. The IP3 receptors are also sensitive to $[Ca^{2+}]$: at low level, the receptors are activated and increase even more the Ca^{2+} efflux from the ER, but at high Ca^{2+} levels they are inhibited. This feedback mechanism, known as Ca^{2+} induced Ca^{2+} release (CICR), determines the emergence of cytoplasmic $[Ca^{2+}]$ oscillations. Specific features of Ca^{2+} dynamics, such as period, amplitude, waveform and baseline levels, width of the spikes and degree of response sustainability, depend on cell type. Organelles, like mitochondria, and the agonist/external signal that elicit the Ca^{2+} response can fine-tune CICR [53]. Frequency-encoding of Ca^{2+} oscillations in nonexcitable cells, following an increase in the stimulatory signal, have been reported [40]. Further studies are needed for understanding whether and how frequency-encoding could be a way to encrypt on-off signals to downstream Ca^{2+} target effectors [47].

In some neurons, Ca^{2+} participates not only as a second messenger, linking neuronal firing to different downstream processes (i.e., gene transcription, phosphorylation of transcription factors and other proteins), but also seems to be involved in the information-encoding mechanism about the number of action potentials fired in a burst and, to a lesser extent, the frequency of action potential firing [49]. Bursting is a type of pulsatile dynamic activity characterized by regular or irregular intense activity during brief time lapses (peaks) separated by long time lapses of quiescent or silent activity. In the context of neuronal activity is defined as a short, high frequency train of spikes, and constitutes one of the underlying information-encoding mechanisms by which neurons can compute [49]. Although there are some dynamic differences between electrical firing patterns and calcium responses [51], Ca^{2+}

firing and bursting dynamics are commonly reported and characterized in diverse types of neurons and used as a readout of neuronal activity [54]. Stochastic calcium dynamics has been reported as the main cause of calcium burst fluctuations [55].

Bifurcations from periodic to chaotic regimes can have fundamental implications in biology and medicine. For example, in cardiac mitochondria, complex oscillatory dynamics in key metabolic variables, that arise at the "edge" between fully functional and pathological mitochondrial behavior, can set the stage for chaotic dynamics [56] that could underlie arrhythmias [57–59]. In this context, the mathematical characterization of the transition between periodic fluctuations to aperiodic chaotic dynamics is presented in Subheading 2.2.2.

1.3 Combining Experimental Design with Appropriate Mathematical Tools to Investigate Temporal Patterns in Time Series

As stated in the previous section, there is a wide variety of biological time series with distinct temporal patterns. Given that the focus of this chapter is on biological rhythms and chaos, to enable their detection and characterization along with other dynamic patterns, next we provide some guidelines for combining experimental design with adequate analytical tools. More specifically, we underscore important considerations to be taken into account in the experimental design (e.g., sampling rate, testing duration).

Distinct set of parameters need to be used to distinguish rhythmic time series from chaotic ones. If experimental rhythmic data is obtained by sampling a periodic function (possibly contaminated with random noise), ideally, it can be fully characterized by the following six parameters [60]: (1) mesor or the rhythm's adjusted mean level (Fig. 1a, gray dotted line); (2) period, the duration of a full cycle or time between two consecutive peaks (Fig. 1a, blue line); (3) amplitude, referring to the height of the wave, basically the distance between the mesor and the peak (Fig. 1a, brown arrow); (4) phase, referring to the displacement between the oscillation and a reference angle (Fig. 1a, green arrow), such as the environmental light-dark cycle [60]; (4) waveform, the shape of the wave (e.g., sinusoidal (Fig. 1a–c), square (Fig. 1d)), and (4) prominence, denoting the strength and endurance of a rhythm. This last parameter corresponds to the proportion of the overall variance accounted for by the signal (signal-to-noise ratio) [60]. If the signal is a sum of more than one rhythm, each rhythm will present its own distinct set of parameters (Fig. 1c), and demixing the rhythms becomes challenging. As for chaotic time series, they can be characterized in a lagged phase space by their strange attractor with fractal properties (*see* Box 1), and sensitivity to initial conditions (*see* Subheading 2.1.9).

Studies aimed at characterizing temporal patterns should be designed in such a way that both experimental and data analysis protocols can balance the trade-off between constraints in both, experimental and analytical demands. In these studies, data is presented as time series, which can be defined as a collection of numerical observations arranged in a natural order [61]. Basically, this usually implies that each experimental observation is associated with a particular instant or interval of time. It is generally assumed that the time values are equally spaced [61], thus the sampling rate (i.e., number of data points collected per unit of time) should be constant. In general terms, the longer the time series and the higher the resolution, the easier to discriminate between separate components with closely similar periods, and to estimate components with longer periods, thus improving the ability to better differentiate rhythms from trends [60, 62].

However, often there are constraints that limit the duration of an experiment, the temporal resolution of data collection (sampling interval), and number of independent samples studied (e.g., number of animals or cell cultures). Importantly, insufficient sampling can lead to aliasing, that is, identification of spurious (alias) rhythms (Fig. 1b, cyan line) or total lack of detection (Fig. 1b, pink triangles) because of poor (i.e., insufficiently frequent) sampling of an actual rhythmic process [60]. The accuracy of parameter estimation and the ability to detect differences between time series, depends upon the sampling rate. Refinetti [63] showed that the accuracy of the temporal resolution determines, in turn, the analytical precision of many well used methods such as autocorrelation (Subheading 2.1.5), Fourier (Subheading 2.1.6) and Enright analyses for circadian period detection in datasets composed of pure waves (cosine as well as square). This represents an improvement compared to other analysis such as acrophase counting, which tolerates accuracies fivefold lower than the data resolution [63]. Nevertheless, Glynn et al. also showed that Lomb–Scargle periodogram (a Fourier based method) is capable of successfully dealing with irregular sampling [64].

When sampling rate is appropriate, the minimum duration required for a time series to attain statistically significant period estimation of an oscillation, depends on other factors such the analytical method available and the signal-to-noise ratio [62]. Deckard et al. (2013) created a decision tree to recommend five different algorithms based on their ability to distinguish periodic from nonperiodic profiles in synthetic data [65]. Sampling across at least two periods is preferable, as periodicity means that values are repeated at regular intervals, which can only be verified with two full periods [64, 66–71]. However, most algorithms require longer time series (see below).

Obtaining very long time series (i.e., weeks, months, years) presents methodological challenges, such as equipment failure or personnel limitations, that hinder the need of keeping consistent experimental conditions throughout the study. Animal housekeeping conditions (e.g., cleaning, feeding, reproduction) can introduce unintentional impact on temporal dynamics.

Additionally, the animals' physiological state or cell culture characteristics changes over long periods of time, for example, through aging or adaptation. A rule of thumb for determining the optimal length of a time series needed for parameter estimation is that it should cover a stretch of time much longer than the longest characteristic time scale that is relevant for the system under study [72]. For instance, in periodic time series 5 to 10 cycles are usually needed for parameter estimation (*see* details for each method in Subheadings 2.1.5, 2.1.6, and 2.1.9). When acquiring a long time series is not possible, other approaches combining experimental results with mathematical modeling are often applied (*see* examples in [73, 74]). For ascertaining the presence of chaos much stricter requirements are necessary, that include very long, high resolution, time series with tens or hundreds of cycles. Consequently, depending upon the phenomenon under analysis or biological setting, mathematical modeling of the system becomes an important tool for obtaining the necessary high-quality time series [56, 75].

In summary, the process involved in understanding the underlying dynamics of a biological system (Fig. 2) begins by recognizing the potential importance of fluctuations over time, followed by an adequate experimental design, which demands the coordinated planning of the experiment and analytical strategy.

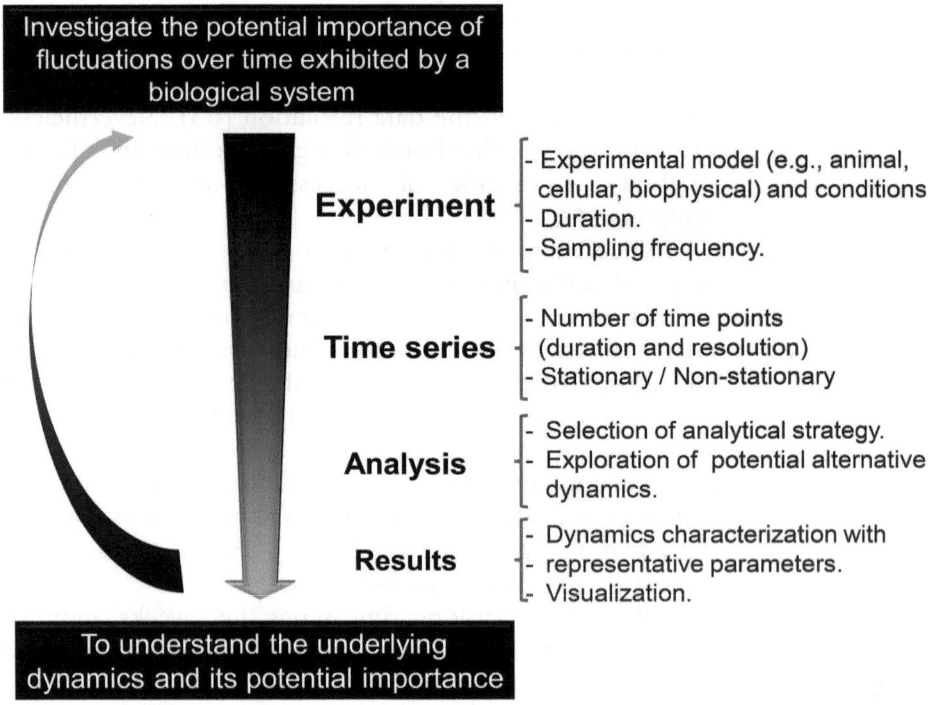

Fig. 2 Coordinated experimental and analytical design to investigate the underlying dynamics and its potential importance in a biological time series

2 Methods

2.1 Informative Metrics in Time Series Analysis

Physiological rhythms and chaos can be difficult to characterize experimentally. For instance, deciding the most appropriate method for determining oscillatory rhythms in biological time series has been a matter of intense debate (for review [46, 60, 65]). Some commonly used analytical methods are autocorrelation (Subheading 2.1.5), Fourier (Subheading 2.1.6), cosinor, maximum entropy spectral analysis (MESA), Enright's method, linear regression of onset, interonset averaging, acrophase counting [60, 63, 72, 76], and, more recently, wavelet (Subheading 2.1.10) [30, 77–80]. Each of these methods has different assumptions which make them valid under certain conditions, therefore may provide different results when applied to the same dataset [62, 63]. In general, these assumptions comprise three basic aspects that refer to the quantity and quality of data. First, different methods differ in their sensitivity to under sampling and short duration (i.e., few cycles) and, as stated in the previous section, the general rule of "more data the better" should be applied. Second, noise levels present in biological time series can interfere with parameter estimation. For example, autocorrelation (Subheading 2.1.5) and Fourier (Subheading 2.1.6) analyses may underperform with respect to the Enright's periodogram method in very noisy datasets [62, 63], where signals that are in reality chaotic or fractal can be confused with insignificant noise. This should be considered in the exploratory phase of data analysis. In this context, probability distributions can help distinguish random noise from fractals (Subheading 2.1.4). Moreover, phase space reconstruction is a useful method for visualization (Subheading 2.1.7 and Box 1) and a standard procedure when studying, potentially, chaotic time series. Third, the majority of time series analyses assume a cyclic model contaminated with an additive stationary noise process (for an in-depth analysis, *see* [72]), thus their performance varies according to their sensitivity to nonstationary trends and long-term correlations [76]. As a caveat, stationarity is frequently not the case in biological time series [76]. Formally, a signal is called stationary if all joint probabilities of finding the system at some time in one state and at some later time in another state are independent of time within the observation period [72]. This definition implies that all parameters that are relevant to the system's dynamics have to be fixed and constant during the observation period [72]. In practical terms, this usually means that the data should exhibit no long-run upward or downward trend (*see* signal with linear trend in Table 2, first column) or, otherwise stated, any fluctuation in the average level of the series should be of relatively short duration compared to the length of the series being analyzed [72]. With this in mind, it is often suggested that the linear trend should be removed from the series when possible (*see* Subheading 2.1.2 for methodology) before proceeding with analysis. When the source of the nonstationarity is in the underlying process itself, more adequate scale-dependent representations must be used to separate rhythms from the small-scale dynamics.

In experimental setups, stationarity is assumed when the mean and standard deviation of the time series, or of the intrinsic period and amplitude of the oscillation, do not change over time. Practically, estimates of mean, standard deviation, transition probabilities, correlations, performed on the first and second half of the time series must not differ beyond statistical fluctuations [72]. Importantly, stationarity is also dependent on the sampling frequency and length of the time series evaluated (*see* discussion Subheading 1.3). Time series that are too short may appear nonstationary while a longer time series of the same system may be stationary. In Subheading 2.1.10 we introduce wavelets as an alternative procedure for handling nonstationary time series [81–84].

Given these difficulties, we strongly recommend the strategy of combining different methods to favor detection of rhythms (circadian and ultradian) and chaos, when noise level and stationarity are not determined a priori. Specifically, when studying time series we propose six visualization and analytical methods as a starting point. First, data can be visualized using actograms (Subheading 2.1.1), moving averages (Subheading 2.1.2), histograms and probability distribution of data points or events (Subheadings 2.1.3 and 2.1.4) (Fig. 3), and/or phase space reconstruction (Subheading 2.1.8). Second, autocorrelation (Subheading 2.1.5), power spectrum (Subheading 2.1.7), and wavelet (Subheading 2.1.10) analyses, Lyapunov exponent estimation (Subheading 2.1.9), and/or synchrosqueezing transform (Subheading 2.1.11) can be used depending upon the characteristics of the time series. Wavelet coherence and correlation are described to investigate association and synchronization between different time series (Subheading 2.1.12). It is important to note that most of these tools are conceptually straightforward and informative if their assumptions are valid for the specific time series under analysis. They are included in almost every data processing or statistical software (MATLAB code is provided in the **Note 2**). Mention of, potentially useful, additional analyses are provided in **Note 1**.

2.1.1 Actograms

Since raw representations of time series often provide a level of detail that hinders visual assessment of underlying dynamics, actograms (Fig. 3, grey plot) are a common form of displaying time series, especially for circadian rhythms detection. Actogramas have the potential to provide visual information about the duration of circadian [60] and even ultradian cycles [30].

Actograms are computed by integrating data in bins of a specific size. Basically, the raw time series is divided into consecutive bins, and the result obtained from adding the data in each bin is

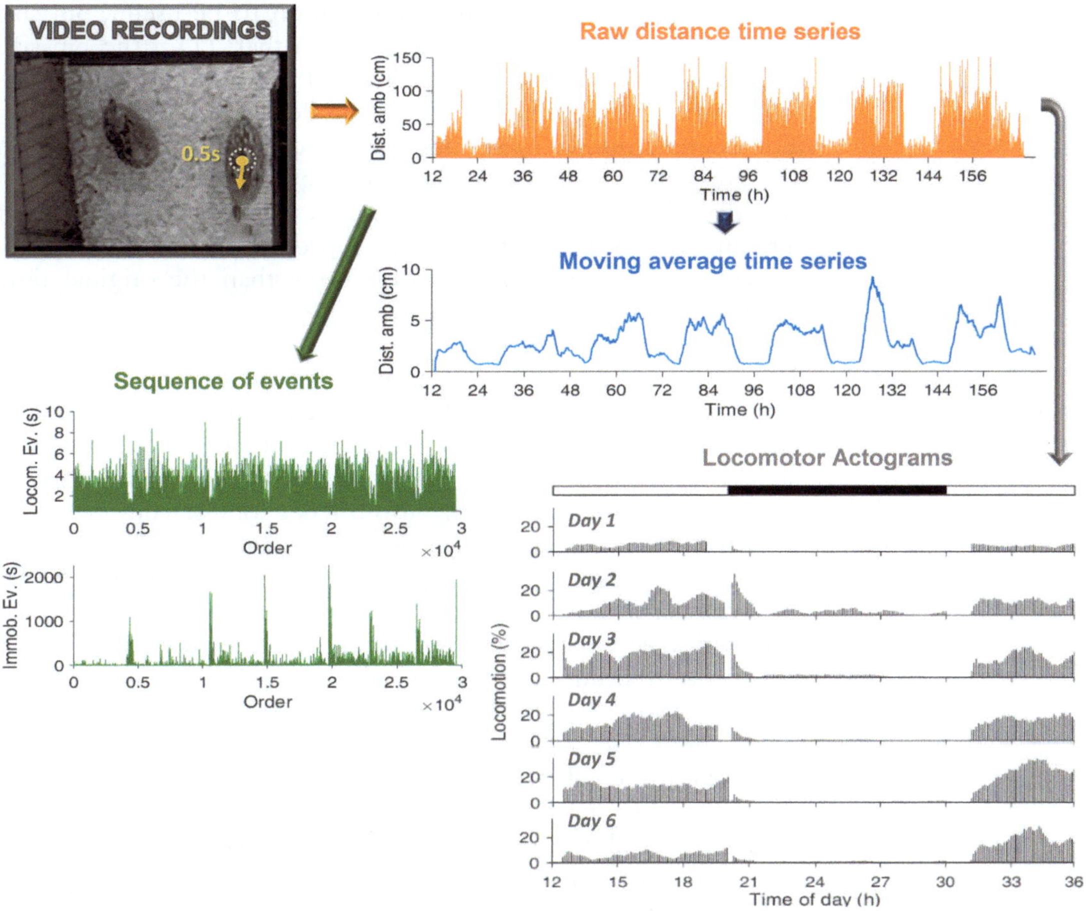

Fig. 3 An example of preprocessing and visualization methods for time series analysis. Time series of distance ambulated by a Japanese quail in their home box can be obtained from video recordings by measuring the displacement of the center of the animal that occurred during the sampling period (here 0.5 s). If the displacement is higher than the 1 cm threshold (white dotted line) the animal is considered to have moved during the period, and the distance ambulated can be plotted as a function of time (orange time series). This raw time series can be processed in different ways. (1) The time series could we smoothed using a moving average algorithm (blue time series), here a 12 h bin was used. (2) A locomotion time series (not shown) of two mutually exclusive states (mobile/immobile) can be estimated from the raw time series using the 1 cm threshold. The lapse the animal stays in any given state (i.e., event) can be estimated and plotted as a sequence of either locomotor (i.e., mobility) and immobility (green plots). (3) Actograms can be constructed from either the distance ambulated time series or the locomotor time. Here a 6 min bin was chosen, and the percent of time mobility during each 6 min period is plotted as a function of time. Twenty-four hour periods are plotted one underneath each other

plotted as a vertical bar. In the example shown in Fig. 3, the original ambulation time series is integrated over a 6 min time interval and represented as vertical bars. Thus, the height of each vertical bar indicates the accumulated time spent ambulating during the 6 min period. Usually, each day is plotted in separate panels, one underneath the other, for comparative purposes [85].

Considerations: For this representation to be useful, bin size needs to be selected appropriately, given that: (1) if the bin size used is too large, important details could be lost; (2) If the bin size is too small, actograms will appear too detailed to be informative. For example, unlike in studies aiming to observe 24 h circadian rhythms where 6 min bin sizes are frequently used [30, 85], in ultradian rhythms with short period cycles, bins sizes of fractions of a minute may be necessary. Once an appropriate bin size is selected, actograms look smoother and less noisy than the original time series.

2.1.2 Smoothing Data: Binning, Moving Average, and Detrending

Data smoothing to favor detection of specific dynamics is a common tool in data processing. Binning, such as that used in actogram construction, is an option that renders smoother and less noisy time series than the original, and, in general, the stationarity assumption holds. The binning processing also offers a solid starting point for applying other methods such as wavelet analysis. However, for analyses focusing on evaluating whether time series exhibit fractal behavior, or if noise is present, the raw rather than the processed time series, with maximum resolution, should be utilized.

Data smoothing can also be achieved by estimating a moving average (also called running mean, Fig. 3, blue time series) for overlapping bins (also called segments or windows). A bin of a fixed size is moved step by step over the original time series, and at each step the mean value of the data within the bin is calculated. Thus, the resulting moving average is a transformed time series (i.e., a subseries) in which each value is an average [86]. As in the case of actograms, once an appropriate bin size is selected, the resulting time series is smoother and less noisy than the original and can be considered stationary. Refinetti [63] showed that filtering the actogram data using a 9 h moving average improved the sensitivity of autocorrelation (Subheading 2.1.5), but not Fourier (Subheading 2.1.6) analysis to detect circadian rhythms in noisy data.

Similarly, moving median (also called running median) or moving standard deviation can also be estimated using overlapping bins. This methodology can be useful for visualizing nonstationary behavior in a time series, such as trends [86], given that changes in mean, median, and standard error over time would be evident. Estimation of mean, median, and standard deviation using non-overlapping bins can also be used for this purpose. For this, a large bin size can be used, for example at tenth or 50th part of the length, N, of the time series (i.e., applied to N/10 or N/50) [72]. If in this way the type of trend can be identified (linear, exponential, etc.), it can be eliminated from the time series in a process referred to as detrending. For this an appropriate function is used to fit the moving average. For a linear trend, for example, a best-fit line to the moving average is calculated [86]. The equation for this line then gives the value of the "trend" at a given time, and can be subtracted from the moving average value [86].

Further considerations: The more data that is included in the moving average (i.e., larger bin size), the greater the smoothing of short-term fluctuations [86]. When bin size is too small, the resulting time series will be very similar to the raw original time series. It should also be taken into account that when smoothing data is necessary, bin size should be selected with caution to avoid deleting important fluctuations present in the time series. It is important to know that many laboratory equipment automatically performs smoothing procedures, thus impacting the data. For example, the wheel running data obtained with ClockLab [87] shown and analyzed in Subheading 2.2.1 was automatically integrated into 1 s bins.

2.1.3 Discretization of Raw Data into Events

Under certain conditions, it is useful to transform continuous raw data into a small number of finite values or events (Fig. 3, green plot), especially in the case of nonstationary time series. For example, fluctuations in membrane potential are frequently transformed into events such as open/closed states of an ion channel; ECG recordings into cardiac interbeat interval; or change in the acceleration or position of a person or animal over time is discretized into steps or ambulation events [30, 88, 89]. In the example depicted in Fig. 3, the following procedure was employed: first, thresholds were determined analytically, and high pass filters were applied. Accordingly, an animal was considered to be ambulating when it moved more than 1 cm in a 0.5 s interval [30]; second, determining the total time of continuous ambulation that is recorded as an ambulation event. This discretization into events have the potential of being more informative than the original noisy signal, although may continue to be nonstationary.

The concept of the existence of discrete physiological or behavioral events is at the base of almost all biological studies. However, the methodological basis for discretization extensively varies between research fields that, in the end, determines the selection criteria for the appropriate method.

2.1.4 Histograms and Probability Distribution of Raw Data and Events

A popular graphical representation of a probability distribution of a continuous variable is a histogram ([86, 90]), where the area of vertical bars represents the probability of certain values in the time series. Specifically, for continuous data, the x-axis shows the range of all possible observable values of the variable partitioned into bins (i.e., classes) of equal or different widths (bin width). Then, the probability is estimated by summing the number of observations that fall within the limits of each bin and dividing it by the total number of observations:

probability = number of observed cases in each class/total number of observations

Depending on the analysis, it may also be desirable that the summation area of the individual bars equals 1, for which the probability density is estimated. For this, the probability values of each class are rescaled by dividing it by the bin width:

probability density = probability/bin width.

In this case, the *y*-axis represents the probability per unit bin width [90], and the bar area (height × width), the probability of occurrence. It is evident that as the bin width becomes smaller and sample size larger, a histogram for a continuous variable gradually blends into a continuous distribution. The resulting smooth curve is called the probability density function (PDF) [86].

Regarding PDF, different type of distributions can be observed in the histograms, for example, raw stochastic time series data may show uniform (Table 1, first row) or Gaussian distributions, while the PDF of a time series of composite sine waves will have distinct peaks (Table 1, second row). As with other simple stationary waveforms, the number of peaks correspond to the periodicity [86]. Unlike time series displaying defined periodicity, raw time series data of a system exhibiting deterministic chaos show a large number and variety of peaks (Table 1, fourth row).

Of note, in addition to histograms for estimating PDF, other methods are available. As a matter of fact, recently Rhee and Gorá (2017) proposed a methodological approach for specifically predicting and estimating probability density functions in chaotic systems [91].

Event duration distribution is a metric widely used in time series analysis (Subheading 2.1.3). In stochastic processes the PDF of event durations decays exponentially, while processes with long-range correlations (fractals) decay as a power law (i.e., linearly in a double logarithmic plot). The most common approach for testing empirical data against a hypothesized type of distribution (e.g., exponential, power law) is to transform *x* and/or *y* axis into a logarithmic scale and fit the data with least-squares linear regression. This provides estimates and standard errors for the slope, in addition to the fraction r^2 of variance accounted for by the fitted line, which is taken as an indicator of the quality of the fit [92]. Although this procedure appears frequently in the literature there are several problems with it and should be avoided especially when axes are transformed to logarithmic scales (for detail review *see* [92]). Instead, goodness of fit and model selection process should be performed on cumulative distribution function (CDF, see below) by comparison of different types of distributions such as power law, lognormal, exponential, stretched exponential, and Gamma distributions [93, 94].

Table 1
Examples of distinct dynamics that can be found in biological time series and their characterization using probability density distributions (PDF), cumulative density functions (CDF), autocorrelation function, power spectrum analysis, lagged phase space plot, and wavelet analysis

	Movement time series and associated PDF	CDF	Autocorrelation function	Power spectrum analysis	Lagged phase space	Gaussian wavelet transform
Noise (Uniform)	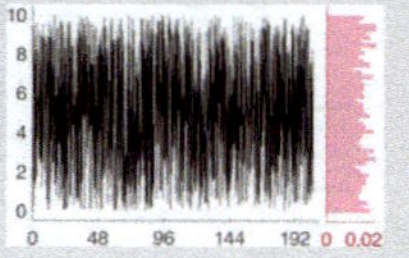	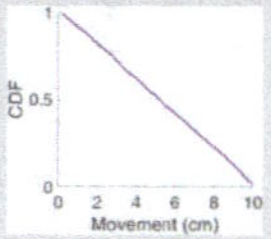	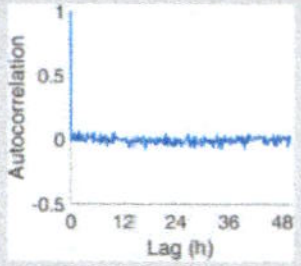	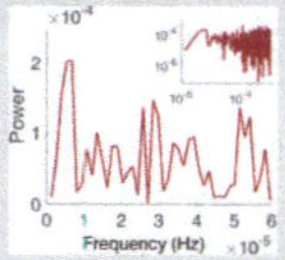	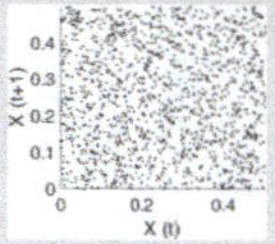	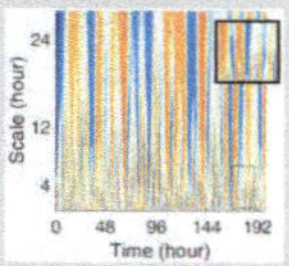
Sum of two sinusoids	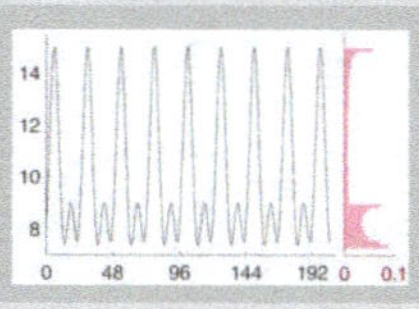	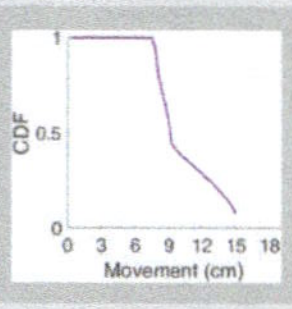	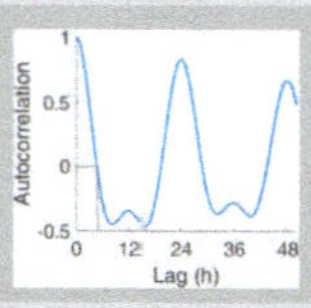	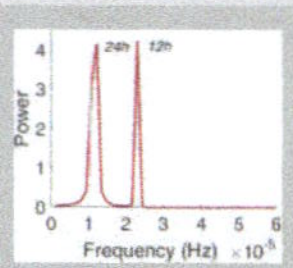	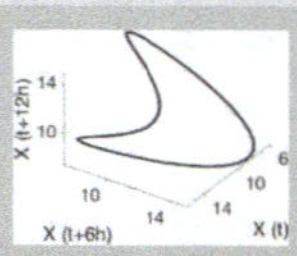	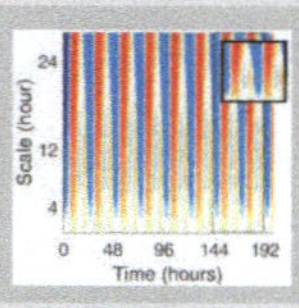
Sum 3 Sinusoids + noise (uniform)	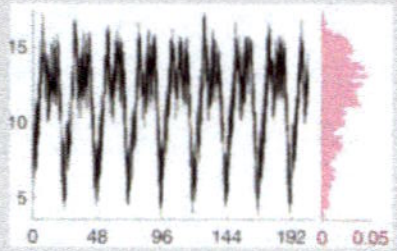	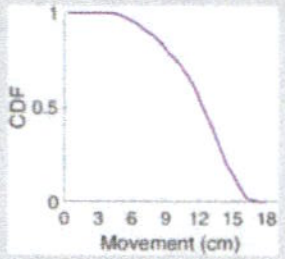	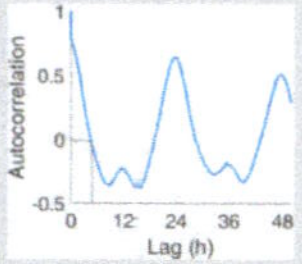	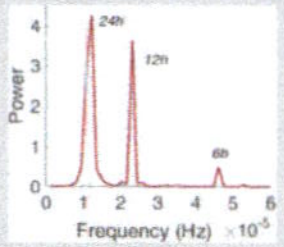	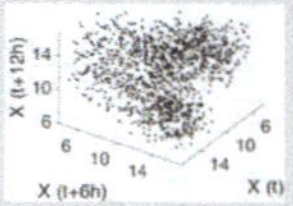	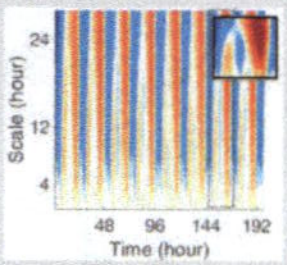
Chaos (Henon)	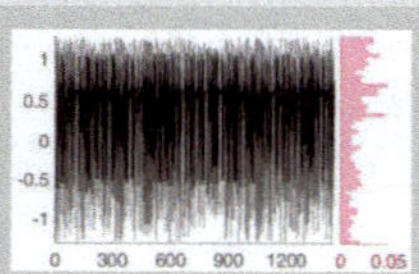	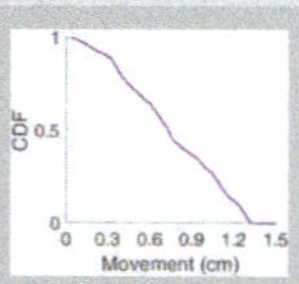	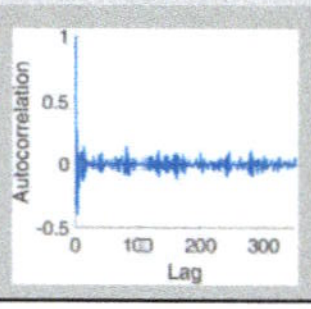	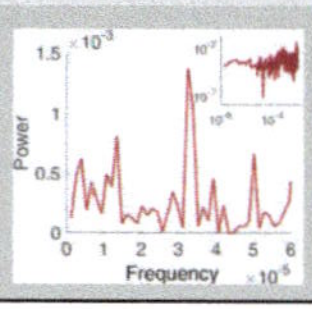	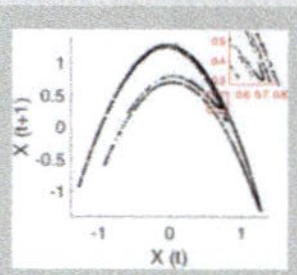	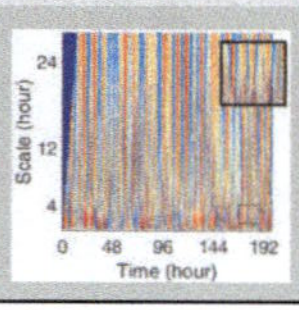

Considerations: when constructing a histogram, selection of bin width is an important factor. Bins that are too narrow will contain few observations per bin and will not provide much insight, whereas, if too wide, will obscure features of the distribution. Several of the rules that have been proposed for the selection of the appropriate bin width (I) are based on the number of observations (N) such as Sturgis's rule, $I = \log_2 N + 1$, or Rice's rule $2 \times N$ 1/3. Although these rules can be useful, different possible bin width should be explored to correctly represent data. Another consideration for bin width choice, is when data are unevenly distributed over the frequency distribution, as frequently is the case for chaotic [86] and fractal time series. This leads to discontinuities (areas where the probability is zero). This can be avoided by using data-adaptive techniques that allow different bin widths depending on local peculiarities [86]. This is done in such a way that narrow bins are used for ranges with a high number of cases, while wide bins are used for ranges that have only few cases.

The PDF can be easily estimated by constructing a histogram of raw data, or derived variables, such as durations of events. Alternatively, one can construct the cumulative distribution function (CDF), which represents the probability that the variable is less than a particular value, and this is done by a simple rank ordering of the data [92]. Empirically, CDF is a more accurate method for estimating characteristic parameters of certain distributions such as the (fractal) scaling parameter α. This is because the statistical fluctuations in the CDF are typically much smaller than those in the PDF [92]. In the example shown in Fig. 3, in the duration of immobility events, the PDF represents the probability that the duration of an immobility event is within a given range. While for the respective CDF, the y-axis $P(x > a)$ represents the fraction of immobility events whose length is larger than a (s). Tables 1 and 2 show the probability distributions (pink plots in the first column) and cumulative distribution probability (second column) for each of the exemplary time series (shown in black).

When using raw data for PDF, trends such as a linear increase in values over time (Table 2, first row), can lead to histograms that are uninformative given that peaks could appear distorted or not evident at all in a periodic time series. Thus, PDF and CDF of time series that are nonstationary or with trends should be avoided when working with raw data, but rather the distribution of derived variables can be studied such as event duration.

2.1.5 Autocorrelation Estimation and the Correlogram

Autocorrelation is a straightforward technique that produces autocorrelation correlation coefficients between the data vector and itself when sequentially "lagged" out of phase, one-time unit at a time (for detailed explanation *see* [86]). Conceptually, the formula used is the same as that of the correlation coefficient utilized in

Table 2
Examples of dynamics that can be found in recorded time series from biological systems and their characterization using probability density distributions (PDF), cumulative density functions (CDF), autocorrelation function, power spectrum analysis, lagged phase space plot, and wavelet analysis

	Time series and associated PDF	CDF	Autocorrelation function	Power spectrum analysis	Lagged phase space	Gaussian wavelet transform
Sinusoid + linear trend	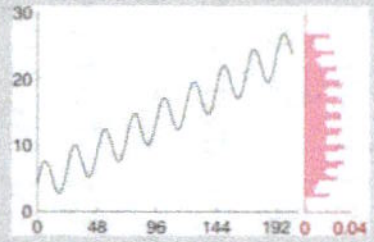	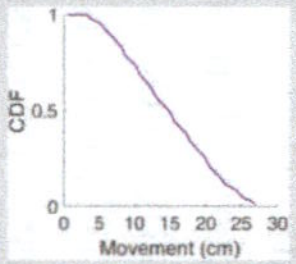	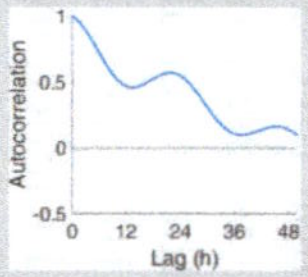	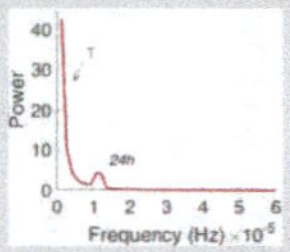	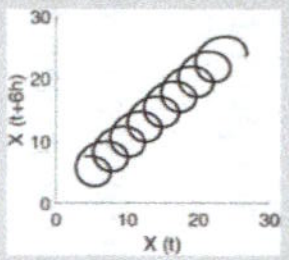	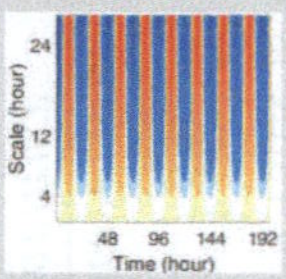
Square waveform	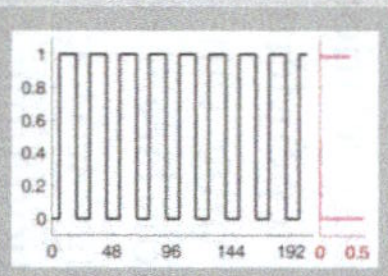	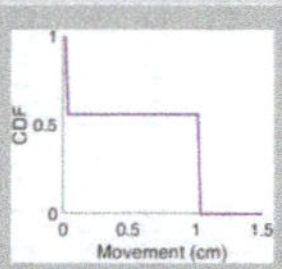	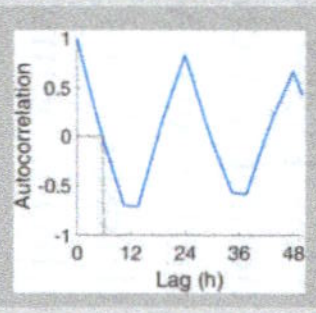	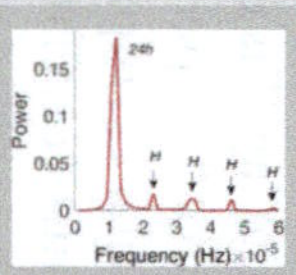	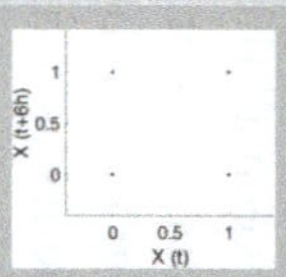	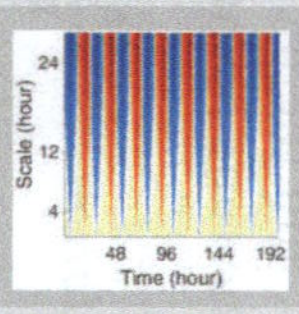
Bursts of random	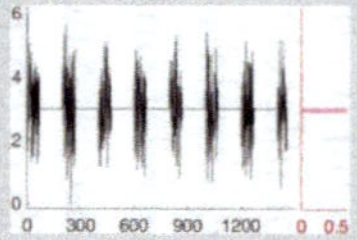	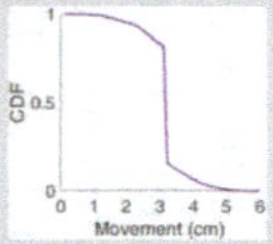	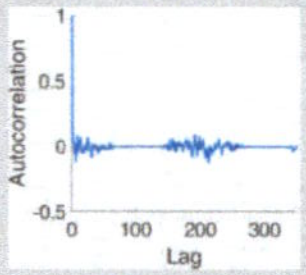	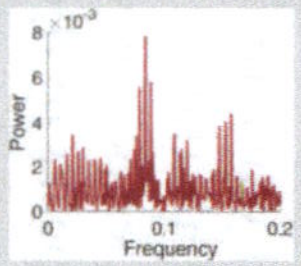	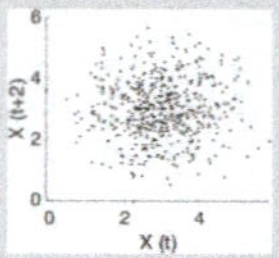	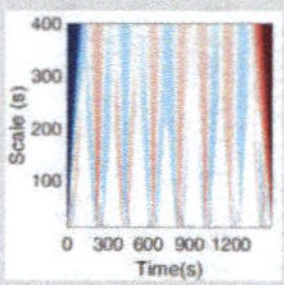
Fractal (Cantor set)	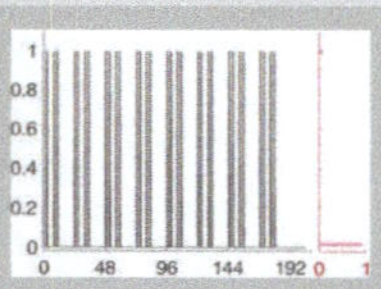	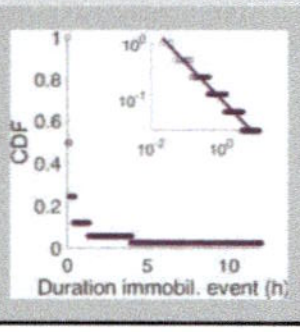	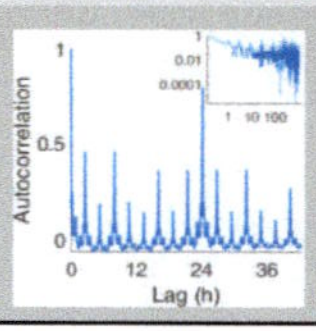	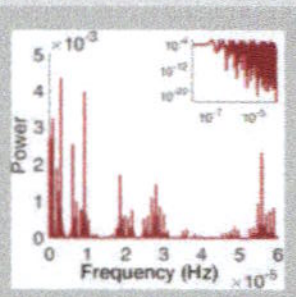	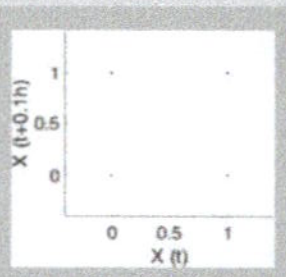	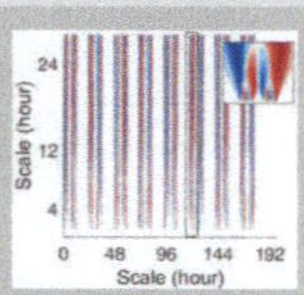

basic statistic between two different variables, with the difference that in autocorrelation it is estimated between the time series and itself after a time lag *T*. Hence, if x_t is the value of the variable x at time point t, and the mean value is x, and x_{t+T} is the value of the variable at the time point *T*:

$$Autocorrelation = \frac{autocovariance}{variance} = \frac{\sum_{t=1}^{N-T}(x_t - \overline{x})(x_{t+T} - \overline{x})}{\sum_{t=1}^{N-T}(x_t - \overline{x})^2}$$

For example, for a lag time of 2 s the correlation coefficient is estimated between all the data points of the original time series and the same points 2 s later.

A correlogram can be constructed by plotting these autocorrelation coefficients as a function of the time lag, *T* [86]. If a time series is periodic, then the autocorrelation function is periodic in the lag *T* [72] as shown in Table 1 (in blue, second row). Specifically, for sinusoidal data the value of the autocorrelation is equal to 1 when the time lag *T* is equal to the period, is −1 when the lag *T* is half the value of the period, and 0 for ¼ and ¾ the value of the period (*see* similar example in Table 1, second row). Thus, recurring peaks in the autocorrelation coefficients indicate that the signal is periodic and provides information on how robust that periodicity might be [35]. When both circadian and ultradian are simultaneously present, peaks indicative of ultradian rhythms could be difficult to detect (such as the smaller peak in the example in Table 1, second row). Mourao et al. (2014) proposed the use of an arbitrary threshold of 10% in order to prevent small peaks in the periodogram from being included in the period estimation analysis of the predominant rhythm [62].

By definition data points that are completely independent of each other render a correlation coefficient of 0, such as in the case of white noise (*see* example of uniformly distributed random values, first row Table 1). Stochastic processes have decaying autocorrelations, but the rate of decay depends on the properties of the process [72]. Fast exponential decay implies that, after a short period of time, data points are practically independent of each other. In the example, high correlation is observed for the first few seconds but completely independent after 5 min. It is evident that a finite decorrelation time can be estimated (*see* details in [95]) where correlations for large time lags compared to the decorrelation time, are negligible due to the fast exponential decay. Interestingly, this decorrelation time can be used as a measure for the memory or persistence of a process. Thus, one also refers to these processes as having short-range or finite memory [95]. Typically, autocorrelation of signals from deterministic chaotic systems also decay exponentially, approaching 0 for large time lags (Table 1, fourth row). Hence, autocorrelations are not characteristic enough to distinguish random from deterministic chaotic signals [72].

A second class of autocorrelation structures can also be distinguished with respect to the form of their decay for large time lags [95]. Long-range correlated processes, such as the case of pink noise (*see* definition Subheading 2.1.7) show a linear decay in a double logarithmic plot as a power law (Table 2, fourth row). Thus, a characteristic time scale does not exist, and the system is, theoretically, considered to have infinite memory. In biology, the concept of long-range correlation has gained considerable power, given that most biologists will intuitively agree that systems have long-term memory (*see* examples [30, 88, 96]). It should be noted that autocorrelation is not the method of choice for testing long-range correlations in biological time series, and other methods should be considered such as Detrended Fluctuation Analysis [97] and, although important, is not the focus of this Chapter.

Autocorrelation estimation is, on the one hand, useful to realize the temporal pattern of stationary data. On the other hand, autocorrelations in time series provide important insights since many statistical metrics are designed for temporally independent, memoryless, data (as opposed to correlated data) [86] where the autocorrelation is 0. We will further address this point for phase space reconstruction (Subheading 2.1.8, Box 1).

Considerations: Although, in noisy datasets, the autocorrelation method can be more sensitive for detecting circadian rhythms than Fourier Analysis, it is highly affected by trends and nonstationarities [72] as shown in Table 2 (first row). In addition, the estimation is only reasonable when the lag T is small compared to the total length of the time series ($T \ll N$) [72]. For periodic data, this method yields an estimated period with resolution that depends on the sampling interval and is best applied to records with at least 4 cycles and a short sampling interval [79]. However, when high levels of noise are present more cycles may be needed to improve period estimation, and other methods such as power spectrum analysis (Subheading 2.1.7) may produce better results [62].

2.1.6 Harmonic Analysis

Two main topics in functional analysis theory have had a great impact in signal processing: analysis and synthesis of functions. The former refers to breaking down the signal into elementary components that better describe the characteristic features of a particular signal, while the latter informs signal reconstruction from the components. Harmonic analysis refers to a branch of mathematics concerned with the representation of functions or signals as the superposition of basic waves and encompasses a diversity of analyses, including Fourier, Hilbert, and Wavelet [81, 82].

Of particular importance, in science and engineering, the process of decomposing a function into oscillatory components, by means of the Fourier transform, is often called **Fourier analysis**,

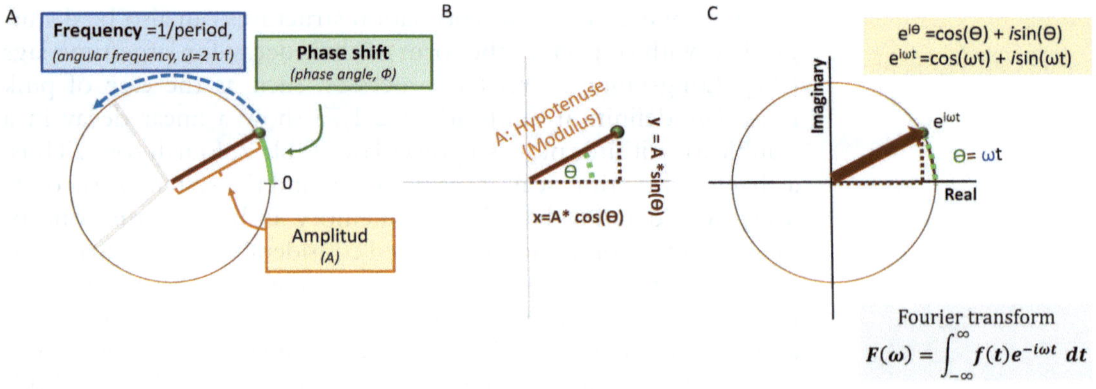

Fig. 4 Relating periodic oscillations to circles and conceptual framework of Fourier analysis. (**a**) Frequently, mathematicians represent periodic sinusoidal oscillations as circles, given by its repetitive nature, as depicted in panel **a**. The radius of the circle is associated with the amplitude of the oscillation. The starting point (with respect to the 0 coordinate) is the phase, and can be considered as an angle, thus the name *phase angle*. The time to complete one circle is the *period* (frequency = 1/period) which is here expressed as an *angular frequency*. (**b**) Basic trigonometry states that the length of the vector, also called *modulus*, can be described knowing the *x*, *y* coordinates or one of the coordinates and the angle (θ). Thus, considering that the hypotenuse (from now on referred to as modulus) is the amplitude (*A*), "*y*" is *A**sin(Θ) and "*x*" is *A**cos(Θ). In our example, we are considering a phase shift = 0 for simplicity (*see* [86] for an in-depth description of this analogy). (**c**) In Fourier transform, the exponential $e^{i\omega t}$ can be regarded as a vector with unit magnitude, rotating in a complex plane at a rate of ω in the direction shown. The magnitude of the unity vector, $|e^{i\omega t}| = 1$, that is, the amplitude is standardized to 1. The oscillation is represented as imaginary numbers, the *x*-axis represents the real part, and the *y*-axis the imaginary part. The angle Θ is equal to the angular frequency multiplied by the time ($\theta = \omega t$), and $\cos(\omega t)$ and $\sin(\omega t)$ are just the projection of this vector on the real (*x*-axis) and imaginary (*y*-axis) axes in this diagram. According to the trigonometry shown in panel (**b**), $e^{i\omega t} = \cos(\omega t) + i\sin(\omega t)$. The formal definition of the Fourier Transform is presented in the gray square. The weighting for each frequency component at ω is $F(\omega)$ which results from adding together (the integral) of the weighted sum of $e^{i\omega t}$ components multiplied by the time series ($f(t)$) at time t, for all time points

while the operation of rebuilding the function from these pieces is known as Fourier synthesis. Generally speaking, the Fourier transform (FT) measures the similarity of a signal with a particular set of analyzing functions, the complex exponentials *exp(iωt)* (Fig. 4). Given a signal, the output of the transform is a complex valued function of a single variable, the angular frequency ω. Such complex function is obtained multiplying the signal with the conjugate complex exponential of frequency ω and integrating the result.

$$F(\omega) = \int_{-\infty}^{\infty} f(t) e^{-i\omega t} dt$$

In Fig. 5 we have broken down the process of computing the FT, for the case of angular frequencies. For comparative purpose, *see* Figs. 8 and 11 in Subheading 2.1.10, for the equivalent process with respect to the wavelet transform.

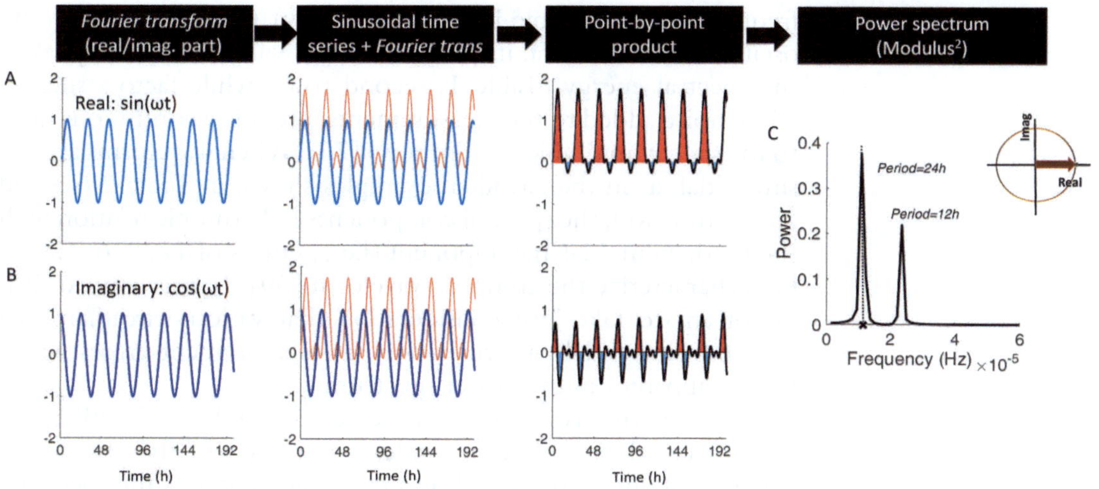

Fig. 5 Breaking down a working definition of the Fourier transform. In the first column, the real (**a**) and imaginary (**b**) parts of the *Fourier transform* are shown for a frequency, ω, of $1.15 \times 10^{-5}\,t$ Hz, equivalent to a 24 h circadian period. In the second column each part (in blue) of the transform is superimposed over the sinusoid time series (in orange). In the third column, the result of the point-by-point multiplication of each part in the time series is shown, with areas under the curve that are positive or negative, in red or cyan, respectively. The sum of this multiplication is positive in the case of the real part, but zero for the imaginary part (note the equal amount of red and cyan areas). Thus, the result is a vector with a positive real part and an imaginary part (inset in **c**). (**c**) The square modulus of this vector is plotted (dotted line marked with an *x*) for the specific frequency assessed. This process is repeated for a broad range of frequencies, and the resulting power spectrum is shown in (**c**)

If the signal is periodic, the FT has a simpler form, the Fourier series, where the sinusoids that decompose the signal are harmonics of a fundamental frequency. The discrete version of the FT can be evaluated quickly on computers using fast Fourier transform (FFT) algorithms, which is the method of choice for performing the power spectrum analysis (Subheading 2.1.7) on biological time series.

2.1.7 Power Spectrum Analysis for the Analysis of Rhythms

Power spectrum analysis is a well-established method for the study of rhythmic processes [61], which is based on Fourier analysis, meaning that any periodic waveform can be exactly described by a combination of pure cosine and sine waves of different amplitudes and frequencies (for review and detailed description of Fourier analysis [61, 82]).

In this context, the power spectrum is defined as the squared modulus of the Fourier transform; it is the square amplitude by which the frequency *f* contributes to the signal being analyzed [72]. For white noise (i.e., independently distributed random numbers, zero autocorrelation, *see* Subheading 2.1.5) equal power is observed for all frequency bins (Table 1, first row) [72]. Thus, the slope of the power density function (power as a function of frequency, f) is 0 (Table 1, inset in first row). Contrarily, if there are oscillations in the data, its period ($=1/f$) will show up as a peak in the spectral energy (Table 1, second row); while factors such as white noise added to the measurements, provide a continuous floor to the spectrum (Table 1, third row) [72]. When waveforms are not sinusoidal, as in the circadian examples shown in Table 2 (second and third rows), the spectral components in harmonic relation with the fundamental 24-h component (i.e., periods of 12, 8, 6, 4.8...) help characterize the complex waveform [60]. In this context, it is important to take into consideration that various circadian data have nonsinusoidal patterns as well as measurement errors which will be apparent in the power spectrum.

For deterministic chaotic time series (Table 1, fourth row), sharp spectral lines may be evident, but even in the absence of added white noise there will be a continuous part of the spectrum [72]. As in the case of white or uniform noise (Table 1, first row), this continuous part of the spectrum is visualized as fluctuations occurring over a broad range of frequencies (compare insets in Table 1, first and forth row, fifth column). Thus, without additional information it is impossible to infer from the spectrum whether the continuous part is due to noise on the top of a (quasi)periodic signal or to chaos; *see* [72] for an in-depth analysis of the important relation between the power spectrum and autocorrelation function. Importantly, for certain cases the power spectrum can show inverse linearity on a double logarithmic plot (*see* inset, Table 1, fourth row) proportional to $1/f^{\beta}$. Generally speaking, this power law, describes colored noise depending on the value of β or the spectral exponent (e.g., $\beta = 0$, 1 or 2 for white, pink, or Brownian noise, respectively) [96, 98]. Thus, a stationary self-similar (fractal) stochastic process, with long-range correlations (*see* section autocorrelation function Subheading 2.1.5), follows pink noise (also called $1/f$ noise) if its power spectral density function is inversely proportional to the frequency ($S(f) \approx 1/f$) [96, 99]. For Brownian noise, the high value of the slope ($\beta = 2$) represents higher energy (i.e., power) at lower frequencies.

Different estimators of the power spectrum, such as the Walsh periodogram and the Lomb–Scargle (LS)] periodogram are classical methods for identifying periodicity in time series data. The (LS) method was developed in the field of astrophysics [70, 100] as a Fourier style method, but was designed to deal with data that exhibit irregular sampling, which is typical of observational data in astronomy. It measures the correspondence to sinusoidal curves and determines their statistical significance [64].

Considerations: First, studies have shown that Fourier analysis has been shown to combine accuracy and precision, and therefore under the correct experimental conditions is superior to autocorrelation and Enright's analysis methods [63]. Although Fourier analysis is robust against changes in amplitude [72], factors such as high noise levels in a dataset [60, 63] along with nonstationarity, may limit its potential for detecting periodicity. In particular, trends appear in the plot as high values of power at low frequencies [72] as shown Table 2 (first row, effect of trend is marked with an arrow and the letter T).

Second, it is important to consider that, as stated previously, when waveforms are not sinusoidal, the spectral components in harmonic relation with the fundamental component help characterize the complex waveform [86]. Thus, the presence of spikes at harmonics in the power spectrum is not a proof of whether ultradian rhythms are present [80]. This is due to the fact that some waveforms, such as a square wave with 24 h period (Table 2, second row), will have spikes at all harmonics (in the example, 12 h, 8 h, 6 h, etc.), even when that signal involves no ultradian periods (*see* discussion in Leise et al. [80], and Glynn et al. [64]).

Third, as in the case of autocorrelation analysis, the power spectrum determines frequencies present *globally* in the signal. For this reason, they do not provide the proper tool for the problem of determining ultradian frequencies present at particular time intervals. Specifically, this analysis should be avoided if the period can differ during, say, subjective day and night for an animal, or when the circadian period changes from day to day [80] (*see* Subheading 2.1.10 wavelet analysis for a more appropriate method).

Fourth, for periodic data, the resolution of the analysis depends on the number of cycles contained in the data and as such requires a relatively long record (typically at least 10 cycles) [101]. Moreover, the periodogram is computed only at frequencies up to 0.5 cycles per sampling interval, the "Nyquist Frequency" [102].

Finally, in theory, the "power" of a rhythm should be concentrated at a single point in the spectrum at the corresponding frequency, however, circumstantially, can be found to "leak" into neighboring frequencies. This implies a limit to resolution and translates into an inability to detect differences between the periodic components of different time series. If biological data is sampled every hour over a 10-day period, the limit of resolution is 0.1, thus cannot distinguish periodic components whose periods differ by less than about 2.5 h.

2.1.8 Lagged Phase Space Plots, Embedding, and Attractor Reconstruction

Conceptually, a phase space is defined as an abstract space in which the coordinates represent the variables needed to specify the phase (or state) of a dynamical system at any particular time [1, 3, 86]. Although phase space plots are constructed using time series, time is only evident by the trajectory given by the sequence of plotted points [86]. In particular, a lagged phase plot compares values of the time series to later measurements within the same data

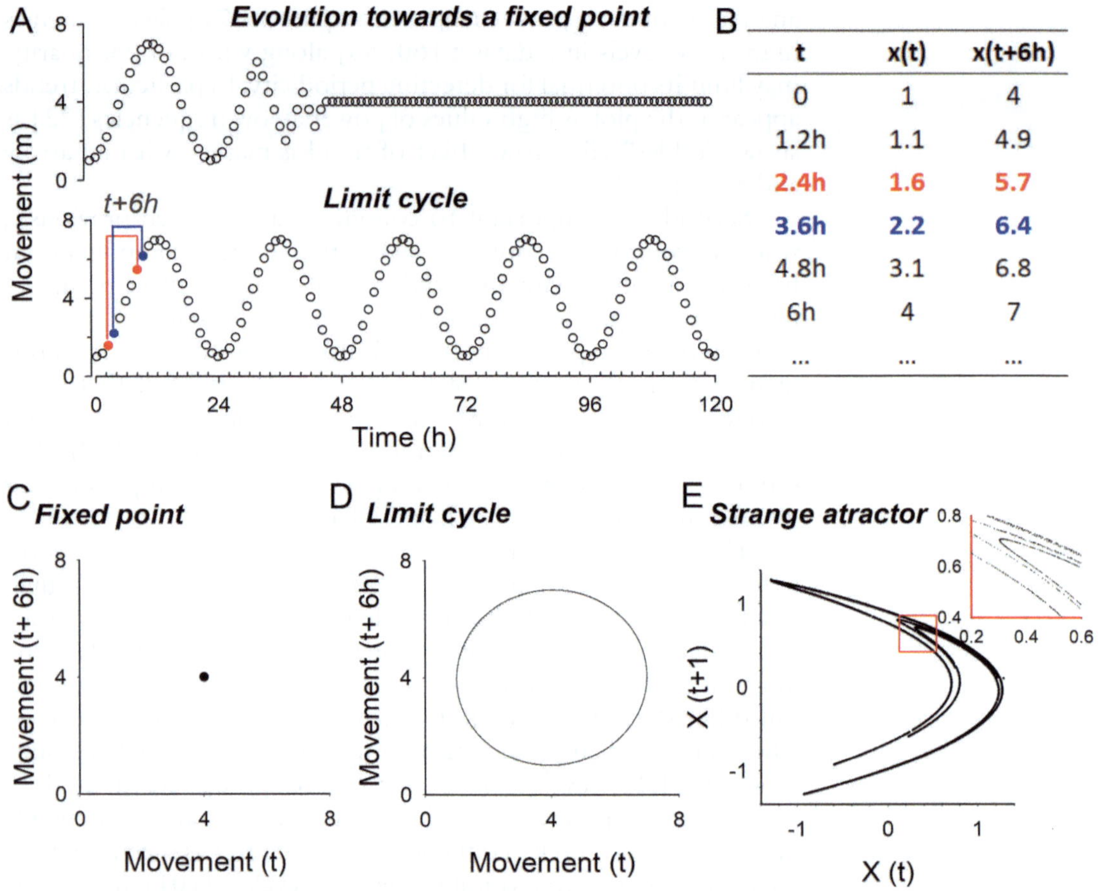

t	x(t)	x(t+6h)
0	1	4
1.2h	1.1	4.9
2.4h	1.6	5.7
3.6h	2.2	6.4
4.8h	3.1	6.8
6h	4	7
...	...	...

Fig. 6 Lagged phase space plots and examples of different types of 2D attractors. (**a**) Examples of a system dynamics that evolves toward a fixed point (constant values) and a limit cycle (the same circadian oscillation presented in Fig. 1a). A time lag, T, of 6 h (equivalent to ¼ period) is represented by the colored brackets. The time series data of the oscillation is shown in table format in (**b**) as $x(t)$. The column $x(t + 6\text{ h})$ represents the phased time series with a time lag. Colored numbers are a reference to the values of the colored circles shown in (**a**). Note that the resulting lagged phase plane plot from plotting the column $x(t + 6\text{ h})$ as a function of $x(t)$ is shown in panel **d**. (**c–e**) Three examples of different types of attractors. (**c**) A fixed point in phase space represents a time series that does not change over time. (**d**) The sinusoidal time series shown in Fig. 1a and Table B is represented in phase space as a limit cycle. Since for a periodic oscillation, autocorrelation is 0 at the lag time T equal to ¼ of the period (*see* Subheading 2.1.5), this lag was used for phase space reconstruction. (**e**) A chaotic time series, such as the Henon equations ($x_{n+1} = 1 + ax_n^2 + by_n$; $y_{n+1} = x_n$; parameters $a = -1.4$, $b = 0.3$) describes a strange attractor. A zoom of the area within the red square shows a fractal appearance of the attractor

[86] (Fig. 6a). The x-axis being the value of the time series at time t, and the y-axis the respective value of the same time series after a time lag $T\,(t + T)$ (Fig. 6b–d). More dimensions may be necessary to represent the data, and, when that is the case, the z-axis would be the time series after two time lags $(t + 2T)$. In this framework, the number of dimensions analyzed or plotted is called the *embedding dimension*. It is important to note that although plots can only have

up to three embedding dimensions given visual limitations, it is possible, and mathematically useful, to construct theoretical spaces with more than 3 embedding dimensions. The criteria for selection of lag and embedding dimension is based on:

1. Abarbanel [103] proposed that an optimal time delay must be: (1) a multiple of the sampling time, (2) sufficiently large for data points to be practically independent of each other, thus the autocorrelation is equal to 0 (for example in sinusoidal data this represents a time lag T equal to the ¼ part of the period, *see* Subheading 2.1.5, and Table 1), and (3) not too large, that any connection between points are lost. This is especially important in the case of chaotic time series, due to the characteristic exponential growth of small errors (*see* sensitivity to initial conditions in Subheading 2.1.8). Although autocorrelation analysis (Subheading 2.1.5) could be used to assess the independence of data points of a time series (Fig. 7a) for different potential time lags, the method of choice is the average mutual information (Fig. 7b). Mutual information, like autocorrelation (Subheading 2.1.5), measures the extent to which values of the variable after a time lag, $T(x(t + T))$, are related to values at time t, $x(t)$ [86]. However, mutual information uses probability to assess correlation, and refers to the amount of information (in bits) that can be learned about the value $x(t + T))$ knowing $x(t)$. When values at $t + T$ are completely independent of those at time t, then the average mutual information between them is 0. Thus, by plotting average mutual information as a function of lag time (T), the first minimum function can be selected as a lag time for phase space reconstruction (Fig. 7b, red arrow).
2. The number of embedding dimensions needs to be sufficient to determine the complete unfolding of the geometrical structure (i.e., attractor, *see* Box 1). In other words, they should be sufficient to undo all overlaps and make orbits unambiguous (for in-depth explanation *see* [103]). When the attractor is completely unfolded, points on the plot are neighbor less, meaning, points lying close to one another in the phase space are because of their dynamics and not due to errors introduced when using too few dimensions. A method frequently used for calculating the embedding dimension is the false-nearest neighbor technique (Fig. 7c). For a certain lag T (estimated by mutual information) the percentage of false nearest neighbors (computed over the entire attractor) is plotted as a function of the number of embedding dimensions. Thus, for phase space reconstruction the lowest embedding dimension with close to 0 false nearest neighbors is selected (Fig. 7c, red arrow).

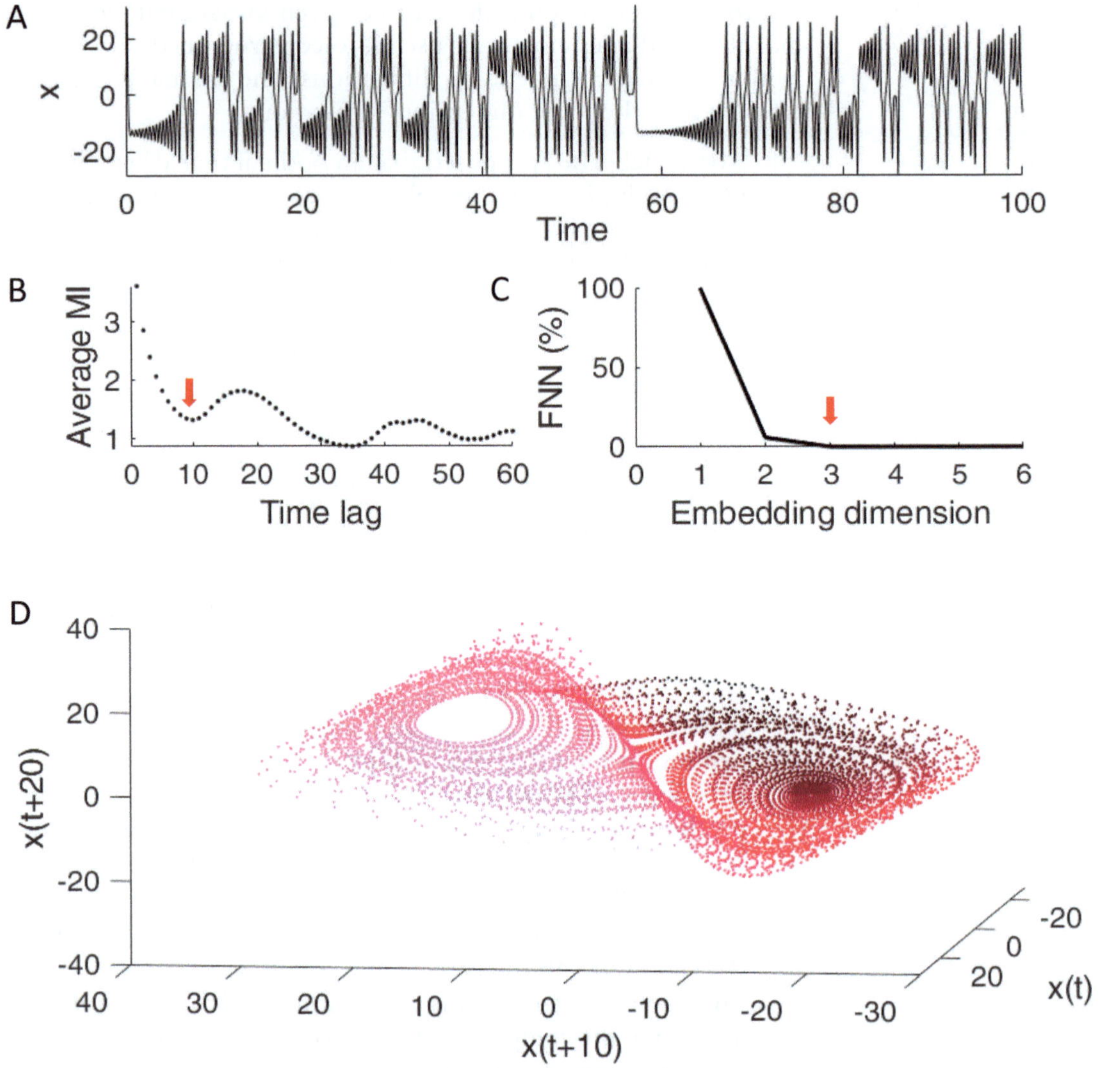

Fig. 7 Example of phase space reconstruction. (**a**) The chaotic time series of the *x*(*t*) component from the Lorenz model (*see* code in **Note 2** and details in [103]). (**b**) Average mutual information (MI) for the *x*(*t*) time series shown in "**a**" as a function of time lag, τ. The first minimum value of this function is at 10, as indicated with a red arrow. (**c**) The percentage of global false nearest neighbors for the *x*(*t*), and a $\tau = 10$ (estimated in "**b**") as a function of the dimension. As indicated with the red arrow, an embedding dimension of 3 is necessary to completely unfold the attractor. (**d**) Resulting phase space plot is shown. Color coding represents the *x*(*t*), *x*-axis values. Code is available in **Note 2**

Overall, and as shown in Fig. 7, performing a lagged phase space plot is a three step process, first, the appropriate time lag T needs to be estimated with a method such as Average Mutual Information (Fig. 7b); second, considering this time lag, the number of embedding dimensions is estimated (Fig. 7c); third, data is represented accordingly as either a 2 or 3 dimensional lagged phase space plot (Fig. 7d).

Once the time series is embedded in the lagged phase plot, a dynamic interpretation can be performed. Data for stationary oscillatory time series will appear in a restricted region of the plot forming circular-like shape. The complexity of the shape is associated with the complexity of the time series. For example, a simple sinusoidal wave can be represented by 2 embedding dimensions (lag $T = 1/4$ period) and will appear as a circle in phase space. However, a time series that is composed for 2 or more sinusoidal ways (as in the case of a circadian rhythm plus ultradian rhythms) needs more than two dimensions to unfold and could appear as smaller circles within the larger circles. On the other hand, in chaotic data a much more complex shape will be evident. These shapes deflect the attractor of a system (*see* Box 1), a fundamental concept for understanding dynamical systems.

Considerations: First, linear trends in data, will cause a drift in values (Table 2, first row), and may even result in the impossibility of defining even simple attractors (*see* Subheading 2.1.2 for methodology to eliminate linear trend). Second, phase space plots will not be informative if the lag time is not selected appropriately, resulting in time points that are autocorrelated (autocorrelation is not 0), thus not independent of each other [86]. For example, if data points (x) exhibit high positive correlation at a given lag T, when $x(t + T)$ is plotted as a function of $x(T)$ a 45° straight line-relation will be observed and not the actual shape of the attractor. Moreover, autocorrelated data introduce several complications in determining Lyapunov exponents and the correlation dimension (*see* Subheading 2.1.9). Third, if the embedding dimension is too low, the attractor will not be completely unfolded. Hence, points that are quite far apart from each other in their dynamic trajectories will be plotted near each other (false neighbors). Such an error could be mistaken for some kind of random behavior even when no noise is present [103].

Box 1 What is an Attractor?

The dynamics of biological systems can change over time acquiring new states. A notable example would be that after death the heart stops beating, thus the variables describing it would remain fixed at a given value. However, in living animals within a given period of time, the oscillatory-like heart dynamics, in theory, could remain fairly constant at rest or if running at a constant velocity, and be fairly responsive to the state of activity (e.g., running, walking, stressed) after which the heart rate eventually goes back to resting behavior. To investigate the dynamics of a system it is often useful to plot the time series as a function of itself after a time lag, T, as

(continued)

explained in Subheading 2.1.8. In this representation, an attractor is the phase space of point or points that, over a time course (iterations), attract all trajectories emanating from some range of starting conditions (the basin of attractors) [1, 3, 86]. In the example of the beating heart, starting conditions could be the rate of someone running, or sitting. Moreover, attractors can be stable or unstable upon perturbation.

The three most notorious types of attractors are shown in Fig. 6. The name of the attractors represents their shape in the phase lagged plot. Hence, a fixed point shows as a single point in the phase plot and represents a time series of constant values (Fig. 6c). A limit cycle appears as a circle (or, e.g., an ellipse, depending upon the waveform of the periodicity) describing a closed phase space orbit, that represents a periodic time series (Fig. 6d, Table 1). Strange attractors, have fractal orbits, and are characteristic of chaotic time series (Fig. 6e). Of note, an infinitely long time series of white noise would become a "scatter plot" that would completely fill the space (Table 1).

Conceptually, unlike homeostasis, a homeodynamic system will dynamically transform a dynamic state into another through instabilities at bifurcation points (*see* Subheading 1). Accordingly, a system's dynamics can evolve from a stable fixed point to a limit cycle (representing periodic dynamics), and, eventually, to a strange attractor (i.e., chaos) [1] and, ultimately, back and forth transitions between these states.

2.1.9 Lyapunov Exponent

The Lyapunov exponent is an important metric for characterizing chaotic dynamics, which exhibits sensitivity of initial conditions and long-term unpredictability [86]. In 1908, Henri Poincaré in his book "Science et méthode" [101] reportedly emphasized that, in chaotic systems, slight differences in initial conditions eventually can lead to large differences, making predictions for all practical purposes "impossible" [86]). In popular culture this has been associated with "The Butterfly Effect" apparently from a 1972 paper entitled "Does the Flap of a Butterfly's Wings in Brazil Set Off a Tornado in Texas" [112].

The Lyapunov exponent measures the time rate at which nearby orbits (or trajectories) diverge (positive Lyapunov exponent) or converge (i.e., negative Lyapunov exponent) from each other in phase space after a small perturbation [103–105]. It is important to note that there will be the same number of Lyapunov exponents as the number of dimensions of the reconstructed lagged phase space, also referred to as the Lyapunov spectrum. Any system containing at least one positive Lyapunov exponent is defined to be chaotic, with the magnitude of the exponent reflecting the time scale at which system dynamics becomes unpredictable [104]. In contrast, periodic or stationary motion will show all negative exponents [104]. Thus, the presence of a positive exponent is sufficient for diagnosing chaos and represents local instability in a particular direction. However, it is important to note that for the existence of an attractor, the overall dynamics must be dissipative, that is, globally stable, and the total rate of contraction must out-weigh the total rate of expansion. Thus, even when there are several positive Lyapunov exponents, the sum across the entire spectrum is negative.

Considerations: As explained in detail by Wolf et al. (1985), accurate exponent estimation requires care in the selection of embedding dimensions as well as time lag. If the dimension is too low the attractor will not be completely unfolded resulting in false neighbors. However, if the embedding dimension chosen is too large, we can expect, among other problems, that noise in the data will tend to decrease the density of points defining the attractor, making it harder to find replacement points needed for implementation of algorithms [105]. Given that the Lyapunov exponents in a chaotic system estimates the rate of divergence of trajectories, sufficiently long (with respect to the length of the fluctuation), high quality, time series are necessary [105]. Wolf et al. (1985) proposed an algorithm for estimating the Lyapunov exponents applicable to experimental data sets, where the underlying equations may be unknown or complex. Expanding on this concept, the algorithm proposed by Rosenstein et al. (*see* **Note 2** for MATLAB code) was designed for smaller data sets [68]. Their algorithm only estimates the largest Lyapunov exponent and is robust to changes in embedding dimension, size of data set, reconstruction delay, and noise level. Although there are fundamental differences between the Wolf [105] and Rosenstein et al. algorithms [68], both track the exponential divergence of nearest neighbors.

2.1.10 Wavelet Analysis

As above mentioned, (*see* Subheading 2.1.6), in the context of functional analysis theory, ***signal analysis*** refers to decomposition of the signal into components that retain meaningful characteristics of the original signal (Subheading 2.1.6, and Figs. 4 and 5), for example, in Fourier analysis the components are produced by the Fourier Transform. These components are coefficients represented by complex numbers (Figs. 4 and 5). As stated previously, power spectrum analysis studies the squared magnitude of these coefficients given by complex numbers, and it is quite successful in detecting constant oscillatory behavior (Subheading 2.1.7). However, if this behavior changes over time in a recorded signal, power spectrum analysis will not help in detecting the source of the change, or when additions to frequency components were made (Subheading 2.1.7). For that matter, a time frequency representation is needed. Although the short term Fourier transform can be used in this case, since the time window is fixed, the localization of a change in signal behavior over time is, in many cases, poor (*see* [81] for a comprehensive description of the short term Fourier transform). Consequently, for these cases, wavelet transform (WT) is the method of choice because it can detect both changes in time and frequency. Basically, the WT generates a representation of the signal into components in the time scale plane. Signals can be enhanced, denoised or filtered by tinkering the elementary components of the wavelet analysis before reconstruction with the wavelet synthesis operator.

Wavelet analysis operator definition. As a starting point, it is important to note that wavelet analysis is not a single, but a family of analyses based on the WT. Generally speaking, the WT operator is defined as a convolution of a signal with a continuously shifting, continuously scalable function, the analyzing wavelet, over the time series, which acts as a measure of correlation between the scaling function and the time series. The result of this convolution scheme are coefficients that represent how well the time series correlates with the analyzing wavelet of a given size (scale) at each time point. In comparison, as presented in Subheading 2.1.7, in the Fourier transform the analyzing functions are complex exponentials $e^{i\omega t}$, and the resulting transform is a function of a single variable, the frequency ω (Figs. 4 and 5). The Fourier transform maps the signal into pure frequency space. On the contrary, as explained in detail below, the WT is a function of both scale (in some cases equivalent to frequency) and time.

The WT operator compares the signal to shifted and compressed or stretched versions of the analyzing wavelet ψ. A dilation operator is the one that stretches or shrinks a function, and in the WT case, the operation made corresponds to a physical notion of scale. By comparing the signal to the wavelet at various scales and positions, a function of two variables is obtained. This two-dimensional representation of a one dimensional signal is redundant; also, if the analyzing wavelet is complex valued, the WT is a complex valued function of scale and position. If the signal and the wavelet are real valued, the WT is a real valued function of scale and position (Fig. 8, first column). For scale parameter $a > 0$, and position b, the WT of a signal *f(t)* with analyzing wavelet ψ is defined as.

$$WT(a, b; f(t), \psi(t)) = \int_{-\infty}^{\infty} f(t) \frac{1}{\sqrt{a}} \psi'\left(\frac{t-b}{a}\right) dt$$

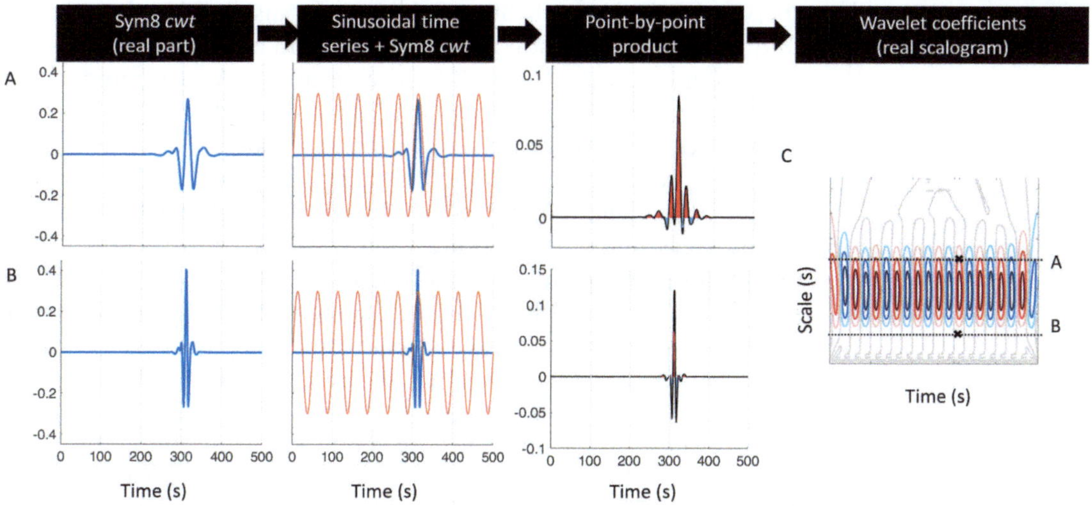

Fig. 8 Schematic representation of the wavelet analysis procedure. In this example a Symlet 8 analyzing wavelets at two different scales are represented in blue (first column), corresponding to (**a**) the 50 s, and (**b**) 20 s time scale. The scaled analyzing wavelet (blue) at time 312 s (first column) is shown superimposed on a sinusoidal time series (orange) in the second column, respectively. The point-by-point product of the time series with the scaled analyzing wavelet, is shown in the third column. The area under the curve is colored in red and cyan for positive and negative areas, respectively. Note that for the 50 s scale, mostly positive values are observed, while for the 20 s scale approximately the same amount of positive and negative values are observed. (**c**) In the last column the real scalogram is shown, with an *x* indicating the scale and time point used in examples. This scale time plot of the all computed wavelet transform coefficients was obtained by integrating the result of the point-by-point product of the time series with the scaled analyzing wavelet, at each scale and time point

where the symbol ′ denotes complex conjugation, an operation necessary if the analyzing wavelet is complex valued. We stress again that the coefficients of the WT not only depend on the position in time and frequency scale, but the choice of the mother wavelet, which gives much more flexibility for detecting features in data than the Fourier transform.

In Fig. 8 we observe how the comparison between the signal and a dilated version of the mother wavelet is made. Since we are interested in detecting rhythms, like sinusoidal oscillations, commonly found in nature, herein we use, as an example, a sine wave with a period of 50 s, and compute a Symlet 8 wavelet transform (note that this is a continuous wavelet transform, *cwt*, as will later be defined). The Symlet family is the least asymmetric Daubechies mother wavelets; the Symlet 8 has eight vanishing moments, a characteristic that makes it very smooth, with a central positive lobe and two small lateral lobes that take negative values. To illustrate the transformation, we chose a specific point in time sampled at 1 s rate, $t = 312$ s, where the sine wave has a peak, and the two selected compressed wavelets of scales 20 (Fig. 8a) and 50 (Fig. 8b) are plotted as a function of time. An inspection of their shapes reveals that scale 50 wavelet is more sustained in time than in frequency in comparison to the scale 20, and both have their central lobes aligned with the peak of the sinusoid. As a reminder, the coefficients are obtained by first computing the point-by-point product of the signal with the shifted and scaled analyzing wavelet (Fig. 8, third column) and integrating the result. That chained operation is called convolution. Hence, the coefficient is the sum of the areas under the product curve with its sign (positive areas depicted in red and negative in blue in Fig. 8). We can see that the scale 50 dilated wavelet matches the sign of the signal quite well (Fig. 8), producing a product signal with a large positive (red) area, therefore the coefficient is large. In the case of the scale 20 coefficient (Fig. 8), the product signal has two negative lobes (blue) and a positive very narrow one, thus the coefficient value should be small. The values of both coefficients are 1.0654 for scale 50 and 0.0047 for scale 20. The last plot of Fig. 8 is a contour plot of the graph of the two parameter function that is the output of the Symlet 8 WT, where the location of the coefficients computed has been marked with an "x" and labeled as A and B.

Scaling and zoom. In our example in Fig. 8, we are able to visualize that the WT provides a mathematical zoom in time, enabling the analysis of signal properties at different time scales (e.g., from seconds to days). Some signals show interesting changes on larger time scales that cannot be envisaged at smaller time scales, or exhibit fractal scale invariance, meaning that the feature is there regardless of the magnification scale. The scale factor introduced by the dilation operator produces an effect on the decomposition, according to which the smaller the scale factor, the more "compressed" the wavelet. Conversely, the larger the scale, the more stretched the wavelet, the less defined the signal features measured by the wavelet coefficients, since only the basic shape is retained. This is a strong trait widely used for denoising and smoothing time series, (*see* [82], for details).

Since the WT maps the signal into a function of time and scale, there is a general correspondence between scale and frequency; at low scales, the wavelet is compressed and able to mimic the regions of high variability of the signal, that is, the ones with high frequency, whereas at high scales, the wavelets are more stretched and detect slow changes or average the details, thus mimicking the regions of low frequency. This corresponds to a general relationship, where scales can be mapped into pseudofrequencies that can be obtained by locating the peak power in the Fourier transform of the mother wavelet, called center frequency, and dividing it for the scale and the sampling interval. Periods are thus estimated as 1/frequency.

Scalogram representation. The redundancy of the transform introduced by applying it to each time-point could make the interpretation of the coefficients challenging, with the added factor of the interpretation of scales, periods, and frequencies. In fact, wavelets were not designed for spectral analysis, but for the detection of singularities, discontinuities, and unusual patterns. The most important feature of the wavelet transform is its ability to track frequency patterns as they evolve in time.

Although, theoretically, signals are continuous in time, the data obtained from real world applications are discrete time series with a specific sampling rate Δt (*see* Subheading 1.2). Thus, the scales and time of the WT are discretized. Since the dilation (stretching and shrinking) operation on the wavelet is multiplicative, both in time and frequency, a common rule of thumb to determine the spacing of scales for the WT is to use the logarithmic scale, where there is a constant ratio between successive elements. Common choices of scales are $2^{1/12}$, $2^{1/16}$, $2^{1/32}$, where 12, 16, and 32 are referred to the number of voices per octave. To construct a logarithmically spaced scale vector, the practitioner must choose the number of voices per octave and total number of octaves. If the sampling period is set to one, the range of scales selected must be greater than one and less than the length of your input signal.

This discretization of the WT of time series $s(k)$, denoted by Ws (j, k), is called continuous wavelet transform (*cwt*), and it is a matrix of discrete values, which graphically corresponds to a scalogram (Fig. 8c). Scalograms are plots of the matrix of wavelet coefficients as a heat map or a contour plot. In particular, the scalogram of the squared modulus of the coefficients is called wavelet spectrogram, in analogy to the short-term Fourier spectrogram, widely used in sound processing. The logarithmic scale discretization in voices and octaves also mimic the classical scales of the spectrogram. If the mother wavelet has complex values, there are four scalograms associated with the cartesian and polar representations of complex numbers, the scalograms of the real and the imaginary part of the coefficients, and modulus and angle. In any scalogram the "*y*" axis is the time scale or pseudofrequency, and the "*x*" axis the actual time vector associated with the original time series (Figs. 8 and 9).

Scalograms are a quite useful source of information. Locating its maxima lines allows it to determine modal components if the maxima lines are localized in scale (horizontal lines in the scalogram), or discontinuities if the maxima lines are localized in time (vertical lines in the scalogram). It is also important to understand that the redundancy in the transform propagates information across scales. For a given point in time, the wavelets centered in that point are increasingly stretched, as a function of the scale [82].

In Fig. 9 we show scalograms of a synthetic signal constructed with pieces of smooth signals and a trajectory of a fractal process. All scalograms show the discontinuities in the regularity of the signal, marking the position of the singularities as vertical lines across the scales, called maxima lines. The first scalogram was made with the Derivative of Gaussian, the second with the Mexican

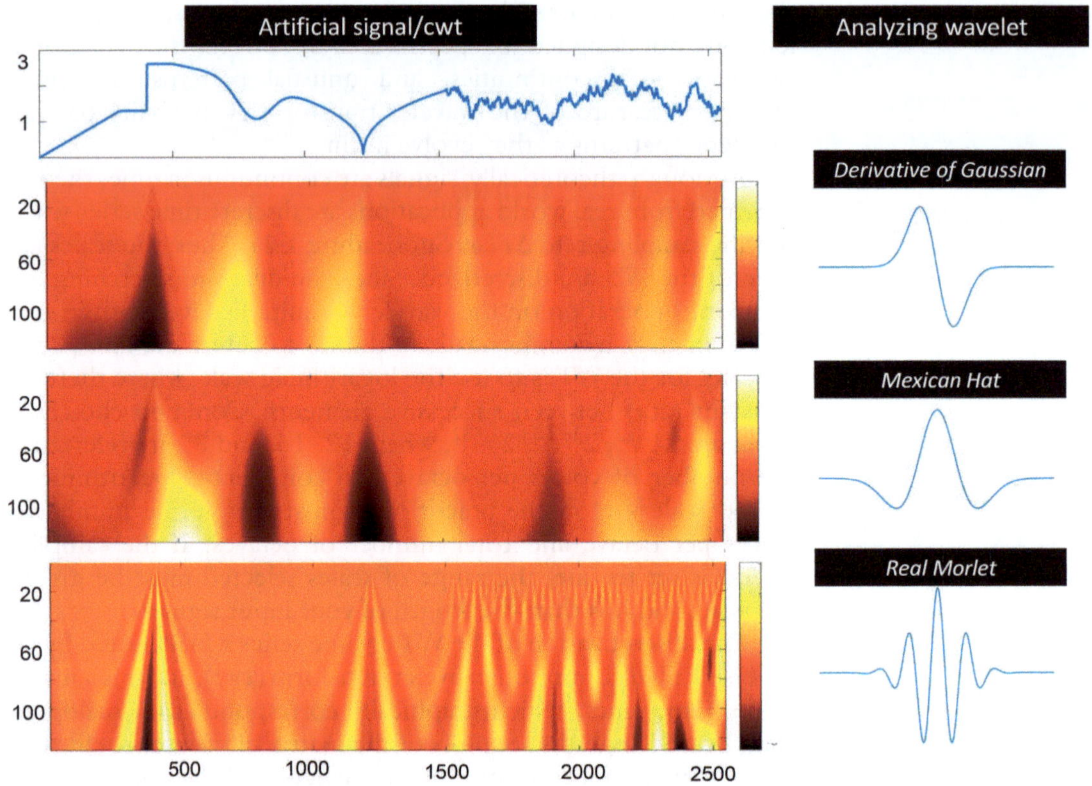

Fig. 9 Synthetic time series of a signal analyzed with different wavelets. The top panel shows the time series analyzed, and subsequent panels depict the absolute values of the coefficients calculated with the different wavelets, from top to bottom: Gaussian wavelet, Mexican Hat, and real Morlet. To the right, insets show the shape of each of the wavelets used in the respective analysis. In the three examples of wavelet analysis, changes in regularity in the signal can be observed, and the discontinuities and variability are well localized in time, although different features are distinctly highlighted depending upon the characteristics of the mother wavelet utilized

hat, the reverse Derivative of Gaussian of order 2, and the third with the real Morlet wavelet. It is important to notice that the localization of the discontinuities is very precise with the Mexican hat, while the real Morlet wavelet detects the rapid variations of the fractal trajectory introduced in the last part of the signal. The pseudocolor in the scalogram must be interpreted carefully since it corresponds to the match or anti match of the wavelet with the shape of the signal.

Our second and third examples correspond to a very simple impulse signal, and a sine wave with constant central frequency. Examining the scalogram of the shifted impulse signal $s_B(t)$ in Fig. 10, it can be seen that the set of *cwt* coefficients is concentrated in a narrow region in the time-scale plane at small scales centered around point $B = 312$. As the scale increases, the set of large *cwt* coefficients becomes wider, but remains centered around point

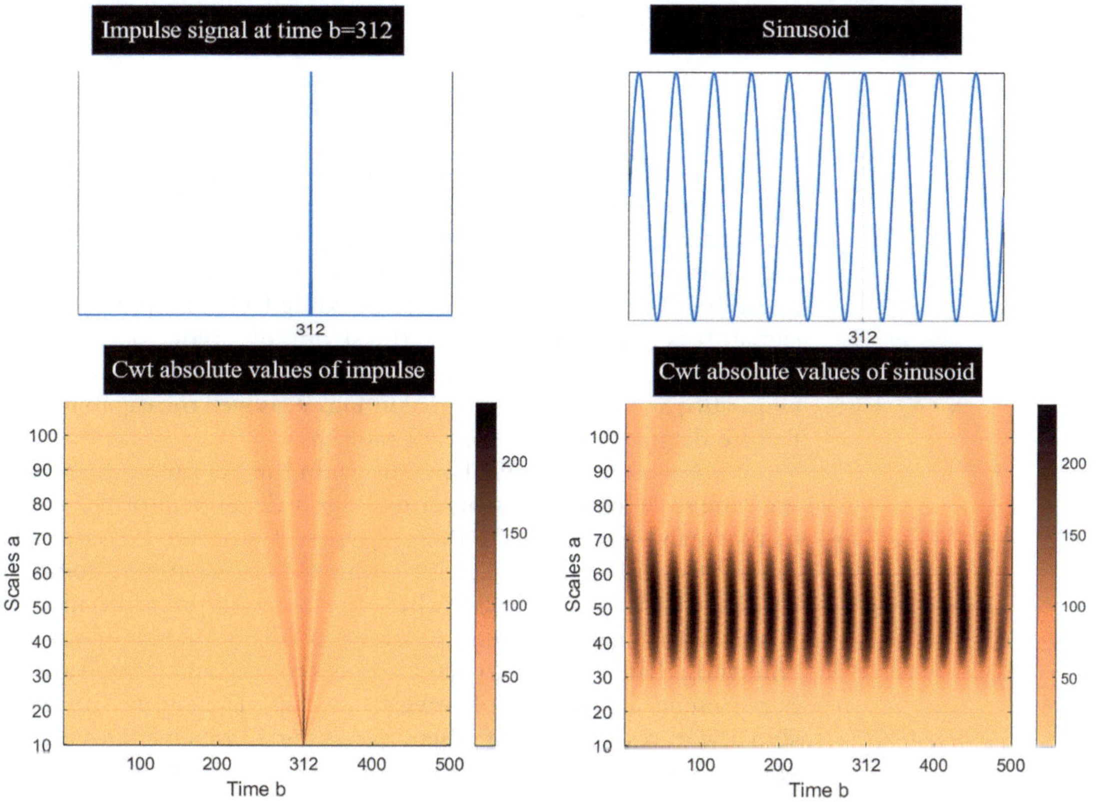

Fig. 10 Wavelet analysis of an impulse (*left panels*), and sinusoidal (*right panels*) functions. Time series are shown at the top panels, while the absolute values of the real valued coefficients are shown in each scalogram. (*left panels*) The impulse, localized at time 312 s, is visualized differently depending on the time scale. Note the resulting cone of influence. (*right panels*) The time series and the Symlet 8 analyzing wavelet is the same as the one used in Fig. 8 (compare this scalogram with Fig. 8c).

$B = 312$. Tracing the border of this region, it resembles an upside-down triangle with a corrugated texture. This region is referred to as the cone of influence of the point $B = 312$ for the Symlet 8 wavelet. The corrugated appearance is due to the lobes that the Symlet wavelet has. A piecewise constant wavelet like the Haar wavelet would produce a smoother triangle, as would a first order Gaussian wavelet.

For a given point, the cone of influence shows you which *cwt* coefficients are affected by the signal value at that point. To understand the cone of influence, assume that you have a wavelet supported on $[-T, T]$ in time. Shifting the wavelet by b and scaling by a result in a wavelet supported on $[-T_{a+b}, T_{a+b}]$. For the simple case of a shifted impulse $s_B(t)$, the *cwt* coefficients are only nonzero in an interval around B equal to the support of the wavelet at each scale. The formal expression of the *cwt* of the shifted impulse is.

$$\begin{aligned} WT(a, b; s_B(t), \psi(t)) &= \int_{-\infty}^{\infty} s_B(t) \frac{1}{\sqrt{a}} \psi'\left(\frac{t-b}{a}\right) dt \\ &= \frac{1}{\sqrt{a}} \psi'\left(\frac{B-b}{a}\right) \end{aligned}$$

For the impulse, the *cwt* coefficients are equal to the conjugated, time-reversed, and scaled wavelet as a function of the shift parameter *b*. This phenomenon is also present in the scalogram of the sinusoid wave in Fig. 10, which shows large coefficients in a broad band around the scale corresponding to the frequency of the signal. It is important to notice the strong difference between the scalograms of the shifted impulse (Fig. 10a) and the sinusoidal wave (Fig. 10b) and the artificial signal in Fig. 9, as well the capability for feature diagnosis that this simple plot has.

Mother wavelets. As we have shown in the examples, the resulting time scale joint representation is highly dependent on the shape of the analyzing wavelet. Different types of wavelets will correlate differently with the time series (compare Gaussian and complex Morlet wavelet in Table 3). There are many different admissible wavelets that can be used to create a WT. While it may seem confusing that there are so many choices for the analyzing wavelet, it represents a strength of wavelet analysis. Depending on what signal features are sought, there are many admissible wavelets to select that would facilitate the detection of that feature. For instance, for detecting abrupt discontinuities in a signal, the Mexican Hat wavelet or the first order Gaussian wavelet (Table 3, fourth column) are appropriate. On the other hand, if the task at hand is to find oscillations with smooth onsets and offsets, a different wavelet may be chosen to better match that behavior. There are several considerations in making the choice of a wavelet, for example, real vs. complex wavelets, continuous vs. discrete, orthogonal vs. redundant decompositions. Briefly, the *cwt* often yields a redundant decomposition (the information extracted from a given scale band slightly overlaps that extracted from neighboring scales) but they are more robust to noise as compared to other decomposition schemes. Discrete wavelets (for definition and uses *see* [79, 82]) have the advantage of fast implementation but, generally, the number of scales and the time invariance property (a filter is time invariant if shifting the input in time correspondingly shifts the output) strongly depends on the data length. If quantitative information about phase interactions between two time-series is required, continuous and complex wavelets provide the best choice (further details can be found in [82]). However, all the wavelets share a general feature: slow oscillations have good frequency and poor time resolution, whereas fast oscillations have good time resolution but a lower frequency resolution. A particular complex continuous wavelet, the Morlet (*see* examples in Table 3, fifth and sixth column), is defined as.

$$psi(t) = pi^{-1/4}\, exp\,(-i2w_0t)\; exp\,(-t^2/2)$$

This wavelet is the product of a complex sinusoid by a Gaussian envelope, where w_0 is the central angular frequency of the wavelet. For the Morlet wavelet, the relation between frequencies and wavelet scales is given by

$$(1/f) = \left((4pia)/\left(w_0 + \sqrt{-i2w_0^2}\right)\right)$$

when $w_0 = 2\pi$ the wavelet scale a is inversely related to the frequency. This greatly simplifies the interpretation of the wavelet analysis and one can replace, on all equations, the scale a by the period (1/frequency) or wavelength.

Applying a complex wavelet to a data series generates a *cwt* of complex value for each combination of s scales (corresponding to frequencies), and translations (corresponding to time). The resulting matrix, often the real part, the imaginary part, the modulus (or magnitude) and the phase angle (recall Fig. 4c) are plotted separately as a function of scale and translation. The magnitude of the complex value at any scale and translation indicates the approximate strength of the frequency corresponding to the selected scale at a time corresponding to the translation [82].

In Fig. 11 the use of a complex *cwt* for analysis is introduced, showing the generation of the real and the imaginary component of the wavelet coefficient. For the scale selected, the real part has a strong correlation with the signal, while the imaginary parte (which is offset at a lag of ¼ the period) has coefficients of almost zero (Fig. 11c). In the scalograms of the real and imaginary parts of the cwt this is observed as a mismatch in colors. The scalogram of the magnitude of the transform (modulus values, Fig. 11d) is the best plot to show the crest of maximum values of the surface, called ridge. The ridge localizes the pseudo frequency of the sinusoid, as the center of the horizontal region marked in brown. Additionally, complex-valued wavelets provide phase information, and are therefore very important in the time-frequency analysis of nonstationary signals.

Advantages of wavelet analysis for rhythm detection. When faced with analysis of nonstationary data, the Morlet wavelet, which is closely related to the familiar tools of Fourier analysis [81, 106], is considered one of the best choices, since it allows for simultaneous estimation of phase, frequency, and amplitude of a particular data set, while simultaneously detrending it, all without the strong parametric assumptions that cause difficulty for traditional methods [107]. Even more importantly, it allows the tracking of the changes in period (Table 3, first row), shape of the curve (Table 3, second row) or loss of periodicity (Table 3, third row) that can happen as a function of time.

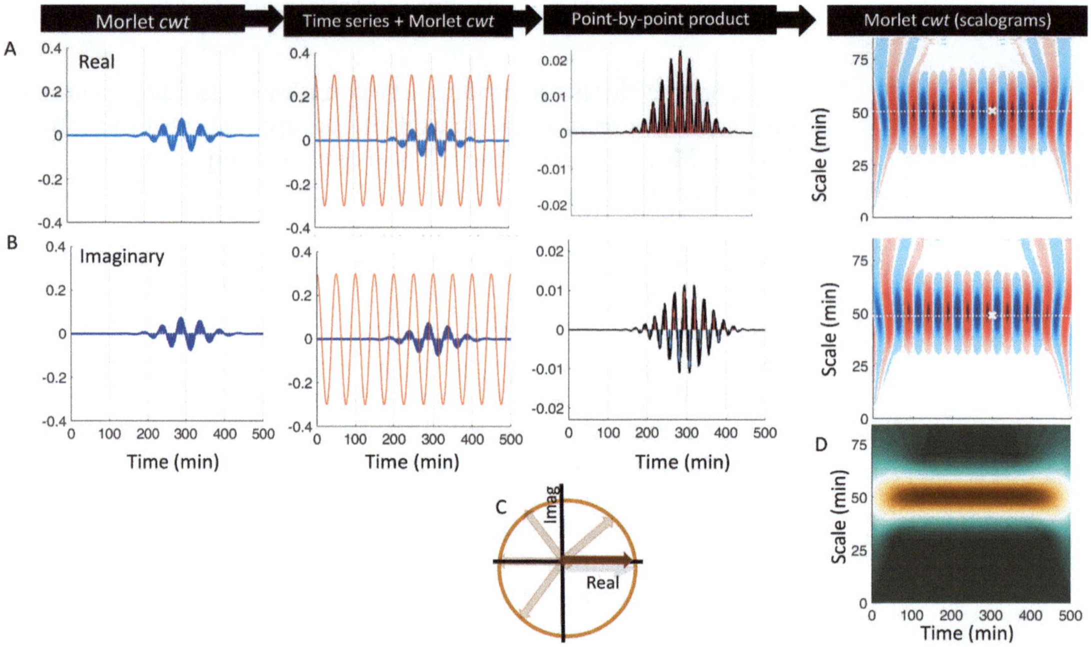

Fig. 11 Representation of the algorithm corresponding to the Morlet continuous wavelet transform (*cwt*). The real (**a**) and imaginary (**b**) parts of the *cwt* using the complex Morlet wavelet at scale 50 at time point 312 min (first column). In the column each part (in blue) is superimposed over the sinusoid time series (in orange). In the third column, the result of point-by-point multiplication of each part of the complex Morlet wavelet (scale 50 and time 312 min) by the time series is depicted. The sum of this multiplication is added and is plotted as a point in the respective scalogram (the specific example is marked with a light gray **x** in the scalogram). In color we see the positive and negative areas of the product of the signal when using the complex Morlet wavelet with the imaginary and real part of the dilated wavelet at scale 50. Also displayed in the last column, is a contour plot of the scalogram of the real part of the transform. (**c**) Schematic representation of the real and imaginary coefficient estimates in **a** and **b**, respectively, shown as a brown arrow (compare with Fig. 4). Since the time series analyzed is sinusoidal the modulus does not change over time (light brown arrows) when the appropriate scaling wavelet is used (scale 50 in this example). (**d**) The modulus scalogram is shown, with near 0 values denoted in green and maximum values in brown. Note that maximum values appear at scale 50 min, which corresponds to the period of the oscillation

By selecting a series of points across time, at which the magnitude of the *cwt* reaches local maxima (the "*cwt* ridges," brown values in Fig. 11d), estimated values of the frequency evolution of the periodic components of the signal can be obtained [108]. The scales of the *cwt* ridge are typically converted back to pseudofrequency or period [109]. Once recovered from the *cwt* scalogram, the *cwt* ridge may be considered as a list of wavelength–time pairs, where each point represents the strongest rhythmic component of the signal at that time point. In some cases, particularly, when noise levels are high and multiple rhythms may be contributing to the signal, simply selecting the global or local maxima at each time point may not be the optimal method for extracting the *cwt* ridge (*see* [110]) for an overview of ridge extraction techniques and their

Table 3

Examples of composite dynamics that can be found in recorded time series from biological systems and their characterization using three different wavelet analyses, namely, the first order derivative of the Gaussian (real scalogram), the complex Morlet (real and modulus scalograms) and synchrosqueezing. For comparison, probability density function (PDF) and power spectrum analysis are also shown

	Time series and associated PDF	Power spectrum analysis	Gaussian cwt (real scalogram) *	Complex Morlet cwt (real scalogram)$^{+}$	Complex Morlet cwt (modulus scalog)	Synchrosqueezing (modulus & ridge)
Change in number of rhythms	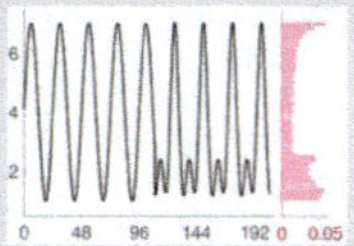	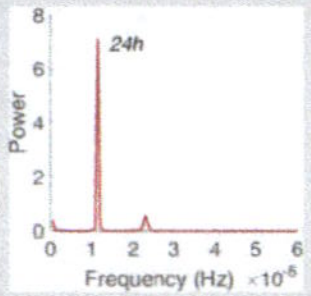	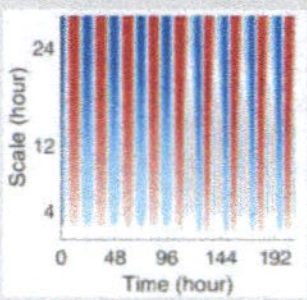	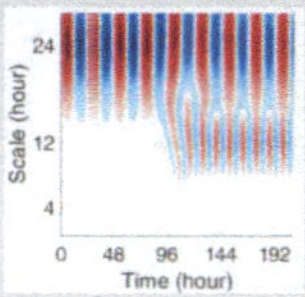	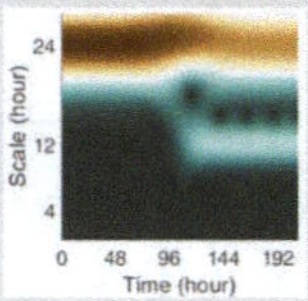	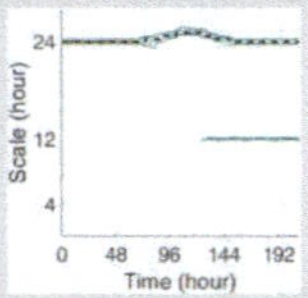
Change in waveform	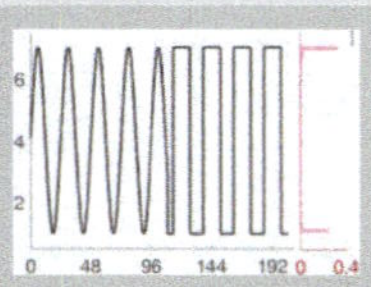	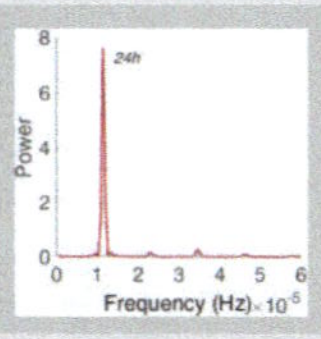	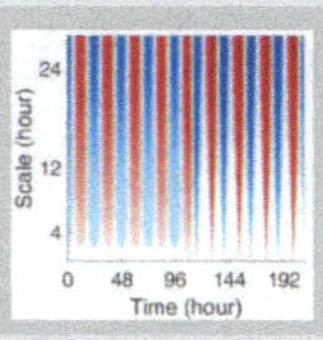	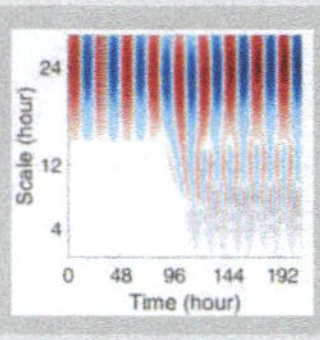	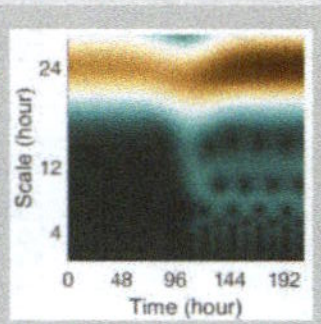	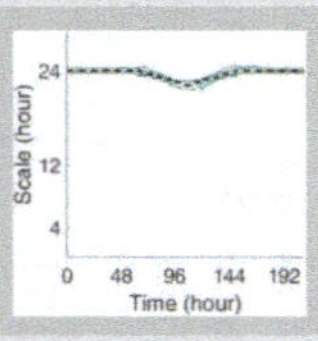
Loss of periodicity	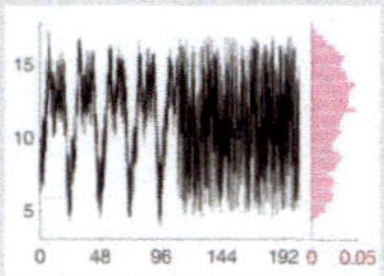	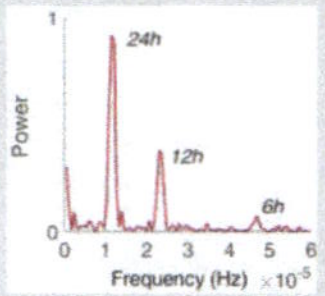	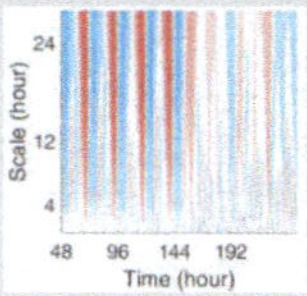	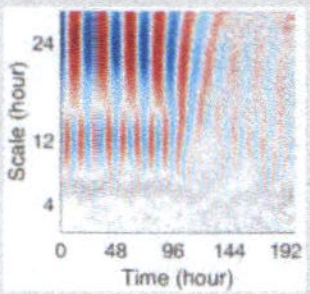	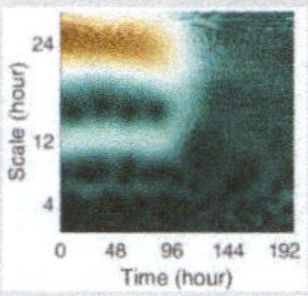	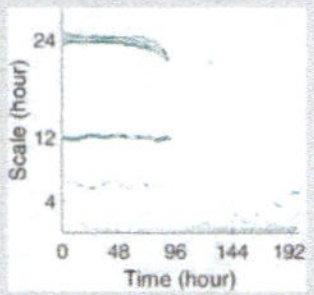

Change in number of rhythms: from a single 24 h rhythms to a series with 2 rhythms 24 h and 12 h (idem Sum 2 Sinusoids, Table 1). Change in waveform: from sinusoidal to a square waveform (idem Square waveform, Table 2). Loss of periodicity: from a sum of 3 sinusoids with noise to only uniform noise (Table 1). *Straight vertical lines in the scale-time plot of the Gaussian cwt indicates discontinuities in the time series. Given the shape of the Gaussian wavelet an upward step corresponds to a negative coefficient (blue), while a downward like step corresponds to a positive coefficient (red). Near zero values are shown in white. In the real part of the Morlet, when the time series is sinusoidal, the positive coefficients (red) coincides with peaks, and negative coefficients (blue) coincide with valleys in sine wave at the corresponding scale. Modulus of the scalogram highlights the period of the rhythms, as maximum values (brown) at the corresponding scale and time. Green values are zero or low values of coefficient. $^{+}$Maximum values in synchrosqueezing are shown in a scale from greens to browns, indicating the period of the rhythm detected. Dotted black lines indicates the ridge

associated issues, *see* [111] for the "crazy climber" algorithm used in [107]). Tracking the rhythm's period by selecting the translation-by-translation maximum algorithm [71, 113] from the *cwt* table provides a robust, rapid, and deterministic method for generating the ridge plot and examining the frequency evolution of the rhythm over time, which enables assessing variations in the dominant period of the signal, and to consider the source of this variability [113].

Considerations: Choosing the correct mother wavelet is essential, not only because it favors detection of desirable aspects, but also because it avoids situations where undesirable, confusing, "leakage" could occur. As in the case of Fourier Analysis, the use of complex wavelets such as Morse and Morlet is based upon the assumption that the time series is a sum of sinusoids of different frequencies. In this context, these methods are very efficient for finding the corresponding sinusoids and estimating their frequency (Table 3). However, if the periodicity is not smooth, but rather spike-like (i.e., a spike train), the resulting wavelet transforms will show leakage into other frequencies. This will result in a scalogram with maximum modulus lines at all the harmonics of the fundamental frequencies, theoretically equivalent to what is observed in the Fourier periodogram (Table 3). Since the choice of the mother wavelet is determined by the researcher, this problem is easily overcome. Specific wavelets have been designed for these specific applications, such as electroencephalographic recordings [83]. Also, these spikes can be considered as singularities, and thus analyzed using orthogonal wavelets such as the Mexican hat or the first-order Gaussian wavelet (Table 3, fourth column). If in doubt, a good starting point is to analyze the data with, for example, the first-order Gaussian wavelet, if variability is observed at both large and small scales (Table 3, fourth column) and if periodic behavior is visually plausible, then apply a complex Morlet or Morse Wavelet (Table 3, fifth and sixth column). An example of a decision tree for data analysis is presented in Subheading 2.2 and Fig. 12.

It has been acknowledged that wavelet analysis is able to decompose physiological time series in components of different frequencies and quantify irregular patterns [114, 115]. The authors suggested the use of the Morlet wavelet scalogram for visual inspection of the time-scale space but also argued that mode extraction depends on the choice of the mother wavelet [114]. Since this choice is arbitrary, they advocate for the use of data-adaptive time-series decomposition techniques, such as singular spectrum analysis (SSA), or Empirical Mode Decomposition (EMD), where the modes are generated by the data itself and are user-independent [114]. Leise [78] also expressed concerns about poor localization of ridges in the time scale plot when the signals are highly nonstationary and noisy, given the frequency smearing that all wavelet

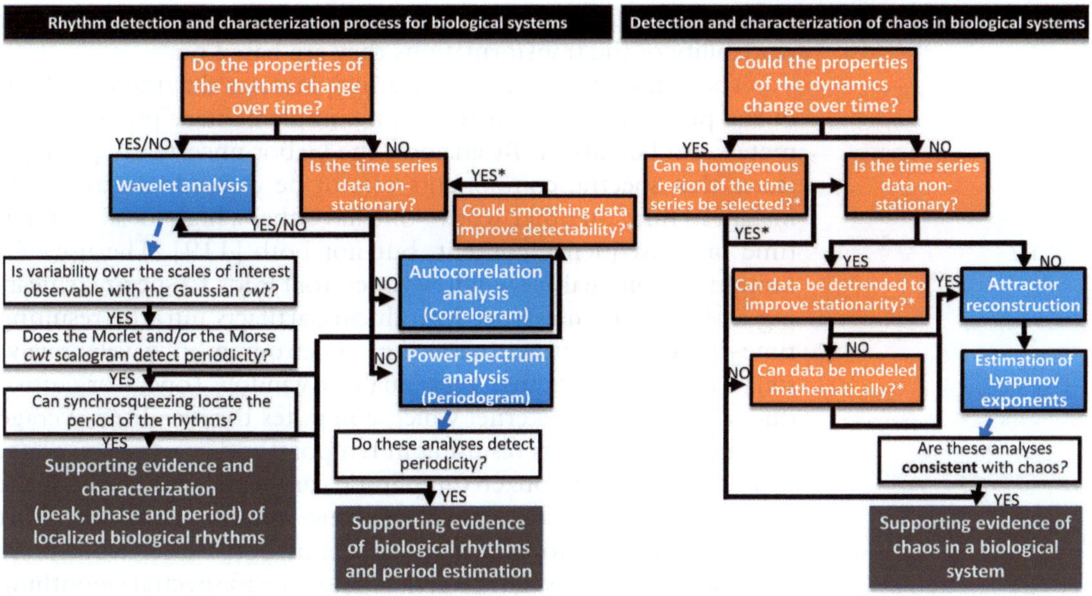

Fig. 12 Flow diagram of a step-by-step decision process associated with detection of rhythms (left panel) and chaos (right panel) in time series from biological systems. Questions associated with selection of the appropriate method of analysis are shown in orange boxes; the method or family of methods are shown in blue. Questions associated with analytical results are shown in white boxes, as indicated by a blue arrow. Final positive results are shown in grey boxes. To simplify the representation, the contrasting negative result (i.e., lack of evidence) is not shown. Arrows indicate the direction of the decision flow through the scheme, according to whether the answer to each question is yes or no. * Steps where the original data is either processed or modeled, and then the resulting time series is fed back into the process for analysis

representations suffer, and also suggests its use only for visual inspection. This problem led recently to the definition and algorithmic design of synchrosqueezed techniques (Subheading 2.1.11) that reduce the frequency leakage and smearing, allowing a sharper mode decomposition. The output of such algorithms is widely used in geologic signal processing and fault detection (*see* [84, 116, 117] and references therein).

2.1.11 Synchrosqueezing

The last few years have witnessed an upsurge of interest in the signal processing community over multicomponent signals. These signals are defined as the super-imposition of amplitude and frequency modes that possess the ability to accurately represent nonstationary signals, which, in practice, are commonly encountered in nonlinear systems as, for instance, analysis of ultradian rhythms [30, 46]. To analyze such signals, analysis operators such as the *cwt* (Subheading 2.1.10) or the short term Fourier transform have attracted overwhelming attention. The effectiveness of these transforms is, however, constrained by the choice of an analysis window which can never be ideal due to the Heisenberg uncertainty principle. To circumvent this issue, reassignment methods were introduced in

[118] and further developed in [116], undeniably improving the readability of the transformations they are based on.

The Heisenberg uncertainty principle states that there is a limit in the precision with which certain complementary physical parameters can be known. By analogy, the Gabor uncertainty principle states that spectral components cannot be defined exactly at any instant in time. In other words, one has either a high localization in time or in frequency content, but not both [119]. The use of a finite-duration analysis window (operator) leads to spectral smearing and leakage, in essence introducing artifacts into the resulting time–frequency representation (i.e., spectrogram in wavelet analysis). This occurs because each analysis window (operator) introduces a convolution kernel which computes the weighted average of neighboring points resulting in temporal and spectral smearing. This implies that a nonzero amplitude can be retrieved, even if the true signal has no component at this time–frequency pair. The short term Fourier transform and the *cwt* thus both suffer from finite localization as well as reduced readability due to spectral smoothing and leakage [120].

To enhance resolution, the synchrosqueezing transform (SST) applies three steps, namely, (1) computation of the *cwt* to ensure varying time-frequency resolution, (2) calculation of the instantaneous frequencies to enhance readability, and (3) frequency reassignment to counter the effect of spectral smearing [118, 121]. Computation of instantaneous frequencies primarily enhances readability but does not affect time-frequency localization as this is imposed by the Heisenberg uncertainty principle. The reassignment method computes the sphere of influence of each analysis window (operator) and reallocates the energy in the time-frequency plane to its center of gravity in the time and frequency domains, thereby improving the readability of the time-frequency picture. Examples of ridge location with synchrosqueezing are shown in Table 3 (last column).

2.1.12 Correlations Between Time Series and Wavelet Coherence

Synchronization is a fundamental phenomenon, described in many biological and physical systems, that arises when there are two or more interacting oscillatory systems. The interactions between coupled oscillators in real systems continuously create and destroy synchronized states, which can be observed as noisy and transient coherent patterns. The statistical detection of temporal and spatial synchrony in networks of coupled dynamical systems is therefore of great interest in disciplines such as geophysics, physiology, and ecology (for examples *see* [1– 3]). *Coherence* is generally defined as the correlation between concurrent time series of a variable measured from several processes (*see* examples of mechanisms that give rise to coherence in biological systems in Lloyd et al. [1]), whereas *synchrony* is referred to as the degree to which their fluctuations behave similarly over time. Usually, the terms synchrony and coherence become interchangeably when used to describe the degree to which different processes evolve in similar ways. Statistical significance of transient coherent patterns cannot be assessed by classical spectral measures and tests, which require signals to be stationary. Synchrony estimators based on nonparametric methods have the advantage of not requiring any assumption on the time-scale structure of the observed signals. Among them, measures of synchrony or coherence based on wavelet transforms have been widely used to detect interactions between oscillatory components in different real systems, i. e. neural oscillations, business cycles, climate variations or epidemics dynamics [122].

In many applications, it is desirable to quantify statistical relationships between two nonstationary signals. In Fourier analysis, the coherency is used to determine the association between two signals, $x(t)$ and $y(t)$. The coherence function is a direct measure of the correlation between the spectra of two time-series [123]. To quantify the relationships between two nonstationary signals, the following quantities can be computed: the wavelet cross-spectrum and the wavelet coherence. The wavelet cross-spectrum is given by.

$$W_{xy}(a,\tau) = W_x(a,\tau)\,W_y(a,\tau)'$$

In this equation the symbol $'$ denotes the complex conjugate. As in the Fourier spectral approach, the wavelet coherence is defined as the smoothed version of the cross-spectrum normalized by the smoothed version of the spectrum of each signal. The smoothing can be obtained by a convolution with a constant-length window, both in time and scale axes. The wavelet coherence is equal to 1 when there is a perfect linear relation at a particular time and scale between the two signals, and equal to 0 if $x(t)$ and $y(t)$ are independent. The advantage of these "wavelet-based" quantities is that they may vary in time and can detect transient associations between analyzed time-series [115, 123, 124]. Since $W_x(a,\tau)$ is a complex number, it can be written in terms of its phase $\phi_x(a,\tau)$ and modulus $|W_x(a,\tau)|$ [125]). The local phase of the Morlet wavelet transform is proportional to the ratio between the imaginary part and the real part of the wavelet transform. The phase of a given time-series $x(t)$ can be viewed as the position in the pseudocycle of the series and it is parameterized in radian ranging from $-\pi$ to π. The phases can then be useful to characterize possible phase relationships between $x(t)$ and $y(t)$ by computing the phase difference. A detailed example is presented in Fig. 17 (Subheading 2.2.1).

A unimodal distribution of the phase difference (for the chosen range of scales or periods) indicates that there is a preferred value of it, and thus a statistical tendency for the two time-series to be phase locked. Conversely, the lack of association between the phase of $x(t)$ and $y(t)$ is characterized by a broad and uniform distribution. To quantify the spread of the phase difference distribution, one can use circular statistics or quantities derived from the Shannon entropy [126].

2.2 Two Cases Studies for Investigating Biological Time Series

When analyzing real biological data, it is often difficult to select the appropriate method of analysis. As stated previously, a combined approach is recommended that exploits the virtues while accounting for the technical limitations in each analytical method. For rhythm detection, we propose a decision tree-like strategy to guide the process (Fig. 12a). As a starting point ask, do the properties of the rhythms change over time? In other words, could certain rhythms be lost or gained at different moments of the time series? If so, the only family of analysis presented herein that could be used is wavelet analysis. If the signal does not present such shifts in dynamic, the following question to be asked is whether data is nonstationary (Fig. 12a, second orange box). If data is nonstationary, a series of wavelet analyses can be performed using a different mother wavelet to accurately detect and characterize period, phase, and peaks of the rhythms. However, if stationary then simpler methods such as power spectrum analysis and autocorrelation analysis can be used for rhythms detection and period estimation (*see* [77–79]). If none of these methods clearly detect rhythms, it is important to rethink the time series used in the analysis. Maybe it was too noisy or not smooth enough, thus smoothing techniques such as a moving average should be implemented to improve detectability of rhythms. The new smoothed time series should be fed back into the analysis process, and the step-by-step process repeated. In the following Subheading 2.2.1, this methodology is applied to mice behavioral data.

Regarding the evidence of chaos in a biological system directly from raw data, the step-by-step flow diagram in Fig. 12b reflects the challenges associated with the strict methodological constraints of the methodology utilized for the analysis. As in the case of rhythm detection, first, it is important to consider changes in dynamics over time. If changes in dynamics are observable, potentially chaotic regions should be selected avoiding regions of transitions between chaotic and nonchaotic states. If this is not possible, mathematical modelling of the system should be considered as an alternative method to obtain a time series representative of the biological system that could be used to test the hypothesis of chaos. Second, are trends present in data? If so, data should be detrended before analysis. As before, if not possible, consider mathematical modelling. In time series that meet the criteria specified in Subheadings 2.1.8 and 2.1.9 the resulting attractors and Lyapunov exponents can be studied, providing supporting evidence of the presence of chaos in that biological system. An example is provided in Subheading 2.2.2.

2.2.1 Wheel Running and Food Intake Behavioral Rhythms in Mice Subjected to Caloric Restriction

In this example, the wheel running and food intake behavior of C57BL/6J mice were evaluated. These time series have been previously described in Acosta-Rodriguez et al., and time series are publicly available [127]. Time series were obtained automatically from a system that not only recorded feeding and voluntary wheel-running activity in mice over a 42 period, but also could control duration, amount, and timing of food availability [127]. The experimental design consisted in allocating the mice individually, in boxes with an unlimited access to a feeder that dispensed pellets and a running wheel. For the first 7-days (starting at day 0) mice were able to feed ad libitum, however after that period, they were subjected to a caloric restriction (CR) protocol that continued for the following 35 days (for details Acosta-Rodriguez et al. [127]). This protocol consisted in 24 h food access but calorie restricted (11 pellets corresponding to 70% of baseline ad libitum levels) fed at the start of the light phase (CR-day). As explained in Subheading 1.2.1, the circadian rhythm in locomotor behavior, such as wheel running, is controlled by the suprachiasmatic nucleus (SCN). Since CR can affect the SCN in the hypothalamus, potentially, it can modulate circadian locomotor activity (for discussion *see* [127]). The respective actogram is shown in Fig. 13a, with feeding data in orange and wheel running in dark grey, as presented in the original publication (Supplementary material in [127]). In the actogram, as well as in the data time series, it is clear that the properties of the time series change over time, especially for food intake. Specifically, implementation of the CR-day protocol leads to a transition from nighttime to daytime feeding in a very localized 2 h time period. Thus, to visualize this transition, the most appropriate family of methods for analysis is wavelet. For the sake of comparison power spectrum (Fig. 13b, c) and autocorrelation (Fig. 13d, e) analyses for the first 5 days and last 5 days are shown for both time series. Given the sparse, nonstationary, nature of the food intake time series, they were preprocessed with moving average of a 1 h window. Note that for the case of the last 5 days of the study, the very localized food intake ("spike-like," see time series in Fig. 14a) rendered a power spectrum with peaks at the harmonics of the fundamental 24 h circadian rhythm (Fig. 13b). While for the case of the much smoother wheel running time series, clearly two peaks appear in the power spectrum, at 24 and 12 h, these being more pronounced the last 5 days in comparison to the first days (Fig. 13c). In the autocorrelation analysis (Fig. 13d, e) only the 24 h circadian rhythm is evident.

As proposed in Fig. 12 (white boxes), a series of wavelet analyses were performed on the food intake (Fig. 14) and wheel running time series (Fig. 15). First, a Gaussian wavelet transform was performed, followed by a Morlet wavelet, and a Morse wavelet. Once visual evidence of variability and potentially periodic dynamics was obtained, peak and phase were estimated using the real part of the Morlet *cwt*.

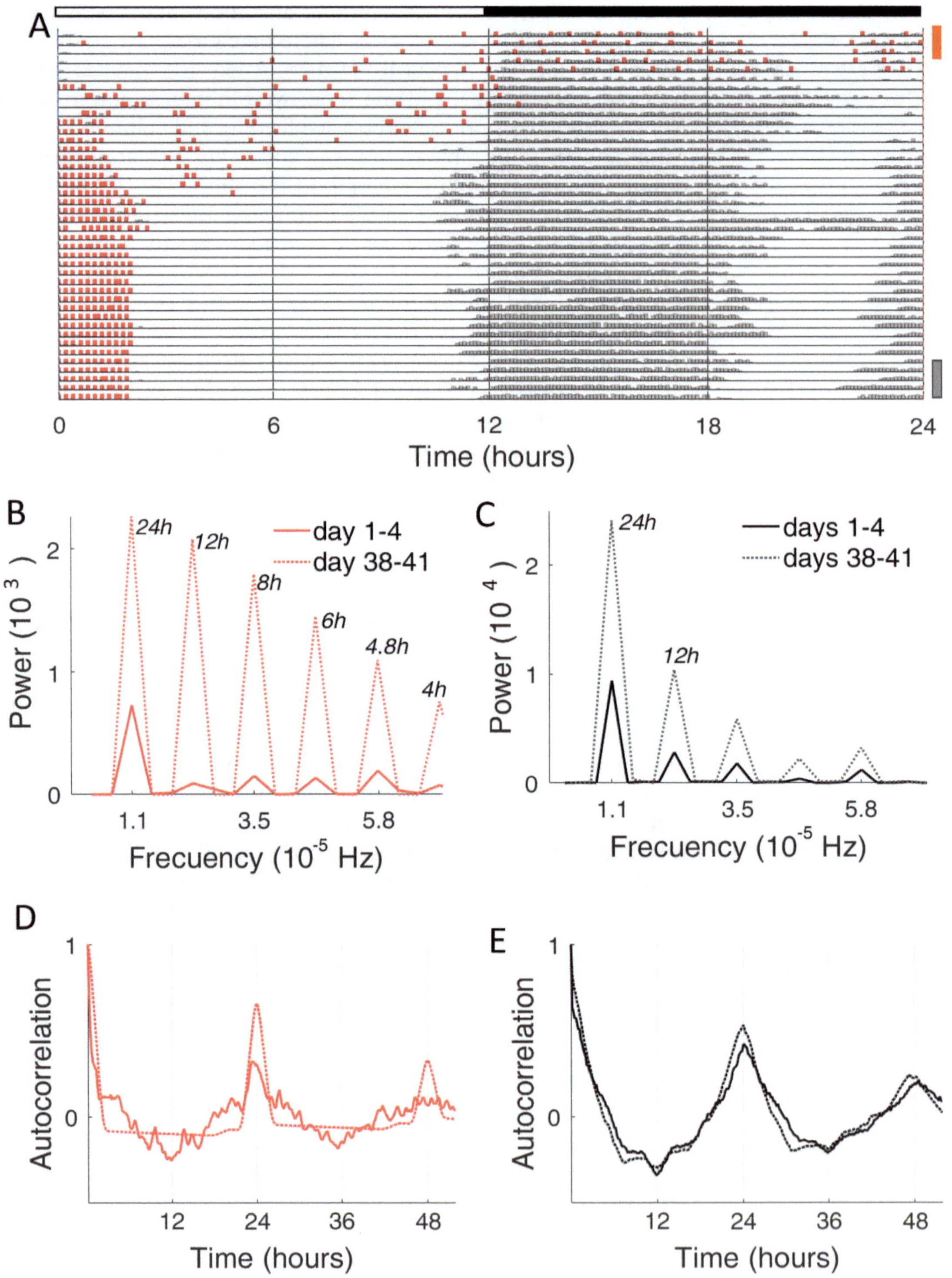

Fig. 13 Traditional visualization from analytical methods, (**a**) actograms, (**b**) power spectrum, and (**c**) autocorrelation, as applied to studying the impact of change in feeding paradigm on scale-dependent dynamics of food intake (orange), and wheel running (gray)

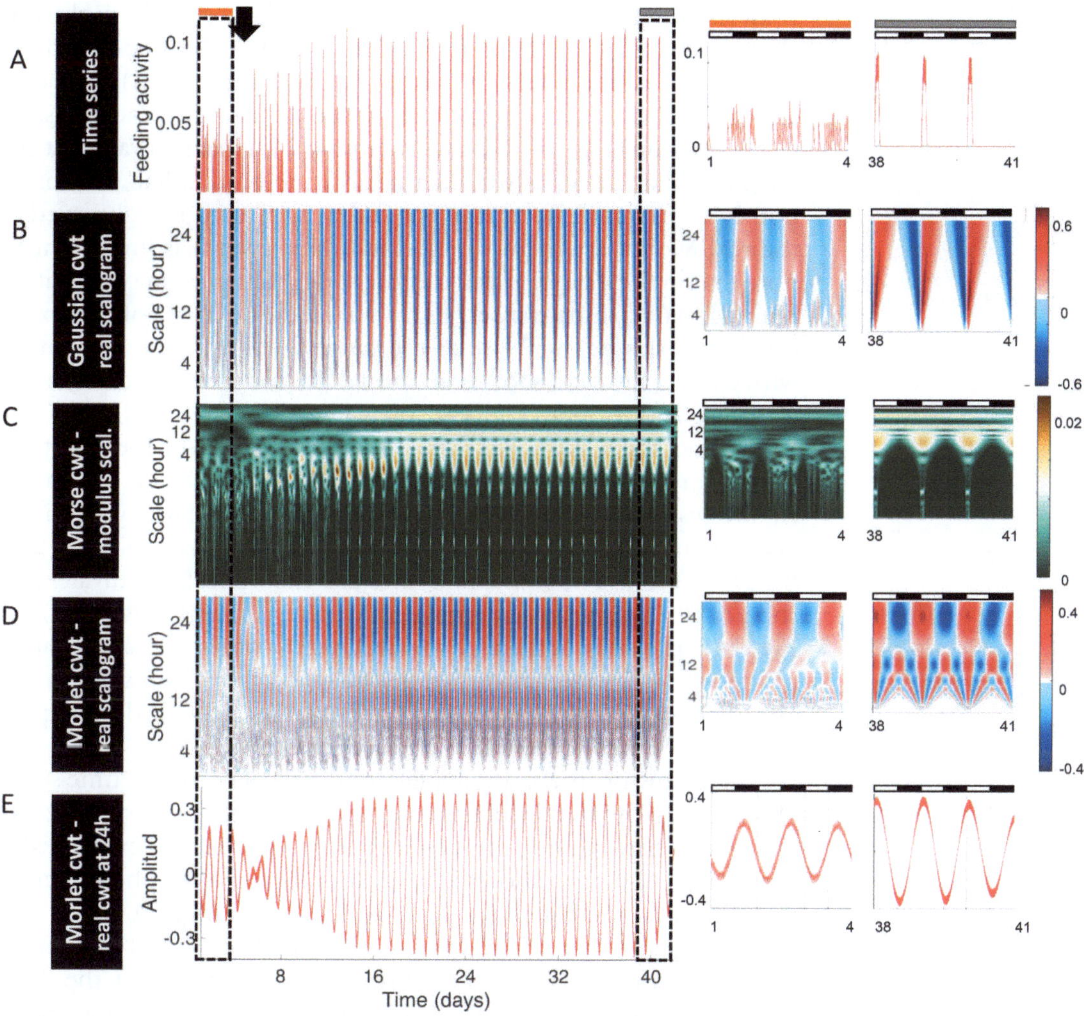

Fig. 14 Wavelet analysis as a tool to assess the impact of change in feeding paradigm on scale-dependent dynamics of food intake. (**a**) Food intake time series for C57BL/6J mice with caloric restriction (CR) during daytime. Corresponding scalograms of the continuous wavelet transform (cwt) of the time series shown in "**a**," using first order Gaussian wavelet (gaus1) (**b**), complex Morse wavelet (modulus is shown) (**c**), and the complex Morlet wavelet (real part is shown) (**d**). Arrow indicates treatment change from ad libitum to CR-day feeding paradigm. Dotted lines with the marked region (orange or gray) shows region amplified that is shown in the respective insets. On top of the insets, the white/black bars indicate the light/dark periods, respectively, of the circadian cycle. Note the loss of complexity for time scales less than 24 h observable especially after day 16 of testing. (**e**) The real coefficients of the complex Morlet cwt for the 24 h scale is plotted as a function of time. Note the shift in the phase of the food intake displayed in insets, from predominantly nighttime to daytime feeding

The most astonishing result presented is the ability of the methodology applied to show, qualitatively and quantitatively, the change in the dynamics of food intake and wheel running elicited by the change in the feeding paradigm. This is particularly evident in the food intake time series (Fig. 14a), where the three wavelet analyses detect the shift from high variability, mostly nighttime (Fig. 14b, feeding toward the localized time—daytime—feeding, specifically during the first hours of the light period). Prior to change in the feeding regime (i.e., when ad libitum), the Gaussian cwt scalogram (Fig. 14b) shows rhythmic 24 h maxima values at the 24 h circadian scale, as well as a bifurcation branch-like phenomenon in maximum values at lower scales, similar to the observed in the real Morlet scalogram (Fig. 14d). However, it is important to note that in the Gaussian wavelet the straight vertical maxima values after day 16 (compare inset, Fig. 14b), indicate a loss of variability for time scales below 24 h, in the range 1–24 h. Thus, evaluation with complex wavelets for estimation of period, and frequency, such as Morse and Morlet (Fig. 14c, d), will only be appropriate for the 24 h time scale. Note that the modulus (Fig. 14c) for this time period (16–42 days), contrary to the first days of the study (0–6 days) shows not only the expected horizontal maxima lines at 24 h circadian period, but also maxima at the harmonics (12 h, 8 h, 6 h, etc.). These maxima lines are theoretically equivalent to the peaks observed in the power spectrum analysis (Fig. 13b). In Fig. 14e, the real coefficients of the Morlet cwt for the 24 h scale is plotted as a function of time. In this plot, the phase shift in the circadian food intake triggered by the change in the feeding regime is evident (compare insets).

The same protocol was applied to the wheel running time series (Fig. 15a). Note that in all the analyses, complexity is observed at all time scales throughout the 42-day period. The Gaussian *cwt* scalogram (Fig. 15b) also shows a bifurcation branch-like phenomenon in maxima values, like in the real Morlet scalogram (Fig. 15d). The modulus scalogram of the Morse *cwt* (Fig. 15c) shows not only a strong 24 h circadian rhythm (brown horizontal line at the 24 h scale) but also a consolidating 12 h rhythm (shifts from a green-white toward a light brown color). In Fig. 15e the real coefficients of the Morlet *cwt* for the 24 h and 12 h scales are plotted as a function of time. In this representation, the consolidation of circadian and ultradian rhythms triggered by the change in feeding regime from ad libitum to caloric restriction, which is evident by the increase of the values of the real coefficients (compare intensity of color in insets Fig. 15d, and values of *y*-axis Fig. 15e).

For increasing the precision in the estimation of period and phase, the synchrosqueezing transform was performed and ridges were detected (Fig. 16a and zoom in Fig. 16b). The ridges (black dotted lines) provide concrete evidence of the localization of two periods, ascertaining the presence of the 24 h and 12 h rhythm. Higher coefficients are also evident in the last days compared to the first day of the study. Inverting the synchrosqueezing transform, the reconstructed time series at the 24 h and 12 h scales is shown in Fig. 16c.

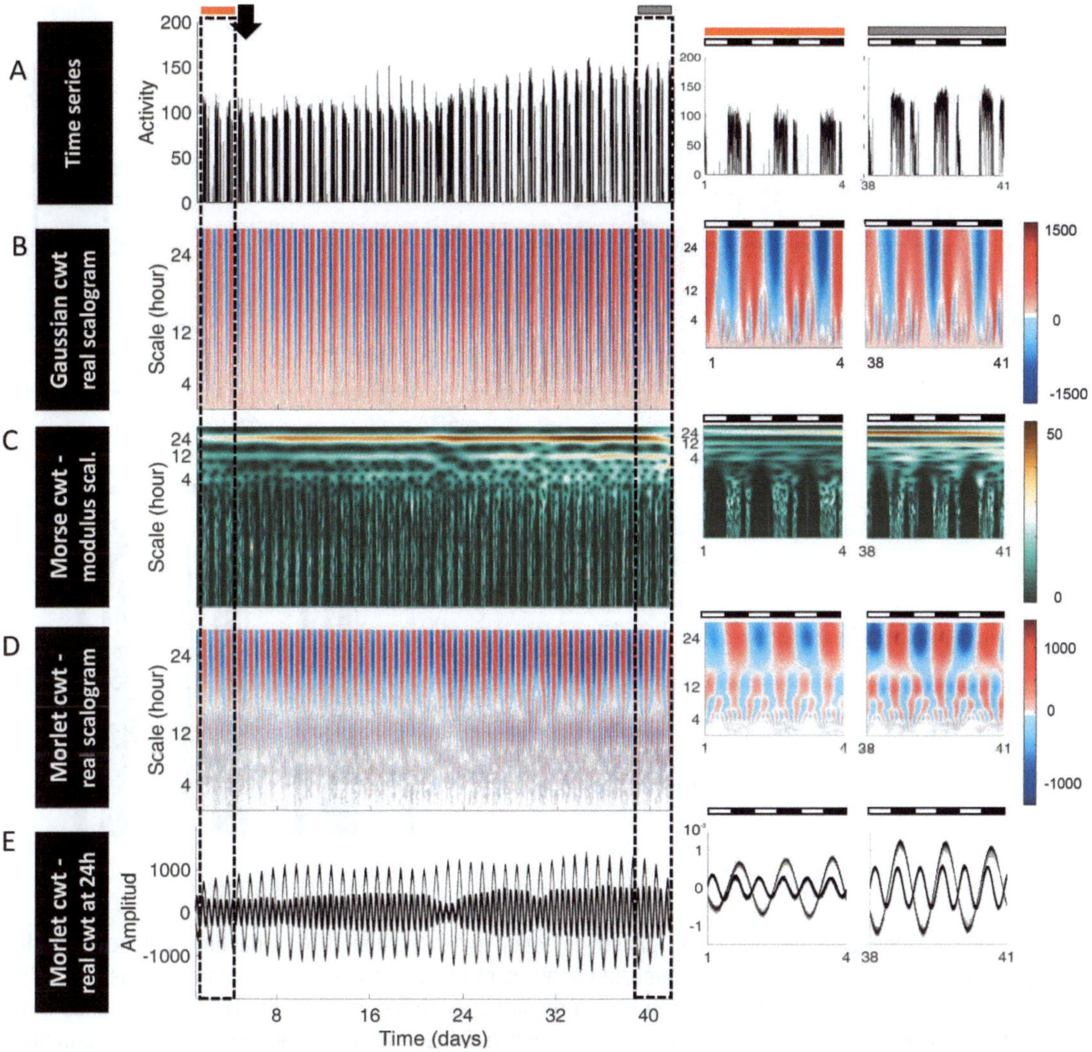

Fig. 15 Wavelet analysis as a tool for assessing the impact of change in feeding paradigm on scale-dependent dynamics in wheel running. (**a**) Wheel running time series for C57BL/6J mice with caloric restriction (CR) during daytime. Corresponding scalograms of the continuous wavelet transform (cwt) of the time series shown in "**a**" using the first order Gaussian wavelet (gaus1) (**b**), complex Morse wavelet (modulus is shown) (**c**), and the complex Morlet wavelet (real part is shown) (**d**). Arrow indicates treatment change from ad libitum to CR-day feeding. Dotted lines with the marked region (orange or gray) show region amplified that is displayed in the respective inset. White/black bars on top of insets, indicate the light/dark periods in the circadian cycle. Note how the 12 h ultradian rhythm consolidates over time, showing larger coefficients over time. (**e**) The real coefficients of the complex Morlet (cwt) for the 24 h and 12 h scales, are plotted as a function of time. Note the increase in the magnitude of the real coefficients after a transition period

Finally, synchronization between the two behaviors was studied with wavelet coherence (Fig. 17). Note that the maxima values (in brown for the 24 h circadian scale) are disrupted after the change in feeding regime from ad libitum to caloric restriction. The arrows at the different time points indicate the phase shift

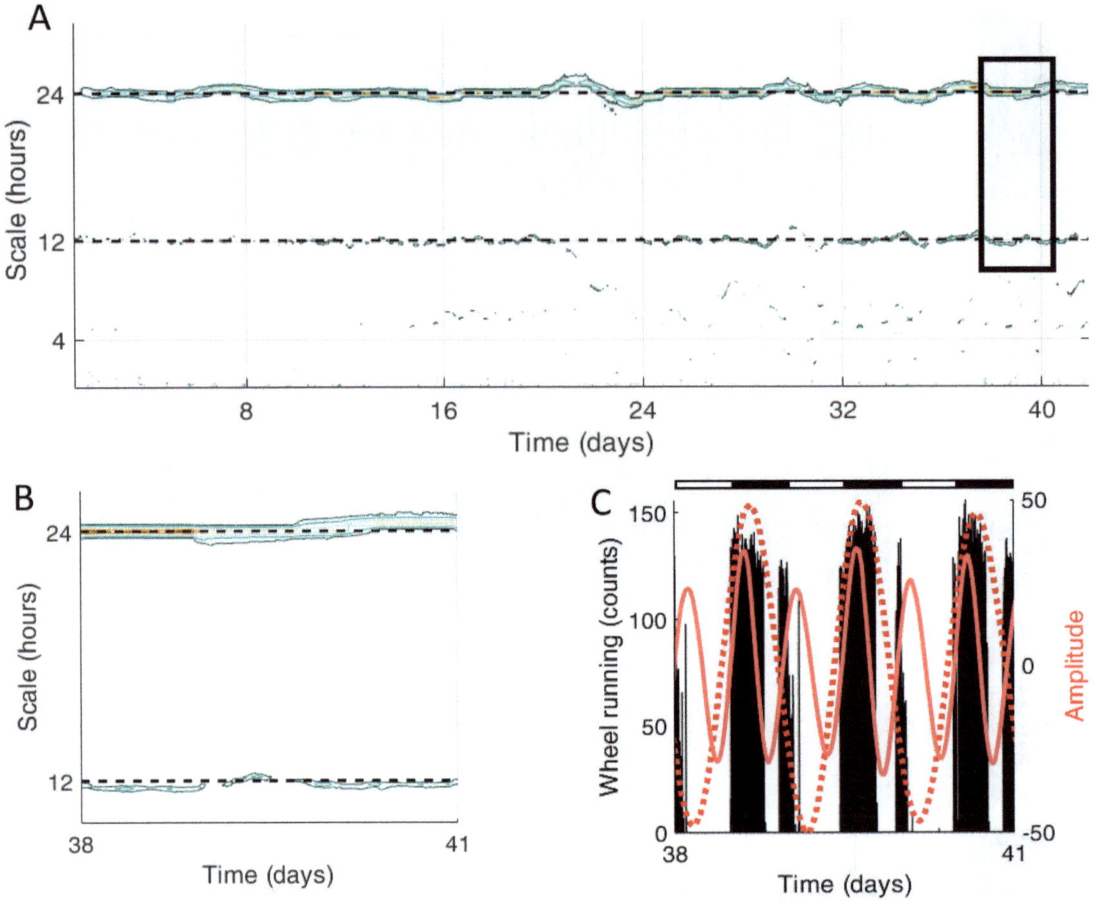

Fig. 16 Synchrosqueezing for ridge and phase detection when switched to a daytime caloric restriction (CR-day) feeding paradigm. (**a**) Synchrosqueezing analysis of the same wheel running time series for C57BL/6J mice switched to caloric restriction (CR) during daytime, as shown in Fig. 15a. Note the increase in the magnitude in the coefficients (brown values) for the 12 h scale after a transition period. (**b**) Amplification of the region shown in the black rectangle in "**a**." (**c**) Inverse transform of synchrosqueezing overset on the original time series for the last part of the study (38–41 days). The real coefficients of the complex Morlet (cwt) for the 24 h (dotted red lines) and 12 h (continuous pink lines) scales are plotted as a function of time

between the two time series. Note the change in the direction of the angle prior and after this disruption. Specifically, at the beginning of the study (mice were being fed ad libitum) where feeding and wheel running both predominated, as expected, during the nighttime period, arrows show an angle close to 0°. However, after the ninth day arrows are pointing in the opposite direction (>180° angle), indicating the phase shift of feeding toward a daytime regime.

2.2.2 Chaos in Calcium Dynamics in a Mitochondrial Model

In this example, calcium dynamics is studied in the context of pathological mitochondrial chaotic dynamics [56] using an experimentally validated computational model of mitochondrial function

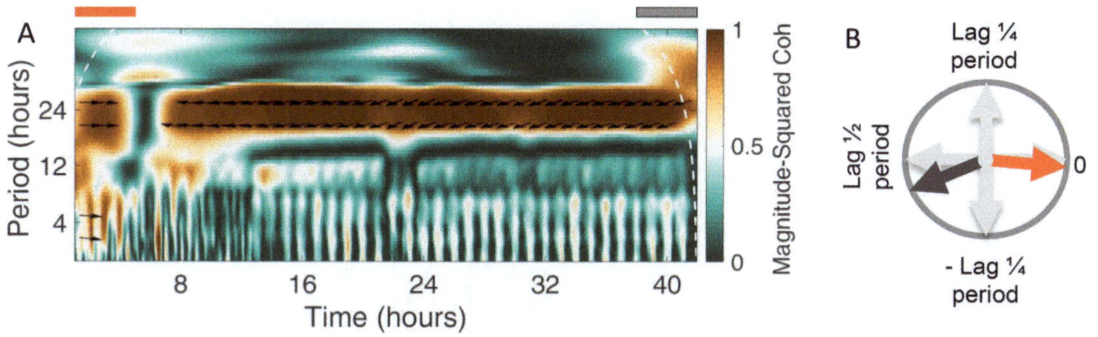

Fig. 17 Wavelet coherence for estimating phase shifts between different behavioral time series. (**a**) Wavelet coherence analysis of the same feeding and wheel running time series for C57BL/6J mice switched to caloric restriction during daytime as shown in Figs. 14a and 15a, respectively. The brown region, corresponding to maximum values, indicates the magnitude-squared coherence. (**b**) Arrows indicate the phase relationship between the two time series. Note the change in direction of arrows from close to 0° to close to 180 ° after a transition period (in green) at the 24 h scale, indicating the shift from predominantly nighttime to daytime feeding

[73] and signaling [74]. Both original publication [56] and time series are open access. In this model, complex oscillatory dynamics in key metabolic variables arise at the "edge" between fully functional and pathological behavior [74], setting the stage for chaos. Under these conditions, a mild, regular sinusoidal redox forcing perturbation triggers chaotic dynamics of the key metabolite Succinate [56].

Given the importance of Ca^{2+} dynamics in physiology (Subheading 1.2.2), herein we evaluated whether this cation also behaves chaotically in the computational model of mitochondrial function. From visual observations of the time series (Fig. 18a) irregular fluctuations in mitochondrial Ca^{2+} are evident. Thus, to establish and characterize chaotic dynamics in Ca^{2+} dynamics in our deterministic model, we performed attractor reconstruction and estimation of the Lyapunov exponent. For attractor reconstruction, the average mutual information function showed a first minimum at 43 s (red arrow, Fig. 18b). This value was used as the time lag for estimation of the embedding dimension with the false nearest neighbor algorithm (Fig. 18c). The percentage of false nearest neighbors approximated 0 only for embedding dimensions equal or above 4 (red arrow, Fig. 18c). Since only 3 dimensions can be represented graphically, the fourth dimension is represented in the color scale. Note how the attractor is not completely unfolded in the three dimensional phase space.

The maximum Lyapunov exponent for this 4-dimensional attractor was estimated using the Rosenstein et al. [68] algorithm. Figure 19 shows a typical plot (solid curve) of logarithm of divergence of trajectory as a function of time. Note that the dashed line

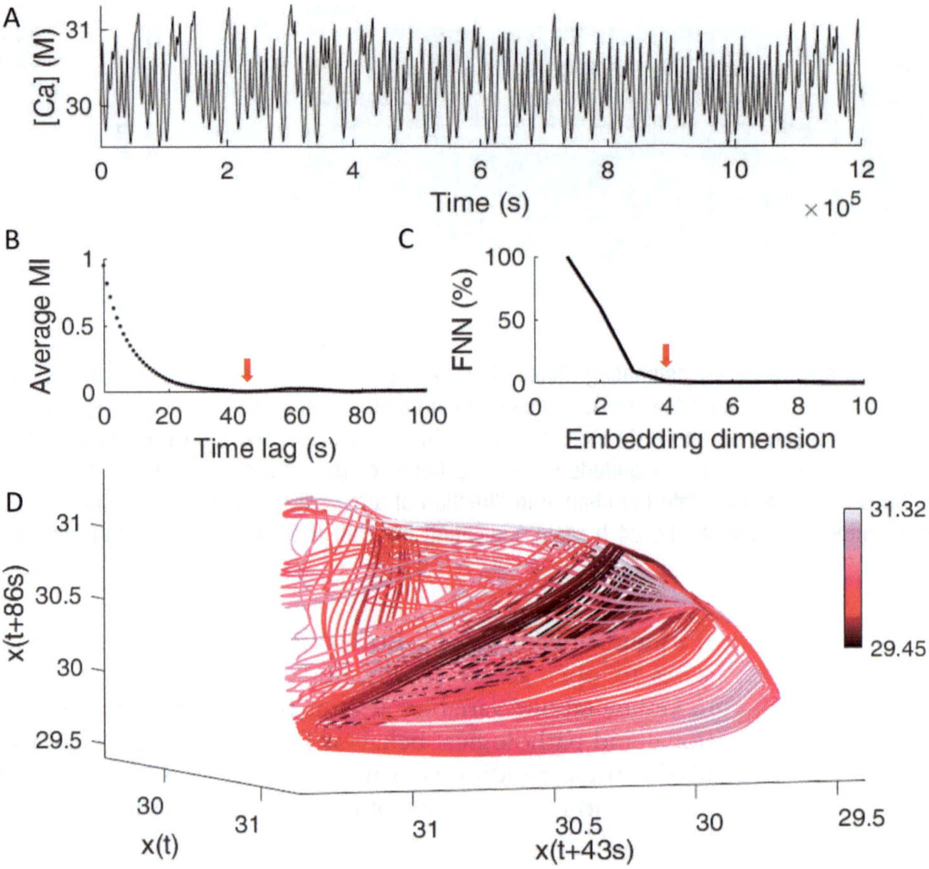

Fig. 18 Phase space reconstruction of mitochondrial calcium concentration. (**a**) Chaotic calcium time series from mitochondrial. (**b**) Average mutual information (MI) for the calcium time series shown in "**a**" as a function of time lag, τ. The first minimum value of this function is at 43 s, as indicated with a red arrow. (**c**) The percentage of global false nearest neighbors for the $x(t)$, and a $\tau = 43$ s (estimated in "**b**") as a function of the embedding dimension. As indicated with the red arrow, an embedding dimension of at least 4 is necessary to completely unfold the attractor. (**d**) Reconstructed attractor of calcium dynamics. Color coding represents a fourth time lag Ca (t+ 43 s), respectively. Model-simulated time series [73, 74] were calculated with [SOD2] = 0.016 mM, Shunt = 0.04, SOD1 = 9.7×10^{-5} mM. External superoxide perturbation: amplitude = 1×10^{-7} mM, period = 30 s [56]

has a slope equal to the theoretical value of the Lyapunov exponent. After a short transition, there is a long linear region that is used to extract the largest Lyapunov exponent. The curve saturates at longer times since the system is bounded in phase space and the average divergence cannot exceed the "length" of the attractor. The resulting positive value of 0.01 indicates sensitivity to initial conditions, a hallmark of chaotic dynamics. Thus, a small perturbation to the system has a large impact on the temporal evolution of the system.

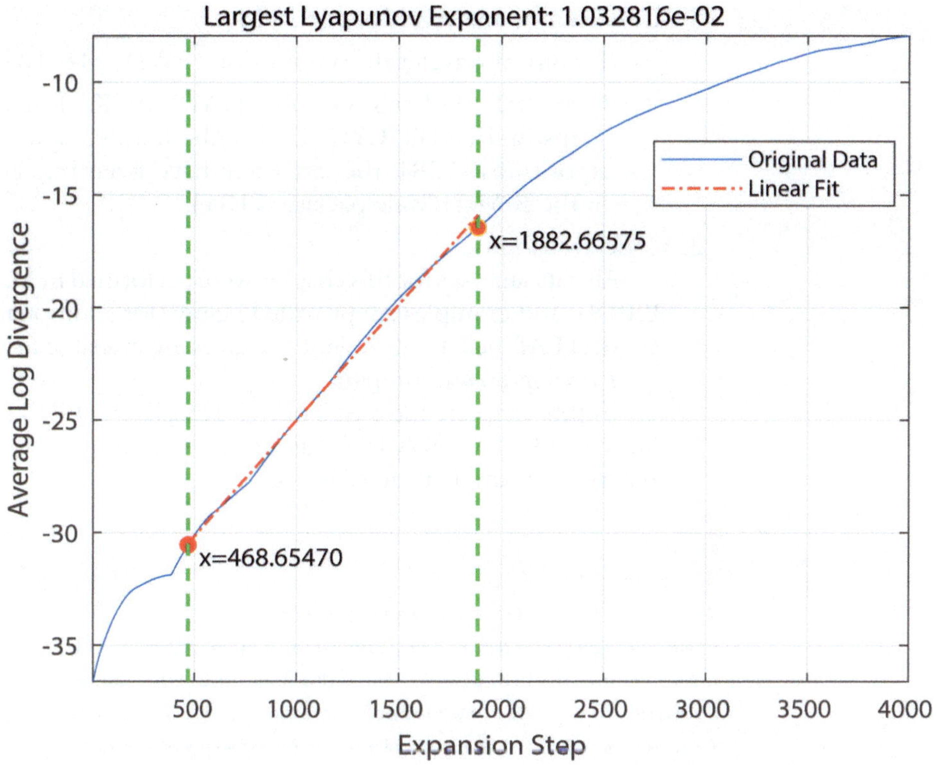

Fig. 19 Maximum Lyapunov exponent estimation for mitochondrial calcium concentration. Output of Rosenstein et al. [68] algorithm implemented in MATLAB R2018a (function: *lyapunovExponent*). Plot of average logarithm of divergence versus time for the time series shown in Fig. 18a.The solid blue curve is the calculated result; the green vertical lines indicate the linear region of this curve. This linear region was fitted (red dashed curve) and the slope is the expected Largest Lyapunov Exponent

3 Notes

1. Other commonly used time series analysis.

   Herein we have focused on a subset of classical and widely used time series analysis methods, however many other analyses exist and can be used as complementary approaches under the appropriate circumstances. Listed below are some examples, and citations for further information on other methods for rhythm detection. As for chaos, in-depth analysis of procedures can be found in [60, 72, 128].

   (a) Enright's method [62, 63, 76].
   (b) Hilbert transform and Empirical mode decomposition (EMD) [30, 129–131].
   (c) Nonparametric circadian rhythm analysis (NPCRA) [100].

(d) Cosinor analysis [76, 115, 124].

(e) Empirical wavelet decomposition (EWD) [84, 132, 133].

(f) COSOPT, Fisher's G test, HAYSTACK, Jonckheere–Terpstra–Kendall CYCLE, Lomb–Scargle, and ARSER algorithms [128]; the last three have been implemented in the R MetaCycle package [134].

2. Code availability.

All data analyses for this chapter were performed in MATLAB R2018a, and examples are provided below (for additional details see MATLAB help). Equivalent packages are available in R.

Power spectrum analysis.

In this chapter, the Fast Fourier Transform (FFT) subroutine of MATLAB R2018a was used to apply power spectral analysis to example time series, x.

```
Y=fft(x);
N = length(Y);
Y(1) = []; %Borra el primer reglon
power = abs(Y)/(N/2); %% absolute value of the fft
power = power(1:N/2).^2; %% take the positve frequency half,
nyquist = 1/2*dt; %Nyquist frequency: is half the sampling %
frequency of a discrete signal processing system
freq = (1:N/2)/(N/2)*nyquist;
```

Phase space reconstruction and Lyapunov exponents.

In the example shown in Fig. 7, the Lorenz system was used [103] with the following code:

```
function df = LORENZ_sysAbarbanel(~, x)

sigma=16;

beta=4;

ro=45.92;


df=[(sigma*x(2))-(sigma*x(1)); ...

            -(x(1).*x(3))+(ro*x(1))-x(2);...

            (x(1).*x(2))-(beta*x(3))];

end
```

```
ICs=[5, 5, 5]; % initial conditions

t=[0, 8000];    % time interval

OPTs = odeset('reltol', 1e-6, 'abstol', 1e-8); %

parameter %settings for ode

sol=ode45(@LORENZ_sysAbarbanel, t, ICs, OPTs);

time=[0:0.01:t(2)];

fOUT = deval(sol,time)';

xdata = fOUT(:,1);
```

For phase space reconstruction, $y(t) = [x(t), x(t + \tau), x(t + 2\tau)\ldots]$, the time lag ($\tau$) value can be determined from the first minimum of the nonlinear correlation function called average mutual information, and can be computed using the Mutual Information computation package in MATLAB [86]. The appropriate embedding dimension can be calculated according to the false nearest neighbor technique to determine the number of dimensions needed for the complete unfolding of the geometrical structure of the attractor (i.e., points should lay close to one another in the phase space due to their dynamics but not their projection), using code [92, 135] in MATLAB. MATLAB R2018a provides an alternate code for phase space reconstruction *phaseSpaceReconstruction*. where both lag and dimensions can be estimated automatically.

```
%%Average mutual information using Peng package
vec1=xdata(1:end-100-1);i=1;
for T=1:100; vec2=xdata(T+1:end-100+T-1);
h(i,1) = mutualinfo(vec1,vec2); i=i+1;
end
plot([1:100],h,'.');
xlabel('Time lag'); ylabel('Average Mutual information')

%% False Nearest Neighbor using Kennel et al. package
[FNN] = knn_deneme(xdata,10,10,15,2)

%%phase space
c=(xdata(21:end)+abs(min(xdata(21:end))))/(max(xdata(21:e
nd))+...
abs(min(xdata(21:end))));
scatter3(xdata(1:end-20),xdata(11:end-
10),xdata(21:end),1,c)
colormap(flipud(cbrewer('seq', 'PuRd', 40,'PCHIP')));
xlabel('x(t)'); ylabel('x(t+10)'); zlabel('x(t+20)')%%%%

%% phaseSpaceReconstruction function
[~,est_lag,est_dim] = phaseSpaceReconstruction(xdata)
```

Wolf et al. [105] provides well documented computational code that more recently was implemented in MATLAB R2018a. Their algorithm is useful for estimation of nonnegative Lyapunov exponents from an experimental time series. Available at: http://www.mathworks.com/matlabcentral/fileexchange/48084-lyapunov-exponent-estimation-from-a-time-series. MATLAB R2018a provides an alternate code for Estimation of the largest Lyapunov exponent *lyapunovExponent* based on the algorithm proposed by Rosenstein et al. [68]. For this algorithm, Ca^{2+} data (Subheading 2.2.2) was transformed so that values ranged in the interval [0 1].

```
fs=10;
dim=4;
ERange=4000;
lyapunovExponent(xdata,fs,'Dimension',dim,'Lag',lag,'Expan-
sionRange',ERange)
```

Wavelets.

MATLAB has a very comprehensive wavelet toolbox (wavelab) that is user friendly. Other paid software, such as Clock Lab also has a wavelet package. Wavelet packages are also available in R. In our example code in MATLAB R2018a, the wavelet to be used is specified (wname) as well as the sampling rate (the time interval between data points in seconds). Noteworthy, the relationship between the wavelet scale (scales_v) and frequency (f; expressed in Hertz) is dependent on the wavelet used and should be calculated with the function scal2frq. The associated period, in hours, can then be estimated (period).

```
%% Complex morlet wavelet

SERIE=x_alim; %time series

sampling_rate=60;

scales_v=6:6:2530;

wname='cmor1-1.5';

f= scal2frq(scales_v,wname,sampling_rate);

period= 1./f/60/60;

cA = cwt(SERIE,scales_v,wname); % complex wavelet
coefficients


%%Gaussian wavelet

  sampling_rate=60;

    sr=1/sampling_rate;

    scales_v=3:3:390;

    wname='gaus1';

    f= scal2frq(scales_v,wname,sampling_rate);

    period= 1./f/60/60;

    SERIE=x_alim; %time series

    c = cwt(SERIE,scales_v,wname,'plot');
```

Correlations between time series and wavelet coherence.

For correlation estimation between two animals (as in the example) or any two different time series of the same duration, first the complex Morlet *cwt* is performed, and complex coefficients are estimated as described previously for both time series (in example, cA_1 and cA_2). Second, the real part of both coefficient matrices are transposed, and then correlated. The result is also a matrix showing the correlation coefficient estimated at all time scales. Normally, only the diagonal (i.e., comparison between animals at the same time scale) is of interest (an example of between different time series where the full correlation coefficient matrix is assessed *see* [136]).

```
cwt1=ctranspose(real(cA_1));

cwt2=ctranspose(real(cA_2));

a=corr(cwt1,cwt2,'type','Spearman');

Correl_alim=diag(a); %%Correlation
```

For wavelet coherence, considering two time series (x_alim and x_run), sampled at 1 min intervals:

```
wcoherence(x_alim,x_run,hours(1/60),'NumScalesToSmooth',16,...
'PhaseDisplayThreshold',0.85);
```

References

1. Lloyd D, Aon M, Cortassa S (2001) Why homeodynamics, not homeostasis? Sci World J 1:133–145. https://doi.org/10.1100/tsw.2001.20
2. Hildebrandt G (1991) Reactive modifications of the autonomous time structure in the human organism. J Physiol Pharmacol 42 (1):5–27
3. Aon MA, Cortassa S (2012) Dynamic biological organization: fundamentals as applied to cellular systems. Springer Science & Business Media, Berlin
4. Edmunds LN (1988) Cellular and molecular bases of biological clocks: models and mechanisms for circadian timekeeping. Springer, New York, NY
5. Refinetti R (2011) Integration of biological clocks and rhythms. Comprehens Physiol 2 (2):1213–1239
6. Devlin PF, Kay SA (2001) Circadian photoperception. Annu Rev Physiol 63(1):677–694
7. Rosbash M, Young M (2009) The implications of multiple circadian clock origins. PLoS Biol 7(3):e1000062
8. Refinetti R (1997) Homeostasis and circadian rhythmicity in the control of body temperature a. Ann N Y Acad Sci 813(1):63–70
9. Chialvo DR (2010) Emergent complex neural dynamics. Nat Phys 6(10):744–750
10. Dunlap JC, Loros JJ, DeCoursey PJ (2004) Chronobiology: biological timekeeping. Sinauer Associates, Sunderland, MA
11. Golombek DA, Rosenstein RE (2010) Physiology of circadian entrainment. Physiol Rev 90(3):1063–1102
12. Schwartz WJ, Daan S (2017) Origins: a brief account of the ancestry of circadian biology. In: Biological timekeeping: clocks, rhythms

and behaviour. Springer, New York, NY, pp 3–22
13. Goldbeter A et al (1997) Biochemical oscillations and cellular rhythms. Cambridge University Press, Cambridge
14. Goodwin C (1965) Oscillatory behavior in enzymatic control processes. Adv Enzym Regul 3:425–437
15. Griffith JS (1968) Mathematics of cellular control processes i. negative feedback to one gene. J Theor Biol 20(2):202–208
16. Griffith JS (1968) Mathematics of cellular control processes ii. positive feedback to one gene. J Theor Biol 20(2):209–216
17. Winfree T (1970) Integrated view of resetting a circadian clock. J Theor Biol 28(3):327–374
18. King DP, Zhao Y, Sangoram AM, Wilsbacher LD, Tanaka M, Antoch MP, Steeves TD, Vitaterna MH, Kornhauser JM, Lowrey PL et al (1997) Positional cloning of the mouse circadian clock gene. Cell 89(4):641–653
19. Konopka RJ, Smith RF, Orr D (1991) Characterization of andante, a new drosophila clock mutant, and its interactions with other clock mutants. J Neurogenet 7 (2–3):103–114
20. Vitaterna MH, King DP, Chang A M, Kornhauser JM, Lowrey PL, McDonald JD, Dove WF, Pinto LH, Turek FW, Takahashi JS (1994) Mutagenesis and mapping of a mouse gene, clock, essential for circadian behavior. Science 264(5159):719–725
21. Zwiebel LJ, Hardin PE, Hall JC, Rosbash M (1991) Circadian oscillations in protein and mrna levels of the period gene of drosophila melanogaster. Biochem Soc Trans 19 (2):533–537
22. Dunlap JC (1999) Molecular bases for circadian clocks. Cell 96(2):271–290
23. Dunlap JC, Loros JJ (2017) Making time: conservation of biological clocks from fungi to animals. Microbiol Spectr 5(3):5–3
24. Takahashi JS (2017) Transcriptional architecture of the mammalian circadian clock. Nat Rev Genet 18(3):164–179
25. Ananthasubramaniam B, Herzel H (2014) Positive feedback promotes oscillations in negative feedback loops. PLoS One 9(8): e104761
26. Forger DB, Peskin CS (2005) Stochastic simulation of the mammalian circadian clock. Proc Natl Acad Sci 102(2):321–324
27. Goldbeter A (1995) A model for circadian oscillations in the drosophila period protein (per). Proc R Soc Lond Ser B Biol Sci 261 (1362):319–324
28. Nieto PS, Condat C (2019) Translational thresholds in a core circadian clock model. Phys Rev E 100(2):022409
29. Risau-Gusman S, Gleiser PM (2014) A mathematical model of communication between groups of circadian neurons in drosophila melanogaster. J Biol Rhythm 29(6):401–410
30. Guzmán DA, Flesia AG, Aon MA, Pellegrini S, Marin RH, Kembro JM (2017) The fractal organization of ultradian rhythms in avian behavior. Sci Rep 7(1):1–13
31. Rijo-Ferreira F, Takahashi JS (2019) Genomics of circadian rhythms in health and disease. Genome Med 11(1):1–16
32. Herzog ED, Hermanstyne T, Smyllie NJ, Hastings MH (2017) Regulating the suprachiasmatic nucleus (scn) circadian clockwork: interplay between cell-autonomous and circuit-level mechanisms. Cold Spring Harb Perspect Biol 9(1):a027706
33. Mohawk JA, Takahashi JS (2011) Cell autonomy and synchrony of suprachiasmatic nucleus circadian oscillators. Trends Neurosci 34(7):349–358
34. Pilorz V, Astiz M, Heinen KO, Rawashdeh O, Oster H (2020) The concept of coupling in the mammalian circadian clock network. J Mol Biol 432(12):3618–3638
35. Dowse HB (2009) Analyses for physiological and behavioral rhythmicity. Methods Enzymol 454:141–174
36. Liu C, Weaver DR, Strogatz SH, Reppert SM (1997) Cellular construction of a circadian clock: period determination in the suprachiasmatic nuclei. Cell 91(6):855–860
37. Wang S, Herzog ED, Kiss IZ, Schwartz WJ, Bloch G, Sebek M, Granados-Fuentes D, Wang L, Li J-S (2018) Inferring dynamic topology for decoding spatiotemporal structures in complex heterogeneous networks. Proc Natl Acad Sci 115(37):9300–9305
38. Izumo M, Pejchal M, Schook AC, Lange RP, Walisser JA, Sato TR, Wang X, Bradfield CA, Takahashi JS (2014) Differential effects of light and feeding on circadian organization of peripheral clocks in a forebrain bmal1 mutant. elife 3:e04617
39. Yoo S-H, Yamazaki S, Lowrey PL, Shimomura K, Ko CH, Buhr ED, Siepka SM, Hong H-K, Oh WJ, Yoo OJ et al (2004) Period2:: Luciferase real-time reporting of circadian dynamics reveals persistent circadian oscillations in mouse peripheral tissues. Proc Natl Acad Sci 101(15):5339–5346
40. Forger DB (2017) Biological clocks, rhythms, and oscillations: the theory of biological timekeeping. The MIT Press, Cambridge, MA

41. Hu K, Scheer FA, Ivanov PC, Buijs RM, Shea SA (2007) The suprachiasmatic nucleus functions beyond circadian rhythm generation. Neuroscience 149(3):508–517
42. Hu K, Ivanov PC, Chen Z, Hilton MF, Stanley HE, Shea SA (2004) Non-random fluctuations and multi-scale dynamics regulation of human activity. Phys A Stat Mech Its Appl 337 (1–2):307–318
43. Goldberger L, Amaral LA, Hausdorff JM, Ivanov PC, Peng C-K, Stanley HE (2002) Fractal dynamics in physiology: alterations with disease and aging. Proc Natl Acad Sci 99(Suppl 1):2466–2472
44. Pittman-Polletta R, Scheer FA, Butler MP, Shea SA, Hu K (2013) The role of the circadian system in fractal neurophysiological control. Biol Rev 88(4):873–894
45. Hu K, Meijer JH, Shea SA, VanderLeest HT, Pittman-Polletta B, Houben T, van Oosterhout F, Deboer T, Scheer FA (2012) Fractal patterns of neural activity exist within the suprachiasmatic nucleus and require extrinsic network interactions. PLoS One 7(11): e48927
46. Wu Y-E, Enoki R, Oda Y, Huang Z-L, Honma K-i, Honma S (2018) Ultradian calcium rhythms in the paraventricular nucleus and subparaventricular zone in the hypothalamus. Proc Natl Acad Sci 115(40): E9469–E9478
47. Carafoli E, Krebs J (2016) Why calcium? how calcium became the best communicator. J Biol Chem 291(40):20849–20857
48. Niggli E, Shirokova N (2007) A guide to sparkology: the taxonomy of elementary cellular ca2+ signaling events. Cell Calcium 42 (4–5):379–387
49. Berridge MJ, Cobbold P, Cuthbertson K (1988) Spatial and temporal aspects of cell signalling. Phil Trans R Soc Lond B Biol Sci 320(1199):325–343
50. Berridge M (1990) Calcium oscillations. J Biol Chem 265(17):9583–9586
51. Sneyd J, Han JM, Wang L, Chen J, Yang X, Tanimura A, Sanderson MJ, Kirk V, Yule DI (2017) On the dynamical structure of calcium oscillations. Proc Natl Acad Sci 114 (7):1456–1461
52. Voorsluijs V, Dawson SP, De Decker Y, Dupont G (2019) Deterministic limit of intracellular calcium spikes. Phys Rev Lett 122(8):088101
53. Dupont G (2014) Modeling the intracellular organization of calcium signaling. Wiley Interdiscip Rev Syst Biol Med 6(3):227–237
54. Gilkey JC, Jaffe LF, Ridgway EB, Reynolds GT (1978) A free calcium wave traverses the activating egg of the medaka, Oryzias latipes. J Cell Biol 76(2):448–466
55. Wakai T, Mehregan A, Fissore RA (2019) Ca2 + signaling and homeostasis in mammalian oocytes and eggs. Cold Spring Harb Perspect Biol 11(12):a035162
56. Kembro JM, Cortassa S, Lloyd D, Sollott SJ, Aon MA (2018) Mitochondrial chaotic dynamics: redox-energetic behavior at the edge of stability. Sci Rep 8(1):1–11
57. Akar FG, Aon MA, Tomaselli GF, O'Rourke B et al (2005) The mitochondrial origin of postischemic arrhythmias. J Clin Invest 115 (12):3527–3535
58. Aggarwal NT, Makielski JC (2013) Redox control of cardiac excitability. Antioxid Redox Signal 18(4):432–468
59. Aon MA, Cortassa S, Akar F, Brown D, Zhou L, O'rourke B (2009) From mitochondrial dynamics to arrhythmias. Int J Biochem Cell Biol 41(10):1940–1948
60. Refinetti R, Cornélissen G, Halberg F (2007) Procedures for numerical analysis of circadian rhythms. Biol Rhythm Res 38(4):275–325
61. Bloomfield P (2004) Fourier analysis of time series: an introduction. John Wiley & Sons, New York, NY
62. Mourão M, Satin L, Schnell S (2014) Optimal experimental design to estimate statistically significant periods of oscillations in time course data. PLoS One 9(4):e93826
63. Refinetti R (1993) Laboratory instrumentation and computing: comparison of six methods for the determination of the period of circadian rhythms. Physiol Behav 54 (5):869–875
64. Glynn EF, Chen J, Mushegian AR (2006) Detecting periodic patterns in unevenly spaced gene expression time series using lomb–scargle periodograms. Bioinformatics 22(3):310–316
65. Deckard A, Anafi RC, Hogenesch JB, Haase SB, Harer J (2013) Design and analysis of large-scale biological rhythm studies: a comparison of algorithms for detecting periodic signals in biological data. Bioinformatics 29 (24):3174–3180
66. De Lichtenberg U, Jensen LJ, Fausbøll A, Jensen TS, Bork P, Brunak S (2005) Comparison of computational methods for the identification of cell cycle-regulated genes. Bioinformatics 21(7):1164–1171
67. Hughes ME, Hogenesch JB, Kornacker K (2010) Jtk cycle: an efficient nonparametric algorithm for detecting rhythmic components in genome-scale data sets. J Biol Rhythm 25 (5):372–380

68. Rosenstein MT, Collins JJ, De Luca CJ (1993) A practical method for calculating largest lyapunov exponents from small data sets. Phys D Nonlin Phenom 65 (1–2):117–134
69. Orlando DA, Lin CY, Bernard A, Wang JY, Socolar JE, Iversen ES, Hartemink AJ, Haase SB (2008) Global control of cell-cycle transcription by coupled cdk and network oscillators. Nature 453(7197):944–947
70. Scargle JD (1982) Studies in astronomical time series analysis. II-statistical aspects of spectral analysis of unevenly spaced data. Astrophys J 263:835–853
71. Cohen-Steiner D, Edelsbrunner H, Harer J, Mileyko Y (2010) Lipschitz functions have l p-stable persistence. Found Comput Math 10 (2):127–139
72. Kantz H, Schreiber T (2004) Nonlinear time series analysis, vol 7. Cambridge University Press, Cambridge
73. Kembro JM, Aon MA, Winslow RL, O'Rourke B, Cortassa S (2013) Integrating mitochondrial energetics, redox and ros metabolic networks: a two-compartment model. Biophys J 104(2):332–343
74. Kembro JM, Cortassa S, Aon MA (2014) Complex oscillatory redox dynamics with signaling potential at the edge between normal and pathological mitochondrial function. Front Physiol 5:257
75. Komendantov O, Kononenko NI (1996) Deterministic chaos in mathematical model of pacemaker activity in bursting neurons of snail, helix pomatia. J Theor Biol 183 (2):219–230
76. Refinetti R (2004) Non-stationary time series and the robustness of circadian rhythms. J Theor Biol 227(4):571–581
77. Leise TL, Harrington ME (2011) Wavelet-based time series analysis of circadian rhythms. J Biol Rhythm 26(5):454–463
78. Leise TL, Indic P, Paul MJ, Schwartz WJ (2013) Wavelet meets actogram. J Biol Rhythm 28(1):62–68
79. Leise TL (2015) Wavelet-based analysis of circadian behavioral rhythms. Methods Enzymol 551:95–119
80. Leise TL (2013) Wavelet analysis of circadian and ultradian behavioral rhythms. J Circadian Rhythms 11(1):1–9
81. Flandrin P (2018) Explorations in time-frequency analysis. Cambridge University Press, Cambridge
82. Mallat S (2011) A wavelet tour of signal processing: the sparse way, 3rd edn. Academic Press, Burlington, MA
83. Addison PS, Walker J, Guido RC (2009) Time–frequency analysis of biosignals. IEEE Eng Med Biol Mag 28(5):14–29
84. Dong S, Yuan M, Wang Q, Liang Z (2018) A modified empirical wavelet transform for acoustic emission signal decomposition in structural health monitoring. Sensors 18 (5):1645
85. Jud C, Schmutz I, Hampp G, Oster H, Albrecht U (2005) A guideline for analyzing circadian wheel-running behavior in rodents under different lighting conditions. Biol Proced Online 7(1):101–116
86. Williams G (1997) Chaos theory tamed. Joseph Henry Press, Washington, DC
87. Clocklab (2020) Clocklab: data collection and analysis for circadian biology. Clocklab, Wilmette, IL
88. Kembro JM, Flesia AG, Gleiser RM, Perillo MA, Marin RH (2013) Assessment of long-range correlation in animal behavior time series: the temporal pattern of locomotor activity of Japanese quail (coturnix coturnix) and mosquito larva (Culex quinquefasciatus). Phys A Stat Mech Its Appl 392 (24):6400–6413
89. Hu K, Ivanov PC, Hilton MF, Chen Z, Ayers RT, Stanley HE, Shea SA (2004) Endogenous circadian rhythm in an index of cardiac vulnerability independent of changes in behavior. Proc Natl Acad Sci 101(52):18223–18227
90. Koks D (2006) Explorations in mathematical physics: the concepts behind an elegant language. Springer, New York, NY
91. Rhee NH, Góra P, Bani-Yaghoub M (2019) Predicting and estimating probability density functions of chaotic systems. Discr Contin Dyn Syst B 24(1):297
92. Clauset A, Shalizi CR, Newman ME (2009) Power-law distributions in empirical data. SIAM Rev 51(4):661–703
93. Kembro JM, Lihoreau M, Garriga J, Raposo EP, Bartumeus F (2019) Bumblebees learn foraging routes through exploitation–exploration cycles. J R Soc Interface 16 (156):20190103
94. Bartumeus F, Giuggioli L, Louzao M, Bretagnolle V, Oro D, Levin SA (2010) Fishery discards impact on seabird movement patterns at regional scales. Curr Biol 20(3):215–222
95. Maraun D, Rust H, Timmer J (2004) Tempting long-memory-on the interpretation of DFA results. Nonlinear Process Geophys 11 (4):495–503
96. Aon M, Cortassa S (2009) Chaotic dynamics, noise and fractal space in biochemistry. In:

Encyclopedia of complexity and systems science. Springer, New York, NY, pp 476–489

97. Peng C-K, Havlin S, Stanley HE, Goldberger AL (1995) Quantification of scaling exponents and crossover phenomena in nonstationary heartbeat time series. Chaos 5 (1):82–87
98. Aon MA, Cortassa S, Lloyd D (2012) Chaos in biochemistry and physiology. In: Encyclopaedia of biochemistry and molecular medicine: systems biology. Wiley-VCH Verlag GmbH & Co. KGaA, Weinheim, pp 239–276
99. Szendro P, Vincze G, Szasz A (2001) Pink-noise behaviour of biosystems. Eur Biophys J 30(3):227–231
100. Lomb NR (1976) Least-squares frequency analysis of unequally spaced data. Astrophys Space Sci 39(2):447–462
101. Poincaré H (1908) Science and method
102. Girling A (1995) Periodograms and spectral estimates for rhythm data. Biol Rhythm Res 26(2):149–172
103. Abarbanel HD, Gollub JP (1996) Analysis of observed chaotic data. Phys Today 49(11):86
104. Shaw R (1981) Strange attractors, chaotic behavior, and information flow. Z Naturforsch A 36(1):80–112
105. Wolf A, Swift JB, Swinney HL, Vastano JA (1985) Determining lyapunov exponents from a time series. Phys D Nonlin Phenom 16(3):285–317
106. Bartnik E, Blinowska KJ, Durka PJ (1992) Single evoked potential reconstruction by means of wavelet transform. Biol Cybern 67 (2):175–181
107. Baggs JE, Price TS, DiTacchio L, Panda S, FitzGerald GA, Hogenesch JB (2009) Network features of the mammalian circadian clock. PLoS Biol 7(3):e1000052
108. Meeker K, Harang R, Webb AB, Welsh DK, Doyle FJ III, Bonnet G, Herzog ED, Petzold LR (2011) Wavelet measurement suggests cause of period instability in mammalian circadian neurons. J Biol Rhythm 26 (4):353–362
109. Torrence C, Compo GP (1998) A practical guide to wavelet analysis. Bull Am Meteorol Soc 79(1):61–78
110. Abid A, Gdeisat M, Burton D, Lalor M (2007) Ridge extraction algorithms for one-dimensional continuous wavelet transform: a comparison. J Phys Conf Ser 76:012045
111. Carmona RA, Hwang WL, Torrésani B (1999) Multiridge detection and time-frequency reconstruction. IEEE Trans Signal Process 47(2):480–492
112. Lorenz EN (1995) The essence of chaos. Taylor & Francis, UK, p 227
113. Carmona RA, Hwang WL, Torrésani B (1997) Characterization of signals by the ridges of their wavelet transforms. IEEE Trans Signal Process 45(10):2586–2590
114. Fossion R, Rivera AL, Toledo-Roy JC, Angelova M, El-Esawi M (2018) Quantification of irregular rhythms in chrono-biology: a time-series perspective. In: Circadian rhythm: cellular and molecular mechanisms. InTech, Rijeka, pp 33–58
115. Fossion R, Rivera AL, Toledo-Roy JC, Ellis J, Angelova M (2017) Multiscale adaptive analysis of circadian rhythms and intradaily variability: application to actigraphy time series in acute insomnia subjects. PLoS One 12(7): e0181762
116. Herrera RH, Han J, van der Baan M (2014) Applications of the synchrosqueezing transform in seismic time-frequency analysis. Geophysics 79(3):V55–V64
117. Kumar CS, Arumugam V, Sengottuvelusamy R, Srinivasan S, Dhakal H (2017) Failure strength prediction of glass/epoxy composite laminates from acoustic emission parameters using artificial neural network. Appl Acoust 115:32–41
118. Daubechies I, Lu J, Wu H-T (2011) Synchrosqueezed wavelet transforms: an empirical mode decomposition-like tool. Appl Comput Harmon Anal 30(2):243–261
119. Auger F, Flandrin P (1995) Improving the readability of time-frequency and time-scale representations by the reassignment method. IEEE Trans Signal Process 43(5):1068–1089
120. Auger F, Flandrin P, Lin Y-T, McLaughlin S, Meignen S, Oberlin T, Wu H-T (2013) Time-frequency reassignment and synchrosqueezing: an overview. IEEE Signal Process Mag 30(6):32–41
121. Thakur G, Brevdo E, Fuˇckar NS, Wu H-T (2013) The synchrosqueezing algorithm for time-varying spectral analysis: robustness properties and new paleoclimate applications. Signal Process 93(5):1079–1094
122. Chavez M, Cazelles B (2019) Detecting dynamic spatial correlation patterns with generalized wavelet coherence and non-stationary surrogate data. Sci Rep 9(1):1–9
123. Cazelles B, Chavez M, Berteaux D, Ménard F, Vik JO, Jenouvrier S, Stenseth NC (2008) Wavelet analysis of ecological time series. Oecologia 156(2):287–304
124. Staff PO (2017) Correction: multiscale adaptive analysis of circadian rhythms and intradaily variability: application to actigraphy time

series in acute insomnia subjects. PLoS One 12(11):e0188674
125. Le Van Quyen M, Foucher J, Lachaux J-P, Rodriguez E, Lutz A, Martinerie J, Varela FJ (2001) Comparison of hilbert transform and wavelet methods for the analysis of neuronal synchrony. J Neurosci Methods 111 (2):83–98
126. Cazelles B, Stone L (2003) Detection of imperfect population synchrony in an uncertain world. J Anim Ecol 72:953–968
127. Acosta-Rodríguez VA, de Groot MH, Rijo-Ferreira F, Green CB, Takahashi JS (2017) Mice under caloric restriction self-impose a temporal restriction of food intake as revealed by an automated feeder system. Cell Metab 26(1):267–277
128. Wu G, Zhu J, Yu J, Zhou L, Huang JZ, Zhang Z (2014) Evaluation of five methods for genome-wide circadian gene identification. J Biol Rhythm 29(4):231–242
129. Huang NE, Shen Z, Long SR, Wu MC, Shih HH, Zheng Q, Yen N-C, Tung CC, Liu HH (1998) The empirical mode decomposition and the Hilbert spectrum for nonlinear and non-stationary time series analysis. Proc R Soc Lond Ser A Math Phys Eng Sci 454 (1971):903–995
130. Rehman N, Mandic DP (2010) Multivariate empirical mode decomposition. Proc R Soc A Math Phys Eng Sci 466(2117):1291–1302
131. Rilling G, Flandrin P (2007) One or two frequencies? the empirical mode decomposition answers. IEEE Trans Signal Process 56 (1):85–95
132. Gilles J (2013) Empirical wavelet transform. IEEE Trans Signal Process 61 (16):3999–4010
133. Liu W, Chen W (2019) Recent advancements in empirical wavelet transform and its applications. IEEE Access 7:103770–103780
134. Wu G, Anafi RC, Hughes ME, Kornacker K, Hogenesch JB (2016) Metacycle: an integrated r package to evaluate periodicity in large scale data. Bioinformatics 32 (21):3351–3353
135. Kennel MB, Brown R, Abarbanel HD (1992) Determining embedding dimension for phase-space reconstruction using a geometrical construction. Phys Rev A 45(6):3403
136. Kurz FT, Kembro JM, Flesia AG, Armoundas AA, Cortassa S, Aon MA, Lloyd D (2017) Network dynamics: quantitative analysis of complex behavior in metabolism, organelles, and cells, from experiments to models and back. Wiley Interdiscip Rev Syst Biol Med 9 (1):e1352

Chapter 14

Computational Systems Biology of Morphogenesis

Jason M. Ko, Reza Mousavi, and Daniel Lobo

Abstract

Extracting mechanistic knowledge from the spatial and temporal phenotypes of morphogenesis is a current challenge due to the complexity of biological regulation and their feedback loops. Furthermore, these regulatory interactions are also linked to the biophysical forces that shape a developing tissue, creating complex interactions responsible for emergent patterns and forms. Here we show how a computational systems biology approach can aid in the understanding of morphogenesis from a mechanistic perspective. This methodology integrates the modeling of tissues and whole-embryos with dynamical systems, the reverse engineering of parameters or even whole equations with machine learning, and the generation of precise computational predictions that can be tested at the bench. To implement and perform the computational steps in the methodology, we present user-friendly tools, computer code, and guidelines. The principles of this methodology are general and can be adapted to other model organisms to extract mechanistic knowledge of their morphogenesis.

Key words Systems Biology, Computational Biology, Machine Learning, Morphogenesis

1 Introduction

Elucidating the mechanisms controlling the morphogenesis of complex multicellular systems is a current challenge [1]. The nonlinear interactions and feedback loops between the genetic components of biological regulation and between these and the cellular and tissue biophysical forces prevent us from easily discerning them directly from experimental phenotypes [2, 3]. Indeed, despite the rich literature of developmental and regenerative biology that has discovered many essential genes and their resultant phenotypes when perturbed, model organisms still lack a comprehensive understanding of the regulatory mechanisms controlling their morphogenesis [4–6].

As a fundamental aid for understanding the regulation of morphogenesis, computational systems biology methods can provide rigorous mechanistic hypotheses recapitulating experimental phenotypes and predicting the outcomes of novel experiments [7–

Sonia Cortassa and Miguel A. Aon (eds.), *Computational Systems Biology in Medicine and Biotechnology: Methods and Protocols*, Methods in Molecular Biology, vol. 2399, https://doi.org/10.1007/978-1-0716-1831-8_14,

10]. For this, dynamic mathematical models based on differential equations are ideal to integrate the processes of genetic regulation, signaling, and cellular and biophysical mechanics to precisely predict the behaviors of tissue, organs, and whole organisms during morphogenesis [11–15]. Importantly, this approach can readily implement surgical, pharmacological, genetic, and environmental perturbations in the simulations, which are essential in developmental and regenerative studies.

Crucially, machine learning methods can infer the parameters and specific terms in the equations of mechanistic models directly from formalized experimental phenotypes [16–22]. This approach requires the tight integration of a diverse set of systems biology methods and protocols—from functional experiments to mathematical modeling and machine learning—forming an iterative process toward the discovery and refinement of models explaining experimental phenotypes. Indeed, the mechanisms controlling morphogenesis involve several scales of complexity including gene regulation, cellular behaviors, and tissue and whole-body morphologies [23, 24] that need to be taken into account and integrated in any mechanistic hypothesis.

Here we present an integrated computational systems biology approach for the modeling, inference, and validation of the mechanisms of morphogenesis. We show practical examples and suggest computational tools that can be used to perform each step. The goal of the methodology is to produce a mechanistic systems-level model that can precisely explain the observed spatial and temporal phenotypes from functional experiments, as well as to make testable predictions from novel perturbations. This methodology can be adapted to a diverse set of organisms and morphogenesis experiments.

2 Materials

2.1 Computational Modeling

1. MATLAB, a scientific-oriented programming language developed by MathWorks Inc. The software also provides a user-friendly programming environment with the same name and available for a fee at https://www.mathworks.com/.
2. As a free open source alternative to MATLAB, GNU Octave (written by John W. Eaton and many others) also provides a user-friendly programming environment mostly compatible with the MATLAB language. GNU Octave is freely available at https://www.gnu.org/software/octave/.
3. Runge–Kutta fourth-order (RK4) solver with dynamic timestep. RK4 solvers can solve some problems such as adhesion forces for whole-embryo simulation that are too stiff for linear solvers (e.g., forward Euler) by using a fourth order polynomial

approximation instead of assuming local linearity. Dynamic step solvers offer additional robustness by allowing the solver to automatically detect simulation steps where numerical error exceeds a given threshold and resimulate those steps using a smaller timestep. For the whole-embryo simulations, we used the dynamic-step, RK4 solver ROWMAP [25] with the default parameters, but solver parameters like minimum step size and error tolerances can be configured if necessary.

2.2 Machine Learning

1. A C++ environment, which provides a very efficient programming language and highly optimized compilers essential for computationally intensive tasks such as machine learning. We regularly use Microsoft Visual Studio as a user-friendly programming environment together with their C++ compiler. The free community edition is freely available at https://visualstudio.microsoft.com/vs/. In addition, we use the GNU C++ compiler (Free Software Foundation, Inc.) for compiling in Linux environments, typically found in high performance servers and clusters.
2. To implement graphical user interfaces and facilitate cross-platform compatibility of the software, we use the Qt libraries (The Qt Company Ltd.), which is freely available at https://www.qt.io. This very useful library also provides efficient and cross-platform classes and functions for multithreading, database access, and file reading and writing.
3. A C++ linear algebra library for performing basic mathematical operations during the simulation of models. We use Eigen (Gaël Guennebaud, Benoît Jacob, and others), which is freely available at https://eigen.tuxfamily.org/.
4. For high performance computers, we use the Message Passing Interface library that is specifically available in a given cluster computer. This facilitates the implementation of system-agnostic parallel computer code.

2.3 Validation

1. MoCha, a tool for the characterization of unknown components and pathways. The tool is freely available at https://lobolab.umbc.edu/mocha/ and is compatible with UNIX systems. An installer is included that compiles the program and downloads and preprocess the necessary data files. Due to the large database of protein interactions that MoCha uses, its installation requires 430 Gigabytes of disk space.
2. Access to the STRING (Search Tool for the Retrieval of Interacting Genes/Proteins) database [26], which can be done automatically through MoCha. The STRING database is a comprehensive repository of known and predicted protein–protein interactions. The latest version comprises more than 5000 organisms, 14 million proteins, and 6 billion links.

3 Methods

3.1 Computational Modeling at the Systems Level

Computational systems biology models are essential for understanding the mechanisms controlling complex developmental phenotypes [7, 27]. A diverse set of formalisms have been proposed for abstracting biological systems of morphogenesis, including discrete cell models [28], cellular automata [29], graph grammars [30], and membrane-computing [31]. However, dynamical systems based on differential equations remains the most versatile method to model developmental and regenerative systems [24, 32, 33]. Differential equations can describe the development of tissues and patterns in time and space and predict the signaling mechanisms in a single [8] or multiple spatial dimensions [34]. The advantage of these systems is in their capacity to integrate controlling mechanisms of gene regulation with spatial signaling and biophysical forces. Furthermore, experimental perturbations can be directly translated to dynamical system models to predict a particular phenotype. Surgical manipulations can be implemented by changing the state of the system, whereas genetic perturbations can be translated to changes in the equation parameters such as the production rate constants. These features make dynamical systems an ideal approach for modeling morphogenesis.

Different computational tools are available for the mathematical modeling of biological systems [9]. However, complex phenotypes including gene regulation, tissue dynamics and growth, and experimental perturbations such as amputations are at the forefront of current modeling research. As a result, general purpose programming languages and environments such as MATLAB, Python, and C++ are common alternatives that give the most versatile approach in which implement models of morphogenesis. In addition, particular programming libraries for the simulation of biological tissue dynamics can aid in the implementation of developmental and regeneration models using general purpose programming languages [35–39].

To illustrate the programming of dynamical models of morphogenesis, here we show a simple example of how to simulate the original reaction–diffusion system proposed by Turing in his seminal paper to explain the phenomena of morphogenesis [40]. In addition, we show how to simulate a perturbation in the system to study its pattern regeneration. The original Turing system includes only two morphogen products, X and Y, that represents two interacting chemical species reacting and diffusing in time and space. This modeling approach is continuous and hence does not model specific cells, but a tissue section abstracted as a continuous space where the morphogens react and diffuse. The following two partial differential equations describe the rates of change of each of the morphogens, which dictates their dynamics in space and time:

$$\frac{\partial X}{\partial t} = \frac{1}{128}(16 - X\Upsilon) + \frac{1}{4}\nabla^2 X,$$

$$\frac{\partial \Upsilon}{\partial t} = \frac{1}{128}(X\Upsilon - \Upsilon - 12) + \frac{1}{64}\nabla^2 \Upsilon,$$

where the production of Υ is zero if $\Upsilon \leq 0$, to avoid negative concentrations.

The computational simulation of a dynamical system requires two main tasks: the initialization of the system and the main simulation loop. The initialization of the system sets the initial values of the variables, such as their concentrations, through space at the initial time point in the simulation, $t = 0$. The main simulation loop iteratively updates the variables according to the governing equations and applies any perturbation performed during the simulation. Box 1 illustrates a simple but complete implementation in MATLAB of the simulation of the Turing reaction–diffusion system described above. The code is compatible with both MATLAB and GNU Octave programming environments, but for the latter the user needs first to load the image package with the command `pkg load image`. The simulation shows a developing stripe pattern that can self-regenerate after a perturbation.

Box 1 MATLAB code to simulate a Turing reaction–diffusion system

```
% Simulation parameters
dt = 0.5;
domain = 100;
duration = 20000;
plotperiod = 100;
perturbation = 10000;

% Initialization
X = 3 * rand(domain, domain) + 2;
Y = 3 * rand(domain, domain) + 2;

% Initial state plot
clf; imagesc(X, [2 5]); axis off; colormap jet;
drawnow;

% Simulation loop
for t=1:duration
  % Diffusion with Neumann boundary condition
  Xd = 1/4 * 4 * del2(padarray(X, [1 1], 'replicate'));
  Yd = 1/64 * 4 * del2(padarray(Y, [1 1],
'replicate'));
  % Production
  Xp = 1/128 * (16 - X .* Y);
  Yp = 1/128 * (X .* Y - Y - 12);
```

(continued)

```
    Yp(Y<=0) = 0;

    % Integration
    X = X + dt * (Xp + Xd(2:end-1, 2:end-1));
    Y = Y + dt * (Yp + Yd(2:end-1, 2:end-1));

    % Perturbation
    if t == perturbation
      X(30:70,30:70) = 0;
      Y(30:70,30:70) = 0;
    end

    % Plot
    if mod(t,plotperiod) == 0
      imagesc(X, [2 5]); axis off; drawnow;
    end
  end
```

The program starts by setting the parameters of the simulation, including the time step (*dt*), the size of the area to simulate (*domain*), the duration of the simulation (*duration*), the period for the plotting function (*plotperiod*), and the time point at which the system will be perturbed to study its regeneration (*perturbation*). Next, the simulation variables (the spatial concentrations of products *X* and *Y*) are initialized with random values (*rand* function) in a 2D space representing a tissue, the simulation domain (*see* **Note 1**). The code includes also the plotting of the simulation (*imagesc* function) to show the concentration of *X* in the spatial domain. After the initialization of the system, the simulation main loop (*for* statement) is executed for a particular time duration. The equations governing the system, which represent the rate of change of the products, are divided into the computation of the diffusion term, stored in the *Xd* and *Yd* variables, and the computation of the production term, stored in the *Xp* and *Yp* variables. The diffusion terms need to account for the special case at the borders of the domain, which the example code treats as an impermeable barrier through which no product can cross (*see* **Note 2**). Then, the integration step updates the concentrations of *X* and *Y* by adding their computed rate of change for the duration of the time step (*dt*)—an implementation of the Euler method to numerically solve the system of equations.

The example program also simulates a perturbation by which the developed pattern is partially removed at the center of the domain. For this, the main loop includes a condition (*if* function) followed by a block of code that is executed only at a specific time point during the simulation (set by the *perturbation* variable). At this time point, the concentration of the *X* and *Y* products are reset to zero in a square region at the center of the domain. After this, the

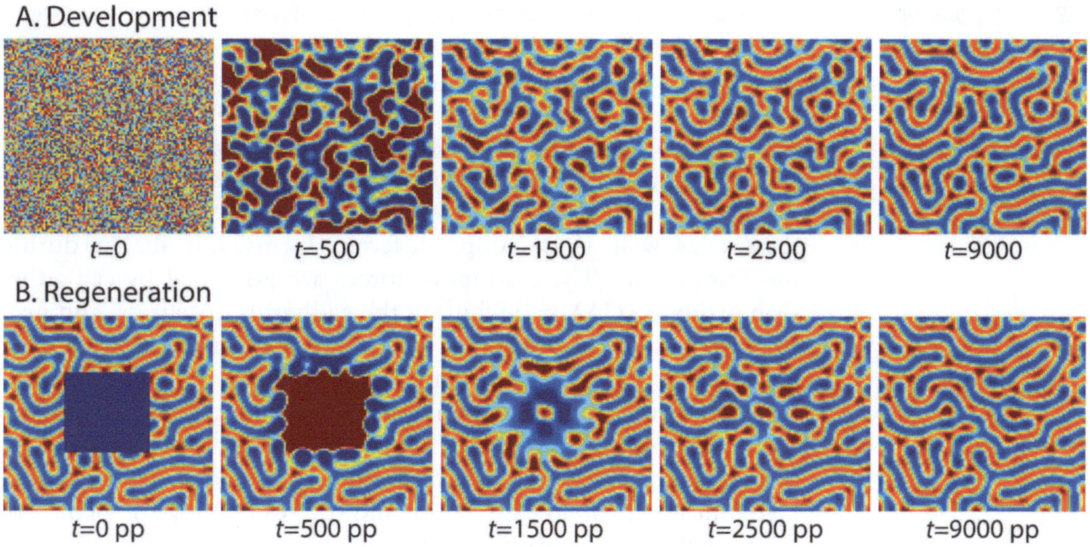

Fig. 1 Simulation of a dynamical mathematical model of morphogenesis for pattern formation and regeneration. (**a**) The development of a stripe pattern starting with a random state. (**b**) Regeneration of the stripe pattern after a perturbation removing an area at the center. Blue colors correspond to low concentration values and red colors to high concentration values. Both patterns reach a steady state by $t = 9000$. The simulation is run in a 100 by 100 grid using arbitrary units. *t* time, *pp.* post perturbation

program continues executing the main loop until reaching the end of the simulation time.

Figure 1 shows the output plots at different time points resulting from the execution of the code. The development plots (Fig. 1a) correspond to the simulation from the initial state until before the perturbation is applied. At $t = 0$, the product concentrations are random with a uniform distribution across the domain. As the simulation advances, a stripe pattern is self-organized due to the reaction–diffusion equations governing the system. By $t = 9000$, the simulation reaches a stable configuration of the pattern, after which the perturbation is applied. The regeneration plots (Fig. 1b) corresponds to the simulation from the perturbation time until the stabilization of the regenerated pattern. At $t = 0$ post perturbation (equivalent to $t = 10000$ in the code), the pattern removal perturbation is applied, as shown by the lack of pattern in a square inside the domain. After this perturbation, the stripe pattern is self-regenerated; however, the resultant regenerated pattern is slightly different than the original, yet it follows the same stripe configuration in terms of band size. By the time the simulation ends, the new stripe pattern has reached a stable state. This simple example illustrates the basic programming components needed to simulate morphogenesis using dynamical systems, a core component of computational systems biology.

3.2 Computational Systems Biology of Whole Embryos

In addition to simulating morphogen dynamics in tissue areas, computational models are essential for the understanding of whole-embryo morphogenesis at the systems level [41]. For this, dynamic cell densities and cell–cell adhesion forces can be incorporated into the models to predict the spatial behavior of tissues and the formation of patterns and shapes, including whole embryos [34]. Cell adhesion forces allow for cells to aggregate into tissues, as well as to adopt different forms and shapes during morphogenesis. These adhesive forces are governed by cell adhesion proteins (CAMs), including the cadherins, integrins, and nectins [42, 43]. Importantly, the temporal and spatial genetic regulation of the expression of these adhesive proteins are essential to orchestrate the formation of specific target morphologies. Computational systems biology methods can be employed to understand the complex feedback loops between gene regulation, cell-adhesion protein expression, and the resultant tissue behaviors to predict the formation of patterns and shapes at the level of a whole embryo.

We illustrate this approach with a whole-embryo continuous model of zebrafish gastrulation integrating the spatial and temporal dynamics of cell density and CAM expression together with the resultant biophysical forces and their genetic regulation [34]. During zebrafish gastrulation a gradient of the morphogen Nodal induces mesendoderm differentiation [44]. In response to this patterning signal, cells will undergo involution, causing the tissue to fold around the germ ring margin and expand up the inner surface of the embryo toward the animal pole. Regulated changes in cell–cell adhesion are essential for these involution dynamics. While all germ layers express similar levels of E-cadherin, N-cadherin expression is upregulated in response to the morphogen Nodal [45]. A specific area of the yolk contains transcriptionally active nuclei that produce Nodal, which then diffuses outward into the blastoderm [46]. The resulting patterning and involution tissue behaviors lead to the formation of the germ layers, where cells unexposed to the Nodal gradient become ectoderm [47] while those exposed to Nodal become mesendoderm [46, 48].

Similar to the Turing system presented in the previous section, the model considers cells as a continuum, but in this case their spatial density can vary dynamically. Hence a variable, u, is used to represent cell density, which can dynamically change due to the movement of the cells through space with some velocity, $\boldsymbol{V}$. Cell growth and apoptosis are ignored in the model, as they are assumed to be negligible during the time period simulated. Cell density is hence defined according to the partial differential equation

$$\frac{\partial u(\boldsymbol{x}, t)}{\partial t} = -\nabla \cdot (u\boldsymbol{V}).$$

Cell velocity is comprised of both dispersion (V_d) and adhesion (V_a) components, such as

$$V = V_d + V_a.$$

Dispersion is proportional to cell density and causes cells to move away from each other into open space. The constant k_p controls the rate of dispersion, resulting in

$$V_d = -k_p \nabla u.$$

The modeling of adhesion forces, which depends on both cell density and CAM concentrations, uses an integral term to define the adhesion strengths between all pairs of locations in space (x and y) that contain both cells and CAMs and are within some sensing radius R of each other. The strength of adhesion depends on the types of CAMs expressed (a_{ij}) and the relative concentration of CAMs ($\frac{c_i(x)}{u(x)}$ and $\frac{c_i(y)}{u(y)}$) in the cells at those locations. A logistic function with crowding capacity k_c limits movement due to adhesion into regions of high cell density. The function $\omega(r)$ describes how the adhesive force varies with the radial distance (r). For simplicity, we assume $\omega(r) = 1$. Additionally, ϕ is a constant of proportionality related to viscosity, n is the number of CAM types, and η is the direction vector. In this way, the velocity of cells due to adhesion forces are defined as

$$V_a = \sum_i^n \sum_j^n \frac{\phi}{R} K(u, c_i, c_j)(x),$$

$$K(u, c_i, c_j)(x) = \int_0^R \int_S a_{ij} \left[f(u, c_i, c_j)(x, x + r\eta) \right] \omega(r) r\eta \, d\eta \, dr,$$

$$f(u, c_i, c_j)(x, y) = \frac{c_i(x)}{u(x)} \frac{c_j(y)}{u(y)} h(u(y)),$$

$$h(u) = \begin{cases} u\left(1 - \frac{u}{k_c}\right) & \text{if } u < k_c, \\ 0 & \text{otherwise.} \end{cases}$$

The concentration of CAMs, c, can also dynamically vary continuously in space. CAMs are advected by the movement of cells and thus move with the same velocity than the cells that express them (since CAMs are membrane bound). Additionally, CAM expression rates can be regulated by morphogen signals, as dictated by the function R_{c_i}, such that

$$c = (c_1, \ldots, c_n),$$

$$\frac{\partial c_i(x, t)}{\partial t} = -\nabla \cdot (c_i V) + R_{c_i}(c, m_{nod}).$$

The specific regulatory dynamics can be different for each CAM. In particular, this example model uses a Hill function for the regulation of N-cadherin in response to the morphogen Nodal, but other sigmoidal or step functions can be used instead, depending on the specific system being studied. The morphogen Nodal is produced in a specific region of the yolk called the eYSL (m_{eYSL}) at some rate k_e. It subsequently diffuses with rate $D_{m_{nod}}$ and decays with rate λ, resulting in

$$\frac{\partial m_{nod}(\boldsymbol{x}, t)}{\partial t} = -\nabla\cdot(m_{nod}\boldsymbol{V}) + \nabla\cdot(D_{m_{nod}} u \nabla m_{nod}) + k_e m_{eYSL} - \lambda m_{nod}.$$

In this way, morphogen diffusion depends on cell density such that the spread of Nodal is limited to moving through the extracellular space between cells in the tissue, as seen in experimental evidence [49], rather than leaking out into other parts of the simulation domain (*see* **Note 3**). m_{eYSL} is represented as a morphogen with no production or decay, such that the factors associated with this region of the yolk are advected with the yolk as the system develops through time, such that

$$\frac{\partial m_{eYSL}}{\partial t} = -\nabla\cdot(m_{eYSL}\boldsymbol{V}).$$

The numerical values of the model parameters were derived from empirical data or estimated based on observing how their values effect the dynamics of the simulation. The system of partial differential equations is discretized using the explicit upwind finite volume method with flux limiting and simulated in a uniform square lattice. Zero-flux boundaries were imposed at the interfaces between the blastoderm and EVL as well as between blastoderm and yolk, to simulate the sealing between EVL cells via apical junctional complexes [50] and the dense cortical yolk cell cytoskeleton [51], respectively. Adhesion prevents cells (and by extension, morphogens) from interacting with the boundary of the simulation domain, so the choice of simulation boundary condition is irrelevant. Special care needs to be taken for computing the nonlocal integral term for adhesion, which is discretized into its angular and radial components using bilinear interpolation. This interpolation is encoded into a precomputed weight matrix representing a kernel, which can then be used with a kernel convolution to efficiently compute the adhesion fluxes. Notably, the kernel convolution must be run twice: once for the x component of the adhesive flux and once for the y component. By symmetry, these two kernels are transposes of each other. In addition, it is important to select an appropriate grid size for the numerical simulation. If the grid is too fine, then the time to simulate the model can become prohibitively large. Conversely, if the grid size is too coarse, then even applying bilinear interpolation will not provide sufficient spatial resolution for the circular nonlocal adhesion term. We find from

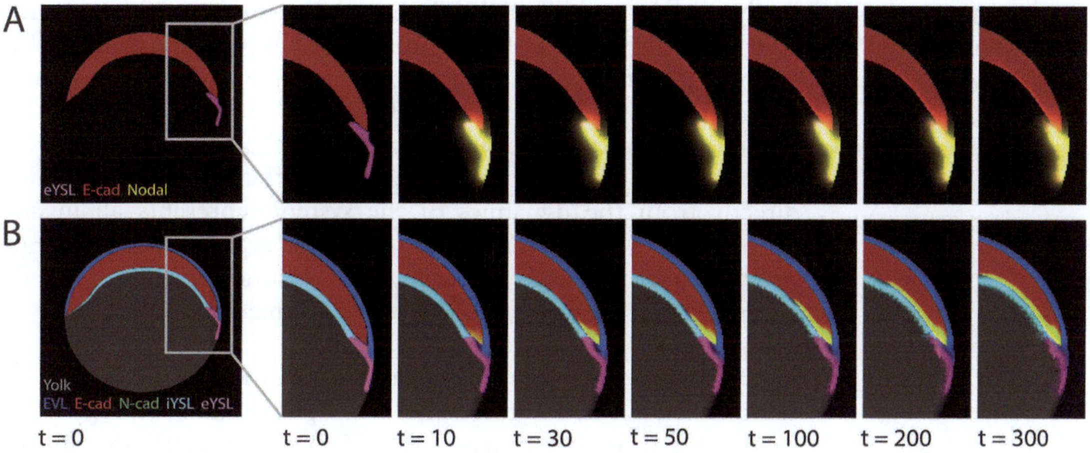

Fig. 2 Simulation of Nodal expression in a whole-embryo, systems-level model of zebrafish gastrulation. (**a**) Nodal (yellow) is expressed in a particular region of the yolk called the eYSL (purple) and diffuses into the blastoderm, which ubiquitously expresses E-cadherin (red). (**b**) Due to the localized expression of Nodal, cells close to the eYSL increase expression of N-cadherin (green), resulting in their involution toward the animal pole (top of the image). Arbitrary units. *t* time

experimentation that for a domain of 25 × 25 units, discretizing into a 250 × 250 grid lattice is tractable on modern hardware. This provides a grid where each square is 0.1 × 0.1 units. Considering that the radius of adhesion is set to 1 unit, this was also sufficient resolution for the adhesion integral.

Figure 2 shows the simulation results of the example whole-embryo model for zebrafish morphogenesis. The initial state with the whole embryo is shown on the left panels, followed by developmental snapshots at different time points (arbitrary units) of the region of interest (gray box)—yet the entire embryo is simulated with the model. Figure 2a visualizes the blastoderm expressing E-cadherin, as well as the eYSL which is fated to express Nodal. Shapes of regions in the initial state are based on experimental images of zebrafish gastrulation [48, 52]. As the simulation progresses, the results show how Nodal is expressed in the eYSL and diffuses outward into the nearby region of the blastoderm, effecting only a small portion of the tissue. As a result of exposure to Nodal, cells increased N-cadherin expression, which then causes their involution and migration toward the animal pole (Fig. 2b). This results in the development of a band of N-cadherin expressing cells close to the yolk, similar to the mesendoderm observed in a zebrafish embryo. This example shows the capacity of a computational systems biology approach to precisely explain the complex tissue dynamics due to regulatory interactions, differential CAM expression, and biophysical forces in a developing embryo.

3.3 Machine Learning of Computational Systems Biology Models

Designing the equations and finding the parameters of a dynamical system that can precisely recapitulate a set of morphogenetic phenotypes is a nontrivial task due to the complexity of the nonlinear interactions of biological regulation [1]. In general, the set of possible combinations of interactions and their parameter space is too vast to be explored manually. Indeed, the inference of models directly from the dynamics of the system represents an inverse problem for which no analytical solution is available [53]. Instead, heuristic methodologies that find the solution can be used for aiding in the design and discovery of mechanistic models of morphogenesis.

Machine learning methods based on heuristic optimization are ideal for the reverse engineering of complex systems biology models directly from experimental data [54]. These methods can efficiently explore the vast space of solutions to a particular problem. Although they cannot guarantee finding the optimal solution, they can perform well enough to find an acceptable model that can recapitulate all the experiments in the input dataset—a hypothesis of the mechanisms governing the observed phenotypes. Evolutionary computation is a popular heuristic methodology based on the stochastic optimization of a population of solutions to a problem [55]. In addition to the reverse engineering of phenotypes of morphogenesis and regeneration [17], we have successfully employed evolutionary computation for discovering solutions to melanoma-like phenotypes [56], morphogenetic designs [57], tensegrity structures [58], and artificial development and differentiation [59].

The main principles behind an evolutionary computation algorithm for the reverse engineering of phenotypes are simple to implement in a general-purpose language. Alternatively, programming libraries providing readily available implementations exist in most modern programming languages [60–62]. The pseudocode for a general evolutionary algorithm for systems biology models of morphogenesis is shown in Box 2.

Box 2 Evolutionary algorithm pseudocode for the inference of systems biology models

```
1. Initialize population with random models
2. Simulate models
3. Calculate error of models with respect target
   phenotype(s)
4. While (best model error > 0) AND (num. generations <
   max. generations)
   (a) Reproduce population of models by crossover and
       mutation
   (b) Simulate children models
```

(continued)

```
    (c) Calculate error of children models with respect
        target phenotype(s)
    (d) Select best models for next generation
        population
5. Return best model found
```

For the reverse engineering of morphogenesis phenotypes, a solution in an evolutionary algorithm represents a model—a system of differential equations—and its fitness represents the ability of the model to recapitulate a given phenotype—such as a gene expression pattern or a particular shape [63]. The input of the algorithm is one or multiple phenotypes, including the initial and final state, and the output is a discovered model that when simulated can recapitulate all the input phenotypes. The method automatically infers the specific equations, parameters, and number of variables needed for the model. The algorithm first automatically generates a set of random models as the initial population. Each random model includes an equation for every morphogen, each with a random number of regulatory terms (modeling the interactions between genes) and with random parameters. The main loop in the algorithm creates new models (reproduction), which substitute the worst models in the population (selection). New models are created by stochastically combining existing ones and adding random changes (mutation). The main loop stops when a satisfactory model has been found or a maximum number of generations have been reached, returning the best model found as output. In this way, the algorithm can be used to find just the parameters of a predefined model or also the specific terms and interactions in the equations (the topology of the regulatory network) by adjusting accordingly the operations performed during model reproduction and mutation. These evolutionary operations can also include adding new variables to the model, which can represent unknown genes predicted by the machine learning as necessary to explain the input dataset of morphogenesis phenotypes [17]. In addition, a hybrid approach can be implemented to include a set of known interactions from experimental evidence that the algorithm can complement with new regulations and product predictions [56].

Figure 3 illustrates the machine learning methodology based on evolutionary computation for the reverse engineering of mechanistic genetic regulatory network (GRN) models that can develop a given expression pattern in a multicellular tissue area or cell culture [64]. The method takes as input an initial expression pattern and a target expression pattern (e.g., a circle shape) and discovers a mechanistic GRN that starting with the input initial expression pattern can develop into the input target expression pattern. The inference method starts with an initial population of

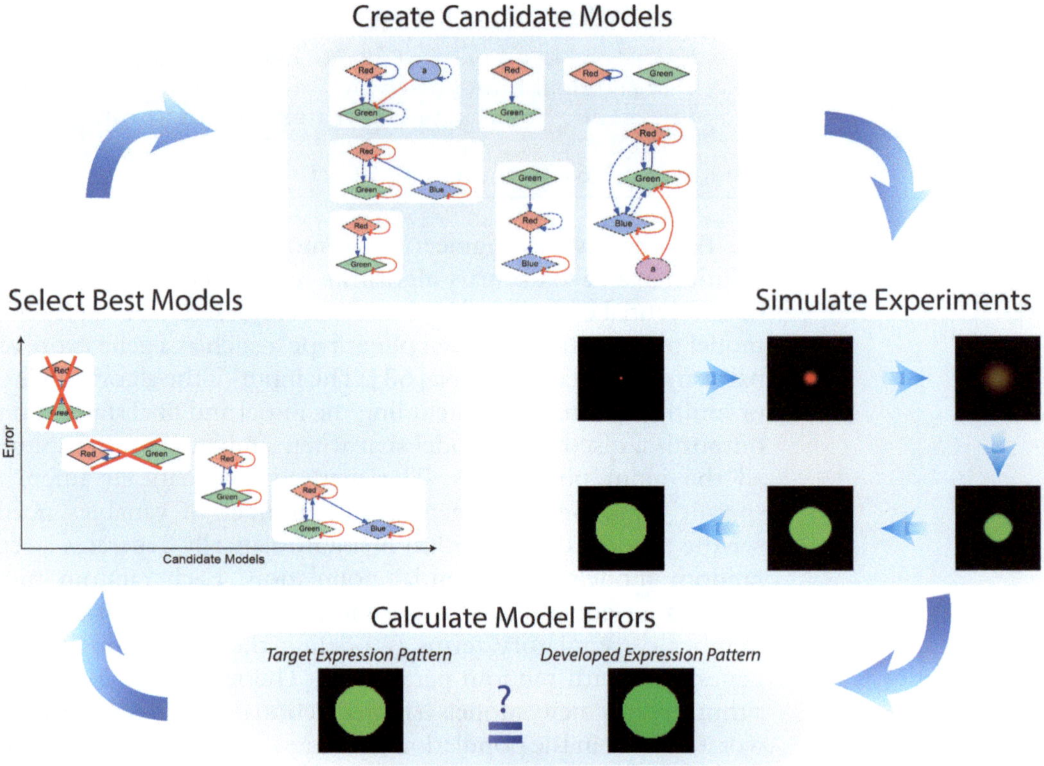

Fig. 3 Machine learning methodology based on evolutionary computation for inferring a mechanistic model that can recapitulate the formation of a particular gene expression pattern. A population of models are iteratively refined and improved by stochastically combining and mutating them, simulating their dynamic behaviors, computing their errors with respect their ability to recapitulate a particular expression pattern, and selecting the best ones among the population. This iterative process is repeated until an acceptable solution is found

simple GRNs with random regulations and parameters. Each candidate GRN model is simulated and its error calculated as the quantitative difference between the developed and target gene expression pattern. Next, the best models—those with the lowest errors—are selected for the next generation, which is combined and randomly mutated to create new candidate models, closing the evolutionary algorithm loop. This evolutionary cycle is repeated until an acceptable solution is found or a predetermined number of generations is reached. In this way, the genetic and signaling interactions (the topology of the network) can be reversed engineered de novo to discover a model that could develop a given gene expression pattern. Multiple executions of the algorithm may discover multiple different models that when simulated can recapitulate equally well the input phenotypes. These models represent different hypotheses that can explain all the experimental input. Crucially, the computational models can aid in the design of new experiments to discriminate between these hypotheses by simulating different perturbations in silico—such as genetic knockouts. The perturbation that maximizes the difference between the predictions from the different models represents indeed the next best experiment to perform at the bench. The following section will further describe methods to validate such inferred models.

3.4 Validating Systems-Level Models with Computational Predictions

A systems-level computational model represents a testable formal hypothesis of the control mechanism explaining a particular phenotype or set of phenotypes observed during morphogenesis. Crucially, new predictions can be formulated in terms of novel phenotypes resulting from new perturbations. Surgical amputations and graftings not included in the input target dataset can be simulated and their resultant spatial phenotypes be predicted. For this, the initial or an intermediate state of the simulation can be altered, and the model run to obtain the predicted phenotype. Genetic perturbations can be also readily implemented. Both loss-of-function and gain-of-function mutations can be simulated by altering the governing equations of the system. For example, the knockout of a gene (or silencing with RNA interference) can be implemented by removing all the production terms in its corresponding equations (or setting their parameters to zero) [17]. Conversely, ectopic gene expression such as with protein-soaked agarose beads can be simulated by adding to a product equation a new spatially dependent source term directly affecting the location of the bead [65]. The biophysical properties of the cell and tissue, such as protein-protein adhesion strengths, can be altered as well in the model by changing specific parameters [34]. The model will then output precise predictions in terms of resultant phenotypes from the novel perturbations, after which the most interesting or extreme can be selected to be validated at the bench.

The presented machine learning methodology can predict both novel regulatory interactions and genes found to be necessary to explain the input phenotypes. These predictions can be characterized and validated at the bench. For this, we have developed the computational tool MoCha [66], which was recently updated to work with the latest version of the STRING database [26]. The tool offers a simple command line interface to input a set of genes and it returns the candidate genes that are known to interact directly or indirectly with all the genes given as input, as illustrated in Fig. 4. The user can limit the search to a specific organism, type of evidence, and confidence, as well as specify a maximum length of the interaction pathways. Box 3 shows an example illustrating the use

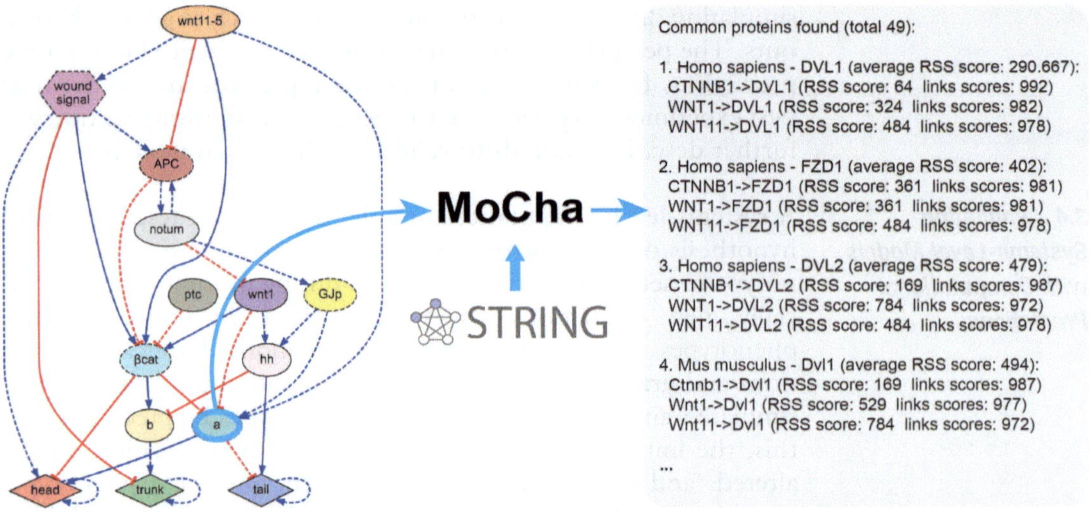

Fig. 4 A reverse-engineered mechanistic model can generate testable predictions and new knowledge. A new unknown gene predicted as necessary by the machine learning methodology can be characterized by our computational tool MoCha, which can find candidate genes with the predicted regulatory interactions among more than six billion known interactions from the STRING database

of the tool for searching all the proteins that directly interacts (1-link pathways) with the products of *ctnnb1*, *wnt1*, and *wnt11* with any type of evidence (type 7) and with a minimum confidence of 900 in any organism ("all").

Box 3 Example execution of MoCha to find genes with particular regulatory interactions. The user input command is in blue and the tool output is in black

```
$ ./mocha.sh 900 1 all 7 ctnnb1 wnt1 wnt11

MoCha: searching in 5090 organisms, 14065928 proteins
with common names and 6183296832 links.

Searching for all the proteins interacting with
ctnnb1, wnt1 and wnt11 in any organism in pathways with
1 links maximum and with link scores of type 7 and equal
or higher than 900. Pathway scores indicate the
residual sum of squares (RSS) of the link scores (lower
is better confidence). Link scores indicate the
STRING confidence of the proteins interaction (higher
is better confidence).
Searching in organism Latimeria chalumnae (taxon id:
7897 num named proteins: 13540 num links: 7487083) ...
 Found 28 links from ctnnb1
 Found 4 links from wnt1
 Found 1 links from wnt11
```

(continued)

```
 Found 0 common proteins
Searching in organism Danio rerio (taxon id: 7955 num
named proteins: 20068 num links: 14116455) ...
 Found 191 links from ctnnb1
 Found 42 links from wnt1
 Found 7 links from wnt11
 Found 0 common proteins

(...)

Common proteins found (total 49):

1. Homo sapiens - DVL1 (average RSS score: 290.667):
 CTNNB1->DVL1 (RSS score: 64 links scores: 992)
 WNT1->DVL1 (RSS score: 324 links scores: 982)
 WNT11->DVL1 (RSS score: 484 links scores: 978)

2. Homo sapiens - FZD1 (average RSS score: 402):
 CTNNB1->FZD1 (RSS score: 361 links scores: 981)
 WNT1->FZD1 (RSS score: 361 links scores: 981)
 WNT11->FZD1 (RSS score: 484 links scores: 978)

3. Homo sapiens - DVL2 (average RSS score: 479):
 CTNNB1->DVL2 (RSS score: 169 links scores: 987)
 WNT1->DVL2 (RSS score: 784 links scores: 972)
 WNT11->DVL2 (RSS score: 484 links scores: 978)

4. Mus musculus - Dvl1 (average RSS score: 494):
 Ctnnb1->Dvl1 (RSS score: 169 links scores: 987)
 Wnt1->Dvl1 (RSS score: 529 links scores: 977)
 Wnt11->Dvl1 (RSS score: 784 links scores: 972)

5. Homo sapiens - FZD2 (average RSS score: 560.333):
 CTNNB1->FZD2 (RSS score: 576 links scores: 976)
 WNT1->FZD2 (RSS score: 576 links scores: 976)
 WNT11->FZD2 (RSS score: 529 links scores: 977)

(...)
```

After the command is executed, the tool outputs within seconds a report with the 49 proteins found to directly interact with the three input products and within the limits specified. The list is sorted in order of combined evidence score (starting with the highest confidence) and it includes information about the organism and particular interaction pathways found. We have used this protocol to characterize a predicted novel gene in planarian regeneration and its capacity to rescue a no-tail phenotype, a prediction that we validated subsequently at the bench [67].

In addition to predicting the phenotype for a given perturbation, reverse-engineered systems biology models can be used to

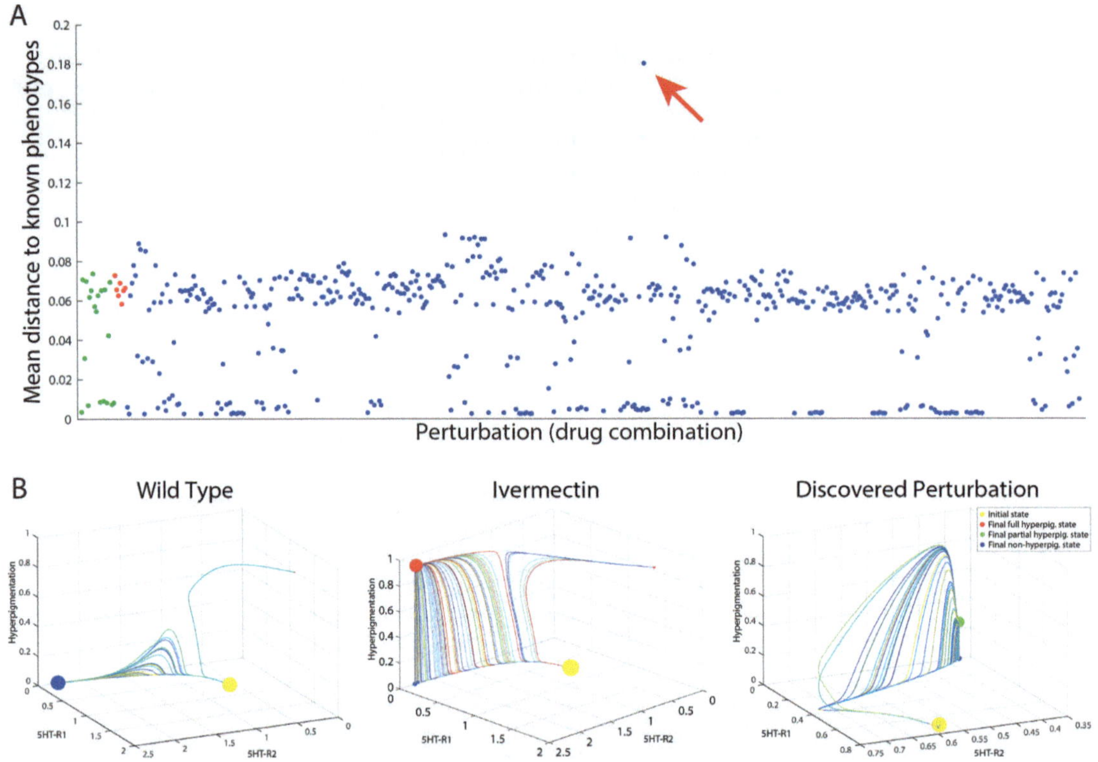

Fig. 5 Computational systems biology models can be used to discover a precise perturbation that results in a particular phenotype of interest. (**a**) Testing in silico all possible one to three drug combinations reveal only one perturbation (combination of drugs) that results in a never-seen-before partially hyperpigmented *Xenopus* tadpole. The red arrow indicates the only combination of three drugs that was predicted to result in the phenotype of interest; green dots correspond to the input dataset, red dots to the validation dataset, and blue dots to novel experiments not previously performed in vivo. (**b**) The phase portraits show the dynamics of the phenotypes obtained in the wild type, when administering Ivermectin, and when administering the discovered combination of drugs. In the first two cases albeit with different probabilities, the stochastic trajectories end in either a low-level pigmentation attractor similar to the wild-type phenotype (blue circles) or a very high pigmentation attractor corresponding to the hyperpigmented phenotype (red circles). In contrast, a bifurcation in the system is observed after applying the discovered novel perturbation, which results in a new intermediate attractor (green circle) corresponding to a never-seen-before partially hyperpigmented phenotype that was subsequentially validated at the bench

discover the specific perturbation that results into a desired morphogenetic phenotype. While the reverse engineering of models from a set of phenotypes represents an inverse problem in need of computationally intensive heuristic methods, the opposite task of finding the phenotype that results from a model is a direct problem involving only the numerical solution of the model equations [3]. Hence, an exhaustive search can be performed to compute the resultant phenotypes from all possible qualitative perturbations to find the one that produces a given phenotype.

Figure 5 shows an example of this approach to find a combination of drugs that could result in a novel phenotype during *Xenopus* morphogenesis [68]. First, a dynamic model was inferred from a set of experiments applying different combinations of drugs during *Xenopus* development. These experiments could result stochastically in two different phenotypes: either the wild-type phenotype with low pigmentation or a melanoma-like phenotype with hyperpigmentation [56]. The stochastic reverse-engineered model could predict precisely the percentage of embryos that would develop the aberrant hyperpigmented phenotype for a given combination of drugs. Indeed, the two experimentally observed phenotypes—wild-type low pigmentation or the hyperpigmented phenotype—represented a stochastic bistable dynamic system. However, an exhaustive simulation of the inferred stochastic model under any combination of one, two, or three drugs revealed a particular combination of three drugs that was predicted to result in a partially pigmented phenotype (Fig. 5a). An analysis of the dynamics of the system demonstrated that this particular drug perturbation produced a bifurcation in the original bistable system, resulting in a new attractor in the intermediate pigmentation region (Fig. 5b, green dot in the discovered perturbation panel). Furthermore, the exact predicted never-seen-before phenotype was validated in vivo at the bench afterward by administering the precise drug cocktail discovered by the algorithm [68].

3.5 Conclusions

In this chapter we have presented an overview of computational systems biology methods toward the understanding of the mechanisms of morphogenesis. Systems biology models based on differential equations represent precise mathematical hypothesis that can include the regulatory, signaling, and biophysical elements sufficient to explain a set of morphological phenotypes. Different levels of complexity can be encapsulated within these models, from morphogens reacting and diffusing in a tissue section to cellular adhesion forces and their regulation at the whole-embryo scale. Crucially, machine learning methods can be employed to infer such models directly from experimental phenotypes. The regulatory elements, their interactions, and parameters of a model can be all reverse engineered, resulting in the automatic formulation of mechanistic hypotheses that can be further validated at the bench. Indeed, mathematical systems biology models can predict the exact phenotypes resulting from novel perturbations. This approach can then be exploited for the discovery of interventions toward target phenotypes—such as healthy states in biomedical applications. In summary, a computational systems biology approach can precisely integrate regulatory and signaling interactions together with biophysical forces toward a comprehensive and mechanistic understanding of morphogenesis.

4 Notes

1. Notice that time and space are continuous in the model defined with the system of partial differential equations. However, they need to be discretized into particular time steps (20,000 steps of 0.5 time units each) and space locations (a 100 by 100 grid), respectively, for their computational simulation. Each space location in the grid corresponds with a location in the tissue and not a single cell.
2. The behavior of the system at the boundary of the domain—the borders of the simulated tissue section—are defined as a boundary condition. A Neumann boundary condition specifies a constant rate of change within the boundary of the domain, which the example sets to 0 to simulate that no species can cross the boundary. An alternative modeling approach is to use a Dirichlet boundary condition, which sets a constant value within the boundary and hence simulates a constant source or sink for the chemical species.
3. Since no cell or morphogens can reach the boundary of the domain, it remains zero for all the variables in the model and hence the boundary conditions are not relevant for this type of whole-embryo models.

Acknowledgments

We thank the members of the Lobo Lab for helpful discussions. This work was supported by the National Institute of General Medical Sciences of the National Institutes of Health under award number R35GM137953. The content is solely the responsibility of the authors and does not necessarily represent the official views of the National Institutes of Health.

Author Contributions: J. K. wrote Subheading 3.2. R. M. wrote Subheading 3.3. D. L. wrote Subheadings 1, 3.1, 3.4, and 3.5. All authors revised and approved the final version of the manuscript.

References

1. Lobo D, Levin M (2017) Computing a worm: reverse-engineering planarian regeneration. In: Adamatzky A (ed) Advances in unconventional computing. Volume 2: prototypes, models and algorithms. Springer International Publishing, Switzerland, pp 637–654
2. Rubin BP, Brockes J, Galliot B et al (2015) A dynamic architecture of life. F1000Res 4:1288
3. Lobo D, Solano M, Bubenik GA et al (2014) A linear-encoding model explains the variability of the target morphology in regeneration. J R Soc Interface 11:20130918
4. McLaughlin KA, Levin M (2018) Bioelectric signaling in regeneration: mechanisms of ionic controls of growth and form. Dev Biol 433:177–189
5. Chiou K, Collins E-MS (2018) Why we need mechanics to understand animal regeneration. Dev Biol 433:155–165

6. Stiehl T, Marciniak-Czochra A (2017) Stem cell self-renewal in regeneration and cancer: insights from mathematical modeling. Curr Opin Syst Biol 5:112–120
7. Sharpe J (2017) Computer modeling in developmental biology: growing today, essential tomorrow. Development 144:4214–4225
8. Herath S, Lobo D (2020) Cross-inhibition of Turing patterns explains the self-organized regulatory mechanism of planarian fission. J Theor Biol 485:110042
9. Bartocci E, Lió P (2016) Computational modeling, formal analysis, and tools for systems biology. PLoS Comput Biol 12:e1004591
10. Kitano H (2002) Computational systems biology. Nature 420:206–210
11. Thieffry D (2007) Dynamical roles of biological regulatory circuits. Brief Bioinform 8:220–225
12. Jiménez A, Munteanu A, Sharpe J (2015) Dynamics of gene circuits shapes evolvability. Proc Natl Acad Sci 112:201411065
13. Economou AD, Ohazama A, Porntaveetus T et al (2012) Periodic stripe formation by a Turing mechanism operating at growth zones in the mammalian palate. Nat Genet 44:348–351
14. Sheth R, Marcon L, Bastida MF et al (2012) Hox genes regulate digit patterning by controlling the wavelength of a Turing-type mechanism. Science 338:1476–1480
15. Prusinkiewicz P, Erasmus Y, Lane B et al (2007) Evolution and development of inflorescence architectures. Science 316:1452–1456
16. Jiménez A, Cotterell J, Munteanu A et al (2017) A spectrum of modularity in multi-functional gene circuits. Mol Syst Biol 13:925
17. Lobo D, Levin M (2015) Inferring regulatory networks from experimental morphological phenotypes: a computational method reverse-engineers planarian regeneration. PLoS Comput Biol 11:e1004295
18. Uzkudun M, Marcon L, Sharpe J (2015) Data-driven modelling of a gene regulatory network for cell fate decisions in the growing limb bud. Mol Syst Biol 11:815–815
19. Jaeger J, Crombach A (2012) Life's attractors: understanding developmental systems through reverse engineering and in silico evolution. In: Soyer OS (ed) Evolutionary systems biology. Springer, New York, pp 93–119
20. Lobo D, Feldman EB, Shah M et al (2014) Limbform: a functional ontology-based database of limb regeneration experiments. Bioinformatics 30:3598–3600
21. Roy J, Cheung E, Bhatti J et al (2020) Curation and annotation of planarian gene expression patterns with segmented reference morphologies. Bioinformatics 36:2881–2887
22. Lobo D, Malone TJ, Levin M (2013) Planform: an application and database of graph-encoded planarian regenerative experiments. Bioinformatics 29:1098–1100
23. Emmons-Bell M, Durant F, Hammelman J et al (2015) Gap junctional blockade stochastically induces different species-specific head anatomies in genetically wild-type Girardia dorotocephala flatworms. Int J Mol Sci 16:27865–27896
24. Durant F, Lobo D, Hammelman J et al (2016) Physiological controls of large-scale patterning in planarian regeneration: a molecular and computational perspective on growth and form. Regeneration 3:78–102
25. Weiner R, Schmitt BA, Podhaisky H (1997) ROWMAP--a ROW-code with Krylov techniques for large stiff ODEs. Appl Numer Math 25:303–319
26. Szklarczyk D, Gable AL, Lyon D et al (2019) STRING v11: protein–protein association networks with increased coverage, supporting functional discovery in genome-wide experimental datasets. Nucleic Acids Res 47: D607–D613
27. Lobo D, Beane WS, Levin M (2012) Modeling planarian regeneration: a primer for reverse-engineering the worm. PLoS Comput Biol 8: e1002481
28. Azuaje F (2011) Computational discrete models of tissue growth and regeneration. Brief Bioinform 12:64–77
29. Plikus MV, Baker RE, Chen CC et al (2011) Self-organizing and stochastic behaviors during the regeneration of hair stem cells. Science 332:586–589
30. Lobo D, Vico FJ, Dassow J (2011) Graph grammars with string-regulated rewriting. Theor Comput Sci 412:6101–6111
31. García-Quismondo M, Levin M, Lobo D (2017) Modeling regenerative processes with membrane computing. Inf Sci (Ny) 381:229–249
32. Eskandari M, Kuhl E (2015) Systems biology and mechanics of growth. Wiley Interdiscip Rev Syst Biol Med 7:401–412
33. Marcon L, Sharpe J (2012) Turing patterns in development: what about the horse part? Curr Opin Genet Dev 22:578–584
34. Ko JM, Lobo D (2019) Continuous dynamic modeling of regulated cell adhesion: sorting, intercalation, and involution. Biophys J 117:2166–2179
35. Germann P, Marin-Riera M, Sharpe J (2019) Ya||a: GPU-powered spheroid models for

mesenchyme and epithelium. Cell Syst 8:261–266.e3
36. Delile J, Herrmann M, Peyriéras N et al (2017) A cell-based computational model of early embryogenesis coupling mechanical behaviour and gene regulation. Nat Commun 8:13929
37. Mirams GR, Arthurs CJ, Bernabeu MO et al (2013) Chaste: an open source C++ library for computational physiology and biology. PLoS Comput Biol 9:e1002970
38. Song Y, Yang S, Lei JZ (2018) ParaCells: a GPU architecture for cell-centered models in computational biology. IEEE/ACM Trans Comput Biol Bioinforma 5963:1–14
39. Ghaffarizadeh A, Heiland R, Friedman SH et al (2018) PhysiCell: an open source physics-based cell simulator for 3-D multicellular systems. PLoS Comput Biol 14:e1005991
40. Turing AM (1952) The chemical basis of morphogenesis. Philos Trans R Soc Lond Ser B Biol Sci 237:37–72
41. Krieg M, Arboleda-Estudillo Y, Puech PH et al (2008) Tensile forces govern germ-layer organization in zebrafish. Nat Cell Biol 10:429–436
42. Maître J-L, Heisenberg C-P (2013) Three functions of Cadherins in cell adhesion. Curr Biol 23:R626–R633
43. Samanta D, Almo SC (2015) Nectin family of cell-adhesion molecules: structural and molecular aspects of function and specificity. Cell Mol Life Sci 72:645–658
44. Schier AF (2009) Nodal morphogens. Cold Spring Harb Perspect Biol 1:–a003459
45. Giger FA, David NB (2017) Endodermal germ-layer formation through active actin-driven migration triggered by N-cadherin. Proc Natl Acad Sci U S A 114:201708116
46. Carvalho L, Heisenberg C-P (2010) The yolk syncytial layer in early zebrafish development. Trends Cell Biol 20:586–592
47. Rodaway A, Takeda H, Koshida S et al (1999) Induction of the mesendoderm in the zebrafish germ ring by yolk cell-derived TGF-beta family signals and discrimination of mesoderm and endoderm by FGF. Development 126:3067–3078
48. Montero J-A, Carvalho L, Wilsch-Bräuninger M et al (2005) Shield formation at the onset of zebrafish gastrulation. Development 132:1187–1198
49. Williams PH, Hagemann A, González-Gaitán M et al (2004) Visualizing long-range movement of the morphogen Xnr2 in the Xenopus embryo. Curr Biol 14:1916–1923
50. Stemmler MP, Koschorz B, Carney TJ et al (2009) The epithelial cell adhesion molecule EpCAM is required for epithelial morphogenesis and integrity during zebrafish epiboly and skin development. PLoS Genet 5:e1000563
51. Bruce AEE (2016) Zebrafish epiboly: spreading thin over the yolk. Dev Dyn 245:244–258
52. Lachnit M, Kur E, Driever W (2008) Alterations of the cytoskeleton in all three embryonic lineages contribute to the epiboly defect of Pou5f1/Oct4 deficient MZ spg zebrafish embryos. Dev Biol 315:1–17
53. Aster RC and Thurber CHCN-J or ABRRQ 8. . A (2012) Parameter estimation and inverse problems. Academic Press, Cambridge, Massachusetts
54. Reali F, Priami C, Marchetti L (2017) Optimization algorithms for computational systems biology. Front Appl Math Stat 3
55. Holland JH (1975) Adaptation in natural and artificial systems: an introductory analysis with applications to biology, control, and artificial intelligence. Michigan Univ. Press, Ann Arbor, Michigan
56. Lobikin M, Lobo D, Blackiston DJ et al (2015) Serotonergic regulation of melanocyte conversion: a bioelectrically regulated network for stochastic all-or-none hyperpigmentation. Sci Signal 8:ra99
57. Lobo D, Fernández JD, and Vico FJ (2012) Behavior-finding: morphogenetic designs shaped by function, In: Doursat, R., Sayama, H., and Michel, O. (eds.) Morphogenetic engineering, pp. 441–472 Springer Berlin Heidelberg
58. Lobo D, Vico FJ (2010) Evolutionary development of tensegrity structures. Biosystems 101:167–176
59. Lobo D, Vico FJ (2010) Evolution of form and function in a model of differentiated multicellular organisms with gene regulatory networks. Biosystems 102:112–123
60. Henry A, Hemery M, François P (2018) φ-Evo: a program to evolve phenotypic models of biological networks. PLOS Comput Biol 14: e1006244
61. Fortin FA, De Rainville FM, Gardner MA et al (2012) DEAP: evolutionary algorithms made easy. J Mach Learn Res 13:2171–2175
62. Mohammadi A, Asadi H, Mohamed S et al (2017) OpenGA, a C++ genetic algorithm library. In: 2017 IEEE international conference on systems, man, and cybernetics (SMC). IEEE, Piscataway, New Jersey, pp 2051–2056
63. Budnikova M, Habig J, Lobo D et al (2014) Design of a flexible component gathering

algorithm for converting cell-based models to graph representations for use in evolutionary search. BMC Bioinformatics 15:178
64. Mousavi R, Konuru SH, Lobo D (2021) Inference of Dynamic Spatial GRN Models with Multi-GPU Evolutionary Computation. Brief Bioinform 22:bbab104
65. Walton KD, Whidden M, Kolterud A et al (2015) Villification in the mouse: bmp signals control intestinal villus patterning. Development:734–764
66. Lobo D, Hammelman J, Levin M (2016) MoCha: molecular characterization of unknown pathways. J Comput Biol 23:291–297
67. Lobo D, Morokuma J, Levin M (2016) Computational discovery and in vivo validation of hnf4 as a regulatory gene in planarian regeneration. Bioinformatics 32:2681–2685
68. Lobo D, Lobikin M, Levin M (2017) Discovering novel phenotypes with automatically inferred dynamic models: a partial melanocyte conversion in Xenopus. Sci Rep 7:41339

Chapter 15

Agent-Based Modeling of Complex Molecular Systems

Mike Holcombe and Eva Qwarnstrom

Abstract

The seamless integration of laboratory experiments and detailed computational modeling provides an exciting route to uncovering many new insights into complex biological processes. In particular, the development of agent-based modeling using supercomputers has provided new opportunities for highly detailed, validated simulations that provide the researcher with greater understanding of these processes and new directions for investigation. This chapter examines some of the principles behind the powerful computational framework FLAME and its application in a number of different areas with a more detailed look at a particular signaling example involving the NF-κB cascade.

Key words Agent based modeling, Computational modeling, Cytoskeleton, FLAME, IL-1, IL1R1 complex, Map kinase, NF-κB, Signal transduction, TILRR

1 Introduction

Understanding the molecular events that drive activation of regulatory systems, and the rules which underpin their control of cell and tissue behavior, provide great challenges for biological scientists. Although much is known about many of these systems, it is likely that we are only aware of a small fraction of the events that contribute to steering cellular responses. Simulation based on accurate and detailed models is a key aspect in search for greater understanding of biology. When modeling and experimental studies go hand in hand much progress can be made. To optimize the value of such interdisciplinary studies, it is of primary importance that the model accurately reproduces the biology and considers parameters such as spatial relationships and the cell environment, known to control cell behavior.

Supplementary Information The online version of this chapter (https://doi.org/10.1007/978-1-0716-1831-8_15) contains supplementary material, which is available to authorized users.

Sonia Cortassa and Miguel A. Aon (eds.), *Computational Systems Biology in Medicine and Biotechnology: Methods and Protocols*, Methods in Molecular Biology, vol. 2399, https://doi.org/10.1007/978-1-0716-1831-8_15,

Agent-based modeling is particularly well suited to reproduce spatially complex and highly dynamic systems and therefore provides an accurate representation of the architectural structure of the cell and its environment. These characteristics are critically associated with regulatory events within the cell and with cell behavior.

Whilst mathematical models that are based on general differential equation provide in depth information of specific regulatory events, they are limited by their lack of spatial representation, which is highly relevant to regulation of complex and dynamic biological systems.

We will illustrate the importance of representing molecular trafficking in a 3 dimensional space on complex regulatory systems using a detailed model of inflammatory receptor-induced activation of the transcription factor NF-κB (nuclear factor kappa B), as an example. In addition, we will briefly describe other applications where the agent-based model also has been central to understanding regulation of a biological system in context of its environment.

2 Materials

The model was developed using the FLAME framework with part using FLAME GPU, a version of the Flexible Largescale Agent-based Modelling Environment (http://www.flame.ac.uk) and modern Graphical Processing Units [1, 2] with the model providing a detailed representation of real time signaling events in a three-dimensional space in live cells.

Agents can move within their environment according to physical laws; can engage, for example bind with other agents; can be broken up into "daughter" agents; and so on (Fig. 1). Usually agents can be in a number of different states—for example, idle, active in some sense, and even dead. Agents can communicate with each other and with their environment. What an agent does at any instant in time depends on the following.

1. Where it is.
2. What state it is in.
3. What messages it receives from the environment or other agents.

Then the agent will do the following.

1. Change its state.
2. Move to a new position.
3. Send a message to other agents or its environment.
4. Transform into another agent or agents through binding or splitting.

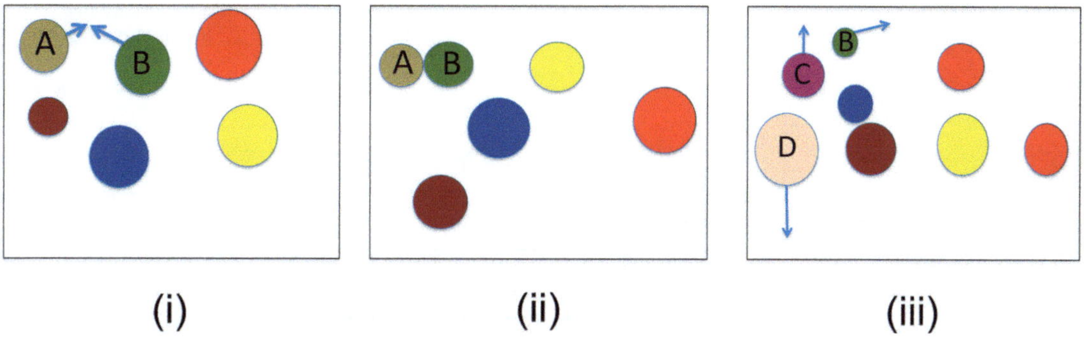

Fig. 1 A cartoon of different stages of agent activity/interaction location, translocation, and movement of agents as molecules. (i) Molecules including A and B are moving around the space. Once A and B are close enough, they will react. (ii) If conditions are suitable the molecules A and B undergo the appropriate reaction and other molecules proceed independently. (iii) In this case molecule A splits into molecules C and D and the agents continue to progress through the space, undertaking other actions where valid. Agent A is deleted from the model run and new agents C and D created

Treating molecules as individual agents endowed with the chemical and spatial properties of the molecule provides an accurate means of creating a highly accurate and realistic basis for a model. An agent-based simulation then involves setting up a model environment replicating the key geometric and chemical characteristics relevant to the system under study, populating it with a suitable, and an appropriate set of different types of agents representing the key molecules involved. This environment and population of agents is then given initial conditions appropriate for the simulations. This means that each agent is given a unique identifier, a unique position and starting state. The shape and internal structures of the environment the agents are located in is defined suitably. The simulation then proceeds to let each agent carry out whatever operation it is permitted to do bearing in mind its internal state.

Software and programming tools.

A number of programming environments and tools are available for building and running agent-based models. Most are based around systems that can be run on desktop computers and offer convenient user interfaces that allow for the design of agents and the simulation context. However, we have seen that the sort of systems we wish to explore will involve a vast number of possibly complex agents and for this a parallel supercomputer is essential. Most of the existing agent-based frameworks are unsuitable since they are mostly based on the programming language Java which is not executable on a supercomputer, these computers usually require programs written in C (or Fortran).

The FLAME (Flexible Large-scale Agent-based Modelling Environment) has been developed precisely to solve these issues (*see* **Note 1**). It provides a set of tools and techniques for building large models in a principled way based on state of the art software

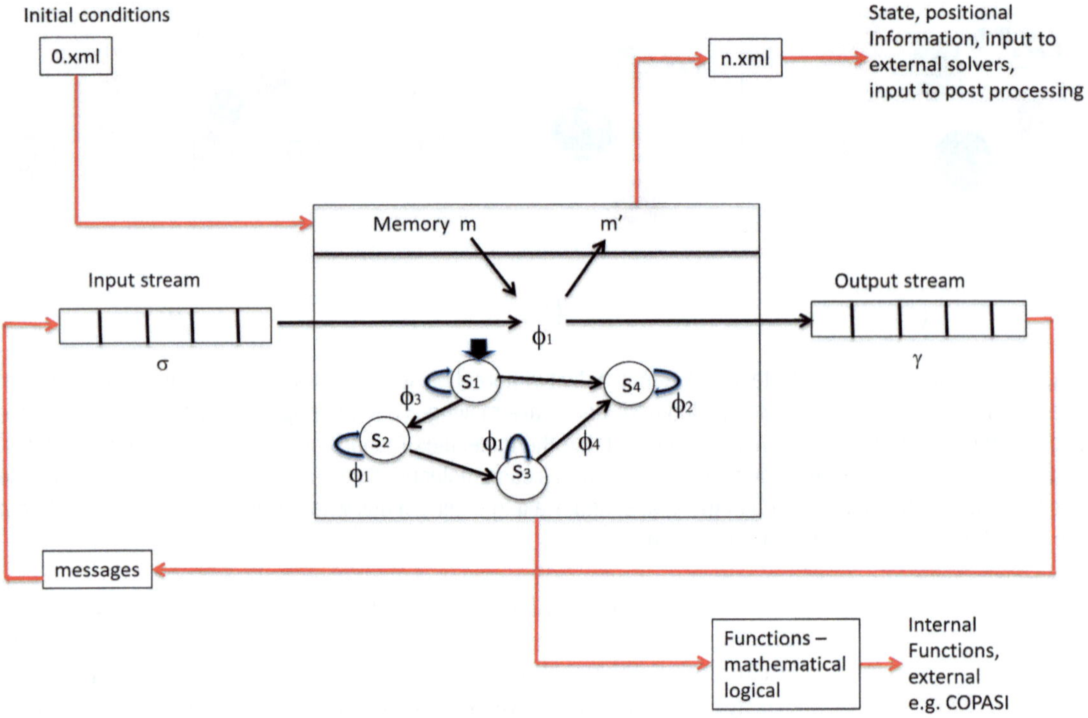

Fig. 2 A conceptual diagram of an X-machine agent model

engineering approaches that will ensure that the resulting model is reliable and trustworthy. Over many years large-scale supercomputer codes have been developed in numerous scientific disciplines. Not all are engineered as well as they could be.

FLAME tries to avoid these problems by using an incremental approach that is precise but intuitive, provides a simple mechanism for specifying agents, tools to analyze and verify the model and automatic procedures for generating highly optimized code that runs automatically on any common supercomputer or desktop. As we will see, it can also be integrated with other programs such as fluid dynamic codes to explore the behavior of agents in dynamic fluid environments such as the blood stream.

FLAME is based on fundamental computational ideas. Each agent is treated as a generalized computing machine (Fig. 2). Thus, each individual agent is an autonomous generalized machine, it has several internal states, receives information—inputs; reads its internal memory; changes state; and generates an output. The input and output mechanism is through a message board that agents have, possibly limited, access to. An agent-based model this comprises many separate machines, often many millions, each updating itself according to the rules and data during any simulation.

During a simulation run, each agent—molecule, cell, particle, and so on— reads relevant messages it has received and proceeds to update its state, move to a new position under whatever laws of motion are in force, sends a message detailing its new position, state, and other relevant information. If two agents are close enough to each other to interact in some way, such as molecules binding, then this can happen with a suitable probability. How close they must be, depend on things like reaction rates and their current state (*see* **Note 2**).

This is a very powerful and fully general computational model that can be used for many types of modeling in biology, social sciences, and so on [3–8]. In silico experiments that use an agent-based model can faithfully describe highly complex biological systems such as regulatory networks that control signaling patterns and cell behavior [3, 9–12].

3 Methods

3.1 Modeling the NF-κB Regulatory Network

The transcription factor NF-κB controls a range of fundamental responses including host defense mechanisms and cell survival [13]. The NF-κB network is highly complex and include multiple pathways, each consisting of a series of tightly controlled steps, which are reliant on molecular translocations, interactions, and activations.

To accurately represent the complex aspects of these events and their impact on network control the agent-based model utilizes a three-dimensional space in which each agent representing for example a cell surface receptor, an intracellular signaling component or a structural molecule has a specific location at any given time and can only interact with other agents within its local vicinity (Fig. 3, Video 1 link) [3, 9–12].

Hence each adaptor protein must move to the location of an activated receptor in order to itself become activated and initiate the signaling cascade. Similarly, proteins such as transcription factors must move to the location of a nuclear import or export receptor in order to translocate between cytoplasm and nucleus. In the nucleus it needs to move into interaction range with a transcription site to trigger the production of new protein agents. These spatial aspects of the agent-based model provide a greater level of detail and realism over more traditional forms of modeling, specifically in functional analysis of biological systems governed by three-dimensional organization. Models which consider the three-dimensional space of the cell and the cell environment also provide reliable predictions for regulatory events in vivo [14].

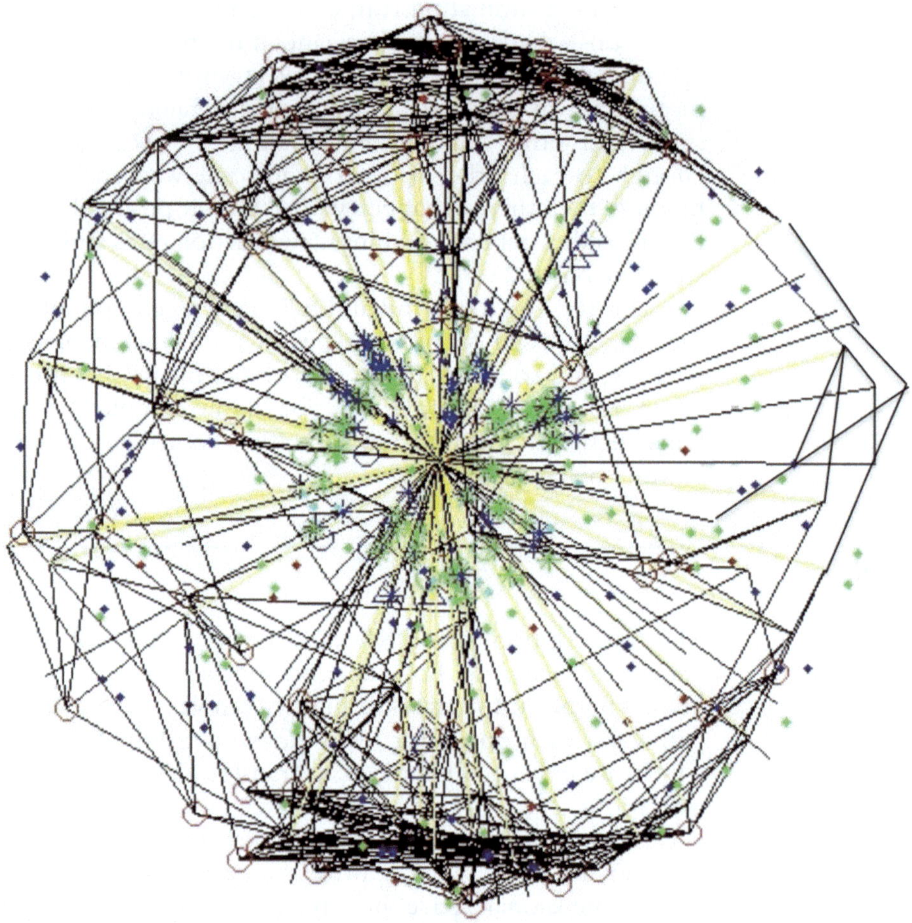

Fig. 3 A still from an animation using FLAME of molecular movement within a stylised cell. Each cellular component (agent) is represented by a sphere. This shows molecules in a cell moving around and with some interacting with the cytoskeleton—the black lines. Supplementary video 1 shows a simulation of this process

Model Development

To build these large-scale simulations a number of important tasks must be carried out. First, the scientific questions being examined must be identified (*see* **Note 3**). This is clearly an important phase that should be led by the biologists. Related to this are the sort of experiment that will be carried out to validate the model, the data that can be collected, and the parameters that can be measured to calibrate and verify the model identified. Development of the model includes three main consecutive steps, model build-up, model expansion and model validation.

Model Build-up

Agent-based models are constructed from three main parts - a description of the agent types including their memory and functions, the implementation of the agent functions, which determines the rule-set for their behavior, and, for each simulation, a starting state including a list of all agents. Execution of the model follows an iterative procedure, where each iteration represents a fixed time step and system update (*see* **Note 4**).

Molecular trafficking is most accurately represented by agent based models developed based on single cell data, where movements within the cell are monitored in real time [15–18]. In this agent-based model, both receptors and regulatory intermediates are represented as agents and the simulation as a whole describes key signaling events occurring in a single cell. The cell is simulated as two concentric spheres, the outer representing the cell membrane and the inner representing the nuclear membrane with the nucleus taking up 5% of the total cell volume (Fig. 3). Proteins can move throughout the cell and move between the compartments through transport receptors that have characteristics which determines the kinetics of movement and enable selection of proteins, which are allowed to move between the compartments.

Most interactions between agents in a signaling cascade involve activation. For example, modeling the IL-1 system, an agent representing an IRAK (IL-1 receptor associated kinase) can be activated by an active MyD88-adapter protein agent when they come in close proximity. Such chains of activation events form the core of the model pathway (*see* **Note 5**).

The next stage is to identify the key agents to be considered, these might be receptors, transcription factors, structural proteins etc. The environment such as the cell or tissue in which the agents exist must be specified. This process is based on prior knowledge and will be determined by the research question. The number of agents can be extensive because of the power of the framework, which makes it possible to include numbers of the various agents which reflects the biology (Table 1). The number and type of agents can easily be adjusted. It is easy to stumble at this point by including far more candidates for agents than are strictly relevant or too few. This needs to be considered carefully as reducing model complexity by reducing the complexity/scale (the number and type of agents) reduces simulation runtime significantly but excessive simplification of the model increases the risk of aberrant system behavior and errors [12]. One benefit of FLAME is that it is easy to test this by adding or removing agents from the model and regenerate the main simulation codes.

The model describes activation of the NF-κB pathway by the cytokine IL-1 (Fig. 4a, b) [11]. It includes key proteins that control activation of the system receptor complex (IL1RI, IL1RAP, and TILRR) and signaling intermediates that regulate the canonical and the noncanonical NF-κB pathways and trigger transcription (Fig. 4a, b). It incorporates branching of signals leading to distinct effects to allow simulations and monitoring of how changes at specific steps propagate downstream through various aspects of the pathway [13] (*see* **Note 6**).

Table 1
Summary of the agent types, location, potential states, and starting numbers

Agent name	Type	Location	Potential states	Starting number
Nuclear Import	Receptor	Nuclear membrane	Active	2500
Nuclear Export	Receptor	Nuclear membrane	Active	200
Il-1R1+ TILRR (WT or mutant)	Receptor	Cell membrane	Active, inactive	Variable depending on experiment, up to 3000
IL1R1	Receptor	Cell membrane	Active, inactive	Variable depending on experiment, up to 3000
Cytoskeleton	Receptor	Cytoplasm	Active and unoccupied. Active and IκB bound, inactive	600,000
Transcription site	Receptor	Nucleus	Active	500
MyD88	Protein	Cytoplasm	Activated by TILRR, activated by ACP, inactive	20,000
IRAK	Protein	Cytoplasm	Active, inactive	20,000
TRAF	Protein	Cytoplasm	Active, inactive	20,000
TAK	Protein	Cytoplasm	Active, inactive	2000
Ras	Protein	Cytoplasm	Active, inactive	20,000
Pl3k	Protein	Cytoplasm	Active, inactive	20,000
Akt	Protein	Cytoplasm	Active, inactive	10,000
IKK	Protein	Cytoplasm	Active, inactive	20,000
IκB	Protein	Cytoplasm, nucleus	Free, bound to NF-κB, bound to actin, being transcribed, to be degraded (pIκB)	Variable, starting endogenous level 50,000
NF-κB	Protein	Cytoplasm, nucleus	Free, bound to IκB	20,000
IL-8	Protein	Cytoplasm, nucleus	Active, being transcribed	Variable. Starting level 0
Caspase 3	Protein	Cytoplasm, nucleus	Active, inactive	2500
NF-κB:IκB Dissociator	Protein	Cytoplasm	Active	500
Actin: IκB Dissociator	Protein	Cytoplasm	Active	70,000
I IκB Phosphorylator	Protein	Cytoplasm	Active	250

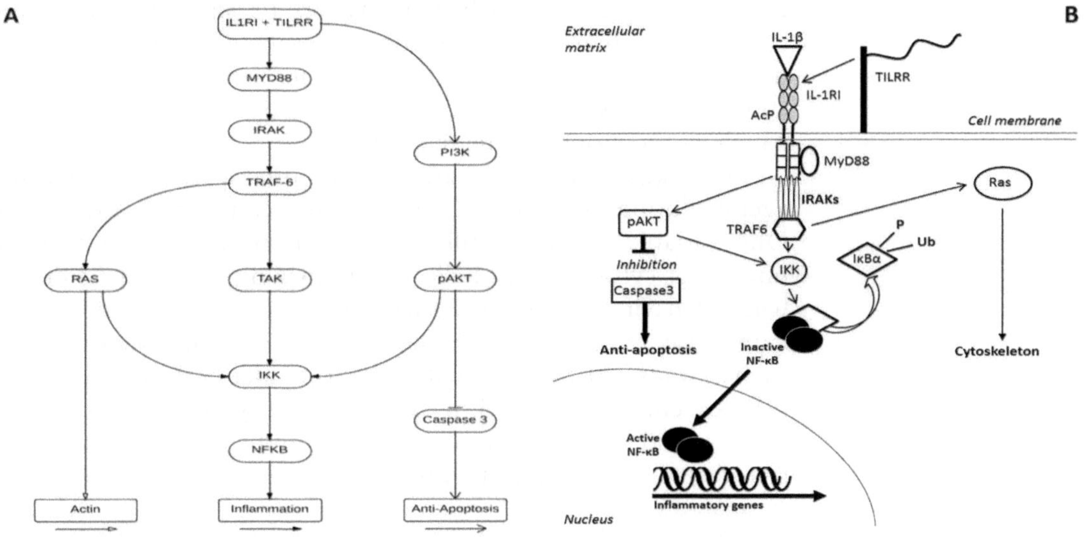

Fig. 4 Schematic outline of the agent-based model. (**a**) Flowchart summarizing the activation cascade represented in the agent-based model of the NF-kB signaling pathway. (**b**) Outline of the biological pathway components represented in (**a**), including showing localization, interactions, and movements represented by agents in the model [11]

Then we must specify the agents using the language XMML a variant of the common web page programming language. The agents each have an internal memory that includes its ID., type, (x,y,z) location, timing information, and other relevant details. The messages each agent can send and receive relate to potential interactions with other agents and the consequences for the internal state and memory.

This agent represents any of the free roaming proteins involved in the signal pathway, MyD88, IRAK, Ras, and so on (Fig. 4). Some Protein agents are confined to the cytoplasm, whilst others such as IκBα and NF-κB can move to and from the nucleus via interaction with nuclear transport receptors. Protein agents can interact with both agents, the rules governing which interactions are permissible are defined in the Protein functions and are determined by the type and state of the Protein agent. For example, a Protein agent representing inactive IKK can only interact with Protein agents representing active forms of upstream regulators (Ras, TAK, and AKT) and will change state to active IKK as a result. Only active IKK can interact with an agent representing a NF-κB:IκBα dimer and will cause the dimer to split into two separate agents representing pIκBα and NF-κB. Protein agents can also change state based on an internal timer, which is used for a variety of processes controlling the pathway. Such timed processes include transcription, in which a new Protein agent is created at the point of transcription but is in a state of being transcribed and unable to interact with other agents until after a set time, when it changes its state to that of the complete protein. Agent functions specify what it can do and under what circumstances in term of its location, internal state and any "messages" it receives from neighboring agents with which its functions may be relevant.

Two agent memory variables are used purely for tracking agents during simulations, to allow for easier data acquisition. One variable, Loc, identifies the localization of each protein, nuclear or cytoplasmic, to allow easy tracking of where agents are without having to compute their coordinates. The other variable named Tag can be used to monitor levels of simulated transfections, such as transfected IκBα agents, which are identical to endogenous IκBα except for the Tag. Endogenous IκBα agents will not contain this Tag and hence the number of Tags can be used to monitor the transfected levels without compromising the basic function of the simulated cell.

Once the model has been defined and the code generated the system can be run. Initially the conditions including the state and position of each agent and the environment are specified using well established biological parameters.

Model Expansion

The initial model includes key events, such as the receptor complex, initial activation steps and gene activity and in the case of complex systems such as the NF-κB, may describe only one branch of the network. Subsequent expansion of the models is made by incrementally increasing the scope and complexity of the model, successively adding regulatory intermediates. Cellular components are included in order depending on their known function general significance to network regulation and their relevance to the specific question. After each expansion predictions from the model are validated experimentally and revisions made to the model in a reiterative process to derive a faithful in silico representation of the biology.

Expanding the model to include representations of structural components of the cell makes it possible to simulate regulation of the NF-κB network in context of cell shape and changes in the cytoskeleton [11]. Our in vitro studies demonstrated that a significant proportion of the NF-κB inhibitor IκBα is sequestered to the cytoskeleton in the resting cell and released during amplified activation through a mechanism controlled by the system coreceptor TILRR. Interaction with cytoskeletal proteins actin and spectrin, was also supported by 3D modeling (Fig. 5) [9, 11].

Model Validation

Validation of the expanded agent based NF-κB model demonstrated that it accurately reproduces cytokine-induced activation profiles monitored in live cells in vitro. This includes comparing system activation profiles from simulations with data from biological experiments in relation to kinetics and concentration of

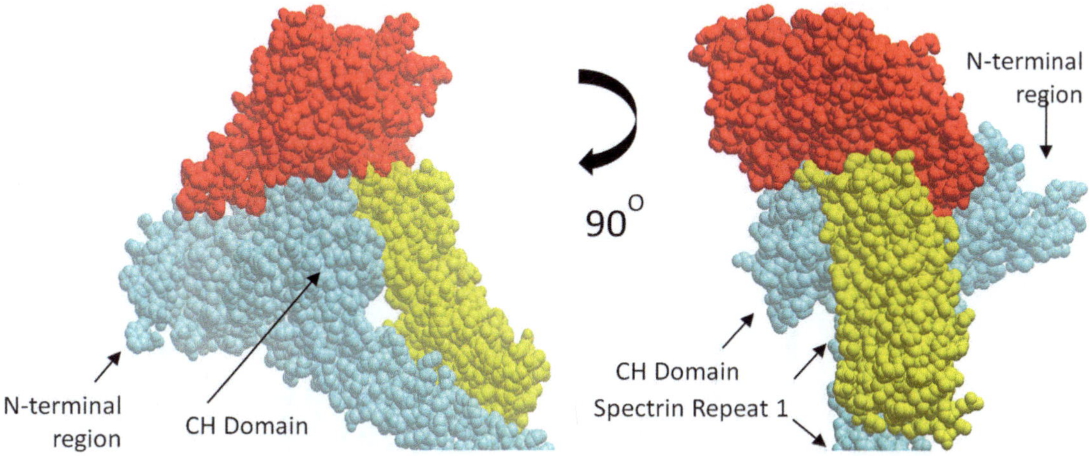

Fig. 5 Space filling representation of the predicted binding interaction of IκBα with cytoskeletal proteins actin and spectrin. Two orientations of the complex are shown to illustrate binding interactions between the three molecules. β-spectrin is shown in blue, actin in red, and IκBα in yellow. 3D protein models were built using multiple-threading alignments and iterative fragment assembly in the de novo I-Tasser Zhang Server and in Swiss-Model. Gramm-X and protein tertiary structure models were viewed and modified in MolSoft ICM Browser, as described in Ref. 11 (*see* Refs. 18–20 in this publication)

a known amplifier of the system, the coreceptor TILRR (Fig. 6a–f). Similarly, a second set of validations, which compared the activation levels in the presence of a wild type TILRR and a negative TILRR mutant demonstrated that the model a faithful represents the biology (Fig. 6g–j).

Simulations to determine the impact of the cytoskeleton on NF-κB regulation, demonstrated pronounced effects on activation during sequestration of the inhibitor at low stimulation levels due to low inhibitor levels, whilst activation levels were barely detectable in the absence of IκBα binding to the cytoskeleton and abundance of free inhibitor (left panel). In contrast high levels of activation were dampened in the absence of cytoskeletal inhibitor binding by the increased amount of free inhibitor (right panel) (Fig. 7). These predictions suggest that the transient inhibitor/cytoskeletal binding provides a mechanism for signal calibration, which enables efficient, activation-sensitive regulation of NF-κB.

3.2 Further Applications for Using Agent-Based Modeling in Biology

Agent-based modeling has been used in a number of research areas. In biology, an early use was in modeling the foraging behavior of social insects, specifically ants [4–7]. More recently, the dynamics of tissue growth and repair, the metabolic basis of bacterial dynamics, the impact of compartmentalization and kinetics on signal specificity and the dynamics of blood flow, have been investigated with this modeling approach. These are discussed next.

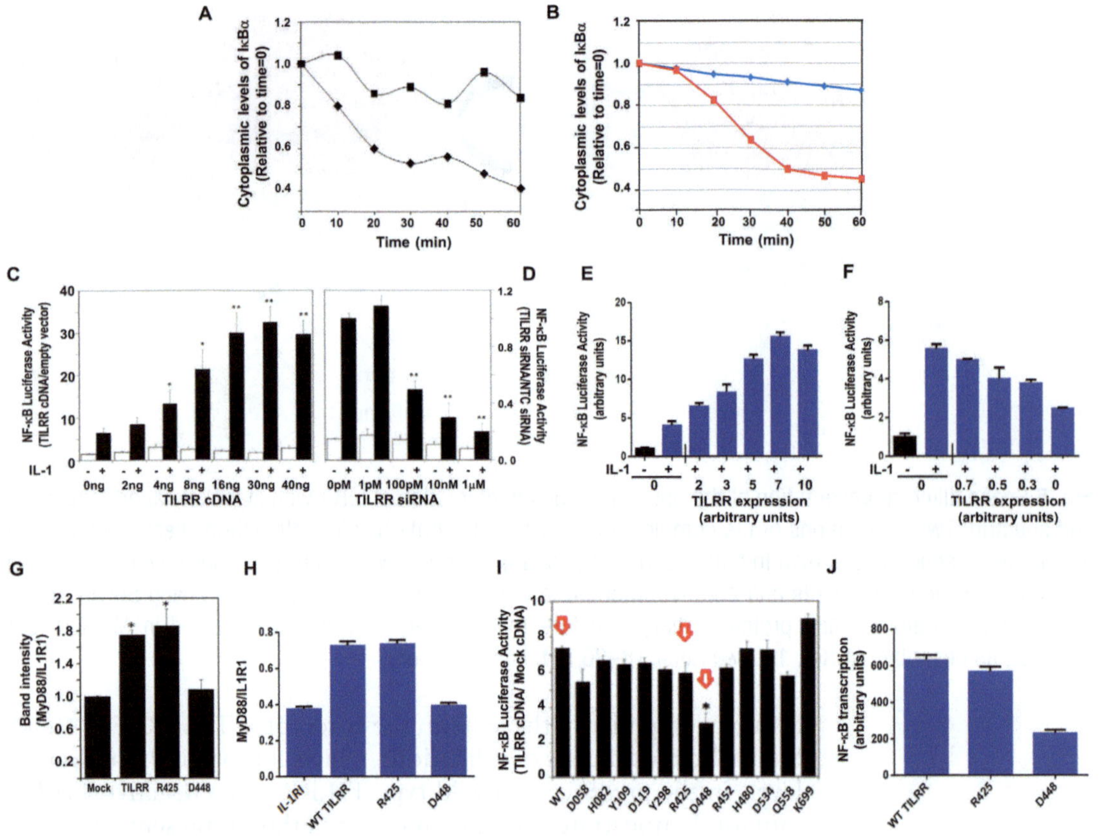

Fig. 6 Comparing model and wet-lab data. The model accurately reproduces activation of inflammatory and antiapoptotic signals, controlled through IL-1RI and the coreceptor TILRR. (**a**) Activation of the IL-1 system causes degradation of the inhibitor IκBα in control cells (◆) which is inhibited by blocking the system coreceptor (■). (**b**) Outputs from the model agree with the biological data (control ■) and reduced effects following inhibition of the receptor complex (◆). (**c**, **d**) TILRR cDNA increases (**c**) and TILRR siRNA decreases (**d**) IL-1 activation, both in a concentration dependent manner, which is faithfully reproduced in simulations (**e**, **f**). A dominant negative mutation at TILRR residue, D448, reduces recruitment of the MyD88 adapter to the IL1R1 complex, inflammatory genes, whilst mutation of residue R425, known not to impact MyD88 regulation, has no impact on adapter recruitment (**g**). Similarly, MyD88 controlled gene activity is reduced by the D448 mutation but unaffected by the control mutant (**i**). The events demonstrated in in vitro experiments shown in **g** and **i** are accurately reproduced by the model in **h** and **j** respectively. Wet-lab experiments (Black **c**, **d**, **g**, **i**); Simulations (Blue **e**, **f**, **h**, **j**)

The Dynamics of Tissue Growth and Repair

A critical player in epithelial tissue regeneration is the TGF-beta network and Transforming Growth Factor TGF-1 in particular. Previous investigations both in vitro and in vivo seemed to indicate that during reepithelialization it acts as a proliferation inhibitor for keratinocytes [19–21]. In previous modeling work, a 3D agent-based model, based on rules at the cellular level governing injury induced emergent behavior, a model component simulating the expression and signaling of TGF-β1 at the subcellular level, and the incorporation of physical solver to resolve the mechanical forces at a multicellular level (Fig. 8, Video 2 link). The model is used to

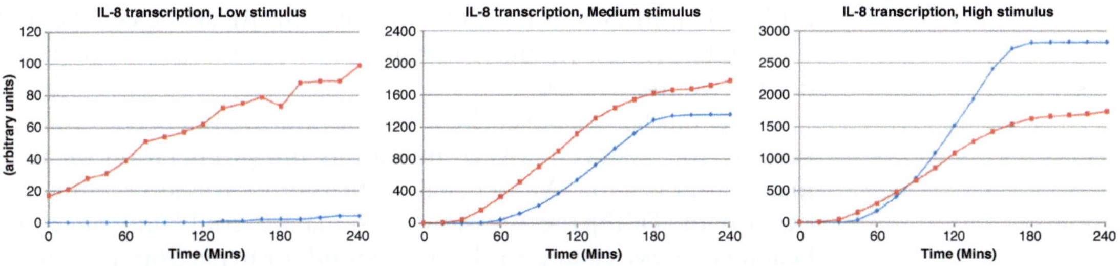

Fig. 7 IL-8 gene activity at low, medium, and high stimulation, in the presence (Red) and absence (Blue) of cytoskeletal sequestration of the NF-κB inhibitor. Simulations show that in the presence of cytoskeletal binding of the inhibitor, a low stimulus produces a measurable level of transcription (Red left graph), and that releases the inhibitor from the cytoskeleton during high stimulus (Red right graph) prevents amplified, aberrant activation of the system.

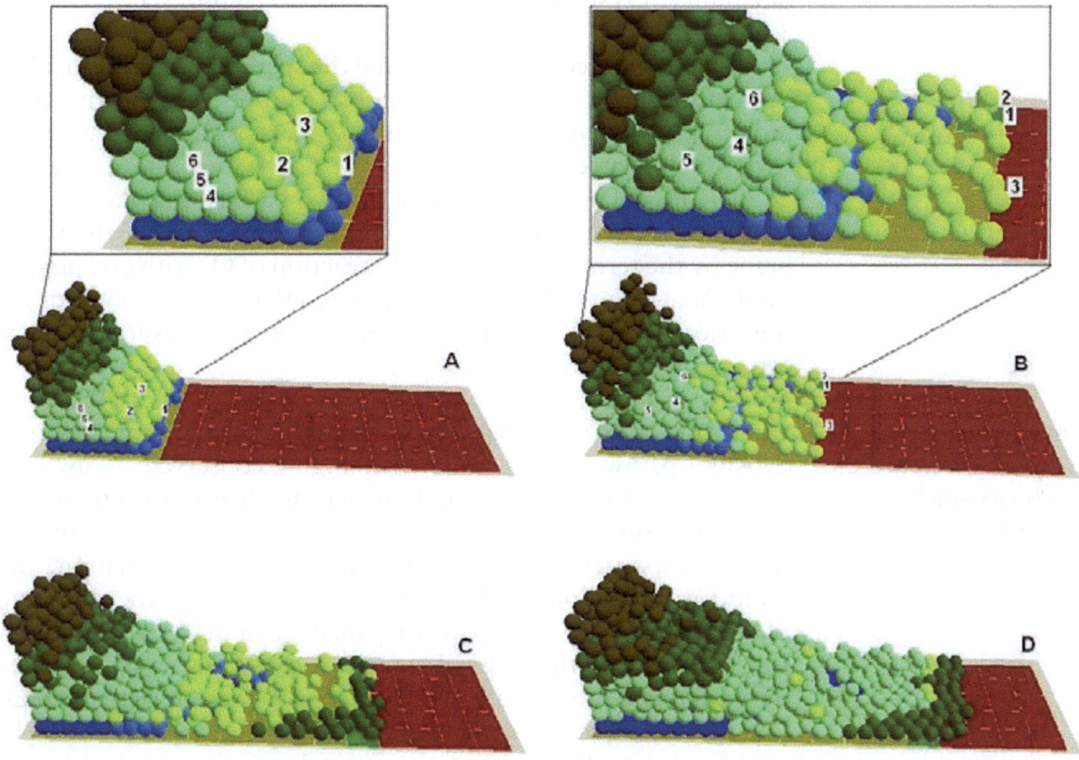

Fig. 8 In virtuo investigation of the functions of TGF-β1 during epidermal wound healing at subcellular level. The virtual wound with normal proliferation and migration rates were simulated for the cells with high TGF-b1 expression levels were labelled with yellow colour. In the integrated model different colors were used to represent keratinocyte stem cells (blue), TA cells (light green), committed cells (dark green), corneocytes (brown), provisional matrix (dark red), secondary matrix (Green), Basal Membrane tile agent (light purple). Supplementary video 2 shows a simulation of stem cells in a tissue culture dividing, differentiating into transit-amplifying cells before their final differentiation into epithelial cells

explore hypotheses of the functions of TGF-β1 at the cellular and subcellular levels on different keratinocyte populations during epidermal wound healing.

The model supports TGF-β1 playing an important role in keeping the balance between migration and proliferation for normal wound healing. Model analysis further indicated that any disruption of TGF-β1 expression or signaling could influence the healing process leading to chronic wounds or hypertrophic wounds as indicated by subsequent biological experimentation.

The Metabolic Basis of Bacterial Dynamics

The bacterium *E. coli* conserves energy by aerobic respiration involving two terminal oxidases *Cyo* and *Cyd.* In environments with different O_2 availabilities the expression of the genes encoding the alternative terminal oxidases, the cydAB and cyoABCDE operons, are regulated by two O_2-responsive transcription factors, ArcA (an indirect O_2 sensor) and FNR (a direct O_2 sensor) (Fig. 9, Video 3 link) [22].

An agent-based model simulated the spatial consumption of O_2 in an individual cell grown in chemostat cultures. The individual O_2 molecules, transcription factors, and oxidases are treated as agents within a simulated *E. coli* cell.

The model implies that there are two barriers that dampen the response of FNR to O_2, that is, consumption of O_2 at the membrane by the terminal oxidases, and reaction of O_2 with cytoplasmic FNR. Analysis of FNR variants suggested that the monomer-dimer transition is the key step in FNR-mediated repression of gene expression.

The Impact of Compartmentalization and Kinetics on Signal Specificity

Signal transduction through the Mitogen Activated Protein Kinase (MAPK) pathways is evolutionarily highly conserved. Many cells use these pathways to interpret changes to their environment and respond accordingly. The pathways are central to triggering diverse cellular responses such as survival, apoptosis, differentiation, and proliferation. Though the interactions between the different MAPK pathways are complex, they maintain a high level of fidelity and specificity to the original signal. In this study an agent based computational model was used to address multicompartmentalization in relation to the dynamics of MAPK cascade activation. The model suggests that multicompartmentalization coupled with periodic MAPK kinase (MAPKK) activation may be critical factors for the emergence of oscillation and ultrasensitivity in the system. Further, it establishes a link between the spatial arrangements of the cascade components and temporal activation mechanisms and predicts that both parameters contribute to fidelity and specificity of MAPK mediated signaling (Fig. 10; Video 4 link) [23].

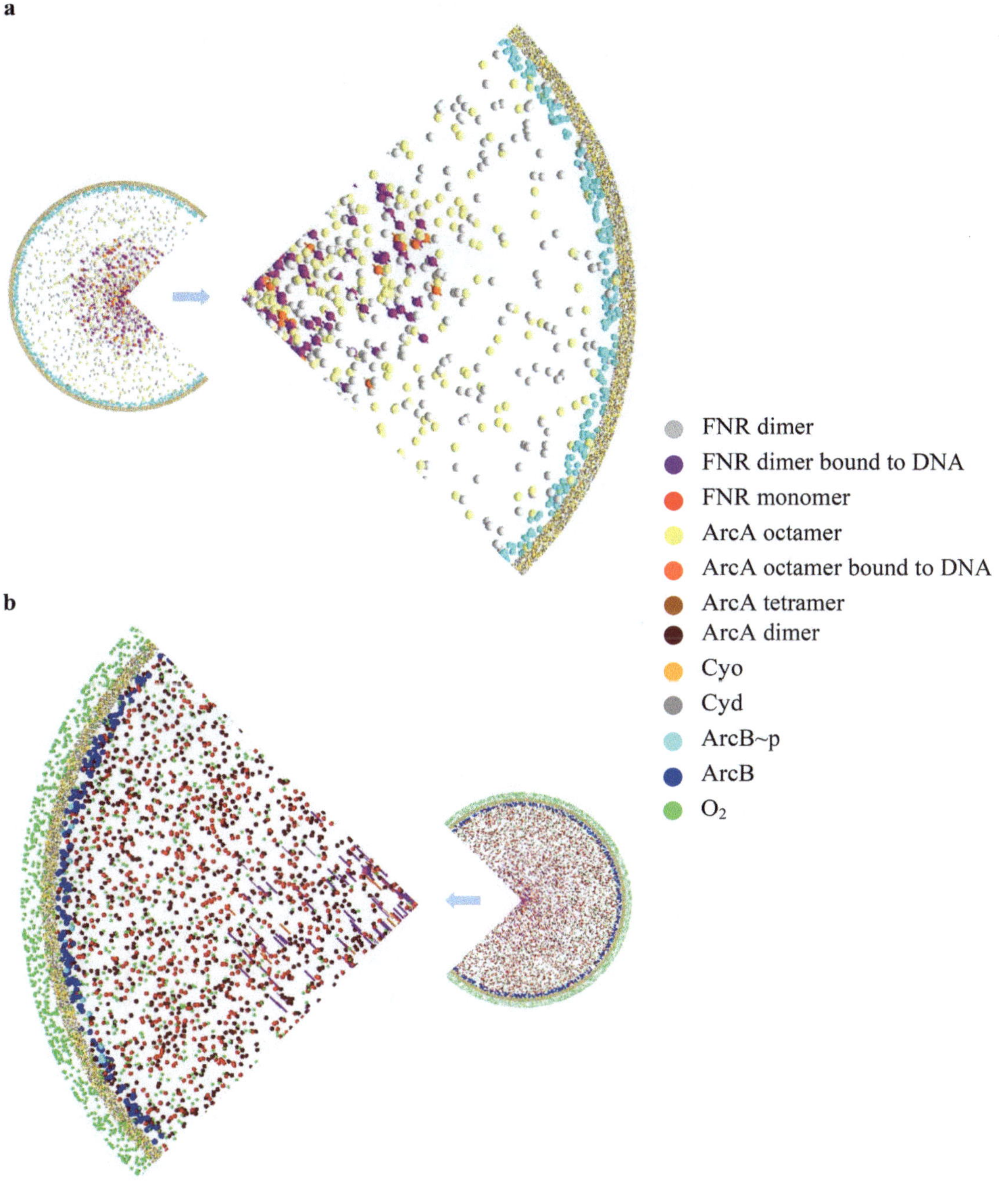

Fig. 9 Initial and final states with no O_2 and with excess O_2. Supplementary video 3 shows a simulation of this process in virtuo

The Dynamics of Blood Flow

Another example, which also incorporates fluid flow modeling, looks at how suitably designed nanoparticles could be used to deliver drugs directly to the brain [24, 25]. The vascular system in the brain can transport a very restricted range of material across this interface and most proteins cannot be absorbed from the blood

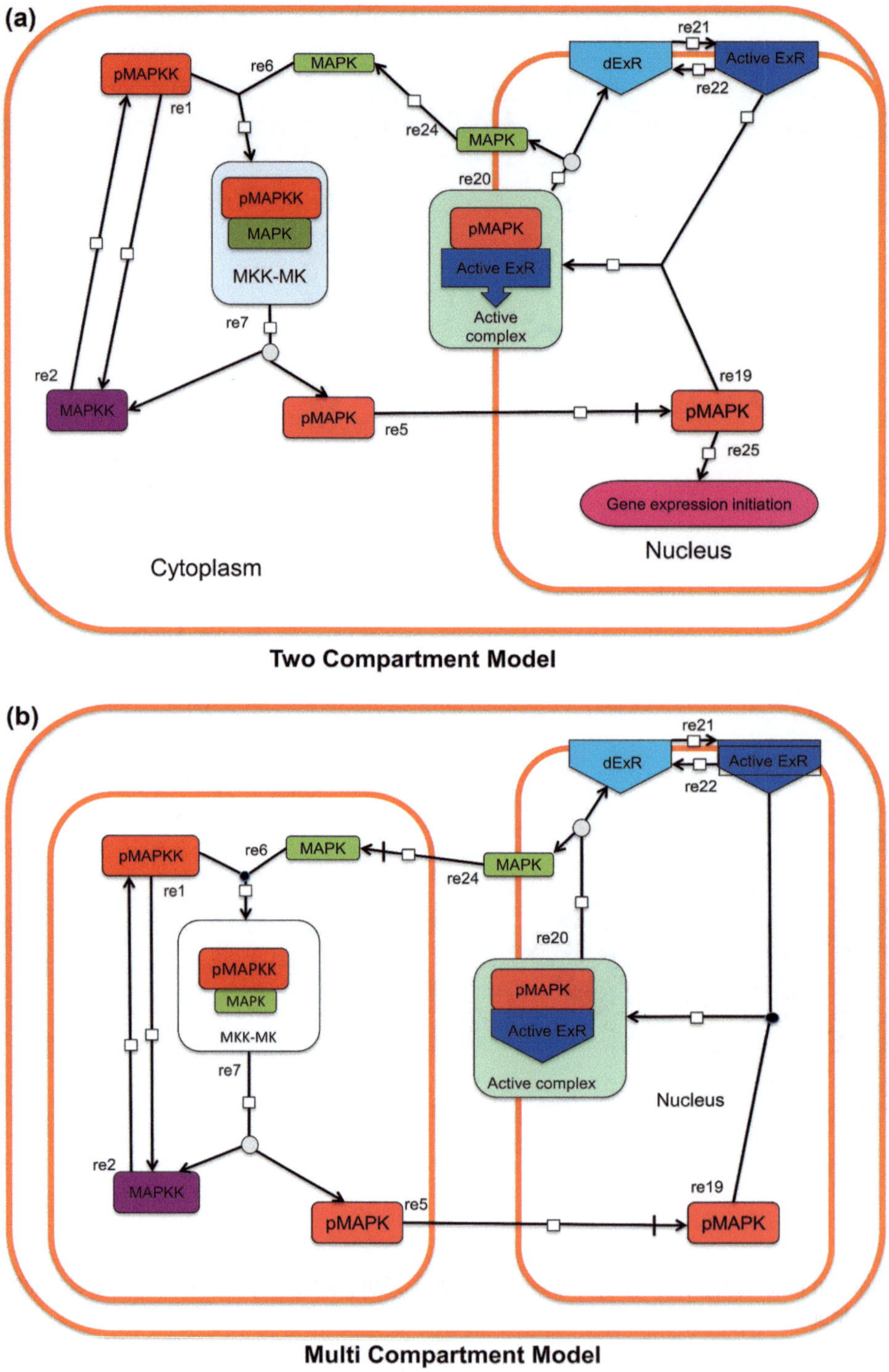

Fig. 10 Schematic for both a two compartment model and a multicompartment models with screenshots of simulations of Map Kinase activity. Supplementary video 4 describes the important role of compartmentation in the interaction of MAPKK and MAPK

Fig. 11 A snapshot of a simulation looking down the blood vessel. Supplementary video 5 shows a simulation through the vessel

flow. By modeling in detail the limited types of gaps and fenestrations together with the flow of blood cells and other particles through the bloodstream, it was possible to identify the sort of particles and the conditions in which transport is possible (Figs. 11 and 12, Videos 5 and 6).

Naturally, the turbulent blood flow environment is a critical aspect and needs the combination of both agents—the particles and blood cells and the fluid dynamics within which they exist. Amongst many insights one is that the red blood cells actually assist suitably shaped nanoparticles in transporting over the barrier. Also, nanoparticle size can selectively target tumor tissue over normal tissue. A simulation snapshot is shown in Fig. 13.

Once the model has been specified there are a few tools that can help in checking out the way the model will run. Dependency state graphs are automatically generated and will show how the model will operate, which agent communicates—receives or sends—messages/data to which other agent and when. Other diagrams can also inform on how the simulation will run (*see* Fig. 14).

These simulations generate a lot of data and tools to manage and analyze the data such as HDF formats; DAIKON, which can be used to check for faulty invariants in the code and tools for visualizing and summarizing data are becoming increasingly available.

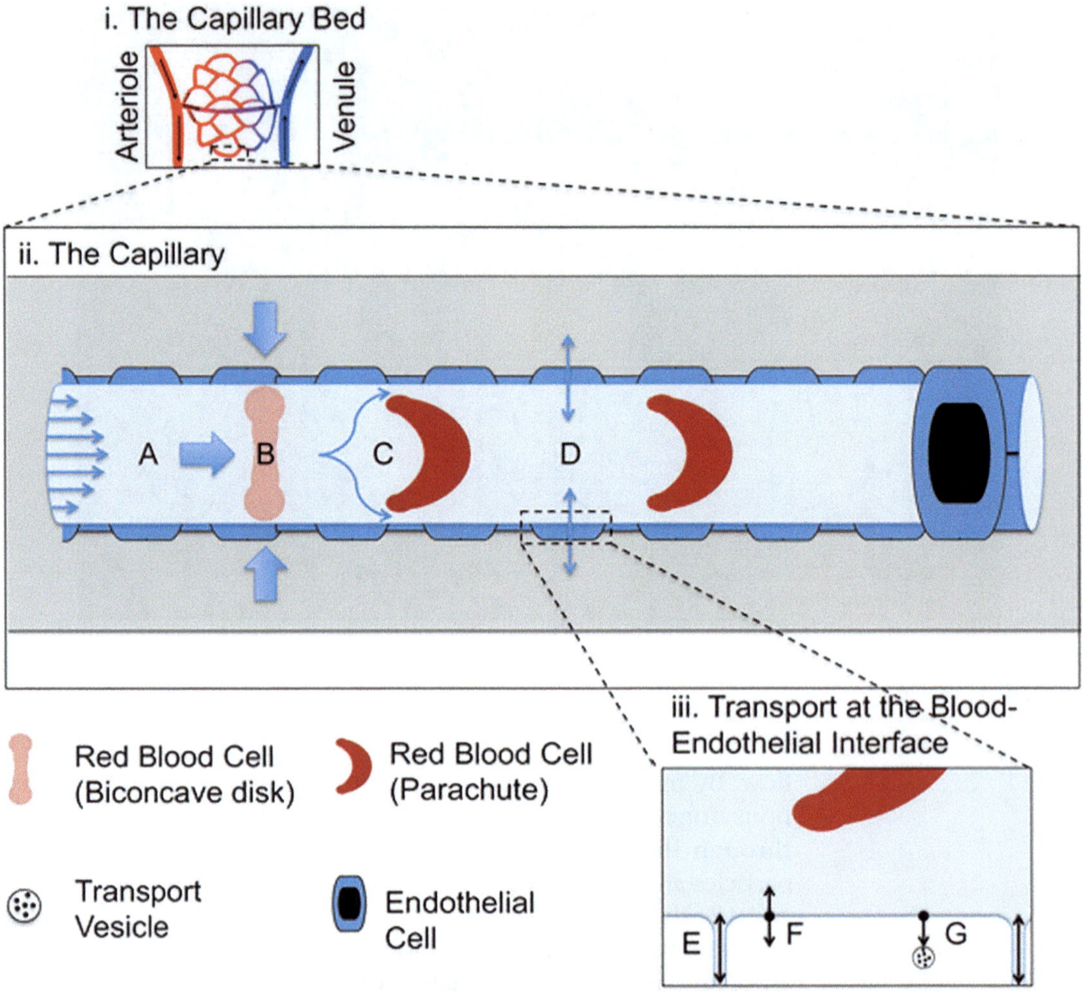

Fig. 12 A lateral view of the simulated particle flow along the blood vessel. The model includes the effect of laminar flow on red blood cells and the behavior of particles at cellular junctions. Supplementary video 6 shows the simulation from the side of the vessel

3.3 Conclusion

The use of agent-based modeling and powerful frameworks such as FLAME within which complex models can be defined, analyzed, verified, and implemented for large-scale supercomputing environments has transformed systems biology. We can now investigate in great detail many biological phenomena and use simulations to examine conjectures, validate against detailed experimental data, and make predictions. The models are also easily maintainable since the FLAME framework has been based on best software engineering practice for large applications.

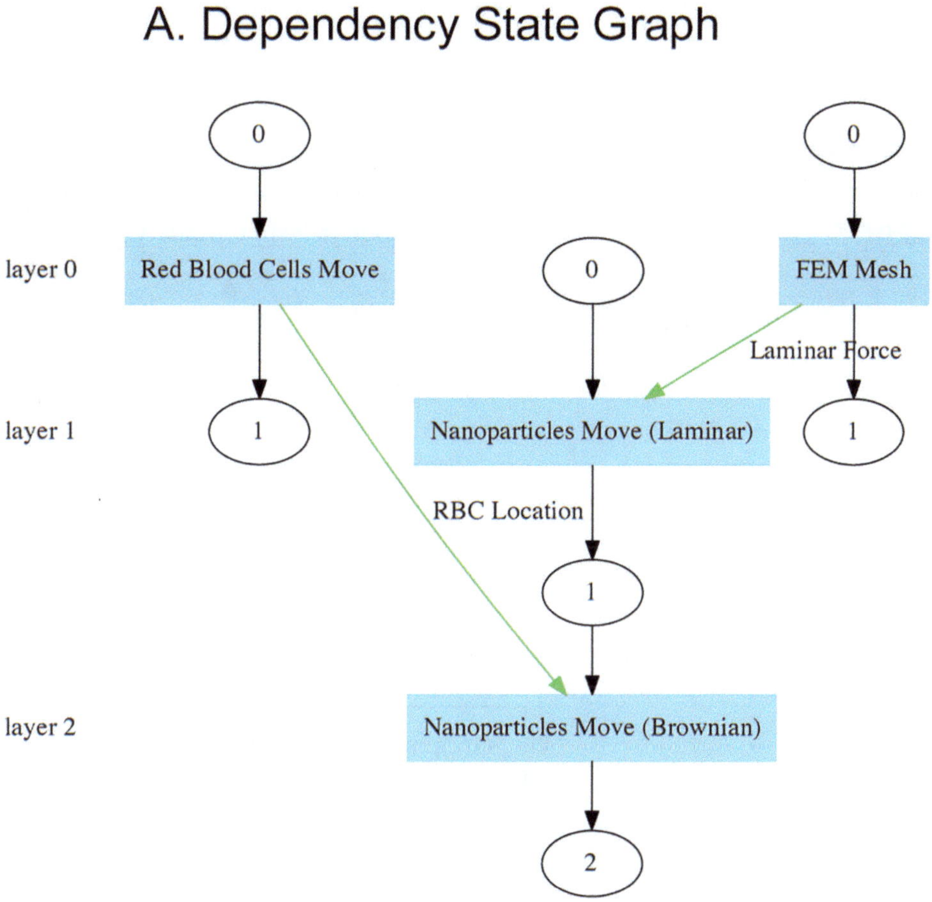

Fig. 13 Software architecture of the example in 4.4 [24]. This state graph demonstrates the dependency of functions on both previous functions and messages for parallelization of the core model. Blue processes are core functions while green ones are optional

4 Notes

1. The use of FLAME as a platform for agent-based modeling provides the researcher with a variety of tools and frameworks that can support the design of high quality and verifiable simulation codes for most computer architectures including GPU and hybrid architectures. Detailed advice can be found at www.flame.ac.uk
2. Models descriptions are formatted in XML (Extensible Markup Language) tag structures to allow easy human and computer readability, enabling easier collaborations between developers writing applications that interact with model definitions.

   The model XML document has a structure that is defined by a schema. The schema of the XML document is currently located at:

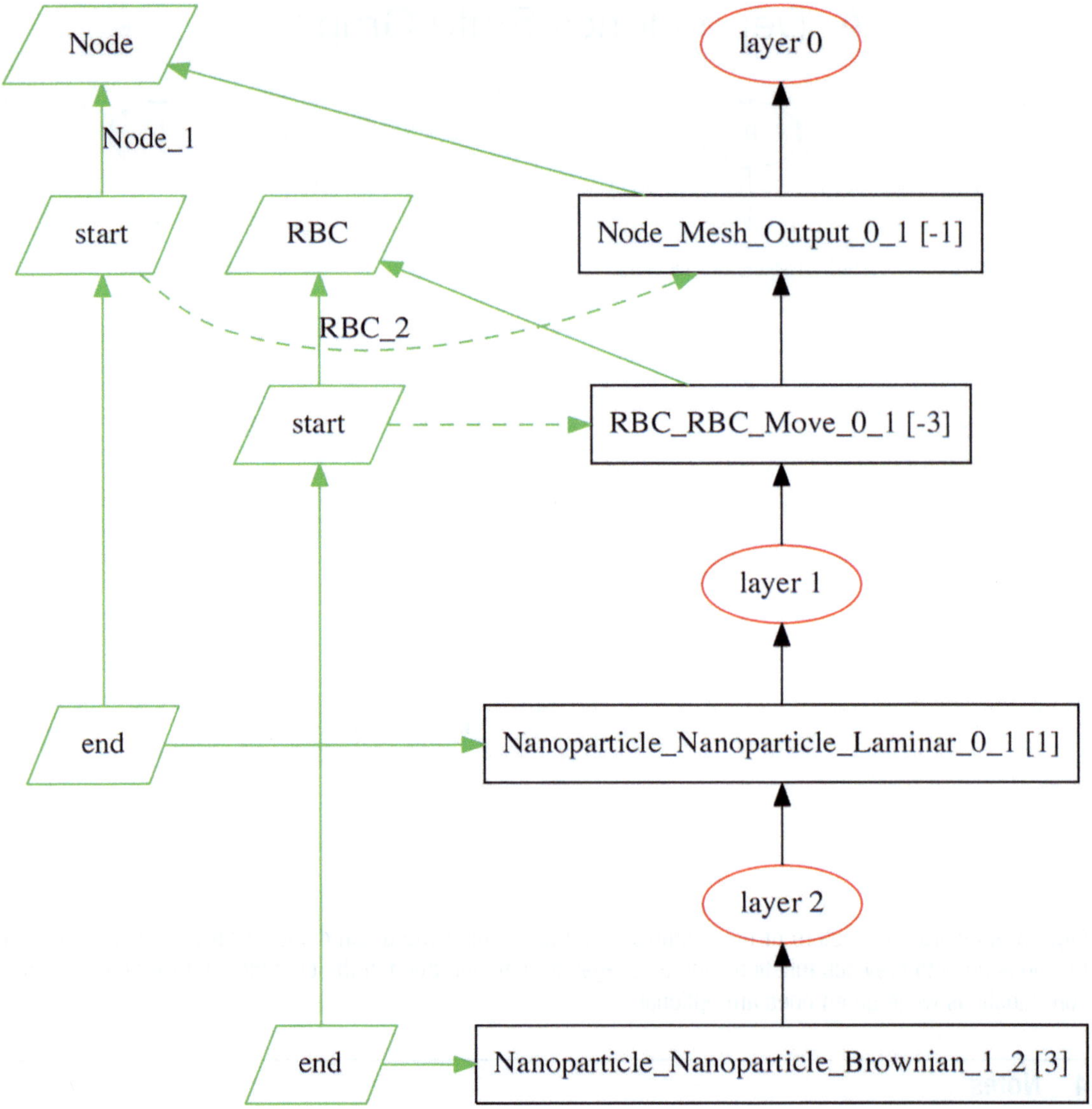

Fig. 14 Dependency state graph and scheduler process order for example 3.4. The process graph shows the order in which FLAME prioritizes the functions to reduce the lag from using the message passing interface

http://flame.ac.uk/schema/xmml_v2.xsd

This provides a way to validate the model document to make sure all the tags are being used correctly. This can be achieved by using xml command line tools like XMLStarlet and xmllint or by using editors that can have xml validation built-in like Eclipse. The start and end of a model file should be formatted as follows.

```
<xmodel version="2" xmlns:xsi="http://www.w3.org/2001/XMLSchema-instance" xsi:noNamespaceSchemaLocation='http://flame.ac.uk/schema/xmml_v2.xsd'>        <name>Model_name</name>
```

```
<version>the version</version> <description>a
description</description> <!-- more... --> </
xmodel>
```

Where name is the name of the model, version is the version, and description allows the description of the model. Models can also contain other models (enabled or disabled), environment, and so on.

The basic concept of agent is defined in a simple format.

Agent Memory: Agent memory defines variables, where variables are defined by their type, C data types or user-defined data types from the environment; a name; and a description.

```
<memory>
    <variable>
      <type>int</type>
      <name>id</name>
              <description>identity number</
description>
    </variable>
    <variable>
      <type>double</type>
      <name>x</name>
           <description>position in x-axis</
description>
    </variable>
  </memory>
```

Agent Functions: An agent function contains the following.

name—the function name which must correspond to an implemented function name and must be unique across the model.

description

current state—the current state the agent has to be in.

next state—the next state the agent will transition to.

condition—a possible condition of the function transition.

inputs—the possible input messages.

outputs—the possible output messages.

And it contains the following as tags:

```
<function>
 <name>function_name</name>
      <description>function description</
description>
<currentState>current_state</currentState>
<nextState>next_state</nextState>
<condition>
```

```
...
</condition>
<inputs>
...
</inputs>
<outputs>
...
      </outputs>
  </function>
```

The current state and next state tags hold the names of states. This is the only place where states are defined. State names must coordinate with other functions states to produce a transitional graph from a single start state to end many possible end states.

The functions are defined in a specific file for each agent.

After every X-machine transition function is accounted for the X-machine is defined. Lastly the messages that can be sent and received need to be well defined also. Each message is defined inside a message tag, is given a name, and any variables it needs to hold. The message defined below refers to the message used in the above X-machine function.

```
<message>
     <name>location</name> <var><type>int</
type><name>id</name></var> <var><type>int</type-
><name>cell_cycle</name></var> <var><type>double</type-
><name>x</name></var> <var><type>double</type><name>y</
name></var> <var><type>double</type><name>radius</name></
var>
</message>
```

The format for the message variables is the same format as the variables defined in the X-machine's internal memory. Many types of messages can be defined with variables for different purposes.

Here is one that sends a message to another agent.

```
#include "header.h"
#include "agent_a_agent_header.h"
/*
 * \fn: int send_message()
 * \brief: Send message.
 */
int send_message()
{
```

```
                        // Send a message of type message_z
containing the id of the agent
                        add_message_z_message(MY_ID);
                    return 0; /* Returning zero means the agent
is not removed */
                }
```

Further details are provided in the FLAME User Manual.

3. As mentioned in Subheading 2, clarity about the research questions being investigated is crucial. It is also important to clearly define the boundaries of the system/s being studied.
4. There must be an integrated and iterative experimental-computational approach as the model is developed, which will require data that may not be available, and experiments may need to be undertaken to acquire them. Using detailed, well controlled biological data, such as single cell readings, as the basis for developing the model greatly improves the accuracy of the model and ultimately its value in predicting biological events and guiding experimental planning. In some successful examples the FLAME model has been built solely by biologists (e.g., [24]). However, in most cases model development is a continuous, multidisciplinary process that calls for a very close relationship between experimentalists and modelers. This is because the conceptual basis of the model is similar to the biological reality, where biologists think in terms of molecules, pathways, properties, and structural elements agent-based modelers should do likewise.
5. Whilst results from conventional biochemical experiments are extremely informative, they do not provide the detail required for in-depth analysis of transient signaling events that are carefully regulated by concentration levels and kinetics as well as by cell shape and cell environment.
6. To optimize the accuracy of the model it is important to consider the biological data in context, to take note of changes in the environment in which the biological event takes place and evaluate the impact this may have on aspects of system control. An example is the role of cell structure on regulation of signal transduction as discussed in "Modeling the NF-κB Regulatory Network" above. This example demonstrates the value of considering results from biological experiments in context and taking note of changes in related systems such as cell attachment and cell shape and shows that these can have significant impact on experimental results.

Acknowledgments

Many scientists have contributed to the development of FLAME and its use in biology. Funding from EPSRC, BBSRC.
Author contributions: Mike Holcombe and Eva Qwarnstrom wrote the text. Simulation examples include work by Mark Pogson, David Rhodes, Salem Adra, Dawn Walker, Phil McMinn, Hao Bai, Aban Shuaib, Gavin Fullstone. Key developers of FLAME include Simon Coakley, Mariam Kiran, Chris Greenough, Paul Richmond, David Worth, and Gemma Poulter.

References

1. FLAME website (2020). http://www.flame.ac.uk. Accessed 30 Dec 2020
2. FLAME GPU website (2020). http://www.flamegpu.com. Accessed 30 Dec 2020
3. Pogson M, Smallwood R, Qwarnstrom E et al (2006) Formal agent-based modelling of intracellular chemical interactions. Biosystems 85:37–45
4. Jackson DE, Holcombe M, Ratnieks FLW (2004) Trail geometry gives polarity to ant foraging networks. Nature 432:907–909. https://doi.org/10.1038/nature03105
5. Jackson DE, Martin SJ, Ratnieks FLW et al (2007) Spatial and temporal variation in pheromone composition of ant foraging trails. Behavioral Ecol 18(2):444–450. https://doi.org/10.1093/beheco/arl104
6. Jackson D, Holcombe M, Ratnieks F (2004) Coupled computational simulation and empirical research into the foraging system of Pharaoh's ant (Monomorium pharaonis). Biosystems 76(1–3):101–112. https://doi.org/10.1016/j.biosystems.2004.05.028
7. Jackson DE, Bicak M, Holcombe M (2011) Decentralized communication, trail connectivity and emergent benefits of ant pheromone trail networks. Memet Comput 3:25–32
8. Holcombe M, Coakley S, Kiran M et al (2013) Large-scale modeling of economic systems. Complex Syst 22(2):175–191. https://doi.org/10.25088/ComplexSystems.22.2.175
9. Pogson M, Holcombe M, Smallwood R et al (2008) Introducing spatial information into predictive NF-κB modelling – an agent-based approach. PLoS One 3(6):e2367. https://doi.org/10.1371/journal.pone.0002367
10. Pogson M (2008) Modelling the intracellular NF-κB signalling pathway. Dissertation, University of Sheffield, UK
11. Rhodes DM, Smith SA, Holcombe M et al (2015) Computational modelling of NF-κB activation by IL-1RI and its co-receptor TILRR, predicts a role for cytoskeletal sequestration of IκBα in inflammatory signalling. PLoS One 10:e0129888. https://doi.org/10.1371/journal.pone.0129888
12. Rhodes DM, Holcombe M, Qwarnstrom EE (2016) Reducing complexity in an agent based reaction model–benefits limitations of simplifications in relation to run time and system level output. BioSystems 147:21–27. https://doi.org/10.1016/j.biosystems.2016.06.002
13. Mitchell S, Vargas J, Hoffman A (2016) Signaling via the NF-κB system. Wiley Interdiscip Rev Syst Biol Med 8(3):227–241. https://doi.org/10.1002/wsbm.1331
14. Smith SA, Samokhin AO, Alfaidi M, Murphy EC, Rhodes D, Holcombe WML, Kiss-Toth E, Storey RF, Yee S-P, Francis SE, Qwarnstrom EE (2017) The IL-1RI co-receptor TILRR (FREM1 isoform 2) controls aberrant inflammatory responses and development of vascular disease. JACC Basic Transl Sci 2(4):398–414. https://doi.org/10.1016/j.jacbts.2017.03.014
15. Carlotti F, Chapman R, Dower SK et al (1999) Activation of NF-κB in single living cells. Dependence of nuclear translocation and anti-apoptotic function on EGFP-RELA concentration. J Biol Chem 274:37941–37949. https://doi.org/10.1074/jbc.274.53.37941
16. Carlotti F, Dower SK, Qwarnstrom EE (2000) Dynamic shuttling of NF-κB between the nucleus and cytoplasm as a consequence of inhibitor dissociation. J Biol Chem 275:41028–41034. https://doi.org/10.1074/jbc.M006179200
17. Yang L, Chen H, Qwarnstrom EE (2001) Degradation of IκBα is limited by a post phosphorylation/ubiquitination event. Biochem Biophys Res Commun 285:603–608. https://doi.org/10.1006/bbrc.2001.5205

18. Yang L, Ross K, Qwarnstrom EE (2003) RelA control of IκBα phosphorylation: a positive feedback-loop for high affinity NF-κB complexes. J Biol Chem 278:30881–30888. https://doi.org/10.1074/jbc.M212216200
19. Adra S, Sun T, MacNeil S et al (2010) Development of a three dimensional multiscale computational model of the human epidermis. PLoS One 5(1):e8511. https://doi.org/10.1371/journal.pone.0008511
20. Sun T, Adra S, Smallwood R et al (2009) Exploring hypotheses of the actions of TGF-β1 in epidermal wound healing using a 3D computational multiscale model of the human epidermis. PLoS One 4(12):e8515. https://doi.org/10.1371/journal.pone.0008515
21. Walker D, Wood S, Southgate J et al (2006) An integrated agent-mathematical model of the effect of intercellular signalling via the epidermal growth factor receptor on cell proliferation. J Theor Biol 242(3):774–789. https://doi.org/10.1016/j.jtbi.2006.04.020
22. Bai H, Rolfe MD, Jia W et al (2014) Agent-based modeling of oxygen-responsive transcription factors in *Escherichia coli.* PLoS Comp. Biol. 10(4):e1003595. https://doi.org/10.1371/journal.pcbi.1003595
23. Shuaib A, Hartwell A, Kiss-Toth E, Holcombe M (2016) Multi-compartmentalisation in the MAPK Signalling pathway contributes to the emergence of oscillatory behaviour and to Ultrasensitivity. PLoS One 11(5):e0156139. https://doi.org/10.1371/journal.pone.0156139
24. Fullstone G, Wood J, Holcombe M et al (2015) Modelling the transport of nanoparticles under blood flow using an agent-based approach. Sci Rep 5:10649. https://doi.org/10.1038/srep10649
25. Fullstone G (2016) Modelling the transport of nanoparticles across the blood-brain barrier using agent-based modelling, Dissertation, University College London, UK

Part VI

Systems Biology in Biotechnology

Chapter 16

Metabolic Modeling of Wine Fermentation at Genome Scale

Sebastián N. Mendoza, Pedro A. Saa, Bas Teusink, and Eduardo Agosin

Abstract

Wine fermentation is an ancient biotechnological process mediated by different microorganisms such as yeast and bacteria. Understanding of the metabolic and physiological phenomena taking place during this process can be now attained at a genome scale with the help of metabolic models. In this chapter, we present a detailed protocol for modeling wine fermentation using genome-scale metabolic models. In particular, we illustrate how metabolic fluxes can be computed, optimized and interpreted, for both yeast and bacteria under winemaking conditions. We also show how nutritional requirements can be determined and simulated using these models in relevant test cases. This chapter introduces fundamental concepts and practical steps for applying flux balance analysis in wine fermentation, and as such, it is intended for a broad microbiology audience as well as for practitioners in the metabolic modeling field.

Key words Constraint-based metabolic modeling, Genome-scale network reconstruction, Wine fermentation, *Saccharomyces cerevisiae*, *Oenococcus oeni*, Metabolic flux

1 Introduction

Wine fermentation is the process whereby grape must is transformed in wine through the biological action of microorganisms. The yeast *Saccharomyces cerevisiae* is the major player in the conversion of sugars present in the grape berry (mainly glucose and fructose) into ethanol and carbon dioxide [1]. This process is known as primary or alcoholic fermentation. Primary fermentation is usually followed by a secondary fermentation, also known as malolactic fermentation (MLF), which takes place in most red wines and some white wines. MLF reduces the acidity of wine and improves flavor complexity and microbiological stability [2–4]. This process is performed by several lactic acid bacteria (LAB), particularly *Oenococcus oeni*, and it usually starts when yeast cells have stopped growing and sugars are almost exhausted in the fermented broth [3]. In this second stage, characterized by high ethanol concentration and low pH, malic acid is transformed into

Sonia Cortassa and Miguel A. Aon (eds.), *Computational Systems Biology in Medicine and Biotechnology: Methods and Protocols*, Methods in Molecular Biology, vol. 2399, https://doi.org/10.1007/978-1-0716-1831-8_16,

lactic acid, which decreases the harsh texture inferred by the former and confers a softer flavor to the wine [2].

While wine fermentation is an ancient process performed throughout thousands of years [1], it has not been until recently that advances in mathematical modeling and bioinformatic and analytical methods have yielded a more comprehensive appraisal of the metabolic phenomena taking place during this process. Availability of genome sequences of different microorganisms has enabled deeper understanding of the physiological features shaping diverse microbial processes, whereby genomic sequences are linked to metabolic functions performed by enzymes [5, 6]. One of the areas that has benefited from the breadth of this data is systems biology. Today, metabolic models reaching genome-scale are available for the yeasts [7] and lactic acid bacteria [8] involved in wine fermentation. They have been constructed using available genomic information from the relevant species. From the prediction of nutritional requirements to the calculation of metabolic flux distributions under different conditions, these models have provided a deeper understanding of the metabolic phenomena involved in wine fermentation [8–14] (Fig. 1). For instance, early work from Sainz et al. [15] successfully predicted glycerol production of

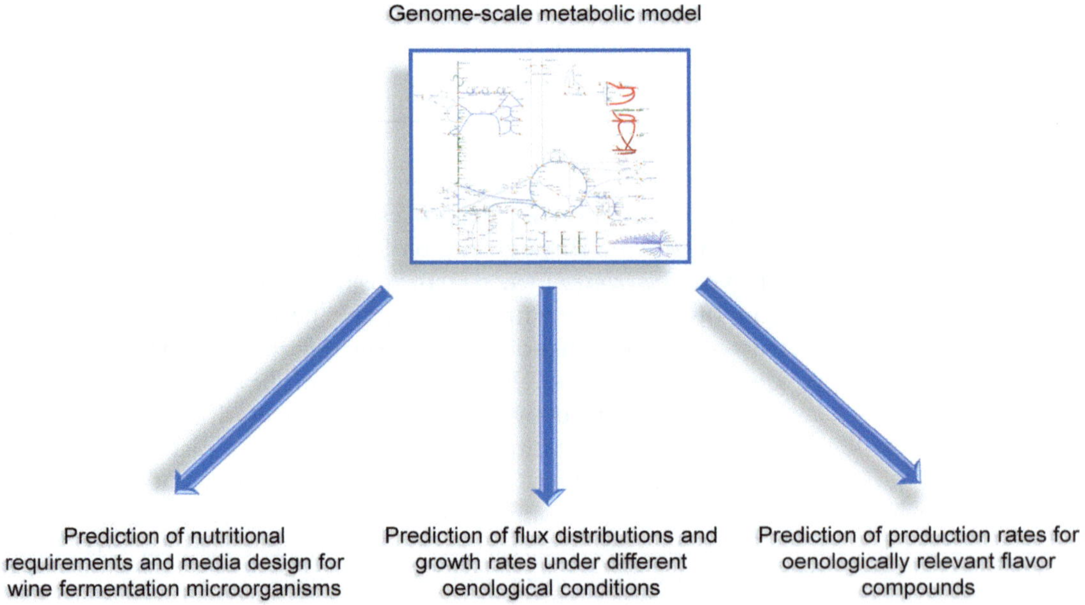

Fig. 1 Applications of genome-scale metabolic models (GEMs) to wine fermentation. Nutritional requirements for many species involved in wine fermentation are difficult and experimentally laborious to determine. Yet microbial genomes of these species are readily available; thus, GEMs can be reconstructed and used to predict essential nutrients for growth. In addition, GEMs are excellent tools for integrating experimentally measured production/consumption rates under different oenological conditions and evaluate their impact on microbial physiology by analyzing the resulting metabolic flux distribution under each scenario. Other uses of GEMs in the wine fermentation context include the prediction of production rates for flavor compounds

S. cerevisiae under enological conditions using a core metabolic model. Subsequent extensions to the model enabled prediction of glycerol production for stuck and sluggish fermentations at various temperatures [16], and even the production of other relevant metabolites in wine at genome scale [17]. Metabolic analysis of the impact of oxygen on the yeast's physiology under enological conditions has also been evaluated using metabolic models [10, 18], complementing and contextualizing previous bioprocess operation studies [19–21]. In the case of *O. oeni*, a recent metabolic model has revealed the energetic consequences of growth on high ethanol concentrations as well as the metabolic mechanisms involved in maintaining cell homeostasis [8, 22]. These and other examples highlight the utility of metabolic models for understanding microbial metabolism in the context of wine fermentation.

A genome-scale metabolic model (usually referred to as GEM) is a mathematical structure generated from a genome-scale metabolic reconstruction (usually referred to as GENRE). The GENRE encodes the stoichiometric matrix that describes the collection of all biochemical reactions encompassing the metabolism of a particular organism. Reconstructions are converted into models by applying assumptions, which can be then translated into mathematical constraints. The steady-state assumption represents the main constraint on metabolic models whereby the production and consumption fluxes of each intracellular metabolite are balanced. Mathematically, the resulting model is described by a system of linear equations that it is typically undetermined, that is, there are infinite solutions describing the same observations; however, it is always possible to compute particular solutions using Linear Programming (LP) optimization methods [23]. Briefly, by defining an objective function and imposing capacity constraints on the fluxes that simulate specific culture conditions or known maximum enzymatic capacities, it is possible to compute the entire flux distribution that achieves the defined goal (Fig. 2). The most famous of such methods is Flux Balance Analysis (FBA) [23], which has been applied in numerous studies to gain a deeper understanding of microbial physiology [24, 25], optimize metabolic production in cell factories [26–28], and even (re)design microbial metabolism [29, 30].

In this chapter, we present a detailed protocol for performing metabolic calculations using genome-scale metabolic models in the context of wine fermentation. More specifically, we illustrate how nutritional requirements can be accurately computed using these models, and also how the specific growth rates and metabolic flux distributions in both, wine yeast and malolactic bacterial cells, can be estimated under the harsh environmental conditions of winemaking. Finally, the chapter provides all the necessary supporting material for reproducing the presented results.

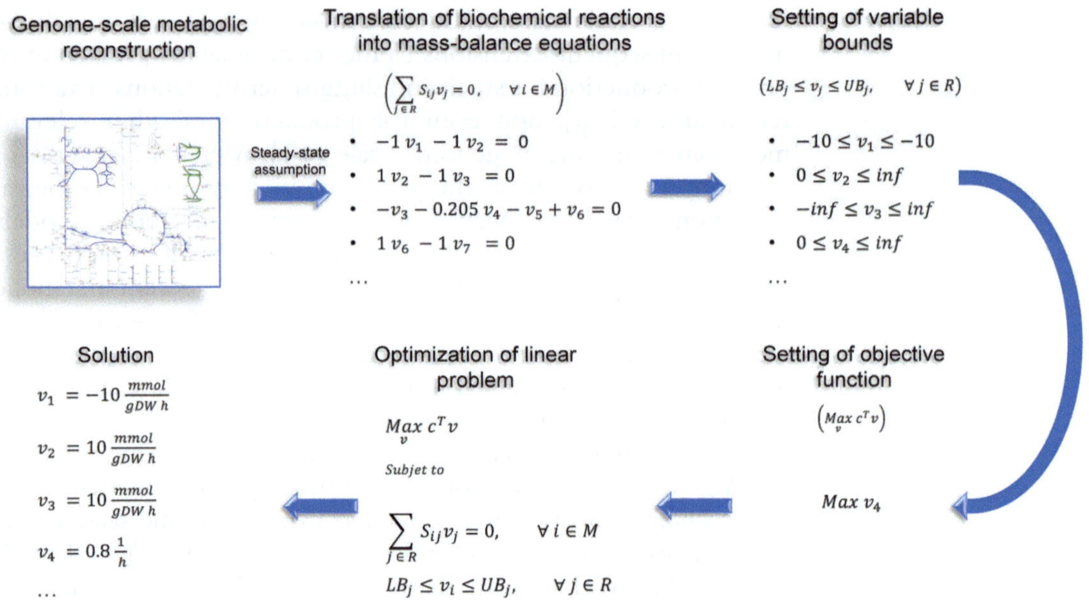

Fig. 2 Schematic representation of the steps for building a genome-scale metabolic model (GEM) from a genome-scale metabolic reconstruction (GENRE). A GENRE contains the collection of all biochemical reactions of cellular metabolism. By applying phenomenological assumptions on the network reactions derived from mass balances (steady state of intracellular metabolites), thermodynamics (reaction reversibility), and observed specific metabolic rates (capacity constraints), a computable model structure (GEM) can be built. Finally, computation of metabolic fluxes requires the definition of an objective function for the network to optimize, for example, biomass growth. The most popular optimization method is called flux balance analysis (FBA) and yields the flux distribution that optimizes the desired biological goal

2 Materials

In the following, we describe the fundamental materials for applying the different methods. A Glossary of key terms and abbreviations is available at the end of the chapter (*see* **Note 1**). Lastly, the different models, files and tutorials presented and employed here can be accessed at https://github.com/SystemsBioinformatics/pub-data/tree/master/protocol_modelling_wine_fermentation

2.1 Metabolic Model

The metabolic network model needs to be of high-quality for the subsequent analyses. In practical terms, high-quality means that the model must be able to generate a positive value through the variable describing the specific growth rate; be mass and charge balanced; avoid free generation of energy through thermodynamically infeasible cycles [31, 32]; and comprehensively describe the relevant metabolism of the species (or strain) under study. We redirect the reader to the detailed protocol for creating a high-quality reconstruction [33], and to the MeMoTe tool for assessing its quality [34].

2.2 Software

Below, there is a list of the fundamental software packages required to run the various protocols described in this chapter. These software include programming environments, software packages and modules. To successfully run the subsequent analyses, the latter software packages need to be appropriately installed in the computing machine.

1. MATLAB Programming Environment (The MathWorks, Natick, MA).
2. The COBRA Toolbox version 3.0 [35].
3. A working version of Python.
4. CBMpy: A Python package to perform constraint-based modeling and analysis (http://cbmpy.sourceforge.net/).
5. EMAF: Enumeration of Minimal Active Fluxes (EMAF) [36].
6. Optimization solvers: CPLEX (IBM ILOG CPLEX Division, Incline Village, NV) and Gurobi (Gurobi Optimization, Inc., Houston, Texas).

3 Methods

3.1 Phenotype Prediction Using Experimental Data

Metabolic networks can be used to predict specific growth rates and flux distributions under different enological conditions such as different grape must compositions, or different culture parameters, like oxygen concentration [10], temperature [37], ethanol concentration [8, 22] or pH (so far not addressed). Unfortunately, genome-scale metabolic models do not have explicit variables for metabolite concentrations; and therefore, the effect of different metabolite concentrations cannot be directly studied. In addition, GEMs do not consider regulatory interactions such as the inhibitory effect of ethanol or pH on cell growth, and thus, the effect of different ethanol concentrations or different pH values in the media cannot be explicitly captured. Despite these limitations, the effect of enological relevant parameters (media composition, oxygen concentration, ethanol concentration, temperature, or pH) can be studied indirectly by performing experiments (in continuous or batch mode) under different conditions and by collecting data that will be used as input to the model. More specifically, specific uptake and production rates can be calculated from data collected in experiments and these rates can be used as inputs for the model. Then, the model can be used to predict, for example, the maximum specific growth rate and flux distribution under particular growth conditions. This is performed using constraint-based modeling methods, the most famous being Flux Balance Analysis (FBA) [23].

All the simulations hereby described, and the following sections rely on this optimization method. FBA is a mathematical formulation that enables the prediction of the flux distribution (i.e., the specific rates for all the reactions in the network) in a metabolic network that achieves a defined objective. Mathematically, FBA is represented by the following linear optimization problem:

$$Max_v\, Z = c^T v \tag{1}$$

Subject to

$$\sum\nolimits_{j \in R} S_{ij} v_j = 0, \forall i \in M \tag{1a}$$

$$LB_j \leq v_j \leq UB_j, \forall j \in R \tag{1b}$$

where v represent fluxes through each biochemical reaction j of the metabolic network composed of i balanced metabolites. The flux variables are in units of $\frac{mmol}{gDW\ h}$, except for the flux representing the growth rate μ which is in units of $\frac{1}{h}$ $\left(corresponding\ to \frac{gDW}{gDW\ h}\right)$. This difference stems from the fact that the stoichiometric coefficients of the biomass equation—representing lipids, DNA, and RNA, among other macromolecules—have units of mmol/gDW so that its flux represents the observed growth rate. In Eq. (1), Z is the objective function and c is a vector containing coefficients (weights) for each of the reaction fluxes to be optimized. In FBA, the objective function usually contains just the growth rate μ, therefore, the dot product $c^T v$ can be expressed just as μ. S is the stoichiometric matrix, where the value in the position i, j represents the stoichiometric coefficient of metabolite i in reaction j. LB_j and UB_j denote capacity constraints and correspond respectively to lower and upper bounds for the rate of reaction j. Lastly, M is the set of all the metabolites in the network and R is the set of all the reactions in the network.

3.1.1 Calculation of Flux Distributions Using Specific Uptake/Production Rates

In this case, experimental data in the form of specific uptake and consumption rates is used as input to constrain the model. Then, the constrained model is used to predict a flux distribution assuming growth rate as the objective function to be maximized (Fig. 3). This type of prediction is usually done using experimental data collected from chemostats, where steady state conditions apply. In this type of experiments, the experimental data collected can be readily incorporated into the model. We note that, in some cases, chemostats are difficult to perform from a practical standpoint as the specific growth rates of some bacteria (e.g., *Oenococcus oeni*) could be very slow. In addition, wine fermentation occurs in batch mode and therefore the data collected under batch cultures is more abundant. Despite their resemblance with wine fermentations, the growth rate of microorganisms during a batch culture changes over time depending on the availability of nutrients and the concentration of compounds that could inhibit growth (e.g., ethanol or lactate); therefore, this analysis is only limited to the exponential phase of batch cultures where external conditions can be considered

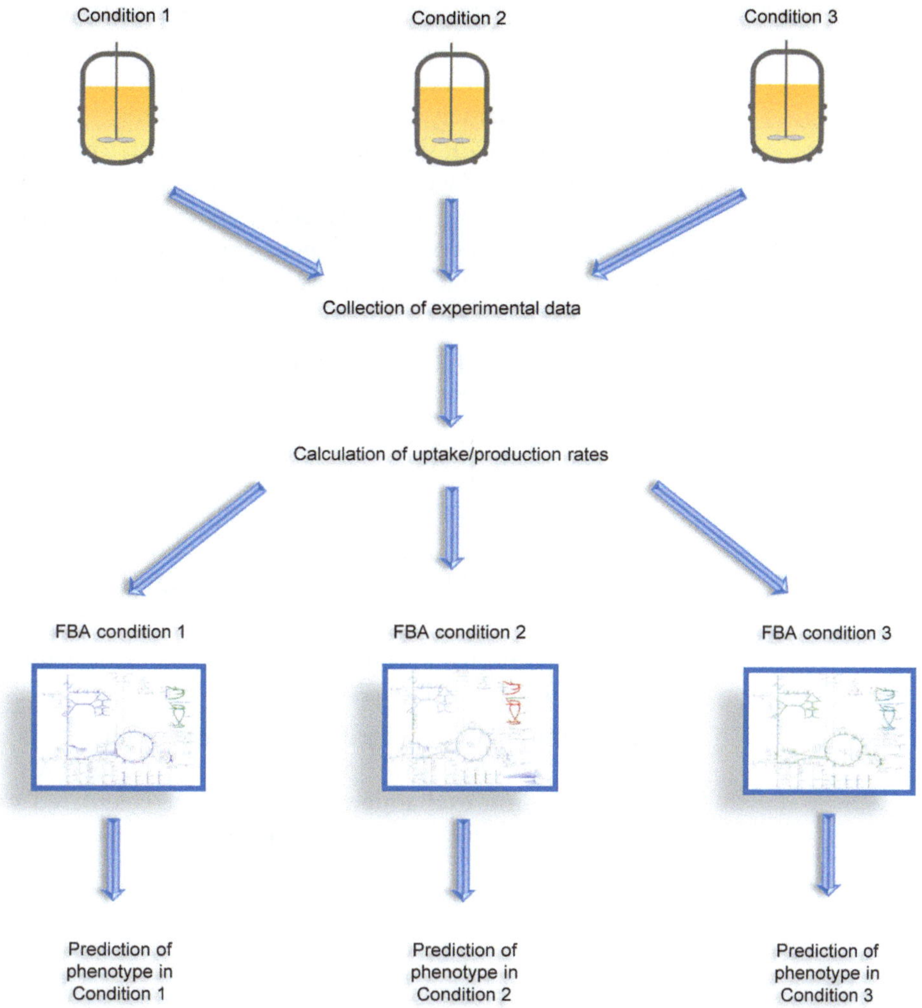

Fig. 3 Illustrative workflow for integrating experimental data into genome-scale metabolic models (GEMs) for studying the effect of oenological parameters on microbial physiology. For example, consider three continuous cultures under different oxygenation conditions. In each culture, metabolites and biomass concentrations are measured and their corresponding specific consumption/production rates are estimated based on the feeding composition and growth conditions. These rates are then incorporated into the model as observed rates, which constrain the range of possible flux values of the model, yielding different metabolic flux distributions

constant (*see* **Note 2**) and the intracellular metabolism can be safely assumed to be in steady state. In this section, we will describe how to perform FBA using data collected from both continuous and batch cultures.

Calculation of Flux Distributions in Continuous Cultures

Metabolite and biomass concentrations are at steady state in continuous cultures, and thus, they can be readily employed to determine the relevant exchange rates and yields under the studied conditions from the inlet and outlet feeds. Metabolite concentrations can be measured using conventional analytical equipment such as HPLC or GCMS. The metabolites to be measured depend on the research question and practical considerations. In our case, the most relevant metabolites correspond to those found in grape must or wine, namely sugars (glucose, fructose), organic acids (malic acid, citric acid, succinic acid, pyruvic acid, formic acid, lactic acid), amino acids, flavor compounds (diacetyl), and vitamins.

Next, we describe how to calculate specific uptake and production rates from data collected from a continuous culture, and how to integrate this data into the model for flux prediction at a specific growth rate.

1. Transform metabolite concentrations in the feed and waste to units of mmol/L.
2. Calculate the metabolite concentrations differences between the waste and the feed.
3. Calculate yields of different metabolites using the difference in metabolite concentrations and the biomass concentrations. These yields will be in units of mmol/gDW.
4. Calculate specific rates for the different metabolites using specific yields and the specific growth rate. These specific rates will be in units of mmol/(gDW h).
5. Incorporate the calculated rates into the model. This is carried out by setting the lower and upper bounds of the corresponding exchange reactions to the calculated values. In particular, the following command must be used:

```
> model = changeRxnBounds(model, rxns, bounds, 'b');
```

6. Perform a flux balance analysis using the COBRA Toolbox v3.0.

```
> model = optimizeCbModel(model);
```

For each condition, FBA computes a flux distribution that maximizes the growth rate under that specific condition.

7. Perform a flux variability analysis (FVA) to assess the robustness of the solution obtained in **step 6**. As solutions returned by FBA are not unique, computed flux distributions must be evaluated for their flexibility by calculating the allowable range for each flux under (sub)optimality. This is conducted by minimizing and subsequently maximizing each reaction. To perform FVA, the following command must be run:

```
> [minFlux, maxFlux] = fluxVariability(model);
```

Complementary to this analysis, a round of random sampling of the flux solution space can be performed for exploring the feasible space using various algorithms [38, 39]. Illustrative applications are found elsewhere [40, 41].

8. Normalize the fluxes to compare flux distributions under different conditions. Typical normalizations involve dividing all the fluxes by the specific uptake rate of the main carbon source or the specific growth rate.
9. Optionally, flux distributions can be drawn in a small metabolic network to summarize the main differences between the simulated conditions. For this task, the platform Escher [42, 43] is useful to visualize the flux difference in the metabolic network.

We illustrate the above workflow with an example using data from Pizarro et al. [37] and the latest genome-scale model of *S. cerevisiae* [7]. A step-by-step tutorial of this workflow is presented in Tutorial 1. In this research, the authors grew the EC1118 yeast strain under anaerobic, nitrogen-limited conditions at 15 °C and 30 °C in continuous cultures. The authors measured the concentrations of nitrogen (ammonium), carbon substrates (glucose), as well as metabolic products (ethanol, pyruvate, succinate, acetate, glycerol, and carbon dioxide), in the feed and waste. Using this information, the specific uptake and production rates of the metabolites can be calculated using **steps 1–4**. The authors reported the specific rates in units $\frac{mmol\ C}{gDW\ h}$, which are common in the wine research field. $\frac{mmol\ C}{gDW\ h}$ can be easily transformed to $\frac{mmol}{gDW\ h}$ by dividing the rate by the number of carbon atoms of the molecule. We present the specific rates in the next table (Fig. 4):

We incorporate these values into the model (**step 5**, Fig. 5).

We compute the flux distributions for each condition (**step 6**, Fig. 6).

And we obtain the specific growth rates (Fig. 7).

The above analysis yielded $\mu = 0.0468$ 1/h and 0.049 1/h, respectively. As expected, the specific growth rates are almost equal to the dilution rates reported in [37], specifically $D = 0.047 \pm 0.000$ 1/h and 0.049 ± 0.002 1/h for cultures at 15 °C and 30 °C, respectively.

In addition, the entire flux distribution can also be obtained (Fig. 8).

As an example, the flux values of the first ten reactions of the yeast genome-scale metabolic model are presented below (Fig. 9):

Next, we perform FVA in each condition (**step 7**, Fig. 10). The goal of this step is to compute the flux range for each reaction in the two studied conditions. Reactions with no

Metabolite	Temperature_15_degrees	Temperature_30_degrees
'Glucose'	-3.56	-4.83
'Ammonium'	-0.1838	-0.1365
'Ethanol'	5.99	7.28
'Pyruvate'	0.033	0.01
'Succinate'	0.0075	0.013
'Acetate'	0.015	0.015
'Glycerol'	0.03	0.03
'Carbon Dioxide'	7.1	8.84

Fig. 4 Specific rates of consumed and produced metabolites from continuous cultures of *S. cerevisiae* at 15 and 30 °C

```
rxns = {'EX_glc__D_e', 'EX_nh4_e', 'EX_etoh_e', 'EX_pyr_e',...
    'EX_succ_e', 'EX_ac_e', 'EX_glyc_e', 'EX_co2_e'};
valuesCondition1 = [-3.56, -0.1838, 5.99, 0.033, 0.0075, 0.015, 0.03, 7.1];
valuesCondition2 = [-4.83, -0.1365, 7.28, 0.01, 0.013, 0.015, 0.03, 8.84];

modelCondition1 = changeRxnBounds(model_condition1, rxns, valuesCondition1, 'b');
modelCondition2 = changeRxnBounds(model_condition2, rxns, valuesCondition2, 'b');
```

Fig. 5 Integration of experimental rates into the model at both temperatures

```
fbaCondition1 = optimizeCbModel(modelCondition1);
fbaCondition2 = optimizeCbModel(modelCondition2);
```

Fig. 6 Execution of Flux Balance Analysis under the studied conditions

```
specificGrowthRateC1 = fbaCondition1.f;
specificGrowthRateC2 = fbaCondition2.f;
```

Fig. 7 Optimal specific growth rates at both temperatures

```
fluxDistributionC1 = fbaCondition1.x;
fluxDistributionC2 = fbaCondition2.x;
```

Fig. 8 Optimal flux distributions at both temperatures

overlapping ranges between conditions hints to parts of cellular metabolism that were differentially affected.

Next, we show the first ten reactions that do not overlap (Fig. 11).

We can also determine which subsystems/pathways are associated to the reactions whose ranges do not overlap (Fig. 12).

We list the first ten most frequent subsystems among the reactions that do not overlap (Fig. 13).

Reaction	Condition_1	Condition_2
'D_LACDcm'	0	0
'D_LACDm'	0	0
'BTDD_RR'	0	0
'L_LACD2cm'	0	0
'r_0005'	0.053458	0.065051
'r_0006'	0.017861	0.021735
'PRMICI'	0.0020171	0.0013431
'P5CDm'	0	0
'r_0013'	0	0
'DRTPPD'	4.6832e-05	4.9003e-05

Fig. 9 Visualization of the flux values for the first ten reactions of the model at both temperatures. Abbreviations denote the following reactions: **D_LACDcm**: (*R*)-lactate:ferricytochrome-c 2-oxidoreductase, **D_LACDm**: (*R*)-lactate:ferricytochrome-c 2-oxidoreductase, **BTDD_RR**: (*R,R*)-butanediol dehydrogenase, **L_LACD2cm**: (*S*)-lactate:ferricytochrome-c 2-oxidoreductase, **r_0005**: 1,3-beta-glucan synthase, **r_0006**: 1,6-beta-glucan synthase, **PRMICI**: 1-(5-phosphoribosyl)-5-[(5-phosphoribosylamino)methylideneamino)imidazole-4-carboxamide isomerase, **P5CDm**: 1-pyrroline-5-carboxylate dehydrogenase, **r_0013**: 2,3-diketo-5-methylthio-1-phosphopentane degradation reaction, **DRTPPD**: 2,5-diamino-6-ribitylamino-4(3*H*)-pyrimidinone 5′-phosphate deaminase

We found that many reactions are associated with the metabolism of amino acids and nucleotides. This again, is consistent with [37], where several genes associated to those subsystems were differentially regulated.

We normalize the fluxes by the nitrogen source (**step 8**, Fig. 14).

We export the fluxes for visualization (**step 9**, Fig. 15).

Finally, we can use Escher to visualize the fluxes (Fig. 16). In the pathway related to the metabolism of nucleotides and amino acids, reactions have different values and the fold change can be readily visualized.

Calculation of Flux Distributions in Batch Cultures

For batch cultures, time courses of metabolites and biomass concentrations need to be available. Based on these time courses, we can calculate specific uptake and production rates that will be incorporated as inputs into the model. However, as mentioned before, in batch mode the specific growth rate as well as specific uptake/production rates (*see* **Note 3**) could drastically change during the culture due to the modification of the extracellular environment. Hence, to apply FBA, a time frame where the intracellular steady state holds must be first found.

```
%we perform a FVA
[minFlux_c1, maxFlux_c1] = fluxVariability(modelCondition1); %condition 1
[minFlux_c2, maxFlux_c2] = fluxVariability(modelCondition2); %condition 2
% we normalize the minimum and maximum fluxes in condition 1 by the uptake rate
% of ammonium
minFlux_c1_norm = minFlux_c1 / abs(specific_uptake_rate_ammonium_c1);
maxFlux_c1_norm = maxFlux_c1 / abs(specific_uptake_rate_ammonium_c1);

% we normalize the minimum and maximum fluxes in condition 2 by the uptake rate
% of ammonium
minFlux_c2_norm = minFlux_c2 / abs(specific_uptake_rate_ammonium_c2);
maxFlux_c2_norm = maxFlux_c2 / abs(specific_uptake_rate_ammonium_c2);
% reactions that do not overlap are those for which the minimum value in
% condition 1 is higher than the maximum value in condition 2 and those for
% which the minimum value in condition 2 is higher than the maximum value in
% condition 1
positions_reactions_not_overlapping = union(find(minFlux_c1_norm>maxFlux_c2_norm),...
    find(minFlux_c2_norm>maxFlux_c1_norm));
%we get the reactions that overlap as the remaining reactions
positions_reactions_overlapping = setdiff(1:length(yeast8.rxns),...
    positions_reactions_not_overlapping);
%we get the reactions that do not overlap
reactions_not_overlapping = yeast8.rxns(positions_reactions_not_overlapping);
%we get the reactions that overlap
reactions_overlapping = yeast8.rxns(positions_reactions_overlapping);
```

Fig. 10 Application of flux variability analysis under each condition and analysis of flux overlap

Reaction	Min_C1	Max_C1	Min_C2	Max_C2
'r_0005'	0.29092	0.29092	0.47657	0.47658
'r_0006'	0.097202	0.097202	0.15923	0.15923
'PRMICI'	0.010977	0.010978	0.0098396	0.0098432
'DRTPPD'	0.00025486	0.00025486	0.000359	0.000359
'r_0015'	0.00025486	0.00025486	0.000359	0.000359
'AATA'	0.047385	0.047386	0.042475	0.04248
'r_0027'	0.047385	0.047386	0.042475	0.04248
'DB4PS'	0.00050973	0.00050973	0.000718	0.00071801
'ADCS'	1.6158e-05	1.7905e-05	2.2761e-05	3.3632e-05
'ADCL'	1.6158e-05	1.7905e-05	2.2761e-05	3.3632e-05

Fig. 11 Visualization of reaction fluxes that differ (differential reactions) under the two conditions (i.e., fluxes do not overlap)

Next, we describe how to calculate specific uptake and production rates using data collected in batch cultures and how to incorporate these rates into the model.

1. Convert all metabolite concentrations to mmol/L and the biomass concentration to gDW/L.
2. Apply the natural logarithmic function to metabolite and biomass concentrations. As metabolite and biomass

```
%we obtain the subsystems associated with the reactions that do not overlap
subsystems = [];
for i = 1:length(positions_reactions_not_overlapping)
    if ~isempty(yeast8.subSystems{positions_reactions_not_overlapping(i)}{1})
        subsystems = [subsystems;...
            yeast8.subSystems{positions_reactions_not_overlapping(i)}'];
    end
end

%we obtain the frequencies
frecuencies = tabulate(subsystems);
%we sort the frequencies
[s,i] = sort(cell2mat(frecuencies(:,2)),'descend');
sorted_frecuencies = frecuencies(i,:);
```

Fig. 12 Generation of subsystems associated with the differential reactions

Subsystems	Frequency	Percentage
'sce01110 Biosynthesis of secondary metabolites'	[44]	[17.8138]
'sce01130 Biosynthesis of antibiotics'	[32]	[12.9555]
'sce01230 Biosynthesis of amino acids'	[28]	[11.3360]
'sce00970 Aminoacyl-tRNA biosynthesis'	[20]	[8.0972]
'sce00230 Purine metabolism'	[11]	[4.4534]
'sce00340 Histidine metabolism'	[9]	[3.6437]
'sce00300 Lysine biosynthesis'	[8]	[3.2389]
'sce00400 Phenylalanine, tyrosine and tryptophan biosynthesis'	[8]	[3.2389]
'Gluconeogenesis'	[7]	[2.8340]
'sce00010 Glycolysis'	[7]	[2.8340]

Fig. 13 Subsystems associated with the differential reactions between conditions

```
%we normalize by the uptake rate of ammonium for condition 1
fluxDistributionC1_norm = fluxDistributionC1/abs(specific_uptake_rate_ammonium_c1);
%we normalize by the uptake rate of ammonium for condition 2
fluxDistributionC2_norm = fluxDistributionC2/abs(specific_uptake_rate_ammonium_c2);
```

Fig. 14 Flux normalization by the nutrient uptake rate for subsequent comparison of each condition

```
exportMultipleSolutionsToJson(yeast8,...
    [fluxDistributionC1_norm, fluxDistributionC2_norm], 'sol.json')
```

Fig. 15 Export of flux solutions to a JSON file for visualization in Escher

concentrations follow an exponential curve, linear curves are obtained when using this transformation. This facilitates the subsequent analysis.

3. Plot the natural logarithm values versus time.
4. Identify a time frame (hours) where steady-state-like conditions apply. This implies to find a time frame where the specific growth rate, as well as specific uptake/production rates of all the metabolites are constant. Lag and stationary phases should

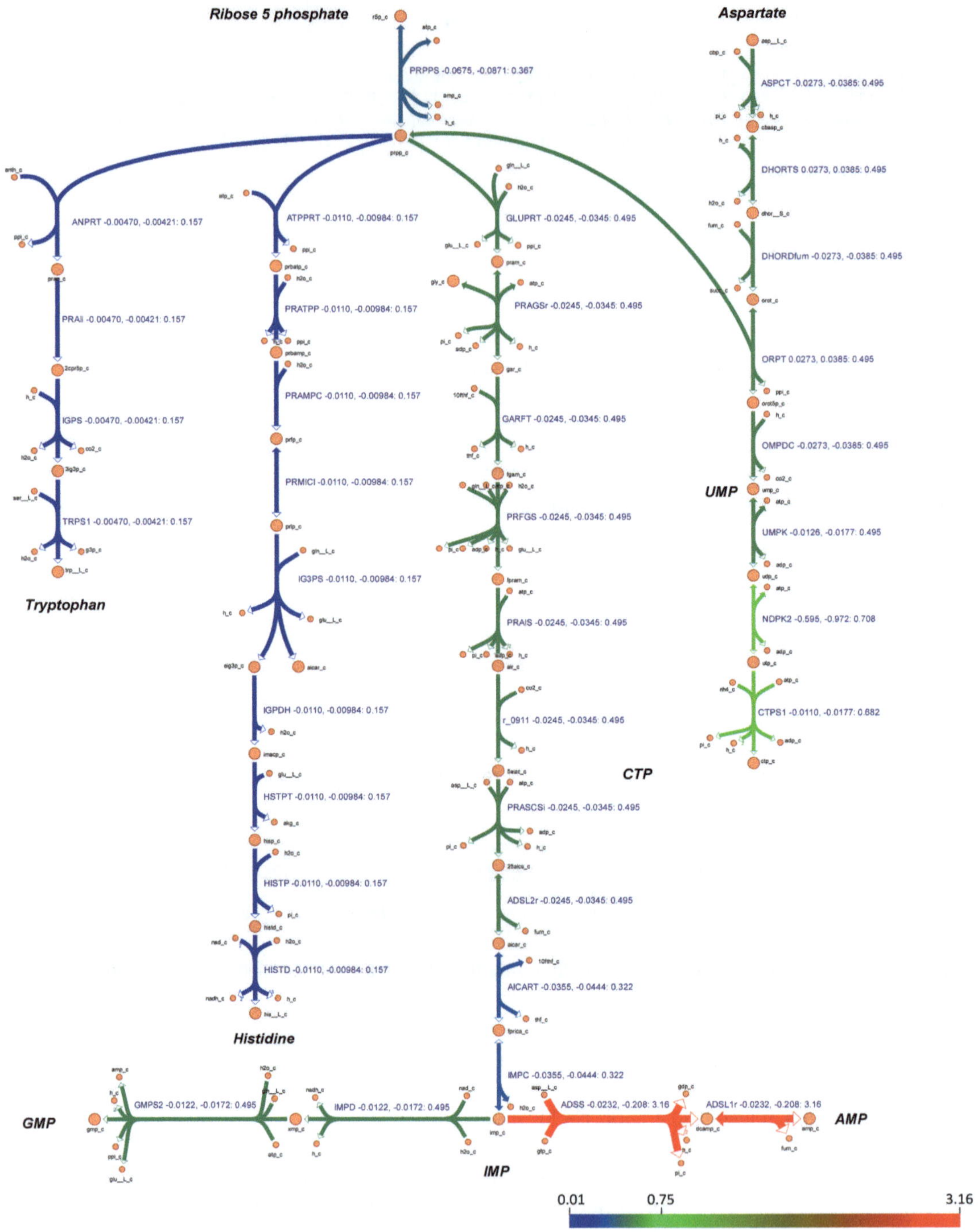

Fig. 16 Flux distributions of the nucleotide biosynthesis pathway of *S. cerevisiae* growing in a nitrogen-limited culture under two different temperatures: 15 and 30 °C. Uptake and secretion rates calculated from cultures at both temperatures were incorporated as inputs in the model. Specific fluxes can be observed in the figure for 15 °C (first value) and 30 °C (second value). Also, the log2 of the fold change can be seen (third value). Reactions in red, green, and blue represent high, medium, and low fold-change, respectively

be discarded in this analysis. As plots are in log-scale, this can be easily carried out by inspecting the slope of the curves. A simple regression should fit the data to a linear curve and reveal the appropriate time frames (*see* **Note 4**). Occasionally, more than one phase may be observed [14, 22]. In such cases, FBA must be applied separately to each time frame.

5. For each time frame identified, calculate yields for each compound. This is done by dividing the change in metabolite concentrations by the change in biomass concentrations. Yields are in units of mmol/gDW.
6. Calculate the specific growth rate in the time frame. This will be in units of 1/h.
7. Calculate the specific uptake/production rates for each metabolite using the calculated yields and growth rate. These uptake/production rates will be in units of mmol/(gDW h).
8. Repeat **steps 4–7** for each time frame identified.

3.1.2 Sensitivity Analysis

Frequently, we want to quantify the extent whereby the lower and upper bounds of the exchange reactions of nutrients and secretion products, affect the specific growth rate. This information can be obtained from the reduced costs of the optimization. In a maximization problem—just like the FBA formulation—the reduced cost is defined as the amount by which the objective function decreases as a result of an increase in the value of a variable by one unit [46]. Therefore, when a reduced cost of a variable is positive - and has a value of a -, the objective function will decrease in a unit as a result of a unitary increase of the analyzed variable, and vice versa.

The formal mathematical description of the reduced cost is (*see* **Note 5**):

$$r_i = -\frac{dZ}{dv_i}$$

One observation that can help the reader to get an intuitive understanding of reduced costs is the following: The reduced cost is always zero for a variable that does not hit the capacity constraints (lower or upper bounds). If the reduced cost is different from zero, then the variable must have hit a capacity constraint. For example, let us consider the case where we compute the FBA solution that maximizes the specific growth rate of *S. cerevisiae* in a nitrogen-limited chemostat with ammonium as the only nitrogen source. If the reduced cost for the reaction that provides ammonium in the model is different from zero, it means that the uptake rate of ammonium has hit the defined capacity constraint. Therefore, an increase in the maximum uptake rate of ammonium will result in an increase in the specific growth rate.

Another important observation is that when the specific growth rate is the objective function, the reduced cost for the flux variables have units of $\frac{\frac{1}{h}}{\frac{mmol}{gDW\ h}} = \frac{gDW}{mmol}$. Therefore, for the variables representing uptake of secretion of metabolites, the reduced costs can also be interpreted as the yield of biomass with respect to those metabolites.

Sometimes, we may not be interested in knowing the effect of each one of the reactions in our network but only how active reactions (i.e., with nonzero flux) affect our objective. In this case, scaled reduced costs are appropriate [48]. The formal mathematical description of scaled reduced cost is

$$R_i = -\frac{r_i * v_i}{Z}.$$

Note that by rewriting the scaled reduced cost as

$$R_i = -\frac{r_i * v_i}{Z} = \frac{dZ}{dv_i} * \frac{v_i}{Z} = \frac{\frac{dZ}{Z}}{\frac{dv_i}{v_i}} = \frac{dln(Z)}{dln(v_i)},$$

they can be interpreted as the relative change in the objective function with respect to a relative change in a flux variable. This nomenclature resembles control coefficients commonly used in metabolic control analysis [49].

The two main differences between *reduced costs* and *scaled reduced costs* are:

1. While reduced costs give information about all the reactions in our metabolic network, scaled reduced costs give information just about active reactions in the solution vector v. This has some practical implications. As most internal reactions are only constrained using thermodynamic information (based on the Gibbs free energy change ΔG), their capacity constraints (bounds) are going to be either $[-\infty, \infty]$ or $[0, \infty]$ for reversible and irreversible reactions, respectively (*see* **Note 6**). As the reduced cost of a reaction is nonzero when it has hit a constraint, and as it cannot hit $-\infty$ or ∞, this implies that if the reduced cost of internal reaction is different from zero, then its value in the solution should be zero, that is, the reaction is inactive. On the contrary, exchange reactions (uptake rates of nutrients or secretion rates of products) are usually constrained with experimental data, and therefore their bounds are usually nonzero. Therefore, in practice, while reduced costs give information about all the reactions in our network, scaled reduced cost will often give information just about how active exchange reactions affect the objective function. Consequently, if the user wants to answer which of all the reactions in the network is affecting the objective function, then filtering the nonzero reduced costs is a suitable alternative to answer that. Instead, if

the user wants to answer how constraints imposed in the exchange reactions affect the specific growth rate, then filter nonzero scaled reduced cost is the most suitable option.

2. While reduced costs inform about absolute changes, scaled reduced costs inform about relative changes. Even though two reactions can have similar reduced costs, they could have very different values in the solution vector *v*. That implies that a 1% change in one reaction with a high flux could result in a much bigger relative change in the objective function than a 1% change in other reaction with a small flux. As scaled reduced costs inform about relative changes, they show more clearly how percentual changes in flux variables affect the objective function.

Next, we describe how to perform a reduced cost analysis (see Tutorial 2 for more details):

1. Set lower and upper bounds.

```
> model = changeRxnBounds(model, rxns, bounds, 'b')
```

2. Perform a FBA simulation.

```
> fba = optimizeCbModel(model)
```

3. Get reduced costs and fluxes.

```
> reduced_costs = fba.w
 > fluxes = fba.x
```

4. Get scaled reduced costs

```
> scaled_reduced_costs = (fba.w .* fba.x)/fba.f
```

5. Interpret costs.

Following the example for the data presented in [37], we will perform a reduced-cost analysis. We analyze here the first experimental condition of low temperature of growth ($T = 15$ °C).

First, we set the bounds (**step 1**, Fig. 17).

We solve the linear problem (**step 2**, Fig. 18).

We get the reduced costs and specific fluxes (**step 3**, Fig. 19).

We get the scaled reduced costs (**step 4**, Fig. 20).

Finally, we display the results (Fig. 21).

We interpret the costs (**step 5**). By inspecting the reduced costs, we can conclude that:

```
rxns = {'EX_glc__D_e', 'EX_nh4_e', 'EX_etoh_e', 'EX_pyr_e',...
    'EX_succ_e', 'EX_ac_e', 'EX_glyc_e', 'EX_co2_e'};
valuesCondition1 = [-3.56, -0.1838, 5.99, 0.033, 0.0075, 0.015, 0.03, 7.1];

model_condition1 = changeRxnBounds(model_condition1, rxns, valuesCondition1, 'b');
```

Fig. 17 Integration of experimental rates calculated from a culture growing at 15 °C

```
fbaCondition1 = optimizeCbModel(model_condition1);
```

Fig. 18 Flux balance analysis at 15 °C

```
%fluxes
fluxesC1 = fbaCondition1.x;
%reduced costs
reducedCostsC1 = fbaCondition1.w;
```

Fig. 19 Determination of optimal flux values and reduced costs

```
scaledReducedCosts = -(fluxesC1.*reducedCostsC1)/fbaCondition1.f;
%we find those scaled reduced costs that are different from zero
positions_src_not_zero  = find(scaledReducedCosts);
%filter out those that are less that a tolerance
tolerance = 1e-8;
higher_than_tolerane = find(abs(scaledReducedCosts(positions_src_not_zero))>tolerance);
positions_src_not_zero = positions_src_not_zero(higher_than_tolerane);

t = table(yeast8.rxns(positions_src_not_zero),...
    num2cell(scaledReducedCosts(positions_src_not_zero)),...
    num2cell(reducedCostsC1(positions_src_not_zero)),...
    num2cell(fluxesC1(positions_src_not_zero)),...
    'VariableNames',{'Reaction','Scaled_reduced_cost','reduced_cost','specific_flux'});

disp(t)
```

Fig. 20 Calculation of scaled reduced costs

Reaction	Scaled_reduced_cost	reduced_cost	specific_flux
'EX_nh4_e'	[1.0000]	[0.2548]	[-0.1838]

Fig. 21 Visualization of reactions with a scaled reduced cost different from zero

1. Ammonium is the only nutrient that is limiting the specific growth rate. This was expected as the experiments were performed under nitrogen-limited conditions and ammonium was the only nitrogen source. Also, this is consistent with the fact that we see a scaled reduced cost of 1, which means that among all the active fluxes, ammonium has "full control" over the objective function.

2. As the reduced cost is 0.2548, that means that we should observe a decrease in the objective function (specific growth rate) of 0.2548 if we were to increase the specific uptake rate of ammonium (v_{NH4}) in 1 unit. Remember that the equation for the exchange reaction of ammonium is "1 nh4[e] <=> " meaning that the uptake is represented by negative values. Consequently, the uptake of ammonium increases as the value of v_{NH4} becomes more negative. As an increase in v_{NH4} yields a lower uptake rate, it makes sense that we would see a decrease in the objective function when increasing v_{NH4}.
3. As the reduced cost is actually the derivative of the objective function w.r.t. the variables $\left(\frac{dZ}{dv_i}\right)$, the reduced cost is only valid for infinitesimal variations of the variables w.r.t. the constraints imposed. Therefore, in many occasions it will not be possible to increase the variable in 1 whole unit. Instead, we should increase the value of the variable in a small number and we should see a proportional decrease in the objective function. For, example, let us suppose that we increase the value of v_{NH4}, which is currently -0.1838, in 10^{-6}. Then, we should observe a decrease of 0.254×10^{-6} (i.e., 2.54×10^{-7}) in the specific growth rate. We can corroborate this with a simple calculation (Figs. 22 and 23).
4. Finally, the units of the reduced cost are $\frac{\frac{1}{h}}{\frac{mmol}{gDW\,h}} = \frac{gDW}{mmol}$. Therefore, the reduced cost can also be interpreted as the yield of biomass w.r.t. the limiting nutrient $yield_{\frac{biomass}{sustrate}}$. In the sourced paper [37], the reported $yield_{\frac{biomass}{sustrate}}$ was $14.17\ \frac{gDW}{g\ NH4}$. This equals to $14.17 \frac{gDW}{g\ NH4} \times 18.039 \frac{gNH4}{mol\ NH4} \times \frac{1\ mol}{1000\ mmol} = 0.2556 \frac{gDW}{mmol\ NH4}$, which is almost equal to the reduced cost predicted by the model.

```
%we create a model with a small perturbation in the uptake rate of ammonium
model_small_variation = changeRxnBounds(model_condition1, 'EX_nh4_e', -0.1838+1e-6, 'b');
%we perform a FBA
fbaCondition1_B = optimizeCbModel(model_small_variation);
%we calculated the difference in the objective function between both simulations
difference = fbaCondition1.f - fbaCondition1_B.f;
fprintf('The decrease in the specific growth rate is :%4.3e\n',difference)
```

Fig. 22 Creation of another model with a small decrease in the uptake rate of ammonium

```
The decrease in the specific growth rate is :2.548e-07
```

Fig. 23 Decrease in the specific growth after a small change in the maximum uptake rate of ammonium (lower bound of the corresponding exchange reaction)

3.2 Determination of Nutritional Requirements and Comparison with Experimental Data

Nutritional requirements for microorganisms involved in wine fermentation can be readily determined using GEMs. As mentioned in the Introduction, GEMs encompass all the metabolic reactions of a particular microorganism, including the synthesis of macromolecules from building blocks. All cells need to synthesize macromolecules such as proteins, lipids, and DNA. However, the machinery used to synthesize these macromolecules differs between species due to its genomic content. This unique genomic content results in a unique set of enzymes which, in turn, catalyzes a unique set of reactions. When a particular gene responsible for coding an enzyme that catalyzes the biosynthesis of a certain building block is missing in the genome, then the cell must incorporate that building block from the environment. When a gene is missing in the genome, the associated metabolic reaction will also be missing in the genome-scale metabolic model. In this way, the model will predict that the cell cannot synthesize that building block and that it needs to be uptaken from the environment to achieve growth. This is the logic by which genome-scale metabolic models can be used to predict nutritional requirements (Fig. 24).

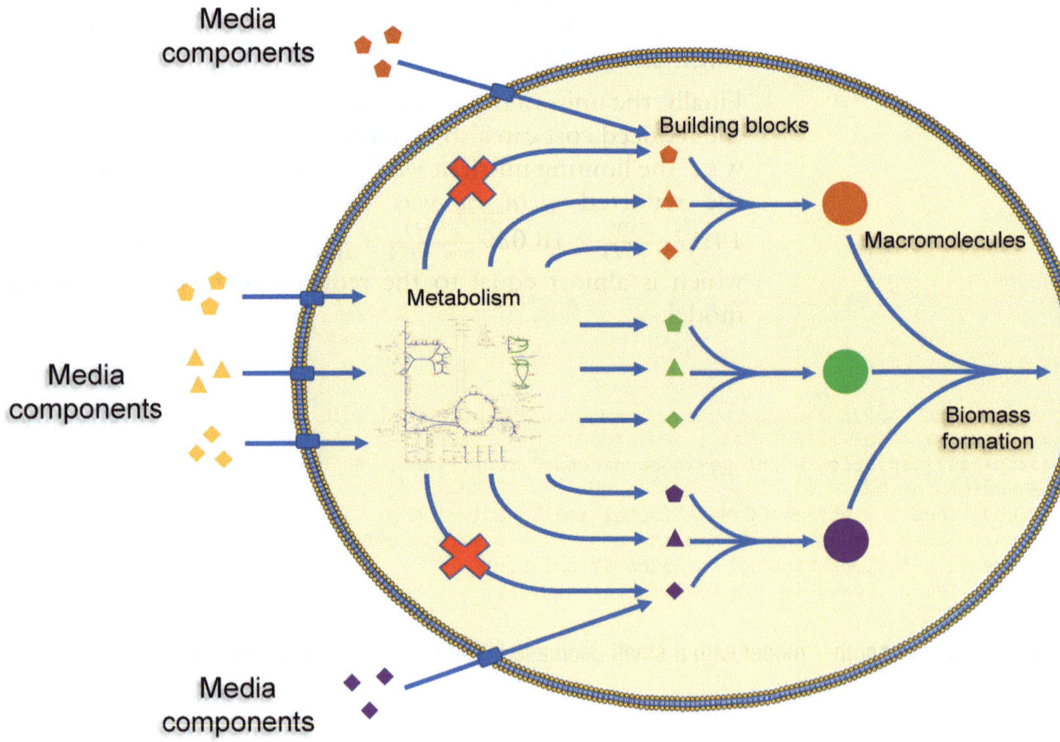

Fig. 24 Genome-scale metabolic models (GEMs) are convenient for predicting nutritional requirements of microbial cells. GEMs contain a detailed description of the synthesis of macromolecules (e.g., proteins) from building blocks (e.g., amino acids). Many of these building blocks are synthesized by the enzymatic machinery of the cell, which is readily captured by GEMs. However, there are other building blocks that need to be taken up from the environment as there may be missing enzymes in the relevant pathways

3.2.1 Minimal Media Determination

In this section, we describe how to use GEMs to list the nutrients which are minimally required to generate biomass; in other words, how to obtain the set of nutrients with minimal cardinality that can sustain growth. For this task, we describe the application of the algorithm EMAF (Enumeration of Minimal Active Fluxes) [36]. This algorithm solves a Mixed-Integer Linear Programming (MILP) problem where the objective function is the minimization of the number of exchange reactions that enable the uptake of nutrients constrained to the mass balances under steady-state. In addition, this algorithm classifies nutrients in two categories: those that cannot be replaced with other nutrients (required) and those that can (interchangeable).

Formally, the problem solved by EMAF is the following:

$$Min_{v,z} \sum\nolimits_{k \in R_{ex}} z_k \quad (2)$$

Subject to

$$\sum\nolimits_{j \in R} S_{ij} v_j = 0, \forall i \in M \quad (2a)$$

$$LB_j \leq v_j \leq UB_j, \ \forall j \in R \quad (2b)$$

$$v_k \leq UB_k * z_k, \ \forall k \in R_{ex} \quad (2c)$$

$$z_k \in \{0, 1\}, \ \forall k \subset R_{ex}$$

$$v_k \in R^+, \ \forall k \in R_{ex}$$

$$v_j \in R, \forall j \in R \backslash \{R_{ex}\}$$

where v are the fluxes through the biochemical reactions of the metabolic network and z are binary variables associated to the exchange reactions that enable the uptake of user-defined nutrients. All the variables associated to fluxes are in units of $\frac{mmol}{gDW\ h}$, except for the variable representing the specific growth rate μ, which is in units of $1/h$. S is the stoichiometric matrix, where the value in the position i, j represents the stoichiometry of metabolite i in reaction j. LB_j and UB_j are values representing the lower and upper bounds for the rate of reaction j, respectively. M is the set of all the metabolites in the network, R is the set of all the reactions in the network and R_{ex} is a subset of R that corresponds to the exchange reactions that describe the uptake of the defined nutrients. Notably, R_{ex} does not necessarily corresponds to the total set of exchange reactions in the network. The challenge here is to find the minimal set of nutrients that can sustain growth given a particular medium composition. In that case, the user has to define R_{ex} as the set of exchange reactions associated with that specific medium composition. Note that results may vary, depending on the simulated media composition. To avoid different output results, R_{ex} has to be defined as the entire set of exchange reactions in the network.

The logic behind this mathematical formulation is the following. Binary variables and fluxes through exchange reactions are associated through Eq. (2c). When the binary variable z_k is zero, then the flux through the exchange reaction v_j must also be zero. When the flux through an exchange reaction v_j is positive, then the binary variable z_k must be one. The formulation minimizes the number of active binary variables z, and, as these variables are associated to the fluxes v of exchange reactions, the formulation will find the minimum feasible number of exchange reactions.

In addition, in the second phase, this algorithm predicts which nutrients can be replaced by others, providing alternative nutrients that can be used to form the minimal media.

Next, we describe how to predict a minimal medium composition. Tutorials 3 to 4 illustrate the steps for performing this analysis.

1. Set an in silico medium where the metabolic model is able to generate biomass. If a chemically defined medium has never been experimentally created for the studied species, then all the exchange reactions can be set to negative values in order to allow the uptake of all the nutrients. We suggest setting all the lower bounds to −10 mmol/(gDW h). Even though this value could be several orders of magnitude higher than the actual uptake rates (especially true for the vitamin uptake rates), in this analysis we are only interested in knowing if reactions are needed or not. Therefore, the actual values of the lower bounds are not particularly relevant. This analysis is also useful in cases when a chemically defined medium is employed. In this case, the interest is to determine which nutrients are required for growth and which ones are not. In this second scenario, if this is done from scratch, first, the compounds in the experimental medium must be mapped to the exchange reactions in the model. Let us define R_{ex} as the entire set of exchange reactions in the model and R_{media} as a subset of R_{ex} corresponding to the exchange reactions for the compounds in the medium. The lower bound of all the exchange reactions in R_{ex} must be set to zero and then, we suggest setting the lower bounds of the exchange reactions in R_{media} to −10 mmol/(gDW h).
2. Perform FBA using the medium formulation to check that the metabolic model is able to generate biomass. Obtain the value of the growth rate, which is done with the following commands,

```
> fba = optimizeCbModel(model)
      > growth_rate = fba.f
```

3. Set the constraints that must be achieved for the desired medium formulation. In particular, it is important to define a constraint for the growth rate. The value for the growth rate constraint depends on the purpose. For example, if the purpose is to determine what are the nutrients required to sustain a specific growth near to the maximum value, we suggest to set a lower bound for the growth rate at 90% of the optimal FBA value (**step 2**). On the contrary, if the purpose is to search for alternative nutrients to sustain growth at any level, 1% could be used. Other constraints in between can also be employed.
4. Define the set of exchange reactions that will be used to create the binary variables for the MILP problem. Let us define R_{ex} as the set of exchange reactions in the genome-scale metabolic model and R_{emaf} as a subset of R_{ex} that represents the set of exchange reactions used by EMAF to create the binary variables that will be minimized. EMAF creates one binary per exchange reaction belonging to R_{emaf}. Exchange reactions that are not in the set R_{emaf} will not be considered for the minimization problem. We suggest defining R_{emaf} as the set of exchange reactions used to set the medium in **step 1** (R_{media}).
5. Create specific input files to run EMAF. The following command line creates all the directories and input files needed to run EMAF.

```
> createInputsForEMAF(model, biomassRxnID, baseDir, model-
File, constraints_ids, constraints_lb, constraints_ub, posEX)
```

The inputs are:

model: a COBRA structure containing the genome-scale metabolic model.

biomassRxnID: a string with the identifier of the growth rate identifier.

baseDir: the path where the EMAF inputs and output are going to be stored.

modelFile: a string with the name of the model.

constraints_ids: a cell array with the list of reaction identifiers used to additionally constrain the model. This list is defined in **step 3**.

constraints_lb: a double array with the lower bounds associated with the reaction identifiers in the cell array constraints. This array is defined in **step 3**.

constraints_ub: a double array with the upper bounds associated with the reaction identifiers in the cell array constraints. This array is defined in **step 3**.

```
% we load the medium formulation created by Terrade and Mira de Orduña [2]
% and we allow the nutrients to be consumed in the model
[model, mediaExchangeRxns, nutrients] = setMediaFromExcelFileWithRxnIDs(model, ...
    'wineMediaFormulations','Terrade_metacyc');
% we also allow the uptake of carbon sources for which we
% ignore if they can sustain growth
c_sources = {'glycerol_ex_', 'sucrose_ex_', 'b_D_glucose_ex_',...
    'b_D_galactose_ex_', 'a_D_galactose_ex_', 'trehalose_ex_', 'cellobiose_ex_',...
    'melibiose_ex_', 'b_D_fructose_ex_', 'L_arabinose_ex_'};
model = changeRxnBounds(model, c_sources,10,'u');
```

Fig. 25 Medium formulation and incorporation of corresponding maximal uptake rates into the model

```
%we solve a FBA
fba = optimizeCbModel(model);
fprintf('The specific growth rate is:%2.2f 1/h\n',fba.f)
```

Fig. 26 Flux balance analysis

```
The specific growth rate is :4.11 1/h
```

Fig. 27 Specific growth rate obtained with the medium formulation according to ref. 38

```
% we identify the reaction ID for the growth rate
growth_rate = model.rxns(model.c~=0);
% we create three vectors to especify reaction ids, lower and upper bounds.
% In this case, there is only one constraint so the vectors have length 1
constraints_ids = growth_rate;
constraints_lb = 0.01*fba.f;
constraints_ub = 1000;
```

Fig. 28 Specification of constraints for EMAF

posEX: a double array containing the list of positions for the exchange reactions defined in **step 4**.

Run EMAF in python with the command:

```
> python runMedia3.py
    > python pushRunMedia3.py
```

Next, we illustrate how to run EMAF for *Oenococcus oeni* using the genome-scale metabolic model reported in [8] and a chemically defined medium composition [50]. First, we setup the in silico medium (**step 1**, Fig. 25).

We verify biomass formation (**step 2**, Figs. 26 and 27).

We set up the constraints (**step 3**, Fig. 28).

```
% we specify which is the set of exchange reactions that we want to
% mimimize
exchangeRxns = union(mediaExchangeRxns, c_sources);
% .. and their positions in the model.
posEX = getPosOfElementsInArray(exchangeRxns',model.rxns);
```

Fig. 29 Specification of exchange reactions to be minimized by EMAF

```
% we specify the folder where the inputs and results are goind to be stored
baseDir = pwd;
%we specify the name of the model
modelFile = 'iSM454';
%we create the inputs for EMAF
createInputsForEMAF(model, growth_rate, baseDir, modelFile, ...
    constraints_ids, constraints_lb, constraints_ub,posEX)
```

Fig. 30 Generation of inputs for EMAF

```
outputFilePath = ['./emaf/media_search_results-(' modelFile '_irrev.xml).csv'];
[required, alternatives] = readEMAFoutput(outputFilePath);

for i = 1:length(required)
    fprintf('%2.0f) %s \n',i,required{i})
end
```

Fig. 31 Results generated by EMAF

We specify the set of reactions to minimize (**step 4**, Fig. 29).

We export the inputs for EMAF (**step 5**, Fig. 30).

In the command console or in the Python programming environment (e.g., Anaconda), we go to the directory where the inputs were exported (*baseDir*) and run the scripts to execute EMAF (**step 6**).

```
> python runMedia3.py
      > python pushRunMedia3.py
```

We interpret the results (**step 7**, Figs. 31, 32, 33, and 34).

In conclusion EMAF found that:

1. L-Arginine, L-cysteine, L-histidine, L-isoleucine, L-leucine, L-methionine, L-phenylalanine, L-serine, L-threonine, L-tryptophan, L-tyrosine, L-valine, manganese, phosphate, nicotinamide ribonucleotide, oleate and pantothenate are needed to sustain a minimum growth rate of 0.04 1/h.

```
1) R_L_Arg_ex_
2) R_L_Cys_ex_
3) R_L_His_ex_
4) R_L_Ile_ex_
5) R_L_Leu_ex_
6) R_L_Met_ex_
7) R_L_Phe_ex_
8) R_L_Ser_ex_
9) R_L_Thr_ex_
10) R_L_Trp_ex_
11) R_L_Tyr_ex_
12) R_L_Val_ex_
13) R_Mn_ex_
14) R_P_ex_
15) R_nicotinamida_RNP_ex_
16) R_oleate_ex_
17) R_panthothenate_ex_
```

Fig. 32 Required nutrients identified by EMAF

```
for i = 1:size(alternatives,1)
    fprintf('GROUP%2.0f:\n',i)
    alternatives_group_i = strsplit(alternatives{i},',');
    for j =1:length(alternatives_group_i)
        fprintf('%2.0f) %s \n',j,alternatives_group_i{j})
    end
end
```

Fig. 33 Printing of alternative groups identified by EMAF

2. At least one the following amino acids has to be chosen to sustain a minimum growth rate of 0.04 1/h: L-glutamate or L-glutamine.
3. At least one of the following carbon sources has to be chosen to sustain a minimum growth rate of 0.04 1/h: galactose, fructose, glucose, cellobiose, melibiose, sucrose or trehalose.

3.2.2 Omission Simulations and Comparison with Experimental Data

In this type of simulation, the model is used to predict if the cell is able to generate biomass when a certain nutrient or nutrients are omitted from the medium (*see* **Note 7**). If omission experiments have been previously performed, then model predictions can be compared against the experimental data and we can judge how accurate are model predictions. In this section, we assume the availability of experimental data and the availability of a chemically

```
GROUP 1:
 1) R_L_Gln_ex_
 2) R_L_Glu_ex_
GROUP 2:
 1) R_trehalose_ex_
 2) R_cellobiose_ex_
 3) R_melibiose_ex_
 4) R_b_D_glucose_ex_
 5) R_b_D_galactose_ex_
 6) R_L_arabinose_ex_
 7) R_sucrose_ex_
 8) R_a_D_galactose_ex_
 9) R_b_D_fructose_ex_
10) R_b_D_ribopyranose_ex_
```

Fig. 34 Alternative groups identified by EMAF

defined medium where the microorganism is able to grow. Otherwise, the protocol described in Subheading 3.2.1 must be employed first.

Next, we describe the steps to perform this analysis. Tutorial 5 illustrates the steps for performing this analysis.

1. Set an in silico medium where the metabolic model is able to generate biomass. If this is done from scratch, first, the compounds in the experimental medium must be mapped to the exchange reactions in the model. In addition, appropriate lower bounds must be defined for each of the mapped exchange reaction. Ideally, these lower bounds should be calculated using experimental data. Estimations based on the maximum amount that can be consumed may also be used. Let us define R_{ex} as the whole set of exchange reactions in the model and R_{media} as a subset of R_{ex} corresponding to the exchange reactions for the compounds in the medium. The lower bound of all the exchange reactions in R_{ex} must be set to zero, and then, the lower bounds of the exchange reactions in R_{media} must be set to the lower bounds determined by the user.
2. Verify that the model is able to generate biomass using the medium composition set by performing FBA.
3. Set a threshold growth rate value to discern between growth or no growth. All the predicted specific growth rates below the threshold will be considered as if there was no growth. Common values vary between 10% and 30% of the growth rate obtained in the default medium using FBA (**step 1**). A strict threshold of zero may also be used (*see* **Note 8**).

4. Create a list of in silico simulations where in each simulation a single medium compound will be tested. Because reaction identifiers can vary between GEMs, the exchange reaction associated with the omitted medium compound must be identified by the user. Usually, one medium component is associated with one exchange reaction. This is the case for most compounds. However, one single medium component can also be mapped to multiple exchange reactions. This usually happens for salts which are combinations of more than one ion.
5. Create a loop where in each iteration one single medium component is omitted. In each iteration, the medium is set as done in **step 1** and then the tested nutrient is removed from the in silico medium by setting the lower bound of the corresponding exchange reactions to zero. Then, FBA is performed to predict whether the cell is able to grow in absence of the omitted nutrient. Using experimental data, the prediction results can be classified into true positives, true negatives, false negatives or false positive. True positive results are defined as those results where the model predicted growth in absence of the omitted compound and the experimental data also shows that there is growth in that condition. True negative results are defined by no growth in absence of the omitted nutrients in both cases: in silico and in vivo. False positive as growth in silico but not in vivo. False negatives as no growth in silico but growth in vivo. If the specific growth rate is higher than the threshold set in **step 3**, then the prediction result will be classified as growth. Other results are classified as no growth.
6. Count true positives (TP), true negatives (TN), false positives (FP) and false negatives (TN) results.
7. Calculate performance metrics using the following equations:

$$Sensitivity = \frac{TP}{(TP + FN)} \tag{3}$$

$$Specificity = \frac{TN}{(TN + FP)} \tag{4}$$

$$Precision\ (PPV) = \frac{TP}{(TP + FP)} \tag{5}$$

$$Negative\ predicted\ value\ (NPV) = \frac{TN}{TN + FN} \tag{6}$$

$$Accuracy = \frac{TP + TN}{TP + TN + FP + FN} \tag{7}$$

$$F - score = \frac{2(precision \times sensitivity)}{precision + sensitivity} \tag{8}$$

3.2.3 Addition of Alternative Carbon Sources and Comparison with Experimental Data

In this type of simulation, the model is used to predict if the cell is able to generate biomass when an alternative carbon source is used as a replacement. We assume here that the medium composition has only one carbon source that sustains growth. This analysis follows the same procedure than the analysis in Subheading 3.1.1, except that instead of omitting a certain nutrient from the medium, the main carbon source is replaced by an alternative carbon source and the model predicts whether there is growth or not in this new condition. Usually, the predictions can be compared with experimental data obtained from Biolog phenotype arrays or API tests. Results using these tests can be readily obtained without time-consuming experiments. However, in some cases, the results from these tests differ from conventional cultures in flasks. While the latter tend to be more reliable, they come at a higher time cost. Finally, it is worth noting that the same analysis can be performed to test alternative nitrogen or phosphorus sources.

Next, we enumerate the steps to perform this analysis. **Steps 1–3** and **6–7** are the same than for the previous analysis. However, we will intentionally repeat them here in order to keep the readability.

1. Set an in silico medium where the metabolic model is able to generate biomass. If this is conducted from scratch, first, the compounds in the experimental medium must be mapped to the exchange reactions in the model. In addition, appropriate lower bounds must be defined for each of the mapped exchange reaction. Ideally, these lower bounds should be calculated using experimental data. However, also estimations based on the maximum amount that can be consumed can be used. Let us define R_{ex} as the whole set of exchange reactions in the model and R_{media} as a subset of R_{ex} corresponding to the exchange reactions for the compounds in the medium. The lower bound of all the exchange reactions in R_{ex} must be set to zero and then, the lower bounds of the exchange reactions in R_{media} must be set to the lower bounds determined by the user.
2. Verify that the model is able to generate biomass using the medium composition set by performing FBA.
3. Set a threshold growth rate value to discern between growth or no growth. All the predicted specific growth rates below the threshold will be considered as if there was no growth. A strict threshold of 0 can also be used. Note that solvers sometimes return values which are different but very close to zero. Hence, if the user wants to use 0 as a threshold, it is convenient to set the threshold to 10^{-6}, which is below experimental growth rates and above typical numerical tolerance.
4. Create a list of in silico simulations where in each simulation the ability of the cell to grow in the presence of an alternative carbon source will be tested. Because reaction identifiers can

vary between GEMs, the exchange reaction associated with each carbon source must be defined by the user.

5. Create a loop where, at each iteration, the presence of a different carbon source is tested in the in silico medium. In each iteration, first, the medium is set in the same way that it was performed in **step 1**. Second, the original carbon source is removed from the medium by setting the lower bound of the corresponding exchange reaction to zero. Third, the alternative carbon source is added to the in silico medium by setting the lower bound to a negative value (for example −10 mmol/(gDW h)). Then, FBA is performed to predict whether the cell can grow in the presence of the alternative carbon source. Using experimental data, the prediction results can be classified into true positives, true negatives, false negatives or false positive. If the growth rate is higher than the threshold set in **step 3**, then the prediction result will be classified as growth. Other results are classified as no growth.
6. Count true positives (TP), true negatives (TN), false positives (FP) and false negatives (TN) results.
7. Calculate performance metrics using the Eqs. (3)–(8).

3.3 Prediction of Flavor Compounds Production

Wine is a very complex mixture of flavor compounds. Some flavor compounds come from grapes. Others are generated by microorganisms during the fermentation. GEMs can be used to predict which flavor compound can be produced in specific medium conditions. However, this analysis should be performed carefully due to the following:

1. Pathways that synthesize flavor compounds are not always known and therefore could be missing in the studied GEM. Furthermore, even though some pathways that synthesize flavor compounds are known, they are not always incorporated into GEMs because of the original scope of the model. Thus, the user may have to check the genome and model of the studied species before doing this kind of prediction to ensure the presence of the pathways synthesizing relevant flavor compounds.
2. The biosynthesis of flavor compounds typically differs greatly for different metabolites. For example, some flavors compounds, such as lactic acid in *Oenococcus oeni*, are directly linked to the central metabolism and, therefore, it is straightforward to understand the conditions that favor their production. For other flavors, we just do not know why cells produce them. This presents both a weakness and an opportunity. Indeed, the absence of knowledge does not allow to directly describe and model such pathways, however, GEMs can be employed as prospective tools to explore possible biosynthetic routes involved in their production.

3. Care must be taken when setting the constraints for simulating specific conditions. GEMs are basically a set of mass balance equations. Therefore, if a user wants to predict the secretion of a product, the user must identify all the reaction rates that affect that production. For example, in *Oenococcus oeni* if the lactate production rate is to be predicted, the specific uptake rate of L-malate must be set in the model as this is the immediate precursor of L-lactate. Note that it may also be important to set specific production rates for other products to obtain more realistic results.
4. Experimental specific production rates of flavor compounds may be lower by several orders of magnitude than the input rates set in the model. Therefore, the relative variation of fluxes may be quite high, resulting in a limited relevance of this analysis. To avoid numerical artifacts, a stringent numerical tolerance is needed when solving the optimizations.

Next, we describe how to perform prediction of flavor compounds.

1. Verify that the metabolic network was created evaluating the presence of all the relevant biosynthetic pathways of the flavors under analysis. If the model has not been curated considering that scope, the user has to curate the network in order to assess the presence of those pathways. Next, verify that the network is able to produce the flavor compounds under study.

```
> model = changeObjective(model, rxnID)
        > fba = optimizeCbModel(model)
```

2. Set bounds for relevant constraints. Usually, accurate prediction of flavor compounds production depends strongly on the availability of accurate uptake/production rates as inputs, and on the researcher's understanding of the metabolic network and biosynthetic pathway functioning.
3. Perform FVA to evaluate the specific production rate range of the flavor compound.

```
> [minFlux, maxFlux] = fluxVariability(model)
```

3.4 Conclusions

This chapter introduces fundamental concepts and practical steps for applying constraint-based methods, namely flux balance analysis, for modeling wine fermentation using genome-scale metabolic models. As shown here, application of these methods offers valuable insights about the metabolism of yeast and bacteria growing under enological conditions. Complemented with appropriate experimental data, model predictions are ultimately a source of rational hypothesis generation and improvement of our understanding of microbial physiology.

4 Notes

1. Glossary
   - (a) FBA: Flux Balance Analysis. This is an optimization method whereby a reaction flux is (typically) maximized under steady state.
   - (b) FVA: Flux Variability Analysis. This is an optimization method used to determine the maximum allowable flux range under (sub)optimal conditions.
   - (c) GEM: Genome-scale Metabolic Model. Mathematical structure that describes the metabolism of an organism under specific environmental conditions and that is used in FBA to compute metabolic fluxes.
   - (d) GENRE: Genome-scale Network Reconstruction. It is the collection of biochemical reactions describing the metabolism of a particular organism.
   - (e) LAB: Lactic Acid Bacteria. Group of microorganisms whose main metabolic product is lactic acid. They are commonly used in the production of fermented foods and drinks such as yogurt, cheese and wine.
   - (f) LP: Linear Programming. Refers to a family of optimization problems where both the objective function and constraints are linear. The decision variables are all continuous.
   - (g) MILP: Mixed-Integer Linear Programming. Refers to a family of optimization problems where both the objective function and constraints are linear. As opposed to LP problems, MILP involves both discrete (e.g., binary) and continuous decision variables.
   - (h) MLF: Malolactic Fermentation. It is the LAB-mediated process where the malic acid present in the fermented grape must is transformed into lactic acid.
   - (i) EMAF: Enumeration of Minimal Active Fluxes. Optimization method for determining the minimum set of nutrients required to sustain growth.
   - (j) Volumetric rates: The velocity at which a certain metabolite is consumed or produced in the system per unit of volume. Its units are $\left(\frac{mmol}{L\,h}\right)$. Volumetric rates do not consider the amount of biomass in the system, and therefore, they are not appropriate for comparing between two conditions where the cell concentration is different.

(k) Specific exchange rates: the velocity at which a certain metabolite is consumed or produced in the environment by the cell population. It is said to be biomass-specific because, in contrast to volumetric rates $\left(\frac{mmol}{L\,h}\right)$ or total rates $\left(\frac{mmol}{h}\right)$, they are expressed per unit of biomass, that is, $\left(\frac{mmol}{gDW\,h}\right)$. For example, specific rates can be calculated by measuring metabolites and biomass concentrations in the feed and waste of a continuous culture.

(l) Specific metabolic fluxes: the velocity at which a reaction occurs inside the cell. It is also expressed in units $\left(\frac{mmol}{gDW\,h}\right)$. In contrast to specific exchange rates, they cannot be directly inferred from metabolites and biomass concentrations measurements, and typically they need to be estimated using a metabolic model along with an appropriate mathematical method, for example, flux balance analysis.

2. Although external concentrations of metabolites do change over time in a batch culture, those changes can be neglected as they exert almost no observable effect on cell homeostasis during vigorous exponential growth.

3. To describe a batch culture in a realistic way, kinetic expressions (*see*, e.g., [44] for a comprehensive review) are needed to capture the dependence of the growth rate on nutrient concentrations [15–17, 45].

4. Time frames can also be identified by plotting metabolite concentrations (mmol/L) as a function of biomass concentration (gDW/L) for the same time points. Consecutive sections of the curve that have the same slope correspond to specific time frames as these have the same yield (mmol/gDW).

5. The employed definition of reduced cost follows [46], although other references omit the minus sign [47]. The same occurs with scaled reduced costs. In any case, this difference does not fundamentally affect the analysis but only how they are interpreted.

6. The nongrowth ATP-associated maintenance represents an exception to this observation. The lower bound of this reaction is always nonzero as it represents the energy required just to maintain the basic cellular processes with exception of growth.

7. While this section is somehow similar to the previous, the addressed question is different. In Subheading 3.2.1, the minimal medium is computed whereas in Subheading 3.2.2 the intent is to determine the accuracy and predictive power of the model.

8. Often, optimization solvers return values different but very close to zero. To avoid numerical issues, it is convenient to set the threshold to a low value, for example, 10^{-6}, which is below measured experimental growth rates but above typical solver tolerance (10^{-9}).

Appendix

Tutorial 1

Metabolic modelling of wine fermentation at genome scale

Tutorial to run FBA with data from multiple conditions in continuous mode

Authors : Sebastián N. Mendoza[1], Pedro Saa[2], Bas Teusink[1], Eduardo Agosin[2]

1. *Systems Biology Lab, AIMMS, Vrije Universiteit Amsterdam, The Netherlands.*
2. *Laboratory of Biotechnology, Department of Chemical and Bioprocess Engineering, School of Engineering, Pontificia Universidad Católica de Chile, Santiago, Chile*

E-mail: snmendoz@uc.cl; s.n.mendozafarias@vu.nl

In this example, we use the experimental data reported by Pizarro et al [1] to simulate the flux distributions of Saccharomyces cerevisiae strain EC1118 in nitrogen-limited, anaerobic continuos cultures at two different temperatures 15° and 30°

We load the models. These models are based on the consensus model for *S. cerevisiae*, version 8 [2]

```
load('yeast841_biomass_pizarro_2007_15_degrees')
model_condition1 = yeast8;
load('yeast841_biomass_pizarro_2007_30_degrees')
model_condition2 = yeast8;
```

The specific rates are already reported in the article. However, we will calculate the specific uptake rate of ammonium to ilustrate the procedure

```
%CONDITION 1 = 15°C (c1 denotes condition 1)
NH4_concentration_feed_g_L_c1 = 0.201; %g/L
NH4_concentration_waste_g_L_c1 = 0; %g/L

biomass_waste_c1 = 2.85; %g/L

dilution_rate_c1 = 0.047; %1/h. Equal to the specific growth rate

%CONDITION 2 = 30°C (c2 denotes condition 2)
NH4_concentration_feed_g_L_c2 = 0.201; %g/L
NH4_concentration_waste_g_L_c2 = 0; %g/L

biomass_waste_c2 = 4.00; %g/L

dilution_rate_c2 = 0.049; %1/h. Equal to the specific growth rate

%Molecular weight of ammonium
MW_NH4 = 18.039;
```

STEP 1: Transformation of concentrations to mmol/L

We transform the concentration from g/L to mmol/L

```
%CONDITION 1
NH4_concentration_feed_mmol_L_c1  = (NH4_concentration_feed_g_L_c1 * 1000) / MW_NH4;
NH4_concentration_waste_mmol_L_c1 = (NH4_concentration_waste_g_L_c1 * 1000) / MW_NH4;

%CONDITION 2
NH4_concentration_feed_mmol_L_c2  = (NH4_concentration_feed_g_L_c2 * 1000) / MW_NH4;
NH4_concentration_waste_mmol_L_c2 = (NH4_concentration_waste_g_L_c2 * 1000) / MW_NH4;
```

STEP 2: Calculation of differences

We calculate the concentrations difference between the waste and feed

```
%CONDITION 1
delta_NH4_c1 = NH4_concentration_waste_mmol_L_c1 - NH4_concentration_feed_mmol_L_c1;

%CONDITION 2
delta_NH4_c2 = NH4_concentration_waste_mmol_L_c2 - NH4_concentration_feed_mmol_L_c2;
```

STEP 3: Calculation of yields

We calculate the yield in the waste and feed

```
%CONDITION 1
yield_NH4_biomass_c1 = delta_NH4_c1/biomass_waste_c1;

%CONDITION 2
yield_NH4_biomass_c2 = delta_NH4_c2/biomass_waste_c2;

fprintf('the yield ammonium/biomass in condition 1 is: %4.2f',yield_NH4_biomass_c1)
```

```
the yield ammonium/biomass in condition 1 is: -3.91
```

```
fprintf('the yield ammonium/biomass in condition 2 is: %4.2f',yield_NH4_biomass_c2)
```

```
the yield ammonium/biomass in condition 2 is: -2.79
```

STEP 4: Calculation of specific exchange rates

We calculate the specific rates

```
%CONDITION 1
specific_uptake_rate_ammonium_c1 = yield_NH4_biomass_c1*dilution_rate_c1;

%CONDITION 2
specific_uptake_rate_ammonium_c2 = yield_NH4_biomass_c2*dilution_rate_c2;

fprintf('the specific uptake rate of ammonium in condition 1 is: %4.2f',...
    specific_uptake_rate_ammonium_c1)
```

```
the specific uptake rate of ammonium in condition 1 is: -0.18
```

```
fprintf('the specific uptake rate of ammonium in condition 2 is: %4.2f',...
    specific_uptake_rate_ammonium_c2)
```

```
the specific uptake rate of ammonium in condition 2 is: -0.14
```

STEP 5: Setting of bounds for specific exchange rates

We incorporate the experimental values into the model

```
rxns = {'EX_glc__D_e', 'EX_nh4_e', 'EX_etoh_e', 'EX_pyr_e',...
    'EX_succ_e', 'EX_ac_e', 'EX_glyc_e', 'EX_co2_e'};
valuesCondition1 = [-3.56, -0.1838, 5.99, 0.033, 0.0075, 0.015, 0.03, 7.1];
valuesCondition2 = [-4.83, -0.1365, 7.28, 0.01, 0.013, 0.015, 0.03, 8.84];

modelCondition1 = changeRxnBounds(model_condition1, rxns, valuesCondition1, 'b');
modelCondition2 = changeRxnBounds(model_condition2, rxns, valuesCondition2, 'b');

% we summarize the specific rates in the next table
names = {'Glucose'; 'Ammonium';'Ethanol';'Pyruvate';...
    'Succinate';'Acetate';'Glycerol';'Carbon Dioxide'};
t = table(names,valuesCondition1',valuesCondition2',...
'VariableNames',{'Metabolite','Temperature_15_degrees','Temperature_30_degrees'});
disp(t);
```

Metabolite	Temperature_15_degrees	Temperature_30_degrees
'Glucose'	-3.56	-4.83
'Ammonium'	-0.1838	-0.1365
'Ethanol'	5.99	7.28
'Pyruvate'	0.033	0.01
'Succinate'	0.0075	0.013
'Acetate'	0.015	0.015
'Glycerol'	0.03	0.03
'Carbon Dioxide'	7.1	8.84

STEP 6: Flux Balance Analysis

We solve the linear optimizacion problems and get the flux distributions using FBA

```
fbaCondition1 = optimizeCbModel(modelCondition1);
fbaCondition2 = optimizeCbModel(modelCondition2);
```

In particular, we can obtain the specific growth rate

```
specificGrowthRateC1 = fbaCondition1.f;
specificGrowthRateC2 = fbaCondition2.f;

fprintf('The specific growth rate in condition 1 is: %4.4f',specificGrowthRateC1)
```

```
The specific growth rate in condition 1 is: 0.0468
```

```
fprintf('The specific growth rate in condition 2 is: %4.4f',specificGrowthRateC2)
```

```
The specific growth rate in condition 2 is: 0.0490
```

As expected, these specific growth rates are almost equal to the dilution rates reported by Pizarro et al., that equals 0.047 ± 0.000 and 0.049 ± 0.002 for cultures at 15°C and 30°C, respectively

In addition, the entire flux distribution can also be obtained

```
fluxDistributionC1 = fbaCondition1.x;
fluxDistributionC2 = fbaCondition2.x;

fprintf('The values for the first 10 reactions in condition 1 and 2 are:\n')
```

```
The values for the first 10 reactions in condition 1 and 2 are:
```

```
t = table( yeast8.rxns(1:10),...
    fluxDistributionC1(1:10),...
    fluxDistributionC2(1:10),...
    'VariableNames',{'Reaction','Condition_1','Condition_2'});
disp(t)
```

Reaction	Condition_1	Condition_2
'D_LACDcm'	0	0
'D_LACDm'	0	0
'BTDD_RR'	0	0
'L_LACD2cm'	0	0
'r_0005'	0.053458	0.065051
'r_0006'	0.017861	0.021735
'PRMICI'	0.0020171	0.0013431
'P5CDm'	0	0
'r_0013'	0	0
'DRTPPD'	4.6832e-05	4.9003e-05

STEP 7: Flux Variability Analysis

We perform a Flux Variability Analysis. Timing: 5-30 minutes

```
%we perform a FVA
[minFlux_c1, maxFlux_c1] = fluxVariability(modelCondition1); %condition 1
[minFlux_c2, maxFlux_c2] = fluxVariability(modelCondition2); %condition 2
```

```
% we normalize the minimum and maximum fluxes in condition 1 by the uptake rate
% of ammonium
minFlux_c1_norm = minFlux_c1 / abs(specific_uptake_rate_ammonium_c1);
maxFlux_c1_norm = maxFlux_c1 / abs(specific_uptake_rate_ammonium_c1);

% we normalize the minimum and maximum fluxes in condition 2 by the uptake rate
% of ammonium
minFlux_c2_norm = minFlux_c2 / abs(specific_uptake_rate_ammonium_c2);
maxFlux_c2_norm = maxFlux_c2 / abs(specific_uptake_rate_ammonium_c2);
```

With this can can find which reactions do have an overlapping flux range and which do not.

```
% reactions that do not overlap are those for which the minimum value in
% condition 1 is higher than the maximum value in condition 2 and those for
% which the minimum value in condition 2 is higher than the maximum value in
% condition 1
positions_reactions_not_overlapping = union(find(minFlux_c1_norm>maxFlux_c2_norm),...
    find(minFlux_c2_norm>maxFlux_c1_norm));
%we get the reactions that overlap as the remaining reactions
positions_reactions_overlapping = setdiff(1:length(yeast8.rxns),...
    positions_reactions_not_overlapping);
%we get the reactions that do not overlap
reactions_not_overlapping = yeast8.rxns(positions_reactions_not_overlapping);
%we get the reactions that overlap
reactions_overlapping = yeast8.rxns(positions_reactions_overlapping);

fprintf('The number of reactions that do overlap is :%4.0f\n', ...
    length(reactions_overlapping))
```

```
The number of reactions that do overlap is :3837
```

```
fprintf('The number of reactions that do not overlap is :%4.0f\n', ...
    length(reactions_not_overlapping))
```

```
The number of reactions that do not overlap is : 152
```

```
fprintf('The first 10 reactions that do not overlap are\n')
```

```
The first 10 reactions that do not overlap are
```

```
labels = {'Reaction','Min_C1','Max_C1','Min_C2','Max_C2'};
t = table(reactions_not_overlapping(1:10),...
    minFlux_c1_norm(positions_reactions_not_overlapping(1:10),1),...
     maxFlux_c1_norm(positions_reactions_not_overlapping(1:10),1),...
     minFlux_c2_norm(positions_reactions_not_overlapping(1:10),1),...
     maxFlux_c2_norm(positions_reactions_not_overlapping(1:10),1),...
     'VariableNames',labels);
disp(t);
```

Reaction	Min_C1	Max_C1	Min_C2	Max_C2
'r_0005'	0.29092	0.29092	0.47657	0.47658
'r_0006'	0.097202	0.097202	0.15923	0.15923
'PRMICI'	0.010977	0.010978	0.0098396	0.0098432
'DRTPPD'	0.00025486	0.00025486	0.000359	0.000359
'r_0015'	0.00025486	0.00025486	0.000359	0.000359
'AATA'	0.047385	0.047386	0.042475	0.04248
'r_0027'	0.047385	0.047386	0.042475	0.04248
'DB4PS'	0.00050973	0.00050973	0.000718	0.00071801
'ADCS'	1.6158e-05	1.7905e-05	2.2761e-05	3.3632e-05
'ADCL'	1.6158e-05	1.7905e-05	2.2761e-05	3.3632e-05

We obtain next the subsystems that are related to the reactions that do not overlap

```
%we obtain the subsystems associated with the reactions that do not overlap
subsystems = [];
```

```
for i = 1:length(positions_reactions_not_overlapping)
    if ~isempty(yeast8.subSystems{positions_reactions_not_overlapping(i)}{1})
        subsystems = [subsystems;...
            yeast8.subSystems{positions_reactions_not_overlapping(i)}'];
    end
end

%we obtain the frequencies
frecuencies = tabulate(subsystems);
%we sort the frequencies
[s,i] = sort(cell2mat(frecuencies(:,2)),'descend');
sorted_frecuencies = frecuencies(i,:);
fprintf('The five most frequent subsystems with are:\n')
```

```
The five most frequent subsystems with are:
```

```
t = table(sorted_frecuencies(1:10,1),...
    sorted_frecuencies(1:10,2),...
    sorted_frecuencies(1:10,3),...
    'VariableNames',{'Subsystems','Frequency','Percentage'});
disp(t)
```

Subsystems	Frequency	Percentage
'sce01110 Biosynthesis of secondary metabolites'	[44]	[17.8138]
'sce01130 Biosynthesis of antibiotics'	[32]	[12.9555]
'sce01230 Biosynthesis of amino acids'	[28]	[11.3360]
'sce00970 Aminoacyl-tRNA biosynthesis'	[20]	[8.0972]
'sce00230 Purine metabolism'	[11]	[4.4534]
'sce00340 Histidine metabolism'	[9]	[3.6437]
'sce00300 Lysine biosynthesis'	[8]	[3.2389]
'sce00400 Phenylalanine, tyrosine and tryptophan biosynthesis'	[8]	[3.2389]
'Gluconeogenesis'	[7]	[2.8340]
'sce00010 Glycolysis'	[7]	[2.8340]

We see that a considerable amount of reactions are related to the metabolism of amino acids and nucleotides. This is in agreement with the finding reported in Pizarro et al where they found several differentially regulated genes in those subsystems.

STEP 8: Normalization of fluxes

We normalize the fluxes obtained with FBA

```
%we normalize by the uptake rate of ammonium for condition 1
fluxDistributionC1_norm = fluxDistributionC1/abs(specific_uptake_rate_ammonium_c1);
%we normalize by the uptake rate of ammonium for condition 2
fluxDistributionC2_norm = fluxDistributionC2/abs(specific_uptake_rate_ammonium_c2);
```

STEP 9: Visualization of fluxes

We export the flux distributions to a .JON file so we can visualize them in Escher

```
exportMultipleSolutionsToJson(yeast8,...
    [fluxDistributionC1_norm, fluxDistributionC2_norm], 'sol.json')
```

We visualize with Escher the subsystems of nucleotides and some amino acids such as L-histidine.

In Figure 4, we can visualize the fold change in fluxes between the two conditions. Red indicates a big fold change, green indicates a moderate change and blue indicates a small change. From this image we can track where the changes occur.

References

1. Pizarro FJ, Jewett MC, Nielsen J, Agosin E. Growth Temperature Exerts Differential Physiological and Transcriptional Responses in Laboratory and Wine Strains of Saccharomyces cerevisiae. Appl Environ Microbiol. 2008;74: 6358–6368. doi:10.1128/AEM.00602-08
2. Lu H, Li F, Sánchez BJ, Zhu Z, Li G, Domenzain I, et al. A consensus S. cerevisiae metabolic model Yeast8 and its ecosystem for comprehensively probing cellular metabolism. Nat Commun. 2019;10. doi:10.1038/s41467-019-11581-3

Tutorial 2

Metabolic modelling of wine fermentation at genome scale

Tutorial to perform reduced cost analysis

Authors : Sebastián N. Mendoza[1], Pedro Saa[2], Bas Teusink[1], Eduardo Agosin[2]

1. *Systems Biology Lab, AIMMS, Vrije Universiteit Amsterdam, The Netherlands.*
2. *Laboratory of Biotechnology, Department of Chemical and Bioprocess Engineering, School of Engineering, Pontificia Universidad Católica de Chile, Santiago, Chile*

E-mail: snmendoz@uc.cl; s.n.mendozafarias@vu.nl

In this example, we use the experimental data reported by Pizarro et al [1] to perform a reduced cost analaysis for the metabolic network of *Saccharomyces cerevisiae* strain EC1118 growing in a nitrogen-limited, anaerobic continuos culture at 15°

We load the model. This model is based on the consensus model for *S. cerevisiae*, version 8 [2]

```
load('yeast841_biomass_pizarro_2007_15_degrees')
model_condition1 = yeast8;
```

STEP 1: Setting of bounds for specific exchange rates

We incorporate the experimental values into the model

```
rxns = {'EX_glc__D_e', 'EX_nh4_e', 'EX_etoh_e', 'EX_pyr_e',...
    'EX_succ_e', 'EX_ac_e', 'EX_glyc_e', 'EX_co2_e'};
valuesCondition1 = [-3.56, -0.1838, 5.99, 0.033, 0.0075, 0.015, 0.03, 7.1];

model_condition1 = changeRxnBounds(model_condition1, rxns, valuesCondition1, 'b');
```

STEP 2: Flux Balance Analysis

We solve the linear optimization problem using FBA

```
fbaCondition1 = optimizeCbModel(model_condition1);
```

STEP 3: Obtention of reduced costs

We obtain the fluxes and the reduced costs

```
%fluxes
fluxesC1 = fbaCondition1.x;
%reduced costs
reducedCostsC1 = fbaCondition1.w;
```

STEP 4: Obtention of scaled reduced costs

We obtained scaled reduced costs

```
scaledReducedCosts = -(fluxesC1.*reducedCostsC1)/fbaCondition1.f;
%we find those scaled reduced costs that are different from zero
positions_src_not_zero  = find(scaledReducedCosts);
%filter out those that are less that a tolerance
tolerance = 1e-8;
higher_than_tolerane = find(abs(scaledReducedCosts(positions_src_not_zero))>tolerance);
positions_src_not_zero = positions_src_not_zero(higher_than_tolerane);

t = table(yeast8.rxns(positions_src_not_zero),...
    num2cell(scaledReducedCosts(positions_src_not_zero)),...
    num2cell(reducedCostsC1(positions_src_not_zero)),...
    num2cell(fluxesC1(positions_src_not_zero)),...
    'VariableNames',{'Reaction','Scaled_reduced_cost','reduced_cost','specific_flux'});

disp(t)
```

Reaction	Scaled_reduced_cost	reduced_cost	specific_flux
'EX_nh4_e'	[1.0000]	[0.2548]	[-0.1838]

STEP 5: Interpretation of costs

We interpret the reduced cost.

By looking at the reduced costs, we can conclude that:

1) Ammonium is the only nutrient that is limiting the specific growth rate. This was expected as the experiments were performed under nitrogen-limited conditions.

2) As the reduced cost is 0.2548, that means that we would see a decrease in the objective function (specific growth rate) of 0.2548 if we would increase the variable representiing the specific uptake rate of ammonium (v_{NH4}) in 1 unit. Remember that the equation for the exchange reaction of ammonium is "1 nh4[e] <=> " meaning that the uptake is represented by negative values. Consequently, the uptake of ammonium increases as the value of v_{NH4} goes more and more negative. As Increasing the value of v_{NH4} represents a lower uptake rate, it makes sense that we would see a decrease in the objective function when increasing v_{NH4}.

3) As the reduced cost is actually the derivative of the objective function with respect to the variables $\left(\frac{d\mu}{dv_i}\right)$,

the reduced cost is only valid for infinitesimal variations of the variables with regard to the constraints applied. Therefore, in many occasions it will not be possible to increase the variable in 1 whole unit. Instead, we should increase the value of the variable in a small number and we should see a proportional decrease in the objective function. For, example, let's say that we increase the value of v_{NH4}, which is currently -0.1838, in 1e-6. Then, we should see a decrease of 0.2548*1e-6 (i.e. 2.548e-7) in the specific growth rate. We corroborate that with a simple calculation

```
%we create a model with a small perturbation in the uptake rate of ammonium
model_small_variation = changeRxnBounds(model_condition1, 'EX_nh4_e', -0.1838+1e-6, 'b');
%we perform a FBA
fbaCondition1_B = optimizeCbModel(model_small_variation);
%we calculated the difference in the objective function between both simulations
difference = fbaCondition1.f - fbaCondition1_B.f;
fprintf('The decrease in the specific growth rate is :%4.3e\n',difference)
```

```
The decrease in the specific growth rate is :2.548e-07
```

4) Finally, the units of the reduced cost is $\frac{\frac{1}{h}}{\frac{mmol}{gDW\,h}} = \frac{gDW}{mmol}$. Therefore, the reduced cost can also be interpreted as the yield of biomass with regard to the limiting nutrient $\text{yield}_{\frac{\text{biomass}}{\text{substrate}}}$.In the article, the reported $\text{yield}_{\frac{\text{biomass}}{\text{ammonium}}}$ is 14.17$\frac{gDW}{g\,\text{NH4}}$. This equal to $14.17\frac{gDW}{g\,\text{NH4}} * 18.039\frac{g\,\text{NH4}}{\text{mol NH4}} * \frac{1\text{ mol NH4}}{1000\text{ mmol NH4}} = 0.2556\frac{gDW}{\text{mmol NH4}}$, which is almost equal to the value predicted by the model.

References

1. Pizarro FJ, Jewett MC, Nielsen J, Agosin E. Growth Temperature Exerts Differential Physiological and Transcriptional Responses in Laboratory and Wine Strains of Saccharomyces cerevisiae. Appl Environ Microbiol. 2008;74: 6358–6368. doi:10.1128/AEM.00602-08
2. Lu H, Li F, Sánchez BJ, Zhu Z, Li G, Domenzain I, et al. A consensus S. cerevisiae metabolic model Yeast8 and its ecosystem for comprehensively probing cellular metabolism. Nat Commun. 2019;10. doi:10.1038/s41467-019-11581-3

Tutorial 3

Metabolic modelling of wine fermentation at genome scale

Tutorial to determine minimal nutritional requirements

Authors : Sebastián N. Mendoza[1], Pedro Saa[2], Bas Teusink[1], Eduardo Agosin[2]

1. *Systems Biology Lab, AIMMS, Vrije Universiteit Amsterdam, The Netherlands.*
2. *Laboratory of Biotechnology, Department of Chemical and Bioprocess Engineering, School of Engineering, Pontificia Universidad Católica de Chile, Santiago, Chile*

E-mail: snmendoz@uc.cl; s.n.mendozafarias@vu.nl

In this tutorial, we will show how to run EMAF (MATLAB version) [1] to determine the minimal nutritional requirements of *Oenococcus oeni*

First, we load the model published in [2]

```
load('iSM454.mat')
model = iSM454;
```

STEP 1: Setup of an *in silico* media

```
% we load the medium formulation created by Terrade and Mira de Orduña [3]
% and we allow the nutrients to be consumed in the model
[model, mediaExchangeRxns, nutrients] = setMediaFromExcelFileWithRxnIDs(model, ...
    'wineMediaFormulations','Terrade_metacyc');
% we also allow the uptake of carbon sources
% we ignore if they can sustain growth
c_sources = {'glycerol_ex_', 'sucrose_ex_', 'b_D_glucose_ex_',...
    'b_D_galactose_ex_', 'a_D_galactose_ex_', 'trehalose_ex_', 'cellobiose_ex_',...
    'melibiose_ex_', 'b_D_fructose_ex_', 'L_arabinose_ex_'};
model = changeRxnBounds(model, c_sources,10,'u');
```

STEP 2: Verifying biomass formation

```
%we solve a FBA
fba = optimizeCbModel(model);
fprintf('The specific growth rate is: %2.2f 1/h\n',fba.f)
```

```
The specific growth rate is: 4.11 1/h
```

STEP 3: Setup of growth rate constraints

```
% we identify the reaction ID for the growth rate
growth_rate = model.rxns{model.c~=0};
fba = optimizeCbModel(model);
constraints = struct('rxnList', {{growth_rate}},'values', 0.01*fba.f, 'sense', 'G');
```

STEP 4: Specifying the set to minimize

```
% we specify which is the set of exchange reactions that we want to
% mimimize
exchangeRxns = union(mediaExchangeRxns, c_sources);
```

STEP 5: Running EMAF

```
% we run EMAF in MATLAB
[required, alternatives, all_sets] = EMAF(model, constraints, exchangeRxns');
```

```
Elapsed time is 2.204083 seconds.
number of solutions found:
    20
```

STEP 6: Interpretation

The required nutrients are::

```
for i = 1:length(required)
    fprintf('%2.0f) %s \n',i,required{i})
end
```

```
 1) L_Arg_ex_
 2) L_Cys_ex_
 3) L_His_ex_
 4) L_Ile_ex_
 5) L_Leu_ex_
 6) L_Met_ex_
 7) L_Phe_ex_
 8) L_Ser_ex_
 9) L_Thr_ex_
10) L_Trp_ex_
11) L_Tyr_ex_
12) L_Val_ex_
13) Mn_ex_
14) P_ex_
15) nicotinamida_RNP_ex_
16) oleate_ex_
17) panthothenate_ex_
```

Additionally, one nutrient must be selected for each of the following groups

```
for i = 1:size(alternatives,1)
    fprintf('GROUP%2.0f:\n',i)
    alternatives_group_i = strsplit(alternatives{i},',');
    for j =1:length(alternatives_group_i)
        fprintf('%2.0f) %s \n',j,alternatives_group_i{j})
    end
    fprintf('\n')
end
```

```
GROUP 1:
 1) L_Gln_ex_
 2)  L_Glu_ex_
```

```
GROUP 2:
 1) L_arabinose_ex_
 2)  a_D_galactose_ex_
 3)  b_D_fructose_ex_
 4)  b_D_galactose_ex_
 5)  b_D_glucose_ex_
 6)  b_D_ribopyranose_ex_
 7)  cellobiose_ex_
 8)  melibiose_ex_
 9)  sucrose_ex_
10)  trehalose_ex_
```

In conclusion, EMAF found that:

1) L-arginine, L-cysteine, L-histidine, L-isoleucine, L-leucine, L-methionine, L-phenylalanine, L-serine, L-threonine, L-tryptophan, L-tyrosine, L-valine, manganese, phosphate, nicotinamide ribonucleotide, oleate and pantothenate are needed to sustain a minimum growth rate of 0.04 1/h

2) at least one the following amino acids has to be chosen to sustain a minimum growth rate of 0.04 1/h: L-glutamate or L-glutamine.

3) at least one of the following carbon sources has to be chosen to sustain a minimum growth rate of 0.04 1/h: galactose, fructose, glucose, cellobiose, melibiose, sucrose or trehalose

References

1. Branco dos Santos F, Olivier BG, Boele J, Smessaert V, De Rop P, Krumpochova P, et al. Probing the genome-scale metabolic landscape of Bordetella pertussis, the causative agent of whooping cough. Appl Environ Microbiol. 2017;83: e01528-17. doi:10.1128/AEM.01528-17
2. Mendoza SN, Cañón PM, Contreras Á, Ribbeck M, Agosín E. Genome- Scale Reconstruction of the Metabolic Network in Oenococcus oeni to Assess Wine Malolactic Fermentation. Front Microbiol. 2017;8: 534. doi:10.3389/fmicb.2017.00534
3. Terrade N, Mira de Orduña R. Determination of the essential nutrient requirements of wine-related bacteria from the genera Oenococcus and Lactobacillus. Int J Food Microbiol. Elsevier B.V.; 2009;133: 8–13. doi:10.1016/j.ijfoodmicro.2009.03.020

Tutorial 4

Metabolic modelling of wine fermentation at genome scale

Tutorial to determine minimal nutritional requirements

Authors : Sebastián N. Mendoza[1], Pedro Saa[2], Bas Teusink[1], Eduardo Agosin[2]

1. *Systems Biology Lab AIMMS, Vrije Universiteit Amsterdam, The Netherlands.*
2. *Laboratory of Biotechnology, Department of Chemical and Bioprocess Engineering, School of Engineering, Pontificia Universidad Católica de Chile, Santiago, Chile*

E-mail: snmendoz@uc.cl; s.n.mendozafarias@vu.nl

In this tutorial, we will show how to run EMAF (python version) [1] to determine the minimal nutritional requirements of *Oenococcus oeni*

First, we load the model published in [2]

```
load('iSM454.mat')
model = iSM454;
```

STEP 1: Setup of an *in silico* media

```
% we load the medium formulation created by Terrade and Mira de Orduña [2]
% and we allow the nutrients to be consumed in the model
[model, mediaExchangeRxns, nutrients] = setMediaFromExcelFileWithRxnIDs(model, ...
    'wineMediaFormulations','Terrade_metacyc');
% we also allow the uptake of carbon sources for which we
% ignore if they can sustain growth
c_sources = {'glycerol_ex_', 'sucrose_ex_', 'b_D_glucose_ex_',...
    'b_D_galactose_ex_', 'a_D_galactose_ex_', 'trehalose_ex_', 'cellobiose_ex_',...
    'melibiose_ex_', 'b_D_fructose_ex_', 'L_arabinose_ex_'};
model = changeRxnBounds(model, c_sources,10,'u');
```

STEP 2: Verifying biomass formation

```
%we solve a FBA
fba = optimizeCbModel(model);
fprintf('The specific growth rate is:%2.2f 1/h\n',fba.f)
```

```
The specific growth rate is :4.11 1/h
```

STEP 3: Setup of growth rate constraints

```
% we identify the reaction ID for the growth rate
growth_rate = model.rxns(model.c~=0);
% we create three vectors to especify reaction ids, lower and upper bounds.
% In this case, there is only one constraint so the vectors have length 1
constraints_ids = growth_rate;
```

```
constraints_lb = 0.01*fba.f;
constraints_ub = 1000;
```

STEP 4: Specifying the set to minimize

```
% we specify which is the set of exchange reactions that we want to
% mimimize
exchangeRxns = union(mediaExchangeRxns, c_sources);
% .. and their positions in the model.
posEX = getPosOfElementsInArray(exchangeRxns',model.rxns);
```

STEP 5: Creating inputs for EMAF

```
% we specify the folder where the inputs and results are goind to be stored
baseDir = pwd;
%we specify the name of the model
modelFile = 'iSM454';
%we create the inputs for EMAF
createInputsForEMAF(model, growth_rate, baseDir, modelFile, ...
    constraints_ids, constraints_lb, constraints_ub,posEX)
```

STEP 6: Running EMAF

In the computer console, go to the directory baseDir and type:

> python runMedia3.py

> python pushRunMedia3.py

STEP 7: Interpretation

```
outputFilePath = ['./emaf/media_search_results-(' modelFile '_irrev.xml).csv'];
[required, alternatives] = readEMAFoutput(outputFilePath);
```

The required nutrients are::

```
for i = 1:length(required)
    fprintf('%2.0f) %s \n',i,required{i})
end
```

```
 1) R_L_Arg_ex_
 2) R_L_Cys_ex_
 3) R_L_His_ex_
 4) R_L_Ile_ex_
 5) R_L_Leu_ex_
 6) R_L_Met_ex_
 7) R_L_Phe_ex_
 8) R_L_Ser_ex_
 9) R_L_Thr_ex_
10) R_L_Trp_ex_
11) R_L_Tyr_ex_
12) R_L_Val_ex_
13) R_Mn_ex_
14) R_P_ex_
15) R_nicotinamida_RNP_ex_
```

```
16) R_oleate_ex_
17) R_panthothenate_ex_
```

Additionally, one nutrient must be selected for each of the following groups

```
for i = 1:size(alternatives,1)
    fprintf('GROUP%2.0f:\n',i)
    alternatives_group_i = strsplit(alternatives{i},',');
    for j =1:length(alternatives_group_i)
        fprintf('%2.0f) %s \n',j,alternatives_group_i{j})
    end
end
```

```
GROUP 1:
 1) R_L_Gln_ex_
 2) R_L_Glu_ex_
GROUP 2:
 1) R_trehalose_ex_
 2) R_cellobiose_ex_
 3) R_melibiose_ex_
 4) R_b_D_glucose_ex_
 5) R_b_D_galactose_ex_
 6) R_L_arabinose_ex_
 7) R_sucrose_ex_
 8) R_a_D_galactose_ex_
 9) R_b_D_fructose_ex_
10) R_b_D_ribopyranose_ex_
```

In conclusion, EMAF found that:

1) L-arginine, L-cysteine, L-histidine, L-isoleucine, L-leucine, L-methionine, L-phenylalanine, L-serine, L-threonine, L-tryptophan, L-tyrosine, L-valine, manganese, phosphate, nicotinamide ribonucleotide, oleate and pantothenate are needed to sustain a minimum growth rate of 0.04 1/h

2) at least one the following amino acids has to be chosen to sustain a minimum growth rate of 0.04 1/h: L-glutamate or L-glutamine.

3) at least one of the following carbon sources has to be chosen to sustain a minimum growth rate of 0.04 1/h: galactose, fructose, glucose, cellobiose, melibiose, sucrose or trehalose

References

1. Branco dos Santos F, Olivier BG, Boele J, Smessaert V, De Rop P, Krumpochova P, et al. Probing the genome-scale metabolic landscape of Bordetella pertussis, the causative agent of whooping cough. Appl Environ Microbiol. 2017;83: e01528-17. doi:10.1128/AEM.01528-17
2. Mendoza SN, Cañón PM, Contreras Á, Ribbeck M, Agosín E. Genome- Scale Reconstruction of the Metabolic Network in Oenococcus oeni to Assess Wine Malolactic Fermentation. Front Microbiol. 2017;8: 534. doi:10.3389/fmicb.2017.00534
3. Terrade N, Mira de Orduña R. Determination of the essential nutrient requirements of wine-related bacteria from the genera Oenococcus and Lactobacillus. Int J Food Microbiol. Elsevier B.V.; 2009;133: 8–13. doi:10.1016/j.ijfoodmicro.2009.03.020

Tutorial 5

Metabolic modelling of wine fermentation at genome scale

Tutorial to compare experimental and predicted growth/no growth data

Authors : Sebastián N. Mendoza[1], Pedro Saa[2], Bas Teusink[1], Eduardo Agosin[2]

1. *Systems Biology Laboratory, AIMMS, Vrije Universiteit Amsterdam, The Netherlands.*
2. *Laboratory of Biotechnology, Department of Chemical and Bioprocess Engineering, School of Engineering, Pontificia Universidad Católica de Chile, Santiago, Chile*

E-mail: snmendoz@uc.cl; s.n.mendozafarias@vu.nl

In this example, we use to genome-scale model of *Oenococcus oeni* [1] to compare model's prediction with experimental growth data. In particular, we will use binary data (growth/no growth) for Oenococcus oeni growing on different carbon sources and also when particular nutrients are ommited from the culture medium.

The purpose of this analysis is to assess model's performance.

```
% we load the model
load('iSM454.mat')
model = iSM454;
```

STEP 1: Setup of bounds for specific exchange rates

```
% we load the medium formulation created by Terrade and Mira de Orduña [2]
% and we allow the nutrients to be consumed in the model
[model, mediaExchangeRxns, nutrients] = setMediaFromExcelFileWithRxnIDs(model, ...
    'wineMediaFormulations','Terrade_metacyc');
```

STEP 2: Verifying biomass formation

```
%we solve a FBA
fba = optimizeCbModel(model);
fprintf('The specific growth rate is: %2.2f 1/h\n',fba.f)
```

```
The specific growth rate is: 0.54 1/h
```

STEP 3: Set thresholds

```
threshold_ommited_percentage = 0.1;
threshold_added = 0.001;
```

STEP 4: Define experimental data

```
% we load the experimental data
[n,data] = xlsread('experiments_ooeni');
```

```
% we get the labels of the experiments. These are unique user-defined
% names.
experiments = data(2:end,2);
% we get the media where each experiment was performed
experiments_media = data(2:end,3);
% we get the objective function that will be use in the test
experiments_objetive_function = data(2:end,4);
% Sometimes, experiments came from different sources
% (different research articles, different laboratories, differnt techniques).
% Experiments can be labeled with different strings
% to get assessments for experimetns from a particular source.
% A set is just the collection of experiments from a common source.
experiments_sets = data(2:end,1);
% we get the type of experiment (omission or addition)
experiments_type = data(2:end,7);
% we get the reaction ids
experiments_rxn_ids = data(2:end,6);
% Finally, we get the result of the experiments.
% 1 means it grew and 0 it didn't
experimentalResults = n(:,3);
```

STEP 5: Performing the comparison

```
% we get the list of sets (see lines 21-25 to understand what a set is)
sets = unique(experiments_sets);

% we initialize the variables to count true positives (TP),
% true negatives (TN), false negatives (FN) and false positives (FP)
% for all the sets of experiments
TP_all = 0;
TN_all = 0;
FN_all = 0;
FP_all = 0;

% for each set of experimetns
for i = 1:length(sets)

    % we gathered all the experiments belonging to that set
    pos = getPosOfElementsInArray(sets(i),experiments_sets);

    % we created a file to export the performance metric for that set
    fi = fopen(['results_' sets{i} '.txt'], 'w');

    % We initialize the variables to count true positives (TP),
    % true negatives (TN), false negatives (FN) and false positives (FP)
    % for the set i
    TP = 0;
    TN = 0;
    FN = 0;
    FP = 0;
```

```
fprintf('Results for set: %s\n', sets{i})
fprintf('%-25s\t%-15s\t%-25s\n', 'Nutrient', 'Classification', 'Specific Growth Rate')

% for each of the experiments in the set i
for j = 1:length(experiments(pos))

    % we load the medium where the experiment was performed
    [rxnsMedia, valuesMedia] = getMediumFromExcelFile(...
        'wineMediaFormulations', experiments_media{pos(j)});


    if strcmp(experiments_type{pos(j)},'ommited')
    % if the experiment is an ommission experiment

        % We set the media
        model = setMediaFromRxns(model,rxnsMedia, valuesMedia,nutrients);
        % We set the objective function
        model = changeObjective(model, experiments_objetive_function{pos(j)});
        % We perform and FBA using the medium with all the nutrients.
        % The growth rate obtained with this FBA will be the reference
        % to compare when we ommit the nutrient.
        fbaRef = optimizeCbModel(model);
        % We ommit the nutrient
        rxn_ommited = experiments_rxn_ids{pos(j)};
        if ismember(rxn_ommited, model.rxns)
            model = changeRxnBounds(model, rxn_ommited,0,'u');
        end
        % We perform an FBA using the medium without the ommited
        % nutrient
        fba = optimizeCbModel(model);
        % We define the threshold
        threshold_ommited = fbaRef.f*threshold_ommited_percentage;
        % If the growth rate obtained using the medium with the ommited nutrient,
        % is less than the threshold, then we classify the result as it
        % didn't grow and therefore, we assign a 0 value. Otherwise,
        % the consider that it grew and we assign a value of 1
        if isempty(fba.x) || fba.f<threshold_ommited
            grew = 0;
        else
            grew = 1;
        end


    elseif strcmp(experiments_type{pos(j)},'added')
     % if the experiment is an addition experiment

        % We set the media
        model = setMediaFromRxns(model,rxnsMedia, valuesMedia,nutrients);
        % We set the objective function
        model = changeObjective(model, experiments_objetive_function{pos(j)});
        % We add the nutrient
        rxn_added = experiments_rxn_ids{pos(j)};
        if ismember(rxn_added, model.rxns)
            model = changeRxnBounds(model, rxn_added,10,'u');
```

```
            end
            % We perform an FBA using the medium with the added nutrient
            fba = optimizeCbModel(model);
            % If the growth rate obtained using the medium with the added nutrient,
            % is less than the threshold, then we classify the result as it
            % didn't grow and therefore, we assign a 0 value. Otherwise,
            % the consider that it grew and we assign a value of 1
            if isempty(fba.x) || fba.f<threshold_added
                grew = 0;
            else
                grew = 1;
            end

        end

        % we classify the result by comparing it with the real observation
        if experimentalResults(pos(j))==1 && ~grew
            %FN: it grew in vivo but it didn't in silico
            classification = 'FN';
            FN = FN + 1;
        elseif experimentalResults(pos(j))==1 && grew
            %TP: it grew in vivo and also in silico
            classification = 'TP';
            TP = TP + 1;
        elseif experimentalResults(pos(j))==0 && ~grew
            %TN: it didn't grow in vivo and neither in silico
            classification = 'TN';
            TN = TN  +1;
        elseif experimentalResults(pos(j))==0 && grew
            %FP: it didn't grow in vivo but it did in silico
            classification = 'FP';
            FP = FP + 1;
        end

        % we print the classification result in the file
        if isempty(experiments{pos(j)})
            s = 'None';
        else
            s = experiments{pos(j)};
        end
        fprintf(fi, '%s, %s, %0.3f\n', s, classification, fba.f);
        fprintf('%-25s\t%-15s\t%0.3f\n', s, classification, fba.f);

    end

    % We detemine metric performance for the set of experiment i
    SEN = TP / (TP + FN);
    SPE = TN / (TN + FP);
    PRE = TP / (TP + FP);
    NPV = TN / (TN + FN);
    ACC = (TP + TN) / (TN + TP + FN + FP);
    FSCORE = (2 * (PRE * SEN))/ (PRE + SEN);

    % We add the counts to the general variables
```

```
    TP_all = TP_all + TP;
    TN_all = TN_all + TN;
    FN_all = FN_all + FN;
    FP_all = FP_all + FP;

    % we export the metric performance for set i to a plain text file
    fprintf(fi, '%s\n', '---------');
    fprintf(fi, '%s%2.0f\n', 'TP=', TP);
    fprintf(fi, '%s%2.0f\n', 'TN=', TN);
    fprintf(fi, '%s%2.0f\n', 'FP=', FP);
    fprintf(fi, '%s%2.0f\n', 'FN=', FN);
    fprintf(fi, '%s%2.0f\n', 'TOTAL=', TP+TN+FP+FN);
    fprintf(fi, '%s\n', '---------');
    fprintf(fi, '%s%0.2f\n', 'SENSITIVITY = ', SEN);
    fprintf(fi, '%s%0.2f\n', 'SPECIFICITY = ', SPE);
    fprintf(fi, '%s%0.2f\n', 'PRECISION = ', PRE);
    fprintf(fi, '%s%0.2f\n', 'N.P.V.=', NPV);
    fprintf(fi, '%s%0.2f\n', 'ACCURACY = ', ACC);
    fprintf(fi, '%s%0.2f\n', 'F-SCORE = ', FSCORE);
    fclose(fi);

    % We show the results
    fprintf('%s\n', '---------');
    fprintf('%s%2.0f\n', 'TP=', TP);
    fprintf('%s%2.0f\n', 'TN=', TN);
    fprintf('%s%2.0f\n', 'FP=', FP);
    fprintf('%s%2.0f\n', 'FN=', FN);
    fprintf('%s%2.0f\n', 'TOTAL=', TP+TN+FP+FN);
    fprintf('%s\n', '---------');
    fprintf('%s%0.2f\n', 'SENSITIVITY = ', SEN);
    fprintf('%s%0.2f\n', 'SPECIFICITY = ', SPE);
    fprintf('%s%0.2f\n', 'PRECISION = ', PRE);
    fprintf('%s%0.2f\n', 'N.P.V.=', NPV);
    fprintf('%s%0.2f\n', 'ACCURACY = ', ACC);
    fprintf('%s%0.2f\n', 'F-SCORE = ', FSCORE);
    fprintf('%s\n', '---------');

end
```

```
Results for set: carbon_source_addition
Nutrient                 Classification  Specific Growth Rate
b_D_glucose_ex_          TP              0.848
b_D_fructose_ex_         TP              0.770
trehalose_ex_            TP              1.343
cellobiose_ex_           TP              1.343
D-deoxyribose_ex_        TN              0.000
L_arabinose_ex_          TP              0.689
L-rhammose_ex_           TN              0.000
D_xylose_ex_             TN              0.000
esculin_ex               FN              0.000
salicin_ex               FN              0.000
glycerol_ex_             TN              0.000
D_mannitol_ex_           TN              0.000
L-sorbitol_ex            TN              0.000
_S__malate_ex_           TN              0.000
citrate_ex_              TN              0.000
fumaric_acid_ex          TN              0.000
```

```
D_Mannose_ex             FN              0.000
b_D_ribopyranose_ex_     TP              0.537
melibiose_ex_            TP              1.302
sucrose_ex_              FP              0.770
maltose_ex               TN              0.000
b_D_galactose_ex_        FP              0.814
raffinose_ex             TN              0.000
lactose_ex               TN              0.000
L-sorbose               TN              0.000
---------
TP= 7
TN=13
FP= 2
FN= 3
TOTAL=25
---------
SENSITIVITY = 0.70
SPECIFICITY = 0.87
PRECISION = 0.78
N.P.V.=0.81
ACCURACY = 0.80
F-SCORE = 0.74
---------
Results for set: nutrient_omission
Nutrient                 Classification  Specific Growth Rate
L_Asn_ex_                TP              0.532
panthothenate_ex_        TN              0.000
b_D_ribopyranose_ex_     TN              0.000
Gly_ex_                  TP              0.537
L_Ala_ex_                TP              0.537
L_Val_ex_                TN              0.000
L_Leu_ex_                TN              0.000
L_Ile_ex_                TN              0.000
L_Ser_ex_                FN              0.000
L_Thr_ex_                TN              0.000
L_Cys_ex_                TN              0.000
L_Met_ex_                TN              0.000
L_Asp_ex_                TP              0.537
L_Glu_ex_                FP              0.537
L_Gln_ex_                TP              0.514
L_Lys_ex_                TP              0.531
L_Arg_ex_                TN              0.000
L_His_ex_                TN              0.000
L_Phe_ex_                TN              0.000
L_Tyr_ex_                TN              0.000
L_Trp_ex_                TN              0.000
L_Pro_ex_                TP              0.534
biotin_ex_               TP              0.537
nicotinamida_RNP_ex_     TN              0.000
pyridoxine_ex_           TP              0.537
riboflavin_ex_           TP              0.537
thiamin_ex_              TP              0.537
adenine_ex_              TP              0.536
guanine_ex_              TP              0.537
xanthine_ex_             TP              0.537
cytosine_ex_             TP              0.537
thymine_ex_              TP              0.537
uracil_ex_               TP              0.537
P_ex_                    TN              0.000
Mn_ex_                   TN              0.000
aminobenzoic_acid_ex_    TP              0.537
choline_ex_              TP              0.537
cyanocobalamin_ex_       TP              0.537
folic_acid_ex            TP              0.537
Mg_ex_                   TP              0.537
```

```
Ca_ex_                    TP            0.537
Cu_ex_                    TP            0.537
Fe_ex_                    TP            0.537
Zn_ex_                    TP            0.537
---------
TP=26
TN=16
FP= 1
FN= 1
TOTAL=44
---------
SENSITIVITY = 0.96
SPECIFICITY = 0.94
PRECISION = 0.96
N.P.V.=0.94
ACCURACY = 0.95
F-SCORE = 0.96
---------
```

STEP 6: Calculate performance metrics

```
SEN_all = TP_all / (TP_all + FN_all);
SPE_all = TN_all / (TN_all + FP_all);
PRE_all = TP_all / (TP_all + FP_all);
NPV_all = TN_all / (TN_all + FN_all);
ACC_all = (TP_all + TN_all) / (TN_all + TP_all + FN_all + FP_all);
FSCORE_all = (2 * (PRE_all * SEN_all))/ (PRE_all + SEN_all);

fi = fopen('general_results.txt', 'w');
fprintf(fi, '%s\n', '---------');
fprintf(fi, '%s%f\n', 'TP=', TP_all);
fprintf(fi, '%s%f\n', 'TN=', TN_all);
fprintf(fi, '%s%f\n', 'FP=', FP_all);
fprintf(fi, '%s%f\n', 'FN=', FN_all);
fprintf(fi, '%s%f\n', 'TOTAL=', TP_all+TN_all+FP_all+FN_all);
fprintf(fi, '%s\n', '---------');
fprintf(fi, '%s%0.2f\n', 'SENSITIVITY = ', SEN_all);
fprintf(fi, '%s%0.2f\n', 'SPECIFICITY = ', SPE_all);
fprintf(fi, '%s%0.2f\n', 'PRECISION = ', PRE_all);
fprintf(fi, '%s%0.2f\n', 'N.P.V.=', NPV_all);
fprintf(fi, '%s%0.2f\n', 'ACCURACY = ', ACC_all);
fprintf(fi, '%s%0.2f\n', 'F-SCORE = ', FSCORE_all);
fclose(fi);
```

```
fprintf(['TN = %2.0f\nFP = %2.0f\nFN = %2.0f\nTP = %2.0f\nTOTAL = %2.0f\n' ...
    'SENSITIVITY = %2.3f\nSPECIFICITY = %2.3f\nPRECISION = %2.3f\n' ...
    'N.P.V = %2.3f\nACCURACY = %2.3f\nF-SCORE = %2.3f\n'], ...
    TN_all,FP_all,FN_all,TP_all,TP_all+TN_all+FP_all+FN_all,SEN_all,...
    SPE_all,PRE_all,NPV_all,ACC_all,FSCORE_all);
```

```
TN = 29
FP =  3
FN =  4
TP = 33
TOTAL = 69
SENSITIVITY = 0.892
```

```
SPECIFICITY = 0.906
PRECISION = 0.917
N.P.V = 0.879
ACCURACY = 0.899
F-SCORE = 0.904
```

STEP 7: Interpretation

In conclusion,

1) The model predicted with an accuracy of 80% for growth on alternative the carbon sources

2) The model predicted with an accuracy of 95% for omission experiments

3) The model predicted with an accuracy of 90% for the all the experiments considered

References

1. Mendoza SN, Cañón PM, Contreras Á, Ribbeck M, Agosín E. Genome- Scale Reconstruction of the Metabolic Network in Oenococcus oeni to Assess Wine Malolactic Fermentation. Front Microbiol. 2017;8: 534. doi:10.3389/fmicb.2017.00534
2. Terrade N, Mira de Orduña R. Determination of the essential nutrient requirements of wine-related bacteria from the genera Oenococcus and Lactobacillus. Int J Food Microbiol. Elsevier B.V.; 2009;133: 8–13. doi:10.1016/j.ijfoodmicro.2009.03.020

References

1. Ribéreau-Gayon P, Dubourdieu D, Donèche B, Lonvaud A (2005) Biochemistry of alcoholic fermentation and metabolic pathways of wine yeasts. Handbook of Enology: The Microbiology of Wine and Vinifications. Wiley, New York, pp 53–77. https://doi.org/10.1002/0470010363.ch2
2. Bartowsky EJ (2005) Oenococcus oeni and malolactic fermentation—moving into the molecular arena. Aust J Grape Wine Res 11: 174–187. https://doi.org/10.1111/j.1755-0238.2005.tb00286.x
3. Bartowsky EJ, Francis IL, Bellon JR, Henschke PA (2002) Is buttery aroma perception in wines predictable from the diacetyl concentration? Aust J Grape Wine Res 8:180–185. https://doi.org/10.1111/j.1755-0238.2002.tb00254.x
4. Davis CR, Wibowo D, Eschenbruch R, Lee TH, Fleet GHS (1985) Practical implications of malolactic fermentation: a review. Am J Enol Viticult 36:290
5. Peter J, Chiara MD, Friedrich A, J-x Y, Pflieger D, Bergström A, Sigwalt A, Barre B, Freel K, Llored A, Cruaud C, Labadie K, J-m A, Istace B, Lebrigand K, Barbry P, Engelen S, Lemainque A, Wincker P, Liti G (2018) Genome evolution across 1,011 *Saccharomyces cerevisiae* isolates. Nature 556: 339–344. https://doi.org/10.1038/s41586-018-0030-5
6. Mills DA, Rawsthorne H, Parker C, Tamir D, Makarova K (2005) Genomic analysis of Oenococcus oeni PSU-1 and its relevance to winemaking. FEMS Microbiol 29:465–475. https://doi.org/10.1016/j.femsre.2005.04.011
7. Lu H, Li F, Sánchez BJ, Zhu Z, Li G, Domenzain I, Marci S, Anton PM, Lappa D, Lieven C, Beber ME, Sonnenschein N, Kerkhoven EJ, Nielsen J (2019) A consensus *S. cerevisiae* metabolic model Yeast8 and its ecosystem for comprehensively probing cellular metabolism. Nat Commun 10:3586. https://doi.org/10.1038/s41467-019-11581-3
8. Mendoza SN, Cañón PM, Contreras Á, Ribbeck M, Agosín E (2017) Genome- scale reconstruction of the metabolic network in Oenococcus oeni to assess wine malolactic fermentation. Front Microbiol 8:534. https://doi.org/10.3389/fmicb.2017.00534

9. Quirós M, Martínez-moreno R, Albiol J, Morales P, Vázquez-lima F (2013) Metabolic flux analysis during the exponential growth phase of *Saccharomyces cerevisiae* in wine fermentations. PLoS One 8:1–14. https://doi.org/10.1371/journal.pone.0071909
10. Aceituno F, Orellana M, Torres J, Mendoza S, Slater A, Melo F, Agosin E (2012) Oxygen response of the wine yeast *Saccharomyces cerevisiae* EC1118 grown under carbon-sufficient, nitrogen-limited enological conditions. Appl Environ Microbiol 78:8340–8352. https://doi.org/10.1128/AEM.02305-12
11. Li H, Su J, Ma W, Guo A, Shan Z, Wang H (2015) Metabolic flux analysis of *Saccharomyces cerevisiae* in a sealed winemaking fermentation system. FEMS Yeast Res 15:1–9. https://doi.org/10.1093/femsyr/fou010
12. Varela C, Pizarro F, Agosin E (2004) Biomass content governs fermentation rate in nitrogen-deficient wine musts. Appl Environ Microbiol 70:3392–3400. https://doi.org/10.1128/AEM.70.6.3392
13. Crépin L, Truong NM, Bloem A, Sanchez I, Dequin S, Camarasa C (2017) Management of Multiple Nitrogen Sources during wine fermentation by *Saccharomyces cerevisiae*. Appl Environ Microbiol 83:1–21
14. Vázquez-lima F, Silva P, Barreiro A, Martínez-moreno R, Morales P, Quirós M, González R, Albiol J, Ferrer P (2014) Use of chemostat cultures mimicking different phases of wine fermentations as a tool for quantitative physiological analysis. Microb Cell Factories 13:1–13
15. Sainz J, Pizarro F, Pérez RJ, Agosin E (2003) Modeling of yeast metabolism and process dynamics in batch fermentation. Biotechnol Bioeng 81:818–828. https://doi.org/10.1002/bit.10535
16. Pizarro F, Varela C, Martabit C, Bruno C, Agosin E, Pe JR (2007) Coupling kinetic expressions and metabolic networks for predicting wine fermentations. Biotechnol Bioeng 98:986–998. https://doi.org/10.1002/bit
17. Vargas FA, Pizarro F, Pérez-Correa JR, Agosin E (2011) Expanding a dynamic flux balance model of yeast fermentation to genome-scale. BMC Syst Biol 5:75
18. Orellana M, Aceituno FF, Slater AW, Almonacid LI, Melo F, Agosin E (2014) Metabolic and transcriptomic response of the wine yeast *Saccharomyces cerevisiae* strain EC1118 after an oxygen impulse under carbon-sufficient, nitrogen-limited fermentative conditions. FEMS Yeast Res 14(3):412–424. https://doi.org/10.1111/1567-1364.12135
19. Saa PA, Moenne MI, Perez-Correa JR, Agosin E (2012) Modeling oxygen dissolution and biological uptake during pulse oxygen additions in oenological fermentations. Bioprocess Biosyst Eng 35(7):1167–1178. https://doi.org/10.1007/s00449-012-0703-7
20. Saa PA, Pérez-Correa JR, Celentano D, Agosin E (2013) Impact of carbon dioxide injection on oxygen dissolution rate during oxygen additions in a bubble column. Chem Eng J 232: 157–166. https://doi.org/10.1016/j.cej.2013.07.081
21. Moenne MI, Saa P, Laurie VF, Pérez-Correa JR, Agosin E (2014) Oxygen incorporation and dissolution during industrial-scale red wine fermentations. Food Bioproc Technol 7: 2627–2636. https://doi.org/10.1007/s11947-014-1257-2
22. Contreras A, Ribbeck M, Gutiérrez GD, Cañón PM, Mendoza SN, Agosin E (2018) Mapping the physiological response of Oenococcus oeni to ethanol stress using an extended genome-scale metabolic model. Front Microbiol 9:291. https://doi.org/10.3389/fmicb.2018.00291
23. Orth JD, Thiele I, Palsson BØ (2010) What is flux balance analysis? Nat Biotechnol 28: 245–248. https://doi.org/10.1038/nbt.1614
24. McCloskey D, Palsson BØ, Feist AM (2013) Basic and applied uses of genome-scale metabolic network reconstructions of *Escherichia coli*. Mol Syst Biol 9:661. https://doi.org/10.1038/msb.2013.18
25. Gu C, Kim GB, Kim WJ, Kim HU, Lee SY (2019) Current status and applications of genome- scale metabolic models. Genome Biol 20:1–18
26. Alper H, Y-s J, Moxley JF, Stephanopoulos GÃ (2005) Identifying gene targets for the metabolic engineering of lycopene biosynthesis in *Escherichia coli*. Metab Eng 7:155–164. https://doi.org/10.1016/j.ymben.2004.12.003
27. López J, Bustos D, Camilo C, Arenas N, Saa PA (2020) Engineering *Saccharomyces cerevisiae* for the overproduction of β -ionone and its precursor β -carotene. Front Bioeng Biotechnol 8:1–13. https://doi.org/10.3389/fbioe.2020.578793
28. Bro C, Regenberg B, Fo J, Nielsen J (2006) In silico aided metabolic engineering of *Saccharomyces cerevisiae* for improved bioethanol production. Metab Eng 8:102–111. https://doi.org/10.1016/j.ymben.2005.09.007

29. Saa PA, Cortés MP, López J, Bustos D, Maass A, Agosin E (2019) Expanding metabolic capabilities using novel pathway designs: computational tools and case studies. Biotechnol J 14:1800734. https://doi.org/10.1002/biot.201800734
30. Noor E, Jona G, Bar-even A, Milo R, Antonovsky N, Gleizer S, Noor E, Zohar Y, Herz E, Barenholz U, Zelcbuch L, Amram S (2016) Sugar synthesis from CO 2 in Escherichia coli. Cell 166:1–11. https://doi.org/10.1016/j.cell.2016.05.064
31. Fritzemeier CJ, Hartleb D, Szappanos B, Papp B, Lercher MJ (2017) Erroneous energy-generating cycles in published genome scale metabolic networks: identification and removal. PLoS Comput Biol 13(4): e1005494. https://doi.org/10.1371/journal.pcbi.1005494
32. Saa PA, Nielsen LK (2016) Fast-SNP: a fast matrix pre-processing algorithm for efficient loopless flux optimization of metabolic models. Bioinformatics 32(24):3807–3814. https://doi.org/10.1093/bioinformatics/btw555
33. Thiele I, Palsson BØ (2010) A protocol for generating a high-quality genome-scale metabolic reconstruction. Nat Protoc 5:93–121. https://doi.org/10.1038/nprot.2009.203
34. Lieven C, Beber ME, Olivier BG, Bergmann FT, Ataman M, Babaei P, Bartell JA, Blank LM, Chauhan S, Correia K, Diener C, Dräger A, Ebert BE, Edirisinghe JN, Faria JP, Feist AM, Fengos G, Fleming RMT, García-Jiménez B, Hatzimanikatis V, Wv H, Henry CS, Hermjakob H, Herrgård MJ, Kaafarani A, Kim HU, King Z, Klamt S, Klipp E, Koehorst JJ, König M, Lakshmanan M, Lee D-Y, Lee SY, Lee S, Lewis NE, Liu F, Ma H, Machado D, Mahadevan R, Maia P, Mardinoglu A, Medlock GL, Monk JM, Nielsen J, Nielsen LK, Nogales J, Nookaew I, Palsson BO, Papin JA, Patil KR, Poolman M, Price ND, Resendis-Antonio O, Richelle A, Rocha I, Sánchez BJ, Schaap PJ, Sheriff RSM, Shoaie S, Sonnenschein N, Teusink B, Vilaça P, Vik JO, Wodke JAH, Xavier JC, Yuan Q, Zakhartsev M, Zhang C (2020) MEMOTE for standardized genome-scale metabolic model testing. Nat Biotechnol 38:272–276. https://doi.org/10.1038/s41587-020-0446-y
35. Heirendt L, Arreckx S, Pfau T, Mendoza SN, Richelle A, Heinken A, Haraldsdóttir HS, Wachowiak J, Keating SM, Vlasov V, Magnusdóttir S, Ng CY, Preciat G, Žagare A, Chan SHJ, Aurich MK, Clancy CM, Modamio J, Sauls JT, Noronha A, Bordbar A, Cousins B, El Assal DC, Valcarcel LV, Apaolaza I, Ghaderi S, Ahookhosh M, Ben Guebila M, Kostromins A, Sompairac N, Le HM, Ma D, Sun Y, Wang L, Yurkovich JT, Oliveira MAP, Vuong PT, El Assal LP, Kuperstein I, Zinovyev A, Hinton HS, Bryant WA, Aragón Artacho FJ, Planes FJ, Stalidzans E, Maass A, Vempala S, Hucka M, Saunders MA, Maranas CD, Lewis NE, Sauter T, Palsson BØ, Thiele I, Fleming RMT (2019) Creation and analysis of biochemical constraint-based models using the COBRA toolbox v.3.0. Nat Protoc 14(3):639–702. https://doi.org/10.1038/s41596-018-0098-2
36. Branco dos Santos F, Olivier BG, Boele J, Smessaert V, De Rop P, Krumpochova P, Klau GW, Giera M, Dehottay P, Teusink B, Goffin P (2017) Probing the genome-scale metabolic landscape of Bordetella pertussis, the causative agent of whooping cough. Appl Environ Microbiol 83:e01528–e01517. https://doi.org/10.1128/AEM.01528-17
37. Pizarro FJ, Jewett MC, Nielsen J, Agosin E (2008) Growth temperature exerts differential physiological and transcriptional responses in laboratory and wine strains of *Saccharomyces cerevisiae*. Appl Environ Microbiol 74: 6358–6368. https://doi.org/10.1128/AEM.00602 08
38. Saa PA, Nielsen LK (2016) Ll-ACHRB: a scalable algorithm for sampling the feasible solution space of metabolic networks. Bioinformatics 32(15):2330–2337. https://doi.org/10.1093/bioinformatics/btw132
39. Haraldsdóttir HS, Cousins B, Thiele I, Fleming RMT, Vempala S (2017) CHRR: coordinate hit-and-run with rounding for uniform sampling of constraint-based models. Bioinformatics 33(11):1741–1743. https://doi.org/10.1093/bioinformatics/btx052
40. Dal'Molin C, Quek L, Saa P, Payfreyman R, Nielsen LK (2018) From reconstruction to C4 metabolic engineering: a case study for overproduction of PHB in bioenergy grasses. Plant Sci 273:50–60
41. Dal'Molin CGD, Quek LE, Saa PA, Nielsen LK (2015) A multi-tissue genome-scale metabolic modeling framework for the analysis of whole plant systems. Front Plant Sci 6:4. https://doi.org/10.3389/Fpls.2015.00004
42. King ZA, Dräger A, Ebrahim A, Sonnenschein N, Lewis E, Palsson BO (2015) Escher: a web application for building , sharing, and embedding data-rich visualizations of biological pathways. PLoS Comput Biol 11: e1004321. https://doi.org/10.1371/journal.pcbi.1004321

43. Rowe E, Palsson BO, King ZA (2018) Escher-FBA: a web application for interactive flux balance analysis. BMC Syst Biol 12:84
44. Saa PA, Nielsen LK (2017) Formulation, construction and analysis of kinetic models of metabolism: a review of modelling frameworks. Biotechnol Adv 35(8):981–1003. https://doi.org/10.1016/j.biotechadv.2017.09.005
45. Sánchez BJ, Pérez-Correa JR, Agosin E (2014) Construction of robust dynamic genome-scale metabolic model structures of *Saccharomyces cerevisiae* through iterative re-parameterization. Metab Eng 25:159–173. https://doi.org/10.1016/j.ymben.2014.07.004
46. Palsson BØ (2015) Systems biology constraint-based reconstruction and analysis, 2nd edn. Cambridge University Press, Cambridge
47. Maarleveld TR, Khandelwal RA, Olivier BG, Teusink B, Bruggeman FJ (2013) Basic concepts and principles of stoichiometric modeling of metabolic networks. Biotechnol J 1: 997–1008. https://doi.org/10.1002/biot.201200291
48. Teusink B, Wiersma A, Molenaar D, Francke C, Vos WMD, Siezen RJ, Smid EJ (2006) Analysis of growth of *Lactobacillus plantarum* WCFS1 on a complex medium using a genome-scale metabolic model. J Biol Chem 281: 40041–40048. https://doi.org/10.1074/jbc.M606263200
49. Visser D, Heijnen JJ (2002) The mathematics of metabolic control analysis revisited. Metab Eng 123:114–123. https://doi.org/10.1006/mben.2001.0216
50. Terrade N, Mira de Orduña R (2009) Determination of the essential nutrient requirements of wine-related bacteria from the genera Oenococcus and lactobacillus. Int J Food Microbiol 133:8–13. https://doi.org/10.1016/j.ijfoodmicro.2009.03.020

Chapter 17

Modeling Approaches to Microbial Metabolism

Andreas Kremling

Abstract

Microbial systems are frequently used in biotechnology to convert substrates into valuable products. To make this efficient, knowledge on the specific metabolic characteristics of a system is required as well as a theoretical description that allows researchers to design the system for a profitable use in an industrial application. In this chapter, basics on mathematical modelling approaches are introduced and examples are provided.

Key words Mathematical modeling, Mass balance equation, Stoichiometric networks, Coarse-grained modeling

1 Introduction

In the last decade "systems biology" has become a research field of its own. This is visible as various novel research programs, the founding of new departments as well as offers of new lectures at universities. There are numerous approaches in experimental and theoretical systems biology, so that a number of definitions exist that emphasize specific aspects [1–3]. However, a common characteristic for research in systems biology is an interdisciplinary approach by combining theoretical (e.g., mathematical modeling), experimental (comprehensive omics technology) and computational methods (databases and data analysis).

In this chapter, the focus is on mathematical approaches to microbiological systems. This has the advantage that complexity is reduced to a simpler metabolism, missing compartmentalisation, and a lower number of behavioural patterns that can be modelled. There are many reasons for applying of theoretical methods, among them the understanding of the complex interplay between thousands of components in large biochemical networks or the design of cellular systems as producing factories for a bio-based economy. Here, microbial systems are used for the production of

Sonia Cortassa and Miguel A. Aon (eds.), *Computational Systems Biology in Medicine and Biotechnology: Methods and Protocols*, Methods in Molecular Biology, vol. 2399, https://doi.org/10.1007/978-1-0716-1831-8_17,

valuables (products used as fine chemicals, drugs, fuels, and enzymes) and, on the other hand, they are important for pollutant decomposition. These research activities often are termed systems biotechnology or systems metabolic engineering [4]. Moreover, the understanding of microbial behaviour during interactions with humans (infectious diseases, biofilm formation) is a necessary driver for the improvement of public health.

Mathematical modelling goes hand in hand with molecular biological experiments. The starting point is a biological experiment, an observation or an unexplained phenomenon. For using a mathematical modelling approach, it is required that the system at hand and the problem can be described mathematically. During problem definition, the researcher hypothesises solutions to the problem. In the modelling process, it is important that the model explains, either qualitatively or quantitatively, the data presented. This is possible by predicting future behaviour if different input conditions (environmental stress but also a modified biochemical network by genetic interventions) are chosen in follow-up experiments. Consequently, the model can also be used to make proposals for new experiments that can be tested to help improve the model using an adequate experimental design strategy. The problem is addressed using an iterative sequence of model improvements and new experiments. Despite the fact that the methods described herein have wide applications, the examples presented originate from the field of microbiology. The bacterium *Escherichia coli* is chosen as "model" organism, as one of the most researched microorganisms today.

Depending on the problem definition, an appropriate modelling framework has to be chosen. In Fig. 1, a number of basic scenarios are shown, each one requiring its own approach: (1) If one is interested in the age or size distribution of a cellular population, the alteration of the distribution over time has to be described with a set of equations that consider changes of these properties, for example, by cell growth and cell division after a certain mass or volume is reached. (2) Bacteria like. *Escherichia coli* shows interesting movement patterns like running and tumbling. If an attractant is added to the medium, the frequency of tumbling is immediately reduced and the cells start to swim to the new source. After some time, they return toward their previous pattern, an observation that led to the detection of feedback structures called integral feedback, which for the first time revealed a robust control circuit [5, 6]. (3) Biofilm formation has gained attraction in recent time because biofilms are not absolutely unwanted but rather show a beneficial behaviour in bioprocess engineering. Microbial systems attach to surfaces, excrete components that build extracellular polymeric substances (EPS) with a defined architecture, thereby providing an ideal environment for production. Structural properties like biofilm thickness will have an impact on nutrient supply.

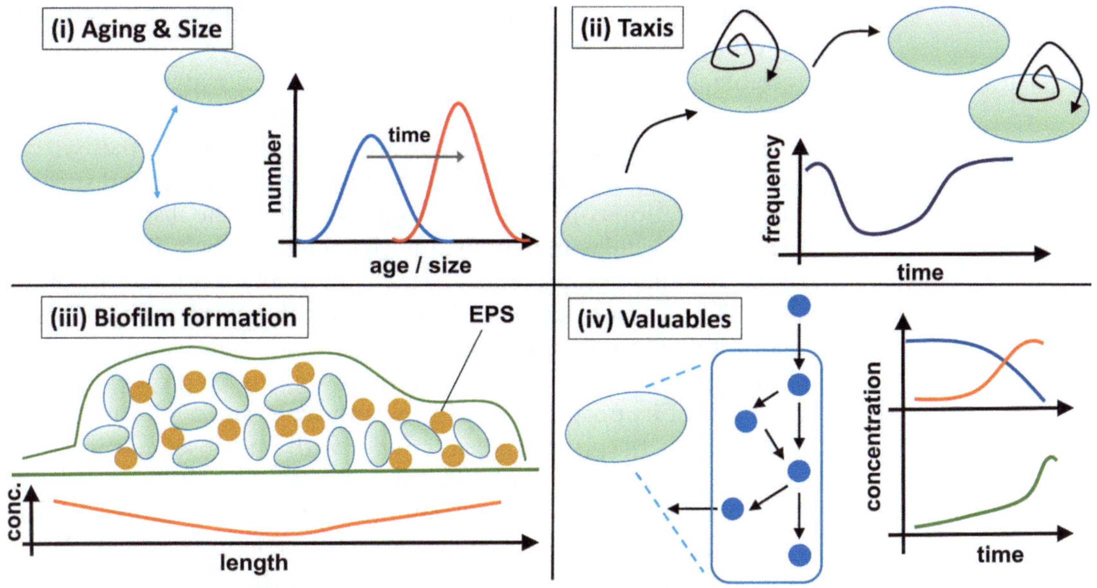

Fig. 1 Examples of behavioural patterns in microbial systems: (i) Cell age and size change due to growth and division, (ii) cell movement is driven by concentration gradients, (iii) biofilm formation leads to mechanical continuum, and (iv) cells as producers of biochemical compounds

Nowadays, sophisticated measurement techniques enable researchers to determine concentration profiles over the depth of the biofilm. (4) Microbial systems are ideal producer and degradation systems; they posses a large repertoire of biochemical pathways for enzymatic conversions. With defined interventions into metabolism and gene expression, not only natural products but also newly designed components and catalysts are made. Strain and process design lead to optimal results for yield and productivity (that is, yield over time) [4, 7].

This very short overview on the diverse and varying metabolic capabilities and behavioural patterns suggest that also the mathematical tools are diverse, and, depending upon the problem formulation, sophisticated. Since an introductory chapter is provided in this book, besides a brief overview on model types, the focus will be on the representation of biochemical networks that is shown on the lower right side of the figure.

The set up of a mathematical model requires the definition of so called state variables (variables of interest that inform about the state of a system) and processes that show an impact on the state variables. Variables and processes are connected with physicochemical basic laws like the fundamental theorems of thermodynamics. One of the most important processes is cell growth and, consequently, mass conservation has to be considered in any case. This is especially true for the conversion of substrates into valuable products, but is also true for cell movement (the energy comes from substrate conversion) and biofilm formation. Models are required to be predictive, that is, able to formulate hypotheses that are testable, first, in simulations, and later, in a real experiment.

Selection of state variables and their properties: Let us start by considering a single cell C. There are several properties that we can assign, like age a, size s, mass m or position p on a defined surface. For applications in biotechnology however, the list has to be extended. Cellular metabolism is characterized by a high number of intracellular components A_i (the sum of all mass fractions of the A_is will be 1) interacting with others and therefore form a densely connected network. So, the goal of our mathematical approach should be to describe the behaviour of our single cell C as a function of its properties and time and we wish to find a function $C(a, s, m, p, A_i, t)$ that is in accordance with our observation. However, modelling is not only the formulation of an equation system, but—more important—a meaningful abstraction of properties and processes to reduce complexity and to focus on the problem definition at hand. F For example, interested in the difference between two carbohydrate systems in our overall population (and we will use carbohydrate transport systems in bacteria as an example later), our function C can be simplified to $C(m, A_1, A_2, \cdots, A_m)$, and we will only describe the mass m of the population and some selected intracellular components $A_1, A_2, \cdots, A_m$.

Besides the reduction of properties, the selection and mathematical description of processes is the second key step in setting up the model. As we know from physics, processes on an atomic level are inherently stochastic but can be successfully described in a deterministic way on a different level. As shown in Fig. 1, properties of interest are quantities like cell age, cell mass or the mass fraction of intracellular components. In Fig. 2, important reduction steps are illustrated. By averaging properties of single cells, an average cell is used in most cases to describe microbial systems (upper row). If we focus on intracellular networks, different representations are also possible and two examples are shown: (1) Based on the detailed knowledge of the system, a model can describe an enzymatic conversion in form of a biochemical reaction eq. $2\ A \rightleftharpoons 3\ B$ (read 2 molecules of A are converted into 3 molecules of B, lower row left) with enzyme E as catalyst. A second example for this type of networks are polymerization processes, for example, to describe the synthesis of proteins. In both cases continuous variables such as the intracellular concentration, given for example in mol/gDW are used. (2) On the other hand, if detailed information is missing, a simple graphical representation given as $A\ B$ (read A influences B, lower row right) might be enough, and the properties of A and B are given as "available"/"not available" or "on"/"off." The two cases described represent a broad spectrum of model reduction possibilities and in literature, often, mixtures of these two cases can be found.

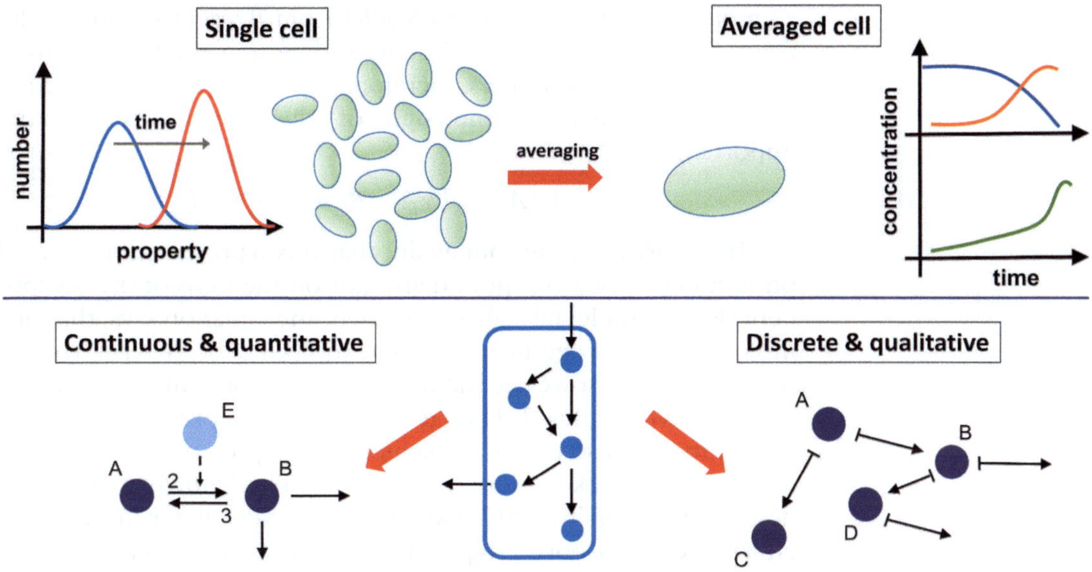

Fig. 2 Reduction of complexity. The single cell level is often too demanding for modelling purposes; by averaging all properties, an averaged cell is considered (top level). For intracellular processes, exemplarily, a continuous/quantitative process and a discrete/qualitative description are shown. Bottom panel, note that a different type of arrow is used here for the representation of the discrete graph. For details, see main text

2 Knowledge Representation and Types of Mathematical Models

In Subheadings 2 and 3 a rough description of available knowledge of microbial systems is given and types of models are introduced in more detail. Exemplary, the database "BioCyc" (biocyc.com) is described. This database includes information from over ten of thousands publications, mainly from *Escherichia coli* [8], *Saccharomyces cerevisiae*, and *Homo sapiens*. A number of tools available enable to computationally predict metabolic pathways and operons in bacterial systems. Data that describe regulatory features and protein networks are included. BioCyc provides tools for navigating, visualizing, and analysing the underlying databases, and for analyzing omics data: a genome browser is available to search genes and information on its position and length. It can display individual pathways and full metabolic maps. Also, it possesses tools for omics data analysis to draw diagrams and maps. Furthermore, metabolic models can be simulated.

As we can see, there is a strong focus on the biochemical conversion of components while other aspects like biophysical or biomechanical properties are not described. Therefore, we also focus on the conversion from substrates outside of the cell into intracellular components that represent the entire biomass. Biochemical reactions, typically, are given on a number base that describes how many substrate molecules are converted into how many product molecules. The following notation is used that says

γ_A molecules from A and γ_B molecules from B react to γ_C molecules from C (stoichiometric coefficients shown in a reaction equation are always positive numbers, therefore $|\bullet|$ is used, since for a further mathematical analysis, the sign of the coefficients becomes important):

$$|\gamma_A| \;\; A + |\gamma_B| \;\; B \rightleftharpoons |\gamma_C| \;\; C$$

It is worth to point out again, that this representation is based on numbers n of a compound and not on the mass of the component. For example, take all $\gamma_i = 1$, then, the equation says, that one molecule of A together with one molecule of B react to one molecule of C; however, the mass m of component C is the sum of the mass of A plus B. Therefore, the type of representation based on numbers is also called a stoichiometric representation.

Now, let us take a more complex scheme, as shown in Fig. 3 and described in detail in [9]. Extracellular lactose is taken up by an enzymatic transport step. The catalyst is enzyme LacY.

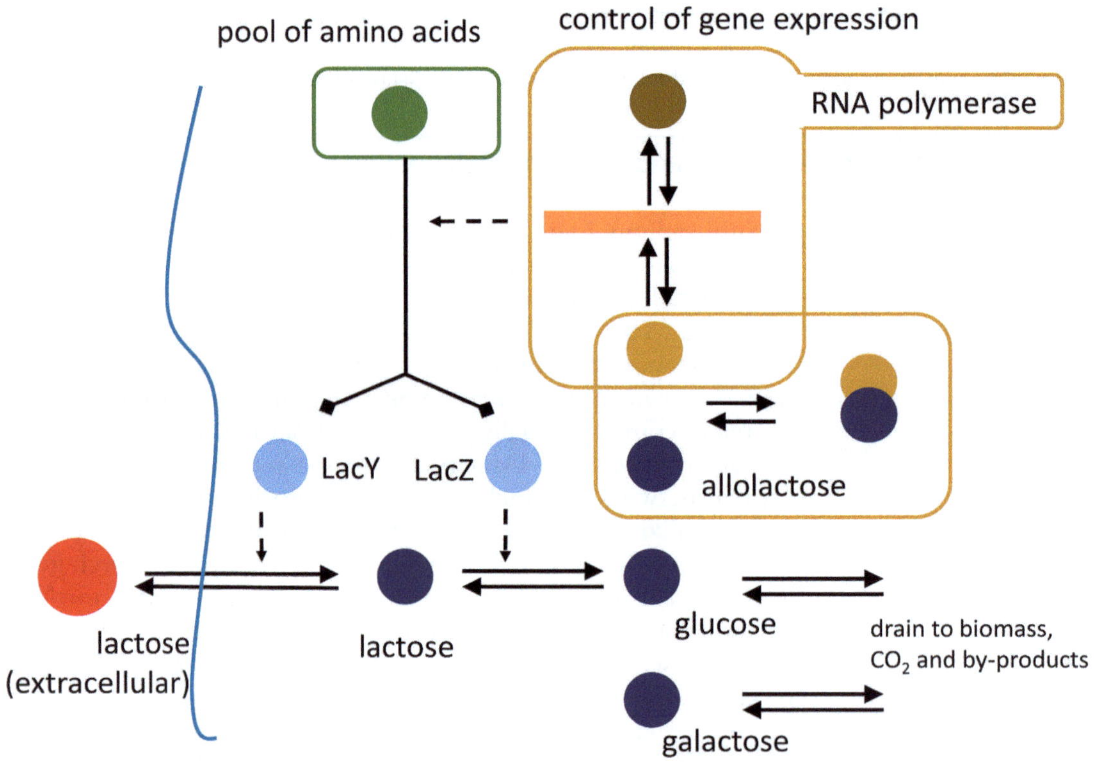

Fig. 3 Reaction scheme for lactose uptake and its control [9]. Lactose is taken up by enzyme *LacY* and further metabolized by *LacZ*. Products of this step are glucose, galactose and, as not further metabolized product, allolactose *ALac*. Intracellular glucose and galactose are further metabolized into precursors for biomass. During these processes, CO_2 as well as by-products like acetate could be formed. *ALac* interacts with repressor *LacI* and form an inert complex, preventing *LacI* to block transcription of the genes by the RNA polymerase. Both, RNA polymerase and *LacI* compete for a binding place on DNA (orange bar). During protein synthesis the pool of amino acids is used as reservoir for enzyme production

Subsequently, the intracellular lactose is further metabolized by enzyme LacZ with products glucose, galactose, and allolactose. To distinguish between mass conversion and catalytic control, dashed lines are used for control. Glucose and galactose are further converted into other metabolites (not shown). However, a by-product, allolactose is not metabolized, but acts as signalling molecule. In the figure, also the synthesis of the two enzymes is shown. Proteins are synthesized from the pool of amino acids. The process is called gene expression and describes how the information on the DNA is transcribed and translated into protein. The process is also controlled by two important proteins, the RNA polymerase and the repressor LacI. If no lactose is available in the medium, the repressor binds to the DNA that codes for the two enzymes. Therefore, the RNA polymerase is not able to read the stored information and gene expression cannot take place. In case that lactose is available, allolactose can inactivate the repressor forming an inert complex that cannot bind to the DNA (shown as blue/ yellow form). But now, the repressor is bound to the complex and cannot interact with the DNA. In this case, the RNA polymerase can start transcribing the information on the DNA.

The following system of four biochemical reactions (r_1 to r_4, represented as Net 1) describe the metabolic mass flow from lactose to the intracellular metabolites ($\emptyset$ means that we do not consider further components in the network):

$$\begin{aligned} \text{Net1}: \quad Lac_{ex} &\rightleftharpoons Lac_{in} \\ Lac_{in} &\rightleftharpoons ALac + Glc + Gal \\ Glc &\rightleftharpoons \emptyset \\ Gal &\rightleftharpoons \emptyset \end{aligned} \tag{1}$$

Furthermore, we can also describe the synthesis of the catalyst *LacY* and *LacZ*. However, before we continue, Equation system (1) given above can be rewritten with the stoichiometric matrix N which includes the stoichiometric coefficients in a systematic manner: each metabolite is represented in a row while the reaction information is given in the columns of N. To make clear that the metabolites on the left side of each equation are consumed and on the right side are produced, we use the negative sign for consumption and the positive sign for production. Matrix N with 5 rows and 4 columns reads therefore:

$$N = \begin{pmatrix} -1 & 0 & 0 & 0 \\ 1 & -1 & 0 & 0 \\ 0 & 1 & -1 & 0 \\ 0 & 1 & 0 & -1 \\ 0 & 1 & 0 & 0 \end{pmatrix}$$

In this way, the stoichiometric matrix is used to represent all metabolic conversions that happen inside the cell. In this representation the influence of the two catalytic enzymes is not given since they only show an influence on the reaction velocity but not on the stoichiometry. Currently, for many bacterial systems, but also for eukaryotic cells, stoichiometric matrices are published with over 1000 reactions and components. A special characteristic here is that, in general, the number of reactions exceed by far the number of components. This information will be important later on when N is used to determine metabolic fluxes in larger systems.

If we now consider the remaining part of the network, we have to describe also the synthesis of the two proteins *LacY* and *LacZ* (Net 2). Strictly formal, a number of amino acids from our pool are converted into proteins; again the influence of the control via the gene expression machinery is not considered (since polymerization is irreversible, we use a simple arrow here).

$$\begin{aligned} \text{Net2}: \quad \gamma_1 \; AA \; &\rightarrow \; LacY \\ \gamma_2 \; AA \; &\rightarrow \; LacZ \end{aligned}$$

Again, to describe the control of protein synthesis, the interaction of RNA polymerase P, repressor R with the DNA binding D and the interaction with inducer $ALac$ have to be described by a set of reaction equations (Net 3):

$$\begin{aligned} \text{Net3}: \quad P \; + \; D &\rightleftharpoons \; PD \\ R \; + \; D &\rightleftharpoons \; RD \\ R \; + \; n \, ALac &\rightleftharpoons \; RALac \end{aligned} \tag{2}$$

The first two reactions describe the competition of the RNA polymerase and the repressor for the binding site on the DNA that is also called control region. Typically, R has a stronger affinity for D and, hence, the RNA polymerase cannot bind to the promoter to start transcription. However, if the amount of inducer *ALac* is increasing, the repressor is bound in an inert complex *RALac*. In this way, more and more polymerase can start transcription and protein is synthesized until a new steady state is reached (this will occur since beside synthesis also degradation will take place).

Up to now, all three networks are not connected. The link between the subnetworks is established by reaction velocities r_i which are necessary to fully describe the biochemical reactions. Reactions kinetics could be very complicated because they cover already a number of single reactions steps that all can be lumped into one equation on the level that we consider here; for example, the first reaction in Net 1 describes substrate uptake of lactose. Enzyme *LacY*, in a series of microconversion steps, first binds the substrate lactose, transports it into the cell and releases it. The field of enzyme kinetics provide a rich set of equations for a high number of different mechanism. A famous example is the Michaelis–Menten equation that can be used here for the first step in Net 1 to describe the reaction velocity in dependence on the substrate and the catalytic device (the reaction is taken as irreversible here):

$$r_1 = k_1 \; LacY \; \frac{Lac_{ex}}{Lac_{ex} + K_1}$$

As already mentioned, processes of signal processing and polymerization described in Net 2 and Net 3 may become difficult to describe since a high number of reaction steps must be considered and knowledge is not available in detail in comparison to metabolic conversion in Net 1. Therefore, an alternative representation is often used when knowledge on specific interactions is scarce. Instead of describing a quantitative stoichiometric conversion, a qualitative interaction is represented by a graph. Here, a graph is composed of components and interactions and only tell us that component *A* in some way influences component *B*; the influence could be positive (+ sign) or negative (− sign). Considering the network above, part of the information can be summarized in the following graph. Note, that we use a different type of arrow to make clear that we do not describe a biochemical conversion but only an interaction:

$$Lac_{ex} \overset{+}{\mapsto} ALac$$

$$ALac \overset{-}{\mapsto} R$$

$$R \overset{-}{\mapsto} LacY$$

$$LacY \overset{+}{\mapsto} ALac$$

If Lactose is available in the medium, it is a driver to produce inducer *ALac*. The inducer shows a negative effect on the repressor because it deactivates it while the repressor itself shows a negative effect on the protein synthesis, here *LacY*. The enzyme itself is also a driver to produce the inducer. Similar to the stoichiometric matrix, we can set up a different type of matrix, the incidence matrix *I* that allows us to summarize the information in a compact way. The incidence matrix describes each interaction with −1 for the starting component and +1 with the target component. For the case at hand, matrix I reads (again the components are represented in the rows while the interactions are represented in the columns):

$$I = \begin{pmatrix} -1 & 0 & 0 & 0 \\ 1 & -1 & 0 & 1 \\ 0 & 1 & -1 & 0 \\ 0 & 0 & 1 & -1 \end{pmatrix}$$

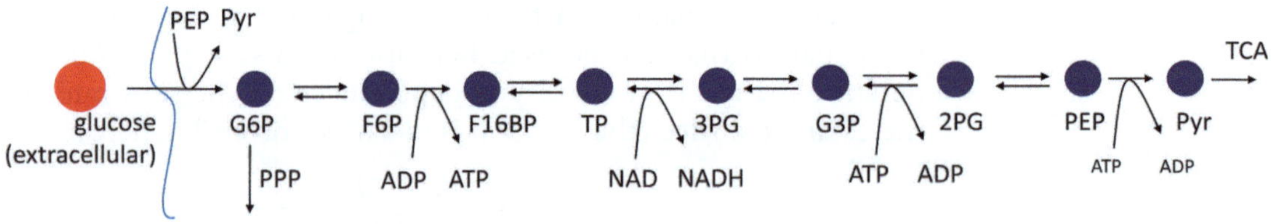

Fig. 4 Set of reactions of the glycolytic pathway. The pathway connects incoming carbohydrate to important metabolites like pyruvate. At the same time, it is the starting point for PPP (pentose phosphate pathway) and TCA (tricarboxylic acid) cycle

Matrix I is used for larger networks to answer the following type of questions: (1) Is there any connection between two selected components in the network? (2) Are there feedback loops in the network (a definition of a feedback loop would be, that we are looking for a connection from a component through the network back to the component itself, however, to be applicable, further restrictions are necessary from a formal point of view)?

To complement this section, in Fig. 4 the complete pathway from uptake of a carbohydrate, here glucose, to metabolite pyruvate is shown. The pathways consists of several steps where the energy of glucose uptake is generated in the sub-system itself; for this reaction phosphoenolpyruvate (PEP) is used as energy source. However, this only works since in one of the reaction steps, 2 mol of triose phosphate (TP) are generated from 1 mol glucose. In this way, also 2 mol of PEP are generated and one mol is used for transport. The pathway also shows that important metabolites for energy generation are involved. ATP is needed in one reaction but produced in two reactions while NADH is produced in one reaction. For the modeling examples later on, we use this pathway and variants of it: For coarse-grained modeling, reactions are lumped together and the glycolytic pathway is represented by a single reaction from glucose to pyruvate. For the stoichiometric approach a detailed model is used and complemented with reactions from pentose phosphate pathway (PPP) and the tricarboxylic acid (TCA) cycle. Finally, a kinetic model for carbohydrate uptake (in this case lactose) describes the interplay between protein synthesis and its control.

To summarize up to here, two possible network representations are provided: a quantitative one that uses the stoichiometric matrix to describe the conversion of substrates into products for single biochemical reaction steps. This format is frequently used in Metabolic Engineering applications to design strains with an increased production of desired valuable components. The second representation is a qualitative one that uses the incidence matrix to describe an interaction between two components in the network to represent interactions between two and more partners. Simple graphs like the one introduced here are not sufficient and must be replaced by so called hyper graphs which are out of scope of this introduction. This type is frequently used to describe signaling networks in medical applications to uncover possible effects of new drugs or of missing proteins due to genetic defects. For further introductory material, a number of textbooks are available [7, 9].

3 Mass Conservation on Biochemical Networks

Based on the results from the preceding section, a deeper look on mass conversion in biochemical reaction networks is taken. A graphical representation like mentioned above is not appropriate here, since the knowledge on the bare interaction does not contain any information on mass conversion nor on the temporal behaviour of our system. In contrast, having available the stoichiometric information of the system, and information on the reaction velocities r_i for each reaction i is enough here. This will allow us to describe the behaviour of the system over time for all components based on the stoichiometric conversion.

However, a problem might be the observation that a system with only a few interacting molecules behaves differently from a system with thousands of molecules for a single species. This is also true for the system of lactose uptake and metabolism since the number of repressor molecules R (10 to 20 molecules per cell) and DNA binding site D (one to two) is very low in comparison to the number of molecules of intracellular glucose for example (1 μmol/gDW $\approx$ 60 $\times$ 10^3 molecules per single cell). A deterministic description is not possible in this case while a stochastic framework is successful here. As the name indicates a stochastic framework will not tell us the exact number n of a certain molecule A at a specific time t in a single cell, but rather a probability that at time t the probability of A to have n molecules is $p_n(t)$. The deterministic equivalent will be the concentration c_A of component A that gives the number of molecules of A in a certain reference volume or—more common in biotechnology—the number of molecules per gram dry weight. As usual, for probabilities, a constraint appears if we sum up over all possible numbers from $n = 0$ to $n \rightarrow \infty$:

$$\int_{n=0}^{\infty} p_n d_n = 1 \tag{3}$$

A relationship between the stochastic framework and the deterministic counterpart can be established if we consider a mean value of the number of molecules in the reference volume V_{ref} that should be equal to the deterministic value:

$$C_A = \int_{n=0}^{\infty} n p_n d_n / V_{ref} \tag{4}$$

Consequently, the output of model simulation studies should give us different information. On the one hand, we expect the change of probabilities $p_n(t)$ over time for the components while in the second case, the time course of the concentration $c_A(t)$ is expected. Working with probabilities requires a sophisticated stochastic framework that is beyond this introductory text.

In the following, we focus on a deterministic framework. From thermodynamic principles mass conversion is the most important one that plays the major role of determining our desired output $c_A(t)$. From our example above, we know that lactose or an other carbohydrate like glucose is taken up and converted into biomass. Possibly, from 1 g of carbohydrate at the beginning of the experiment, not only biomass, but CO_2 and also by-products are produced. Before we formulate a complete set of mass balance equations, a static view on carbohydrate uptake and subsequent metabolism already offers us the possibility for a quantitative view on cellular processes, enabling us to introduce macroscopic parameters like the yield coefficient, an important measure, for example, in biotechnology. A network is considered with only two stoichiometric equations that describe the overall cellular metabolism:

$$S \;\rightarrow\; \beta_{11}\; P \;+\; \beta_{12}\; H_2$$

$$\gamma\;[P\; + \alpha_{21}\; H_2 \;+\; O_2] \rightarrow\; B \;+\; \beta_1\; CO_2 \;+\; \beta_2\; H_2O$$

The equation reads that substrate S is converted into compound P, that is, a representative of central metabolism plus hydrogen. Further P is converted with oxygen into biomass B, CO_2 and water. Stoichiometric coefficient γ is used to describe that the cellular macromolecules are composed of monomers (such as P in the example). On an extended view, substrate S will be glucose $C_6H_{12}O_6$ and P metabolite pyruvate $C_3H_4O_3$ and we focus first on the atomic composition of the network components. To write down the equations in a balanced form, the number of molecules for C, O and H must be equal on the left and the right side. In the following example, we restrict the analysis to these type of molecules, however, taking into account nitrogen N and further molecules is straightforward:

$$\begin{aligned} C_6H_{12}O_6 &\rightarrow\; 2\;\; C_3H_4O_3 +\; 2\;\; H_2 \\ \gamma\;[C_3H_4O_3 \;+\; H_2] \;+\; O_2 &\rightarrow\; 0,6\;\; B^* \\ &\qquad +\; \gamma[0.53\;\; CO_2 +\; 0.8\;\; H_2O] \end{aligned} \tag{5}$$

It is imported to stress here, that substrates like carbohydrates are first degraded into smaller molecules (all these reactions are named catabolic reactions) and subsequently used to build polymeric structures (anabolic reactions). This is exactly described here with the two stoichiometric equations. The second equation deserves closer attention: to be consistent for the number of molecules, the biomass B^* is written $C_{4.1 \times \gamma}H_{7.3 \times \gamma}O_{1.9 \times \gamma}$ with $\gamma = 300$ the number of monomers used to provide a single polymeric compound. If we count the number of hydrogens for the second reaction, we have $300 \times 4 + 300 \times 2 = 1800$ on the left side and on the right side $0.6 \times (7.3 \times 300) + 300 \times 0.8 \times 2 = 1794$ that is only a minor difference due to rounded values.

More formally, we represent each component (in our example, we have seven components) of the network in a vector describing the content on C, H and O atoms. For example, pyruvate is represented as $[3, 4, 3]$ in a column vector. All components are then stored in a matrix K, which has seven columns in the following order: carbohydrate, pyruvate, hydrogen, oxygen, biomass, carbon dioxide, water for each component, and three rows for each atom:

$$K = \begin{pmatrix} 6 & 3 & 0 & 0 & (4.1\ \gamma) & 1 & 0 \\ 12 & 4 & 2 & 0 & (7.3\ \gamma) & 0 & 2 \\ 6 & 3 & 0 & 2 & (1.9\ \gamma) & 2 & 1 \end{pmatrix} \tag{6}$$

As before, the stoichiometric coefficients are displayed in a matrix N with the stoichiometric coefficients in the columns. N has seven rows and two columns:

$$N = \begin{pmatrix} -1 & 0 \\ 2 & -\gamma \\ 2 & -\gamma \\ 0 & -1 \\ 0 & 0.6 \\ 0 & 0.53\gamma \\ 0 & 0.8\gamma \end{pmatrix} \tag{7}$$

To verify that all atoms are balanced we explore the term $K \times N$ and we notice that all entries are zero (as said above, a small difference due to rounding appears for the second reaction).

It is worth to look also at the mass fraction that is converted from glucose to biomass. From the atomic composition, glucose for example, has a molecular mass of 180.2 g/mol ($\approx 6 \times 12 + 12 \times 1 + 6 \times 16$). Assume that we start with 1 mol, that is, we start with 180.2 g carbohydrate and we ask how much biomass we get. Since we get 2 mol of pyruvate in the second reaction, we have to multiply the outcome of the second reaction by $2/\gamma$. Therefore, our output is $1.2C_{4.1}H_{7.3}O_{1.9}$ that corresponds to $\approx$95.2 g. From these two numbers, the yield with respect to biomass from the substrate can be calculated and we get 9.2/180.2 = 0.52 g biomass/g substrate. Furthermore, we calculate that $1.06CO_2$ is produced that corresponds to 46.6 g. So, a large mass fraction of the carbohydrate exit the system as gas.

From this static information, however, it is not possible to infer on the time course of one state variable. At this point, a mass balance equation that describes the change of a compound over time and sums up the material flow into the system and out of the system comes into play. The mass balance equation is a differential equation, that is, an equation for a variable x that also contains its derivative with respect to t, written as $d\,x/dt$. Since we are interested in the mass m of a component, the mass balance reads as:

$$\frac{d\,m}{dt} = J[m] + P[m] \tag{8}$$

In this general equation, term $J[m]$ describes the mass flow into the system while $P[m]$ describes conversion inside the system, for example, by biochemical reactions. In the current form, the equation cannot be applied directly. The reason is the following: for biochemical reaction networks, the term $P[m]$ describes mass conversion by reactions, and we have already seen above, that the reaction velocity strongly depends on the concentration of the reactions partners and not on the mass of the reaction partners. However, it is not possible to write down directly an equation for the concentration c that is defined as the relation between mass m and a reference value, in most cases the volume V of interest: $c = m/V$. If we now consider the change of mass m over time t that is written as $\dot{m} = dm/dt$ we get, by applying the product rule for derivatives:

$$m = c\ V \Rightarrow \dot{m} = \dot{c}\ V + c\ \dot{V} \tag{9}$$

If for example our mass production is zero $\dot{m} = 0$, the concentration could change, if at the same time the volume changes also.

To avoid inconsistencies, it is recommended to start always from the mass balance and to reformulate the mass balance into an equation for the concentration (the resulting equation is not a mass balance in the strict sens but in literature we often find this term). In the next steps we derive equations for four different type of variables of interest that are called state variables since they characterize the state of interest. A reaction system as shown in Fig. 5 is considered with the possibility to feed substrate into the reactor and to take out biomass and medium for subsequent downstream processing.

The volume V of the system changes due to feeding q_{in} (we consider only one feeding) and outflow q from the reactor:

$$\dot{V} = q^{in} - q \tag{10}$$

Next, a substrate S is considered and its concentration c_S is based on the reactor volume V:

$$\dot{m}_S = q^{in}\ c_S^{in} - q\ c_S - r_S^*, \tag{11}$$

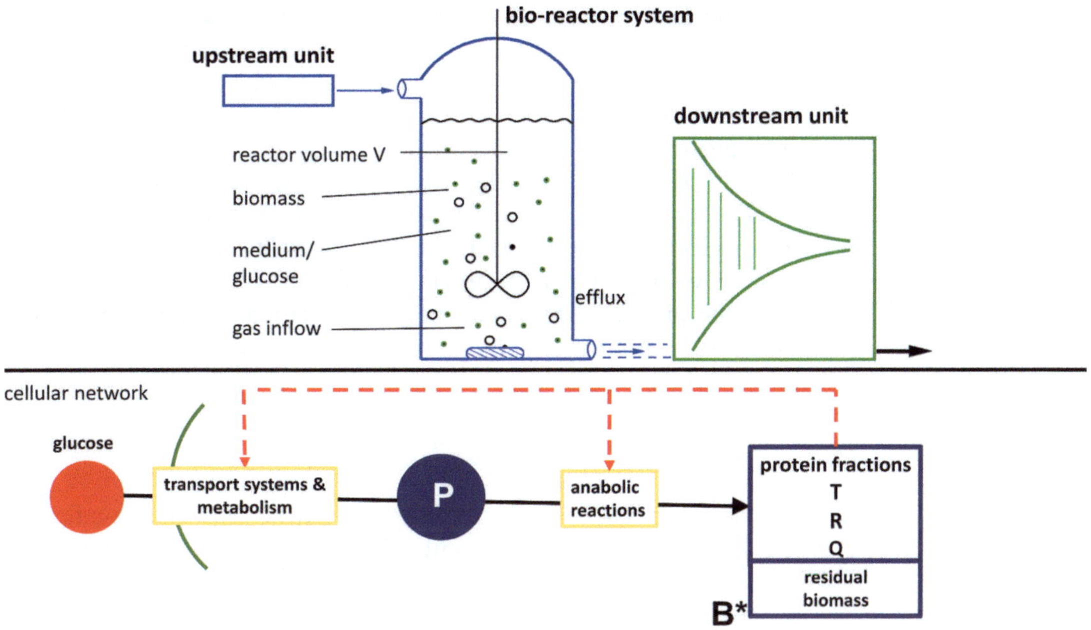

Fig. 5 General scheme of unit with upstream, bioreactor and downstream unit. The system under consideration is the bioreactor unit with biomass and medium. Medium feed with rate q^{in} is shown on the right side; outflow with rate q connects the reactor with the downstream unit. The outline of a coarse-grained model is shown below the modules. Details of the model are given in the main text

with c_S^{in} is the concentration of S in the feeding and r_S^* is the rate of uptake by the cells (given for example in g/h). Since reaction systems are considered, we have already seen that the number of molecules n is a more appropriate variable. Here, fortunately, the relation between mass m and number n is given by a fixed number, the molecular weight w, and we can rewrite the above equation:

$$\dot{n}_S = q_{in}\, c_S^{in} - q\, c_S - r_S^{**}, \tag{12}$$

with the concentration is defined as $c = n/V$ and r_S^{**} is given in mol/h.

In biotechnology, quantity (dry) biomass m_X is of fundamental interest and a mass balance equation is used to describe the course over time by using the specific growth rate μ. By definition, μ describes the change of time of biomass divided by the biomass itself under batch conditions, and is the most important indicator for the quality of the process. From a formal point of view, biomass only can change, if nutrients are taken up; however, this poses the question or problem to know all substances in the medium that are taken up or excreted over time. For a first approach, the specific growth rate μ is used as cumulative parameter and we will show later on in which way this parameter is connected to the biochemical reaction network. The equation for the biomass reads:

$$\dot{m}_X = -q\, c_X + \mu\, m_X \tag{13}$$

The last quantity of interest is a compound or a metabolite M that is inside the cell and therefore, its mass in our reference system (see Figure above) will change if (1) biomass is leaving the reactor and (2) by biochemical reactions that take part inside the cell. Hopefully, it becomes clear that now the reference for the concentration c_M is not the bioreactor, but the cell;therefore, it is an accepted standard to define an intracellular concentration based on a selected quantity of the cell. Normally, one would use the cellular volume which is, however, difficult to measure, and therefore, the biomass dry weight m_X is the standard reference (*see* also **Note 1**). This implies $c_M = n_M/m_X$ and accordingly the mass balance read:

$$\dot{n}_M = -q c_M c_X + \sum_i r_i^{**} \tag{14}$$

with the first term on the right side describes the mass flow of the biomass—with component M inside—out of the reactor and the sum term considers all reactions where M is involved. This second term can be written more precisely since with the stoichiometric matrix N (*see* Subheading 2) all reactions are already defined with the stoichiometric factors n_{ij} and in addition, for each reaction r_j, the velocity can be defined as a function of the concentration of the metabolites of the network. The mass flow by reaction of component M_i is then given by the respective line of the stoichiometric matrix N multiplied with the vector $\underline{r}$ of all reactions in the system:

$$\dot{n}_{Mi} = -q\ c_M\ \ c_X + \underline{n_i}^T \underline{r}^{**} \tag{15}$$

Still, all reactions r_j^{**} in the vector are given in mol/h. Now, we apply the procedure from above to reformulate this equation in terms of the concentration. For the left side, we get as above:

$$\dot{n}_M = \dot{c}_M\ \ m_X + c_M\ \ \dot{m}_X \tag{16}$$

that must be equal to the right hand side of the equation above. Rearranging the left hand side leads to an equation for $\dot{c}_M$ and with some basic calculations using the equation for the biomass from above, we obtain:

$$\begin{aligned} \dot{c}_M &= \frac{1}{m_X}\ \left(-q\ c_M\ \ c_X + \underline{n_i}^T \underline{r}^{**} - c_M\ \ \dot{m}_X\right) \\ &= \frac{1}{m_X}\ \left(\underline{n_i}^T \underline{r}^{**} - \mu\ c_M\ \ m_X\ \right) = \underline{n_i}^T \underline{r} - \mu\ c_M \end{aligned} \tag{17}$$

with $\underline{r}$, the specific rate vector, that is, the rate vector $\underline{r}^{**}$ based on the biomass with units mol/gDW h. If we write down the equations for all metabolites, then all single row vectors are combined in the stoichiometric matrix N:

$$\underline{\dot{c}}_M = N\ \underline{r} - \mu\ \underline{c}_M \tag{18}$$

For the other quantities of interest, we proceed analogously and our final set of equations for biomass and substrate concentration reads:

$$\dot{c}_S = D\,\left(c_S^{in} - c_S\right) - \underline{n}_S^T\,\underline{r}$$
$$\dot{c}_X = (\mu - D)\,c_X \tag{19}$$

with the dilution rate $D = q/V$ and $\underline{n}_S^T$ the stoichiometric coefficients for substrate uptake.

Obviously, the equation system is not yet consistent since the growth rate μ is still present. As already said, the change of mass of all cells should only depend on uptake and excretion of nutrient fluxes. To make the relation clear, we exploit a different way to write down the change of mass of all cells. Assume that in our vector of metabolites the complete mass of the cell is represented as:

$$m_X = \sum_i m_{Mi} \Rightarrow \dot{m}_X = \sum_i \dot{m}_{Mi} = \sum_i \dot{n}_{Mi} W_i \tag{20}$$

Plugging in all terms for $\dot{n}_{Mi}$ from above and comparing with Eq. (13) reveals an expression for the specific growth rate with the property of strict mass conversion:

$$\mu = \underline{w}\;N\;\underline{r} \tag{21}$$

with vector $\underline{w}$ of all molecular weights. This equation, however, is not suitable for large networks since not all rates can be determined with high accuracy; instead, often, an empirical relationship is used, and the specific growth rate is coupled for example to the available main substrate S in the experimental system. In this way, a possible growth law reads:

$$\mu = \mu_{max}\;\frac{c_S}{c_S + K_S} \tag{22}$$

with parameters μ_{max} the maximal growth rate and K_S the half saturation parameter. A different approach is used when coarse-grained models are applied; here the growth rate is coupled to the number of ribosomes R in the cells. In literature, a linear relationship is reported and is named "bacterial growth" law [10]:

$$\mu = k\;c_R \tag{23}$$

To summarize, mass balance equations are set up to allow for a quantitative description of cellular processes; this is necessary especially in systems biotechnology where such models represent the bases for a model-based design. However, mass balance equations have to be rewritten to bring in kinetic information in the reaction equations. In this way a set of differential equations $\dot{c}_i$ is generated for each compound i in the network that is a function of all compounds of the network: $\dot{c}_i = f(c_1, c_2, \cdots)$.

4 Examples for Models for Bacterial Systems

4.1 Coarse-Grained Model

Coarse-grained models are now widely used to describe resource allocation for bacterial systems. Due to their simple structure, simulation studies reveal interesting relationships between the different parts of the proteome. As a basis, we use the reaction network from above that is given by:

$$
\begin{aligned}
r_1 : &\quad S \rightarrow 2\ P + 2\ H_2 \\
r_2 : &\quad 300\ [P + H_2] + O_2 \\
&\rightarrow 0.6\ B^* + 300\ [0.53\ CO_2 + 0.8\ H_2O]
\end{aligned}
\tag{24}
$$

which is shown in Fig. 5. For this reaction network, the stoichiometric matrix for the main components P ($w_P = 88$ g/mol) and B^* ($w_B = 26070$ g/mol) reads:

$$
N = \begin{pmatrix} 2 & -300 \\ 0 & 0.6 \end{pmatrix}
\tag{25}
$$

and for the specific growth rate μ, we get with Eq. (21):

$$
\mu = 2\ r_1\ w_P + r_2\ (0.6\ w_B - 300\ w_P)
\tag{26}
$$

The differential equations read:

$$
\begin{aligned}
\dot{c}_P &= 2\ r_1 - 300\ r_2 - \mu\ c_P \\
\dot{c}_{B^*} &= 0.6\ r_2 - \mu\ c_{B^*}
\end{aligned}
\tag{27}
$$

As a final step to complete the model, the reaction kinetics for r_1 and r_2 has to be fixed:

$$
\begin{aligned}
r_1 &= k_1\ \frac{c_S}{c_S + K_S} \\
r_2 &= k_2\ \frac{c_P}{c_P + K_P}
\end{aligned}
\tag{28}
$$

To study the system, we vary the substrate concentration c_S in the system and solve for the steady-state solution.

The proposed model structure allows us to extend the set of equations by taking into account regulatory characteristics. So, far our macromolecule B^* represents nearly the entire biomass. Proteins are the most abundant part and play an important role for metabolism and its control. Coarse-grained models subdivide the proteome in at least two important fractions: one fraction T represents proteins involved in transport and metabolism while the second fraction R represents the complete transcription and translation apparatus. Depending on the growth situation, the resource B^* has to be spread to the two fractions in order to allow for maximal growth at best. However, the question arise what is the best strategy to allocate the resource to T and R respective, in dependence on the available substrate S. Mathematical optimization provides useful tools to answer the question. We extend our kinetic reaction expression to make them depending on T and R (*see* **Note 2**):

$$r_1 = k_1^j \; c_T \; \frac{c_S}{c_S + K_S}$$

$$r_2 = k_2^j \; c_R \; \frac{c_P}{c_P + K_P} \tag{29}$$

and we formulate the optimization problem to maximize the growth rate μ with allocation of c_T and c_R as follows:

$$\max_{c_R, c_R} \; \mu(r_1, r_2, N, \underline{w})$$

$$s.t.$$

$$0 = 2\, r_1 \; - \; 300\, r_2 \; - \; \mu\, c_P$$

$$0 = 0.6\, r_2 - \; \mu\, c_{B^*}$$

$$c_{B^*} = c_T \; + \; c_R \; + \; c_Q \tag{30}$$

With the last equation we introduce a fixed quantity Q that takes into account that only some part of B^* can be allocated (50%). Except for rescaled parameters k_1, $k_{2'}$ and the fixed fraction for Q no additional parameters are needed (Table 1).

Fig. 6 shows the steady state solution for varying substrate concentrations (upper row left, blue without optimization). The optimal solution for the growth rate is nearly the same as without optimization. The course of the T and R fraction after optimization reveals that the T fraction shows a negative trend for increasing growth rate while R is increasing over the growth rate. This is in good qualitative agreement with data presented in [11]. The plot in the lower row shows that for all substrate concentrations, strict mass conversion is guaranteed and that with increasing substrate concentration also the fraction of P is increasing. For higher growth rates the fraction of P is around 20% while the rest of biomass are macromolecules B^*. With the model the yield coefficient can be estimated with μ/r_1 and a value 0.5 as above is obtained.

Table 1
Additional model parameter for the two models described by Eqs. (28) and (29). Model 1 is without optimization, and model 2 with optimization

	Max	*K*
Model 1: r_1	5×10^{-3}	10^{-5}
Model 1: r_2	6.3×10^{-3}	0.5
Model 2: r_1	400	10^{-5}
Model 2: r_2	8000	2

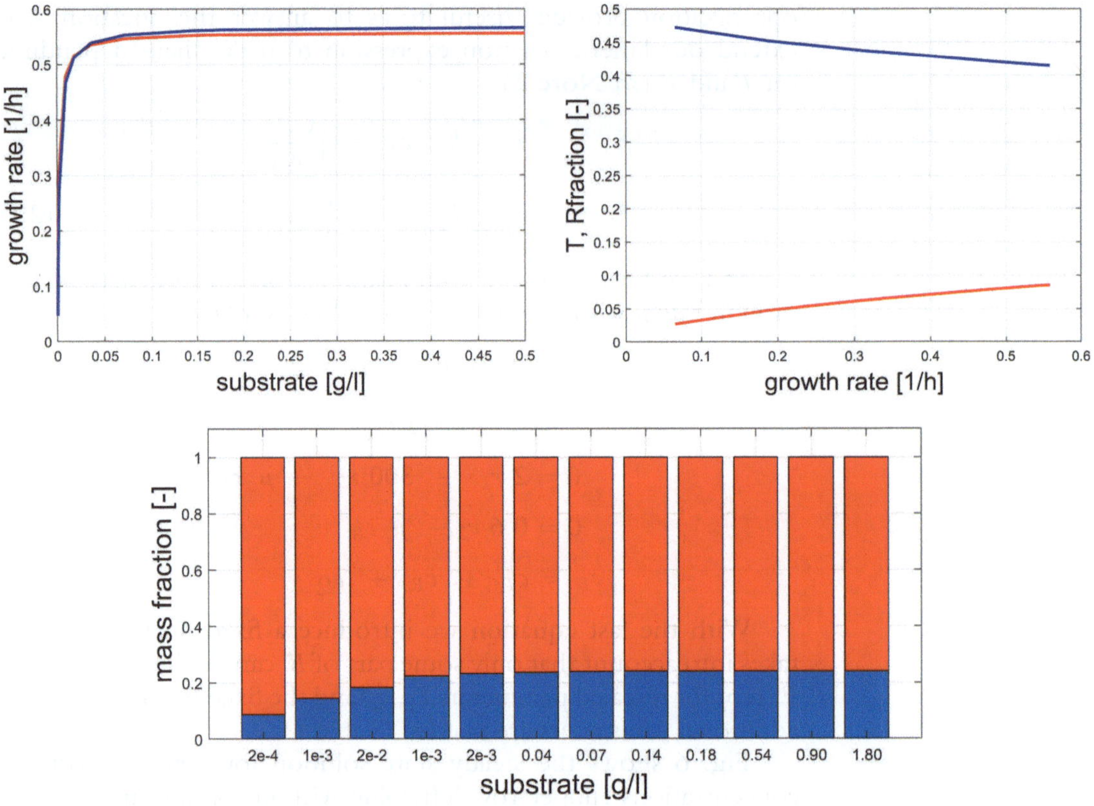

Fig. 6 Simulation study for two model variants. Upper row left: growth rate dependence on substrate concentration (blue without optimization, red with optimization). Upper row right: T (blue) and R (red) fractions as a function of the growth rate μ after optimization. Lower row: mass fraction of P and B^* over substrate after optimization (P blue and B^* red)

4.2 Stoichiometric Model

Currently, stoichiometric models are available for a number of microorganisms and a small example was given in Subheading 2 on lactose uptake. A closer look at Eq. (18) reveals that for systems with hundreds of components a numerical solution depends on a large number of kinetic parameters that are uncertain or even are not known. However, already some decades earlier, a simplified form of this equation has been extensively explored. Two simplifications are done; first, only fluxes with high numerical values are considered (*see* **Note 3**). Since the molar concentration of an intracellular metabolite is rather low, the dilution term $\mu\ \underline{c}_M$ can be neglected. Second, often the system's dynamics is not of great interest and therefore, a steady state situation is considered. Both simplifications lead the following set of linear algebraic equations with known stoichiometric coefficients stored in N and unknown rates in $\underline{r}$:

$$\underline{0} = N\ \underline{r} \tag{31}$$

For metabolic systems, the number of unknown rates is by far higher than the number of metabolites. In this case, there is a set of solution vectors, called the kernel or null space *K*. Matrix *K* has as many rows as *N* possess columns and the number of columns depend on the rank *r* of matrix *N*. Having an infinite number of solutions makes it difficult to select physiologically meaningful ones. Therefore, again, mathematical optimization comes into play and allows to select solutions that besides Eq. (31) fulfill an objective function. To mimic a natural behavior of the system, and based on a number of studies (for example [12]), maximisation of the growth rate is used today as a standard. For the problem at hand, the growth rate has to be expressed as a function of the rate vector. Typically, a reaction is introduced that couple metabolic precursors describing the drain from central metabolism into anabolism. With knowledge on upper and lower bounds ($\underline{r}_{lower}, \underline{r}_{upper}$) of these rates, a general formulation of this linear optimization problem, often called flux balance analysis (FBA), reads:

$$\max_{\underline{r}} \ \underline{c}^T \ \underline{r}$$

$$s.t.$$

$$\underline{0} = N \ \underline{r}$$

$$\underline{r}_{lower} \leq \underline{r} \leq \underline{r}_{upper} \tag{32}$$

A core model for the bacterium *E. coli* is proposed in the COBRA toolbox (a MATLAB toolbox with a number of sophisticated tools for model set-up and analysis [13]) which consists of 95 reactions and 72 metabolites and comprise reactions like glycolysis, pentose phosphate pathway, and respiratory chain. This model is used to reveal basic properties of the network and to show possibilities for strain design for problems in Metabolic Engineering. A simple workflow is shown in Fig. 7.

Fig. 8 shows exemplary simulations studies for the core model with varying glucose and oxygen uptake. Only for high growth rates, a strong acetate production is observed that indicates an imbalance between the carbohydrate flux and oxygen uptake. To satisfy higher energy demand for higher growth rates, the ATP production is increasing with increasing growth rate; the data

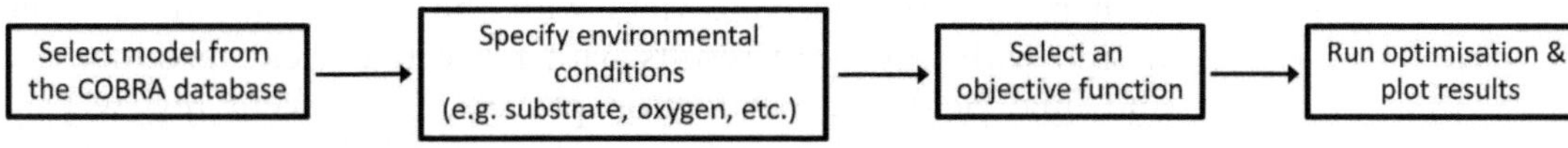

Fig. 7 Simple workflow to generate simulated data with the COBRA toolbox. Step 1: A model is selected from the database stored in the standard installation of the toolbox. Step 2: Environmental conditions are selected (in the example, glucose uptake rate is set to a fixed value while the oxygen uptake rate is varied from a maximal value to zero). Step 3: As objective function, maximisation of the growth rate is selected. Step 4: The optimization is started and the rates for the by-products are plotted

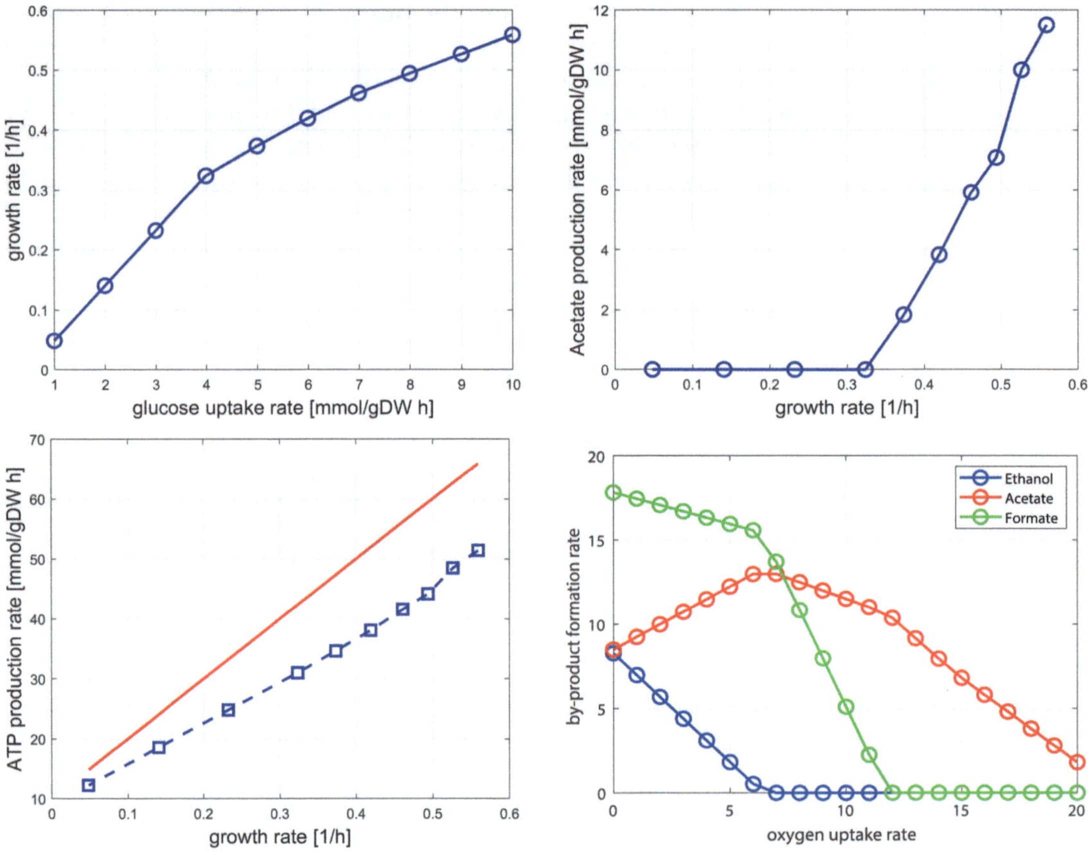

Fig. 8 Simulation study with a core model for *E. coli*. All rates are given in mmol/(gDW h). Upper row left: growth rate dependence on the substrate uptake rate. Upper row right: acetate production as a function of the growth rate. Lower row left: Overall ATP production as a function of the growth rate (blue) compared to literature data (red solid line). Lower row right: dependence of by-product formation on oxygen uptake. All simulation studies were done with the COBRA toolbox

show a good agreement with experimental data (red solid curve) shown in [14]. By decreasing the oxygen uptake rate, the respiratory chain cannot transduce all hydrogen produced in the central pathways into NAD(P)H and ATP. To balance this excess, by-products like formate and ethanol are released to the medium.

Succinate is an interesting compound of economic importance and hence, a number of studies are published analysing the model with respect to succinate yield with the aim of finding a strain design with improved results with respect to yield, productivity, and final yield in a production process. In a recent review [15] theoretical methods to increase succinate yield by genetic modifications were compared. Since there are different ways to generate succinate from glucose, a number of different modifications were proposed in the literature. The model presented above is used to figure out the yield of succinate as a function of the growth rate.

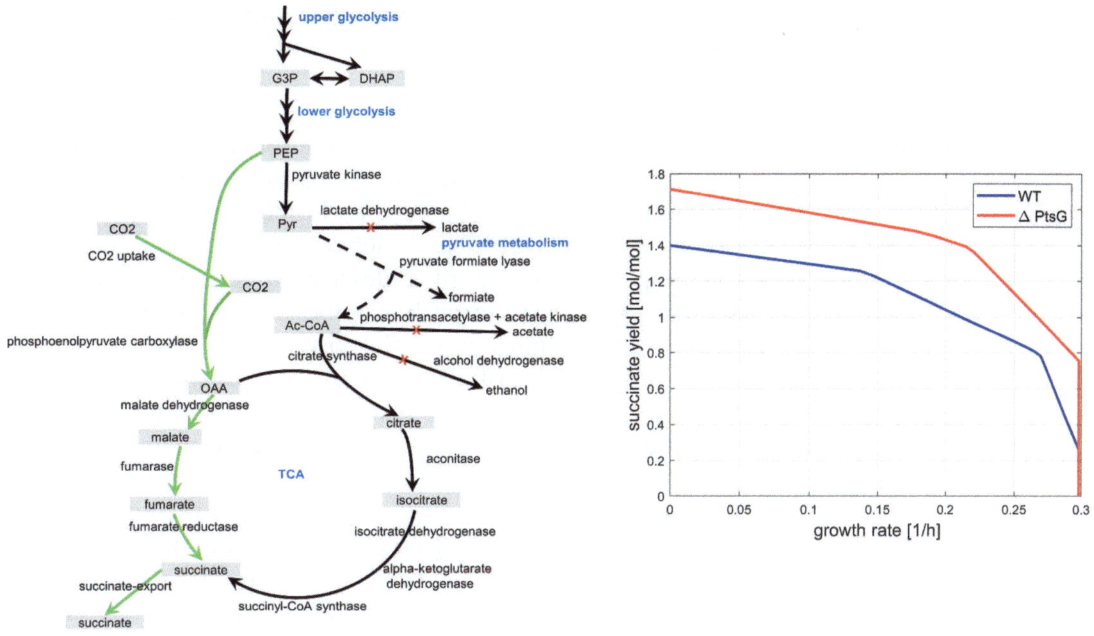

Fig. 9 Left: Optimal route from glucose to succinate. Other by-products are knocked out and the TCA cycle, in part, runs in the opposite direction in comparison to aerobic growth. **Right:** Simulation study with a core model of *E. coli* for the production of succinate under anaerobic conditions; the ΔPtsG strain performs much better than the wild type strain

Figure 9 (right panel) shows the optimal route from glucose to succinate under anaerobic conditions. On the right side, the course of the (optimal) dependence of succinate yield on the growth rate for two model variants is depicted: A mutant strain replacing the main glucose uptake system PtsG (needs PEP as energy source) with a different system (uses a different energy source) performs much better than the wild type strain. However, there is another interesting observation: Central to the study are reactions that link glycolysis to the TCA cycle: Phosphoenolpyruvate carboxylase (Ppc, PEP → oxaloacetate) as well as phosphoenolpyruvate carboxykinase (Pck, oxaloacetate + ATP →PEP + ADP) which are designed as irreversible in the standard model. However, experimental studies reveal a positive influence of modified strains with overexpressed Pck that suggests that the enzyme also operates in the reverse direction. A positive side effect would also be the generation of ATP that could be used for growth. This results in a higher growth rate in comparison to the standard model (data not shown). Of note is that the values shown are based on glucose that is taken up; the simulation shows that on CO_2 is also needed, bringing in more C atoms.

In recent years, FBA has extended into several directions to account for thermodynamic constraints, crowding effects due to a limited cellular volume or integration of other cellular levels like signaling processes. Normally, this results in further constraints that must be fulfilled in addition to the boundary conditions shown above (*see* **Note 4**).

4.3 Kinetic Model

As a last representative of models for microbial systems, a kinetic model is introduced in detail. Control of lactose uptake is a paradigm for genetic control in bacteria and therefore under investigation already for years. This small network was already introduced above. A simpler model variant to describe gene expression, signaling and metabolism presented here, does not consider all known interactions but concentrates on the formation of the protein and describes the influence of the protein back onto the promoter dynamics. To determine feedback structures in larger networks with only qualitative information, the incidence matrix (*see* Subheading 2)—can be exploited. In case that the incidence matrix describes only intracellular processes and links to the environment are eliminated, the nullspace again comes into play and reveals such structures. Here we found vector $\underline{c} = [0\ 1\ 1\ 1]$ that display a loop from allolactose over R and $LacY$ back to allolactose (interactions 2,3, and 4): The protein enhances the formation of the inducer and therefore has a positive influence on its own synthesis. This is used later on for the choice of the kinetic expression in the model.

The scheme describes the promoter D in two states: closed D and open PD, that is, with polymerase occupied. In the following, we will not consider RNApolymerase as state variable and we use D_o for the open complex (Net 3 from above). After RNA polymerase binding, the promoter is open and synthesis of a mRNA and the protein ($LacY$) can take place (Net 2 from above, we omit the dependency from amino acids and use directly D_o as a driving force for gene expression). A third reaction describes protein degradation by dilution. The following reaction scheme with kinetic parameters k_i illustrates the situation:

$$\begin{aligned} P + D &\rightleftharpoons PD\,(D_o) \quad (k_1 LacY) \\ D_o &- D_o + LacY \quad (k_2, k_{2'}) \\ LacY &- \quad (k_3) \end{aligned} \tag{33}$$

In reaction 1 the dependence of the step from protein $LacY$ is reflected in the reaction velocity; in reaction 2 also a basal expression rate is considered independent of D_o ($k_{2'}$). Simple rates are assumed for the reaction rates: The transition is proportional to the number of molecules available. Since the DNA binding site is either open or closed the total number of binding D_t can be used and the system can be described with only two equations (the term for the dilution is not considered for the promoter conformations, since the reactions occur very quickly):

$$\dot{D}_o = k_1\ D\, LacY^2 - k_1^-\ D_o$$

$$= k_1 \ (D_t - D_o) \ LacY^2 \ - \ k_1^- \ D_o$$

$$\dot{LacY} \ = \ k_{2'} \ + \ k_2 \, D_o \ - \ k_3 \ LacY \tag{34}$$

The kinetics for the DNA binding site is far more faster than protein synthesis and metabolic reactions. Therefore, an equilibrium is applied and one obtains for the promoter conformation D_o:

$$D_o \ = \ \frac{k_1 \ D_t \ LacY^2}{k_1 \ LacY^2 + k_2} \ = \ \frac{D_t \ LacY^2}{LacY^2 + K_B} \tag{35}$$

with the binding constant $K_B = k_1^- / k_1$.

The system can now be written with one ordinary differential equation for the protein:

$$\dot{LacY} \ = \ \underbrace{k_{2'} \ + \ k_2 \frac{D_t \ LacY^2}{LacY^2 + K_B}}_{r_{syn}} \ - \ \underbrace{k_3 \ LacY}_{r_{deg}} \tag{36}$$

Since we do not explicitly model the inducer, the system is analyzed by changing the initial conditions for the protein and results are shown in Fig. 10 (left panel). It turns out that there are two possible steady states for larger time t. If the initial value is too low, protein cannot be synthesised; however, for higher values, the system accelerates itself due to the autocatalytic structure and reaches a new steady state. A closer look on the rate of synthesis r_{syn} and rate of degradation r_{deg} of the protein reveals that there are three intersection points (for a steady state, both rates must be equal).

The arrows in Fig. 10 right indicate these steady states. Left of the first steady state, the rate of synthesis is higher than the degradation, meaning that the amount of *LacY* will increase. The same applies for the values at the right of the second steady state. At the

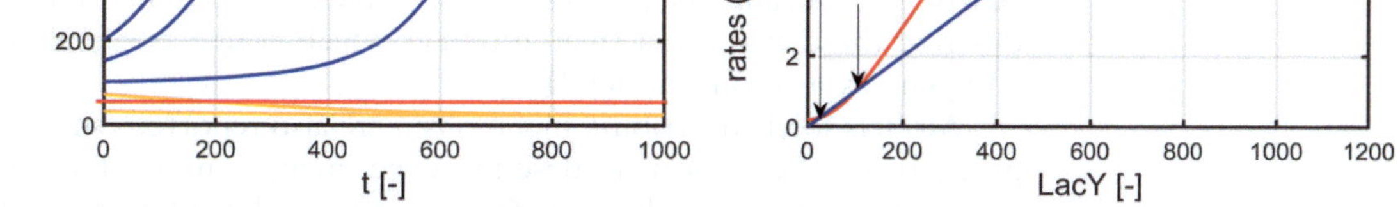

Fig. 10 On the left: Simulation study for different initial conditions for *LacY*. The red line subdivides the region of initial conditions. **On the right:** Rates for synthesis and degradation as a function of *LacY*. The number of intersections represents the number of steady states. Arrows indicate the three steady states

right of the first, and right of the third steady state, the degradation rate is higher. The number of molecules will decrease. Using the deterministic simulation, depending on the initial value, one ends either in the first or third steady state.

The example nicely demonstrates the connection between gene expression and signaling. For deterministic systems the observation of bistability is important for the design of such networks. A simulation study at the single cell level with the same model structure would lead to the observation of bimodality, that is, some cells will express the protein and some will not. This is indeed observed in experimental studies with *E. coli* cultures [16] growing on lactose. For biotechnological application the understanding of such a behavior is necessary for further strain design and optimization.

4.4 Conclusion

Understanding and efficiently exploiting of biochemical networks for biotechnological production is at the core of theoretical studies based on mathematical models for metabolism, gene expression and signaling. This chapter introduces basic concepts for a deterministic description of these networks with stoichiometric equations that are finally translated from mass balance equations into a set of differential equations for concentrations of metabolites inside the cell. It is shown that a number of steps are necessary for leading into an abstract representation of all cellular processes of interest that can be simulated with a computer system. The models presented differ with respect to the number of metabolites/components, type/number of processes under investigation, continuous in time or steady state, and—an aspect that becomes more and more important—the allowance of optimization strategies to describe for example, cellular design principles.

Fig. 11 summarizes different model types according to the mentioned criteria. Models with only a few components like coarse-grained models already have a high predictive power as shown in the first example above mentioned (model 1). Coarse-grained models are also used to explore design principles like resource allocation (model 2). Here, an optimization criterion is used, that is maximisation of the growth rate, that allows us to describe how resources like enzymes and ribosomes have to be allocated during the course of growth. An application of a kinetic model with a moderate number of equations was also shown with lactose uptake and its control. These models are often used to describe the dynamics of small modules which connects signaling and gene expression (model 3).

Models solely describing metabolism are also reported (shaded green box); they normally describe experiments on short times scales like substrate pulse experiments, where the interest is on fast metabolites dynamics [17]. In such experiments, protein will not be synthesized due to the short time window. It is important to stress that optimization is used here also, but in a different way as before. Parameter identification (parameter estimation and

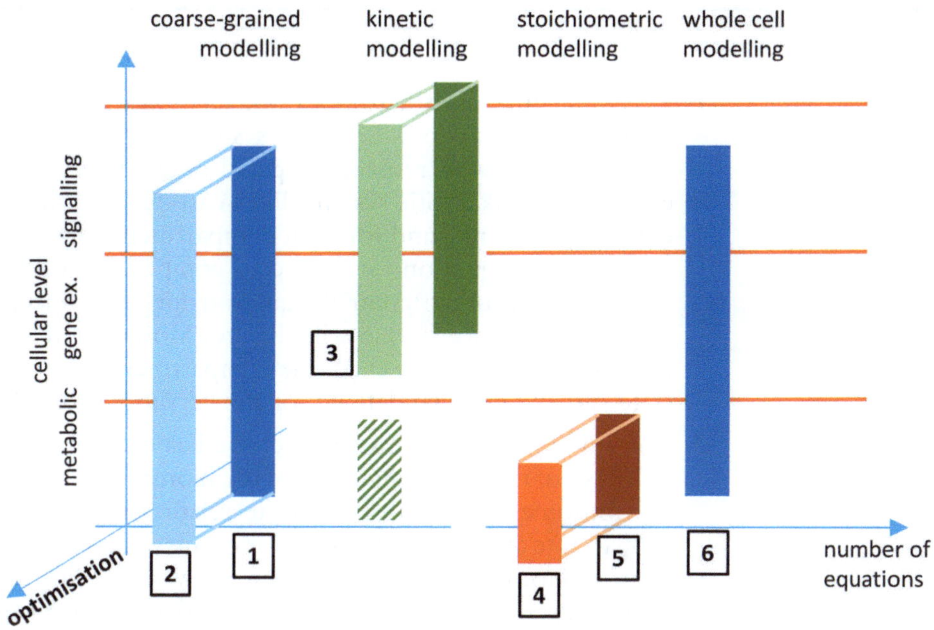

Fig. 11 Summary of model types according to three coordinates: the *x*-axis represents the number of equations; the *y*-axis shows the application of the models with respect to the cellular level (metabolism, gene expression/protein synthesis, and signaling); the *z*-axis differentiates if optimization strategies are used. For details on the model numbers given see explanation in the main text

analysis) as well as model structure optimization, are important tools for calibrating the model. This is reflected in the negative *z*-axis in Fig. 11.

Bigger stoichiometric models are especially reported for bacterial systems and they use optimisation to perform flux distribution that maximises an objective function (model 4). Incorporation of metabolites uptake and excretion measurements can reduce the number of unknown rates and, ideally, all rates can be determined without any optimisation (model 5).

"Whole cell" models are found on the extreme right side of Fig. 11. They describe the overall metabolism, including gene expression and signaling, with a high number of equations [18]. The effort to develop such a model is extremely high and needs also a large number of experiments for calibration and validation.

Mathematical modeling together with optimization methods offer many ways to describe, understand, and exploit cellular systems for industrial production. The model choice strongly depends on the scientific problem, prior knowledge on the system, and on the available data for model calibration. Nowadays, omics technologies provide a rich set of data from different cellular levels to be integrated into distinct model structures. Having detailed models available that can describe a broad range of experimental conditions with high accuracy will pave the way into a new, bio-driven, economy.

5 Notes

1. The procedure used here does not take into account cellular compartmentalization. Bacterial systems can be regarded as homogeneous with respect to spatial organization. However, even simple eukaryotic systems like yeasts already show different compartments, and for each compartment, the set up of a mass balance equation for a selected component will lead to a different type of differential equation for the concentration since the reference system will change. Moreover, if a component is present in more than one compartment, more that one state variable must be defined for the very same component.
2. The simple coarse-grained model used as example relates the synthesis of the protein fraction of the biomass to the total pool of ribosomes c_R because a linear relationship between the growth rate and the pool of ribosomes was observed. However, from a mechanistic point of view only free ribosomes can initiate a translation event. For a more detailed description of gene expression of a single gene, the overall distribution of free and busy ribosomes has to be taken into account. As described in [9], the number of ribosomes on the mature mRNA can be estimated by $n = f\, l/v$ with the initiation frequency f of the ribomsome, l the length of the transcript, and v the velocity of the ribomsomes. The equation can also be used to estimate the number of ribosomes on the nascent mRNA. The same arguments hold true for the RNA polymerase. Here, the number of busy RNA polymerase molecules can be estimated with the velocity and binding frequency of the RNA polymerase and the length of the gene.
3. Most stoichiometric models using FBA have a focus on central metabolic pathways like glycolysis, pentose phosphate pathway, and tricarboxylic acid cycle. In these pathways, fluxes are relatively high in comparison to the dilution term that is growth rate times concentration (to give an example, the uptake rate for glucose is approx. 6 mmol/gDW h; the dilution term with a growth rate $\mu = 0.5/\text{h}$ for a standard metabolite with a concentration of approx. 1 μmol/gDW results in 0.5 μmol/gDW h). However, fluxes in anabolism will be much smaller thus the dilution term becomes more important. It is recommended to estimate the dilution term based on literature data and to decide afterward if the term is small enough to neglect it in further calculations.
4. As can be seen in the example, a stoichiometric flux analysis requires an objective function to figure out possible and meaningful flux distributions. This is based on the observation that the number of unknown fluxes by far exceeds the number of

equations and therefore, a high number of degrees of freedom exist. To circumvent this problem, labeled ^{13}C glucose is used in experiments and—also based on a stoichiometric model—the way the labeled ^{13}C atoms change as they are processed by the metabolic network until it reaches the proteins. In this way, it is possible to estimate the intracellular fluxes in a much more accurate way than the standard method without labeling.

6 Glossary

6.1 Bio-based Economy

Current economy is mainly based on products from petrol. Chemical synthesis of interesting products suffer from high temperature and high pressure. In contrast, in a bio-based economy renewable substrates like agricultural waste or even light are used to produce valuable products with microorganisms under moderate temperature and pressure.

6.2 Intervention Strategy

To improve cellular metabolism to favour the production of a desired product, the reaction system has to be modified to redirect metabolic fluxes. This can be done by either deleting genes on the DNA (knockout) to block specific pathways, to overexpress genes to increase the flux in the pathway, or to bring in new genetic information that result in new proteins or a complete pathway.

6.3 Iterative Cycle of Experimental Investigation and Model Based Analysis

The Design-Build-Test-Learn cycle is a now well established procedure to combine experimental and theoretical methods in systems and synthetic biology. Especially in synthetic biology, the procedure is used to implement completely new modules in cellular systems, that are, small network with a defined functionality like amplifiers, oscillators or controllers.

6.4 Mathematical Model

The term model is used in very different ways in biology and mathematics. In biology, a model is an ideal test object, a special bacterial system, to study for example selected physiological properties. A mathematical model is a set of equations that describes the behaviour of selected quantities of the system under consideration. Such quantities are called state variables because they characterize at best the state of the system also with respect to the aim of the modelling procedure. The selection of state variables should be oriented on the available experimental data. For example, a kinetic model with a high number of uncertain or unknown parameters requires a high information content in form of measured data to calibrate the model adequately. This does not mean that all state variables have to measured; here sophisticated tools for parameter identification can help to determine the expected quality of the parameters given a set of experimental data. Besides state variables, input variables are also part of a model. Input variables allow us to influence and to alter the state variables of the system (in statistics the input variables are called factors or independent variables).

6.5 Network Reconstruction

To set up a model to describe a cellular network different types of information are needed. Following the dogma of molecular biology: DNA—mRNA—protein, the sequence of base pairs represents the information that is finally found in the respective protein. Proteins fulfil different functions in a cell either as enzymes converting metabolites or as structural entities for example as membrane proteins. In biotechnology, metabolic pathways converting substrates into biomass or valuable products are very important and for a high number of organisms stoichiometric networks are available and stored in databases. For new bacterial systems, sequence information is available and a comparison with existing sequences often allow the researchers to assign a function to the sequence/ protein. This expensive procedure to determine the function of a protein, determines cofactors and stoichiometric coefficients, and, in the best case, also kinetic parameters, is called network reconstruction.

6.6 Network Representation (Stoichiometric or Incidence Matrix)

There are two main approaches to represent cellular networks in a very compact way. Both are based on a matrix structure, that is, a mathematical structure with a defined number of rows and columns. In both types, a row represents a component while in a column stoichiometric conversions or interactions are represented. The stoichiometric matrix is used when material conversion by biochemical reactions (normally catalysed by enzymes) is described. The entries indicate the number of molecules of the substrate that are consumed during the process (negative sign) and the number of molecules of the products that are synthesised (positive sign). The stoichiometric matrix is used in mass balance equations. In contrast, the incidence matrix only describes interactions between two components; if there is a directed interaction in the sense that A influences B (but not vice versa) the entry for A is -1 while the entry for B is $+1$. The incidence matrix is used if knowledge on the process is only rough. Main applications are found in medicine.

Acknowledgments

I wish to thank all members of my team at TUM for fruitful discussions.

References

1. Kitano H, editor (2001) Foundations of systems biology. MIT Press, Cambridge, Massachusetts
2. Cortassa S, Aon MA, Iglesias AA, Lloyd D (2002) An introduction to metabolic and cellular engineering. World Scientific, Singapore

3. Breitling R (2010) What is systems biology? Front Physiol 1:159
4. Ko Y-S, Kim JW, Lee JA, Han T, Kim GB, Park JE, Lee SY (2020) Tools and strategies of systems metabolic engineering for the evelopment of microbial cell factories for chemical production. Chem Soc Rev 49:4615–4636
5. Yi T-M, Huang Y, Simon MI, Doyle J (2000) Robust perfect adaptation in bacterial chemotaxis through integral feedback control. Proc Natl Acad Sci U S A 97(9):4649–4653
6. Alon U (2006) An introduction to systems biology: design principles of biological circuits. Chapman & Hall/CRC Press, London
7. Palsson BO (2006) Systems biology: properties of reconstructed networks. Cambridge University Press, Cambridge
8. Keseler IM, Mackie A, Santos-Zavaleta A, Billington R, Bonavides-Martínez C et al (2016) The EcoCyc database: reflecting new knowledge about *Escherichia coli* K-12. Nucleic Acids Res 45(D1):D543–D550
9. Kremling A (2014) Systems biology—mathematical modeling and model analyses. Chapman & Hall/CRC Press, London
10. Scott M, Gunderson CW, Mateescu EM, Zhang Z, Hwa T (2010) Interdependence of cell growth and gene expression: origins and consequences. Science 330(6007):1099–1102
11. Schmidt A, Kochanowski K, Vedelaar S, Ahrne E, Volkmer B, Callipo L, Knoops K, Bauer M, Aebersold R, Heinemann M (2016) The quantitative and condition-dependent *Escherichia coli* proteome. Nat Biotechnol 34 (1):104–110
12. Schuetz R, Kuepfer L, Sauer U (2007) Systematic evaluation of objective functions for predicting intracellular fluxes in *Escherichia coli*. Mol Syst Biol 3(1):119
13. Heirandt L et al (2019) Creation and analysis of biochemical constraint-based models: the COBRA toolbox v3.0. Nat Protoc 14:639–702
14. Shimizu K, Yu M (2019) Regulation of glycolytic flux and overflow metabolism depending on the source of energy generation for energy demand. Biotechnol Adv 37(2):284–305
15. Valderrama-Gomez M, Kreitmayer D, Wolf S, Marin-Sanguino A, Kremling A (2017) Application of theoretical methods to increase succinate production in engineered strains. Bioprocess Biosyst Eng 40:479–497
16. Ozbudak EM, Thattai M, Lim HN, Shraiman BI, Van Oudenaarden A (2004) Multistability in the lactose utilization network of *Escherichia coli*. Nature 427:737–740
17. Chassagnole C, Noisommit-Rizzi N, Schmid JW, Mauch K, Reuss M (2002) Dynamic modeling of the central carbon metabolism of *Escherichia coli*. Biotechnol Bioeng 79 (1):53–73
18. Karr JR, Sanghvi JC, Macklin DN, Gutschow MV, Jacobs JM, Bolival B, Assad-Garcia N, Glass JI, Covert MW (2012) A whole-cell computational model predicts phenotype from genotype. Cell 150(2):389–401

Correction to: Multiscale Modeling of the Mitochondrial Origin of Cardiac Reentrant and Fibrillatory Arrhythmias

Soroosh Solhjoo, Seulhee Kim, Gernot Plank, Brian O'Rourke, and Lufang Zhou

Correction to:
Chapter 11 in: Sonia Cortassa and Miguel A. Aon (eds.),
***Computational Systems Biology in Medicine and Biotechnology: Methods and Protocols*, Methods in Molecular Biology, vol. 2399,**
https://doi.org/10.1007/978-1-0716-1831-8_11

The chapter was inadvertently published with one of the co-author's surname incorrectly spelled as "Sohljoo" instead of "Solhjoo".

The correction has been incorporated by updating the author's name as "**Soroosh Solhjoo**".

The updated version of this chapter can be found at
https://doi.org/10.1007/978-1-0716-1831-8_11

Sonia Cortassa and Miguel A. Aon (eds.), *Computational Systems Biology in Medicine and Biotechnology: Methods and Protocols*, Methods in Molecular Biology, vol. 2399, https://doi.org/10.1007/978-1-0716-1831-8_18,

INDEX

A

B

Sonia Cortassa and Miguel A. Aon (eds.), *Computational Systems Biology in Medicine and Biotechnology: Methods and Protocols*, Methods in Molecular Biology, vol. 2399, https://doi.org/10.1007/978-1-0716-1831-8,

C

D

E

T

W

MIX
Papier aus verantwortungsvollen Quellen
Paper from responsible sources
FSC® C105338

If you have any concerns about our products,
you can contact us on
ProductSafety@springernature.com

In case Publisher is established outside the EU,
the EU authorized representative is:
Springer Nature Customer Service Center GmbH
Europaplatz 3, 69115 Heidelberg, Germany

Printed by Libri Plureos GmbH
in Hamburg, Germany